Non-Stoichiometric Compounds

Surfaces, Grain Boundaries and Structural Defects

NATO ASI Series

Advanced Science Institutes Series

A Series presenting the results of activities sponsored by the NATO Science Committee, which aims at the dissemination of advanced scientific and technological knowledge, with a view to strengthening links between scientific communities.

The Series is published by an international board of publishers in conjunction with the NATO Scientific Affairs Division

A Life Sciences **B Physics**	Plenum Publishing Corporation London and New York
C Mathematical and Physical Sciences **D Behavioural and Social Sciences** **E Applied Sciences**	Kluwer Academic Publishers Dordrecht, Boston and London
F Computer and Systems Sciences **G Ecological Sciences** **H Cell Biology**	Springer-Verlag Berlin, Heidelberg, New York, London, Paris and Tokyo

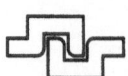

Series C: Mathematical and Physical Sciences - Vol. 276

Non-Stoichiometric Compounds

Surfaces, Grain Boundaries and Structural Defects

edited by

J. Nowotny

and

W. Weppner

Max-Planck-Institut für Festkörperforschung, Stuttgart, F.R.G.

Kluwer Academic Publishers

Dordrecht / Boston / London

Published in cooperation with NATO Scientific Affairs Division

Proceedings of the NATO Advanced Research Workshop on
Non-Stoichiometric Compounds
Surfaces, Grain Boundaries and Structural Defects
Rottach-Egern, F.R.G.
3–9 July 1988

Library of Congress Cataloging in Publication Data

NATO Advanced Research Workshop on Non-stoichiometric Compounds (1988 : Rottach-Egern, Germany)
Non-stoichiometric compounds : surfaces, grain boundaries, and structural defects : proceedings of the NATO Advanced Research Workshop on Non-stoichiometric Compounds, Rottach-Egern, FRG, 3-9 July 1988 / edited by J. Nowotny and W. Weppner.
p. cm. -- (NATO ASI series. Series C, Mathematical and physical sciences ; vol. 276)
"Published in cooperation with NATO Scientific Affairs Division."
Includes index.

1. Crystals--Defects--Congresses. 2. Mass transfer--Congresses. I. Nowotny, Janusz, 1936- . II. Weppner, W. (Werner), 1941- III. North Atlantic Treaty Organization. Scientific Affairs Division. IV. Title. V. Series: NATO ASI series. Series C, Mathematical and physical sciences ; no. 276.
QD921.N379 1988
548--dc19 89-2458

ISBN-13: 978-94-010-6914-4 e-ISBN-13: 978-94-009-0943-4
DOI: 10.1007/978-94-009-0943-4

Published by Kluwer Academic Publishers,
P.O. Box 17, 3300 AA Dordrecht, The Netherlands.

Kluwer Academic Publishers incorporates the publishing programmes of
D. Reidel, Martinus Nijhoff, Dr W. Junk and MTP Press.

Sold and distributed in the U.S.A. and Canada
by Kluwer Academic Publishers,
101 Philip Drive, Norwell, MA 02061, U.S.A.

In all other countries, sold and distributed
by Kluwer Academic Publishers Group,
P.O. Box 322, 3300 AH Dordrecht, The Netherlands.

Printed on acid free paper

TABLE OF CONTENTS

PREFACE

The material in this book is based on invited and contributed papers presented at the NATO Advanced Research Workshop on "Non-stoichiometric Compounds" held in Ringberg Castle, Rottach-Egern (Bavarian Alps), Germany, July 3-9, 1988. The workshop followed previous meetings held in Mogilany, Poland (1980), Alenya, France (1982), Penn State, USA (1984) and Keele University, UK (1986).

The aim of these workshops is to present and discuss up-to-date knowledge in the study of non-stoichiometry and its effect on materials properties as well as to indicate the most urgent research pathways required in this field. Since the subject of non-stoichiometry is interdisciplinary, the workshops bring together solid state physicists and chemists, surface scientists, materials scientists, ceramists and metallurgists.

The present workshop, which gathered 42 scientists of an international reputation, mainly considered the effect of surfaces, grain boundaries and structural defects on materials properties. From discussions during this meeting it emerged that correct understanding of properties of ceramic materials requires urgent studies on the defect structure of the interface region. Progress in this direction requires the development of the interface defect chemistry. This is the task for materials scientists in the near future.

The present proceedings includes both theoretical and experimental work on general aspects of non-stoichiometry, defect structure and diffusion in relation to the bulk and to the interface region of such materials as high tech ceramics, solid electrolytes, electronic ceramics, nuclear materials and high T_c oxide superconductors.

We would like to acknowledge the financial support of the NATO Science Committee and the Max-Planck-Society. Thanks are due to Mr. W. Kernler, Mrs. I. Koch, Dr. N. Nicoloso, Mr. W. Payer, Dr. M. Rekas and Dr. F. Sim for their kind help in the organization the meeting. We would like also to thank all authors for their co-operation in the preparation of these proceedings.

Janusz Nowotny
Werner Weppner

WORKSHOP PARTICIPANTS. First row (sitting) from left: Smyth, Alcock, Stubican, Nowotny, Simkovich, Janowski, Hirschwald; second row: Moya, Naito, Freer, Gleitzer, Sorensen, Weppner, Saito, Tuller; third row: Alcock, Smyth, Petot-Ervas, Catlow, Cormack, de Vries, Matsui, Chiang, Boukamp, Rekas, Karatas; last row: Dechamps, Nicoloso, Schulz, Tasker, Fromm, Eror, Mason, Sarma, Moya, Hennings, Monty, Colson, Pecheur, Borchardt, Werber, Matzke, Dufour.

LIST OF PARTICIPANTS

1. C.B. Alcock, University of Notre Dame, Department of Materials Science, Notre Dame, Indiana 46556, USA
2. G. Borchardt, Technische Universität Clausthal, Institit für Allgem. Metallurgie, 3392 Clausthal-Zellerfeld, FRG
3. B. Boukamp, University of Twente, Department of Chemistry, 7500 AE Enschede, The Netherlands
4 G. Boureau, Université Paris-Sud, Laboratoire de Composés Non-stoichiométrique, Bat. 415, 91405 Orsay Cedex, France
5. C.R.A. Catlow, University of Keele, Department of Chemistry, Staffordshire, ST5 5BG, UK
6. Y.M. Chiang, MIT, Department of Materials Science, Cambridge, MA. 02139, USA
7. J.C. Colson, Université de Burgogne, Faculté des Sciences Mirande Laboratoire de Réactivité des Solides, 21004 Dijon Cedex, France
8. A.N. Cormack, Alfred University, New York State College of Ceramics, Alfred, NY 14802, USA
9. M. Déchamps, Université Paris-Sud, Laboratoire de Chimie du Solide, 91405 Orsay Cedex, France
10. L.-C- Dufour, Université de Burgogne, Faculté des Sciences Mirande, Réactivité des Solides, 21004 Dijon Cedex, France
11. R. Freer, University of Manchester, Materials Science Centre, Machester M1 7HS, UK
12. E. Fromm, Max-Planck-Institut für Metallforschung, Seestr. 72, 7000 Stuttgart 1, FRG
13. N.G. Eror, Oregon Graduate Center, Department of Materials Science 19600 NW von Neumann Dr., Beaverton, OR. 97006, USA
14. C. Gleitzer, Université de Nancy, Laboratoire de Chimie du Solide, 54506 Vandoeuvre Cedex, France
15. D. Hennings, Philips Forschungslabor, 5100 Aachen, FRG
16. W. Hirschwald, Freie Universität Barlin, Institut für Physikalische Chemie, Takustrr. 3, 1000 Berlin 33, FRG
17. J. Janowski, Academy of Mining and Metallurgy, Institute of Metallurgy, 30059 Krakow, Poland
18. C. Karatas, Hacettepe University, Department of Nuclear Engineering, Beytepe, Ankara, Turkey
19. T.O. Mason, Northwestern University, Department of Materials Science and Enginering, Evanston, IL. 60201, USA
20. T. Matsui, Nagoya University, Department of Nuclear Engineering, Furo-cho, Nagoya 462, Japan
21 Hj. Matzke, Joint Research Centre, European Institute for Transuranium Elements, 7500 Karlsruhe, FRG
22. C. Monty, Laboratoire de Physique des Matériaux, CNRS, 92195 Meudon, France
23. F. Moya, Faculté des Sciences et Techniques, Laboratoire de Metallurgie, 13397 Marseille Cedex 13, France
24. K. Naito, Nagoya University, Department of Nuclear Engineering, Furo-cho, Nagoya 462, Japan

25. N. Nicoloso, Max-Planck-Institut für Festkörperforschung, 7000 Stuttgart 80, FRG
26. J. Nowotny, Max-Planck-Institut für Festkörperforschung, 7000 Stuttgart 80, FRG
27. P. Pecheur, Ecole des Mines, Laboratoire de Physique du Solide, 54042 Nancy Cedex, France
28. G. Petot-Ervas, Université Paris-Sud, ISMA, 91405 Orsay Cedex, France
29. M. Rekas, Max-Planck-Institute für Festkörperforschung, 7000 Stuttgart 80, FRG (present address: Academy of Mining and Metallurgy, Institute of Materials Science, 30059 Krakow, Poland)
30. Y. Saito, Tokyo Institute of Technology, Research Laboratory of Engineering Materials, 4259 Nagatsuta, 227 Yokohama, Japan
31. D.D. Sarma, Indian Institute of Science, Bangalore 560012, India
32. H. Schulz, Universität München, Institut für Kristallographie, Theresienstr. 41, 8000 München 2, FRG
33. F. Sim, Imperial College, Department of Chemistry, London SW7, UK
34. G. Simkovich, Penn State University, Department of Materials Science, University Park, PA. 16802, USA
35. D.M. Smyth, Legigh University, Materials Science Centre, Bethlehem, PA. 18015, USA
36. O.T. Sorensen, Riso National Laboratoty, 4000 Roskilde, Denmark
37. V.S. Stubican, Penn State University, Department of Materials Science, University Park, PA. 16802, USA
38. P.W. Tasker, Harwell, Oxfordshire OX11 ORA, UK
39. H.L. Tuller, MIT, Department of Materials Science and Engineering, Cambridge. MA. 02139, USA
40. K.J. de Vries, University of Twente, Department of Chemistry, 7500 EA Enschede, The Netherlands
41 W. Weppner, Max-Planck-Institut für Festkörperforschung, 7000 Stuttgart 80, FRG
42. T. Werber, Technion, Department of Materials Science, Haifa 32000, Israel

I. Non-stoichiometry and Defect Structure

THE CONTROL OF STOICHIOMETRY IN OXIDE SYSTEMS

C.B. Alcock
Department of Materials Science and Engineering
University of Notre Dame
Notre Dame, IN 46556 USA

ABSTRACT. The experimental methods for the control and determination of the metal/oxygen ratio in binary metallic oxides are critically reviewed. The present limitations on classical analytical procedures as well as the newer solid state electrochemical and mass spectrometric techniques are discussed using data from systems already reported in the literature.

Introduction

The solid state scientist who studies the transport properties of oxide systems at high temperatures, needs accurate information concerning the metal/oxygen ratio as a function of both temperature and oxygen pressure. The correlation between these factors has provided the high temperature chemist with problems which have tested his analytical techniques to their practical lmits. In this paper we shall discuss these techniques and outline some of the limitations which have been reached, so that future studies may be made in the light of past experience. The classical analytical procedures are based on volumetric and gravimetric analysis, whilst techniques of more recent development are solid state electrochemistry and mass spectrometry.

Classical Analytical Procedures

Traditional methods of analysis are usually based on volumetric analysis in aqueous solutions, or conversion of the element to be analysed to some well defined solid, followed by a gravimetric determination e.g., sulphur is converted to barium sulphate $BaSO_4$ for gravimetry.

The earliest accurate study of non-stoichiometry is due to Darken and Gurry (1) who first defined the composition of wustite from equilibrium with iron to equilibrium with magnetite. The metal/oxygen ratio after equilibration with a known CO/CO_2 mixture at a controlled temperature was determined by volumetric analysis on quenched samples. Each sample was dissolved in concentrated hydrochloric acid under a CO_2 atmosphere and the solution titrated with standard permanganate solution first to yield the ferrous ion content, and after complete reduction to yield the ferric ion content by difference. This is one of the rare instances where a direct chemical determination of the upper/lower valency ratio for the metallic species in a non-stoichiometric oxide has been made. This is also a rare case where an easily determined range of stoichiometry can be subjected to a well established analytical procedure for the metal. In all other cases

J. Nowotny and W. Weppner (eds.),
Non-Stoichiometric Compounds Surfaces, Grain Boundaries and Structural Defects, 3–10.

cited in this paper, this fortunate state of affairs does not apply.

The high temperature equivalent of the classical gravimetric procedure when applied to the determination of the metal/oxygen ratio takes one of two forms. Either the substance is equilibrated with an oxygen pressure at fixed temperature which is low enough to ensure reduction to the stoichiometric ratio, or it is oxidized to a higher valency oxide which is assumed to attain precise stoichiometry.

In their study of UO_{2+X}, Hagemark and Broli (2) made use of high temperature thermogravimetry to determine the extent of non-stoichiometry when a sample was equilibrated with a gas of known oxygen potential at a fixed temperature. The oxygen potential in the gas phase was fixed by mixing CO and CO_2 in the required proportions. Over a wide range of CO_2/CO ratios the only weight change observed in a sample as constant temperature was that due to buoyancy changes in the gas phase, and it was assumed that under these conditions the oxide sample was exactly stoichiometric. Increases in weight appear as the oxygen pressure increases the metal/oxygen ratio to hyperstoichiometry, but weight decreases occur as hypostoichiometric oxide is formed under extreme reducing conditions. Hagemark and Broli give evidence for the thermodynamic behavior of $UO_{2 \pm x}$ in their study using this criterion.

Bransky and Tallan (3) used thermogravimetry and equilibration with CO/CO_2 mixtures in a study of $MnO_{1 + x}$, but found that the volatility of manganese increased considerably as the stoichiometric ratio was approached. The increasing carbon potential of the CO/CO_2 atmosphere under these circumstances raised the possibility of oxide-carbide solid solution formation. Because of these complications it appears almost impossible to look for hypostoichiometry in MnO using this method. Direct Mn^{2+}/Mn^{3+} ratio determination by chemical analysis does not appear to have been attempted for this system.

It should be noted that these authors assumed from their measurements that the stoichiometric oxide was reached at 800C with a CO/CO_2 ratio equal to one. This was because the departure from stoichiometry was unmeasurable below this limit and was probably less than 1 part per thousand, corresponding to $MnO_{1.001}$. This CO/CO_2 ratio is considerably more oxidizing than the oxygen potential at which manganese metal should be formed, in principle $10^6/1$ at this temperature, and hence there is a wide gap of ignorance in the $MnO_{1 \pm x}$ system. The observed tendency for manganese to evaporate from the sample at this lowest CO/CO_2 ratio, 1/1, which was used, suggests a high metal activity, and hence a range possibly only of hypostoichiometry at lower oxygen pressures.

These results provide a useful guide for studies where a volatile metal, such as manganese, may form part of a non-stoichiometric oxide system. It should be noted that manganese has a vapour pressure of 3×10^{-5} atmospheres at 1000°C. This high volatility is clearly indicated by the relatively low heat vaporization of the element. A number of studies have been made where the oxide can be reduced to metal in hydrogen at high temperature in order to determine the non-stoichiometry of a previously oxidized sample. For example, with Co and Ni because of the low vapour pressures of these metals a weighed sample of pure metal can be oxidized during thermogravimetry, and reduced to regain the initial weight of metal (4) (5). These elements have heats of vaporization of 417.4KJ and 421.4KJ per g. atom respectively compared with 265.6KJ for manganese at 1000°C

This technique was used by Zador and the author (6) to confirm electrochemical studies of non-stoichiometry in $MoO_{2 \pm x}$. There are several aspects of this study which exemplify the earlier procedures. The principlal technique for varying the metal/oxygen ratio was electrochemical using a solid electrolyte rather than gas equilibration. Because of the low vapour pressure of molybendum, oxidized samples could be easily reduced in hydrogen to determine the metal/oxygen ratio, and these gravimetric studies provided a useful check on the electrochemical technique. However, samples which had the greatest departure from stoichiometry in this study indicated another aspect of high temperature thermogravimetry which must be carefully considered. Oxides which are highly hyperstoichiometric, exhibit vaporization of MoO_3 in the monomeric or polymeric forms $(MoO_3)_n$ (7) at high temperatures. It was possible to show in this study that a large increase in oxygen pressure accompanies the changeover at constant temperature from MoO_{2-x} to MoO_{2+x}.

This volatility of higher valency oxides raises experimental difficulties in a number of systems. For example in a study of NbO_2 the only gravimetric method which has been used is to determine the non-stoichiometry by oxidation to Nb_2O_5 (8). Because of the votalility of Nb_2O_5 the oxidation was restricted to 800°C since measurable weight loss occurred at 900°C. Vasil'eva et al (9) found in a study of $VO_{1 \pm x}$ that determination of stoichiometry by oxidation to V_2O_5 must be made under restricted temperature conditions (≤ 500°C), due to V_2O_5 vaporization.

The vaporization of UO_3 (g) must be considered carefully in studies of the uranium-oxygen system, since although UO_3 is not a stable condensed phase at high temperatures and moderate oxygen pressures, the gaseous molecule is formed in significant quantities as the U_4O_9 composition is approached in UO_{2+x} at 1000°C (10).

Vaporization is thus always a matter of concern in gas/solid equilibration studies of non-stoichiometry as a limiting factor in this very flexible technique. On thermodynamic grounds there is little to choose between CO/CO_2 mixtures, H_2/CO_2 mixtures aned H_2/H_2O mixtures for controlling the oxygen potential in the gas phase. However the relative ease of diffusion of hydrogen out of apparently sealed systems at high temperatures, and the carburizing potential of CO/CO_2 mixtures rich in CO are factors limiting the choice of a gas phase at high temperatures.

Another limitation of this technique is that ratios of gas composition in excess of $10^4/1$ are difficult to make under good control. The oxygen potential ranges over which gas/solid equilibration may be used with any accuracy is therefore about 160 K Joule to 600 KJ at 1000°C. This covers the range from Cu/Cu_2O equilibrium at the upper end, to Cr/Cr_2O_3 equilibrium at the lower extreme. Above the upper limit of these gas mixtures it is quite possible to use oxygen/nitrogen gas mixtures, but no successful technique exists below Cr/Cr_2O_3 equilibrium for gas/solid equilibration.

Solid State Electrochemistry

A quite different approach to the measurement and control of non-stoichiometry is the use of solid state electrochemistry with zirconia- or thoria-based electrolytes. In this well-established procedure the oxide sample is usually presented to the face of a solid electrolyte pellet in the form of a sintered pellet. The sample serves as an electrode in an oxygen galvanic cell which incorporates a reference electrode metal/metal oxide mixture of known oxygen potential at the temperature of measurement. The stoichiometry of the oxide pellet can be varied under control by polarizing the cell with an external potential such that oxygen is transported from or to the non-stoichiometric electrode in known amounts. When equilibrium is achieved, a measurement of the cell EMF can be used to establish the oxygen potential of the non-stoichiometric oxide sample. Two independent studies of UO_{2+x} using the electrochemical technique have provided closely agreeing results (11) (12) which also agree satisfactorily with the thermogravimetric study of Hagemark and Broli (loc. cit.). The currents passed in these studies were of the order of a few microamps., over a period of some hours. Amounts of oxygen of the order of one microgram can easily be added or removed from the sample in this manner, and therefore oxide samples of a few hundred milligrams can be successfully cycled across the range of non-stoichiometry.

In open systems where the inert gas atmosphere surrounding the solid state cell has equal access to both electrodes, transport of oxygen from one electrode to the other can occur via the gas phase, which usually contains traces of a chemically active gas such as hydrogen or methane. It is therefore preferable to separate the oxide sample from the reference electrode, and provide a separate atmosphere for each electrode. The use of getters to produce a chemically neutral atmosphere should be minimized because of the unanticipated introduction of reducing gaseous species which usually accompanies oxygen trace elimination in argon, for example. It is better to mount a sample of the electrode material upstream of the electrode to reduce this effect to a minimum by pre-equilibrating the "inert" gas to an oxygen potential close to that of the electrode.

When oxygen is provided to an oxide sample by this procedure of coulometric titration, clearly some diffusion processes must occur either of oxygen or metal in the sample to produce a homogeneous non-stoichiometric oxide. For the oxide to be readily manageable a thickness of at least 2 mm. is normally required. The time to reach equilibrium in such a sample can be calculated from known diffusion data using standard procedures (13) but a rough estimate can be made by using the relationship between mean penetration squared and diffusion coefficient multiplied by time. With a 2 mm. thick sample this suggests that a sample of UO_{2+x}, where it is established that oxygen moves more rapidly than uranium, and Do ~ 10^{-6} cm^2 sec^{-1}, should equilibrate in a few hours. Nakajima and Fujino in their study of UO_{2+x} (loc.cit.) allowed more than 5 hours after each coulometric titration at 1000°C for equilibrium to be established before measuring the oxygen potential via the cell E.M.F. This allowed an adequate period of time to confirm the absence of long-term drift in the E.M.F. With lower diffusion coefficients than this it can be seen that the study of non-stoichiometry in an oxide could be very lengthy.

In a study of symmetrical cells with ZrO_2 based electrolyte, Worrell and Iskoe (14) showed that overvoltage develops at 900°C in cells with electrodes of Cu/Cu_2O, and Fe/FeO and Ni/NiO, in that order when current is passed through the cell. The overvoltage is directly related to the current and whereas a steady overvoltage is achieved with currents up to 80 μA with Cu_2O, and 50 μA with FeO, the stable overvoltage can only be achieved below 10 μA with Ni/NiO as electrode. The growth and dissipation of the overvoltage is linearly related to the current being passed in the cell, and hence is clearly related to diffusion controlled processes in the electrode and, presumably at the electrode/electrolyte interface.

This limitation on the rate of transfer of oxygen and the consequent growth of this extraneous potential due to kinetic effects can be considerably reduced if a liquid metal containing oxygen in solution is used as electrode. Thus the author has used liquid tin through which a controlled H_2/H_2O ratio is bubbled to maintain a constant oxygen content in the electrode. The liquid metal is stirred by the bubbling gas, and thus concentration gradients are only formed at the electrode/electrolyte interface. Using this electrode it was possible to carry out coulometric titrations to liquid sodium through a thoria-based electrolyte at 400°C without undue polarization, at 1 μA current (15).

Improved oxygen transfer across the electrode/electrolyte interface has also been demonstrated for example with electrodes based on Sc_2O_3/UO_{2+x} solid solutions at temperatures as low as 400°C (16). This is presumably due to the relatively high oxygen diffusion coefficient in UO_{2+x} referred to above.

Another approach to this problem, used in the study of $MoO_{2\pm x}$ by Zador and the author, was to use an electrode in powder form within an encapsulating crucible. A platinum wire dipped into the powdered sample through the lid of the capsule, and the cell could be operated as long as an electron conducting path could be maintained between the platinum lead and the electrolyte, which formed the body of the capsule. Rapid equilibration of the sample powder probably occurred through vapour phase transport of oxygen as water vapour or CO_2 within the capsule atmosphere. Such confinement of the sample also helps to minimize any effect due to vaporization of oxides such as MoO_3 in this example. The results of this electrochemical procedure were subsequently confirmed by gas/solid equilibration under optimum conditions to minimize vaporization whilst yielding a readily measurable departure from stoichiometry.

Electrochemical studies of non-stoichiometry may only be accurately carried out during long, cumulative, periods of time when the transport number of oxygen ions in the electrolyte is very close to unity. This is because any electron transport across a cell with a chemical potential gradient between the electrodes, allows oxygen atoms to be transferred from the higher oxygen pressure to the lower. This fact limits the use of solid state electrochemistry to temperatures less than about 1200°C in this application, since measurable permeability of electrolytes from normal commercial sources is found at higher temperatures (17). Furthermore the range of oxygen potentials over which the presently available electrolyte materials have the ion transport number substantially equal to unity narrows considerably with increasing temperature (18).

Mass Spectrometry

Both gravimetric and solid state electrochemical procedures have their limitations at low oxygen potentials less than, say, Cr/Cr_2O_3 at 1000°C, and at high temperatures, i.e., greater than about 1300°C. An alternative approach to the study of non-stoichiometry can be seen in the work of Gilles and co-workers (19) on $TiO_{1 \pm x\ at}$ at 1527°C. Samples covering the wide range of stoichiometry of this phase $TiO_{0.86}$ - $TiO_{1.25}$ were prepared by arc melting, and then placed in the tungsten crucible of the Knudsen effusion cell of a high temperature mass spectrometer. The collision-free molecular beam effusing from the cell at high temperatures contained the vapour species, Ti(g), TiO(g), TiO_2(g) and oxygen both monatomic and diatomic species at much lower pressures.

In such a Knudsen cell operating at high temperatures the weight loss due to effusion can be kept very small i.e., less than a few micrograms from a sample of about one gram, and the ion currents formed in the mass spectrometer can be used to determine the chemical potentials of the metal and oxygen in the sample. For example the Ti^+ and TiO^+ ion currents may be used to determine oxygen potentials relative to some reference state. Following Neckel and Wagner (20) we may write

$$\frac{1}{2} \log \frac{pO_2}{pO_2\ (ref)} = \log \frac{TiO^+}{Ti^+} \cdot \frac{Ti^+\ (ref)}{TiO^+\ (ref)}$$

The monoxide $TiO_{1 \pm x}$ is in equilibrium with the ß solid solution of oxygen in titanium according to the phase diagram (21), and the oxygen potential of the α phase solid solutions have been determined by Komarek and Silver (22). Combining these data with the TiO^+/Ii^+ ratio at the α-phase/monoxide interface makes it possible to use the equation above to map out chemical potentials in $TiO_{1 \pm x}$.

This technique has not been used in the study of non-stoichiometry to the extent which seems possible, and it offers a new method for high temperature studies as well as those where the vapour pressure of the metal could lead to experimental difficulties in the conventional gravimetric procedure. Such difficulties arose, it will be remembered, in the study of manganese oxide at low oxygen potentials.

Conclusion

In conclusion it should be said that the application of thermogravimetry to the determination of the metal/oxygen ratio has been well explored, and information from earlier studies referred to above give clear indications as to its strengths and weaknesses. Solid state electrochemistry has a larger potential than the width of its applications to date appear to indicate. There appears to be no pressing need to push this technique to higher temperatures since this would undoubtedly bring up some new materials problems. However the development of new oxide electrolytes of a higher conductivity than zirconia-based electrolytes (23) appears to be promising in the present applications especially at lower temperatures.

Finally, mass spectrometry has now reached an impressive state of maturity and the use of Knudsen cell studies, especially when these can be combined with solid oxide electrochemical cells (24), would appear to open a promising new avenue for the study of the thermodynamics and kinetics of non-stoichiometric oxides.

References

(1). L.S. Darken and R.W. Gurry, J. Amer. Chem. Soc. 67, 1398 (1945)

(2). K. Hagemark and M. Broli, J. Inorg. Nucl. Chem 28, 2837 (1966)

(3). I. Bransky and N.M. Tallan, J. Electro Chem. Soc. 118, 788 (1971)

(4). B. Fisher and D.S. Tannhauser, J. Chem. Phys. 44, 1663 (1966)

(5). W.C. Tripp and N.M. Tallan, J. Amer. Ceram. Soc. 53, 531 (1970)

(6). S. Zador and C.B. Alcock, J. Chem. Thermo. 2, 9 (1970)

(7). J. Berkowitz, M.G. Inghram and W.A. Chupka, J. Chem. Phys. 26, 842 (1957)

(8). C.B. Alcock, S. Zador and B.C. H. Steele, Proc. Brit. Ceram. Soc. No. 8, 231 (1967)

(9). I.A. Vasil'eva, Zh. V. Granovskaya and I.S. Sukhushina, Russ. J. Phys. Chem. 48, 905 (1974)

(10). M.H. Rand, and T.L. Markin, "Thermodynamics of Nuclear Materials" I.A.E.A. Vienna (1968)

(11). T.L. Markin and R.J. Bones, A.E.R.E. Rep. 4042 (1962), 4178 (1962)

(12). A. Nakamura and T. Fujino, J. Nucl. Mat. 149, 80 (1987)

(13). W. Jost, "Diffusion in Solids, Liquids, Gases" Academic Press N.Y. (1960)

(14). W.L. Worrell and J.L. Iskoe, "Fast Ion Transport in Solids" ed. W. van Gool, North Holland (1973)

(15). C.B. Alcock and G. Stravropoulos, Can. Met. Quart. 10, 257 (1971)

(16). S.P.S. Badwal, M.J. Bannister and W.G. Garrett, "Science and Technology of Zirconia II" ed. N. Claussen, M. Ruhle and A. Heuer, Amer. Ceram Soc. (1984) p. 598

(17). C.B. Alcock and J.C. Chan, Can. Met. Quart. 11, 559 (1972)

(18). T.H. Etsell and S.N. Flengas, Chem. Rev. 70, 339 (1970)

(19). S.A. Heideman, T.B. Reed and P.W. Gilles, High Temp. Sci. 13, 79 (1980)

(20). A. Neckel and S. Wagner, Berichte Bunsen 73, 210 (1969)

(21). O. Kubaschewski, O. Kubaschewski von Goldbeck, P. Rogel and H.F. Franzen "Titanium" Special Issue No. 9, I.A.E.A. (1983)

(22). K.L. Komarek and M. Silver "Thermodynamics of Nuclear Materials p. 749 I.A.E.A. (1962)

(23). T. Takahashi, H. Iwahar and Y. Nagai, J. App. Electrochemistry 2, 97 (1972)

(24). C.B. Alcock, J. Butler and E. Ichise, Solid State Ionics, 314, 499 (1981)

PHASE RELATIONS OF METAL OXIDES BY COULOMETRIC TITRATION

Y. SAITO
Research Laboratory of Engineering Materials
Tokyo Institute of Technology
4259 Nagatsuta-cho, Midori-ku, Yokohama 227
Japan

T. MARUYAMA
Department of Metallurgical Engineering
Tokyo Institute of Technology
1-12-1 O-okayama, Meguro-ku, Tokyo 152
Japan

ABSTRACT. The purpose of this paper is to summarize the results on the phase relations and nonstoichiometry in V_2O_{5-x}, $BaBiO_{3-x}$ and PrO_x investigated by the present authors. The equilibrium oxygen partial pressures over the above oxides have been measured by coulometric titration using yttria-stabilized zirconia as a solid electrolyte. Nonstoichiometry in these oxides has been determined as a function of the oxygen partial pressure. The maximum oxygen deficiency in V_2O_{5-x} is 0.055 at 920K. The $BaBiO_{3-x}$ has three distinct phases in $0<x<0.5$. The existence of the $PrO_{1.667}$ is suggested in the PrO_x. Detailed discussions have been made on the defect structures of the nonstoichiometric oxides based on the relation between the deviation from the stoichiometric composition and the oxygen partial pressure. The results prove that coulometric titration is one of the powerful tool to study the phase relation particularly in the oxides with the small deviation from the stoichiometric dcomposition.

1. Introduction

The phase relation in metal-oxygen system is a basic factor in designing the advanced materials with high performance. Nonstoichiometry in metal oxides has been of great interest because it sensitively affects physical and chemical properties of the materials. The thermodynamic properties and the range of nonstoichiometry of metal oxides have been determined by several techniques of quenching-chemical analysis, thermogravimetric measurements and the tensimetric method. Following the pioneering work by Kiukkola and Wagner [1] in 1957, a considerable number of thermodynamic studies on metal oxides have been carried out by an electrochemical technique using galvanic cells with solid electrolytes (e.g., ZrO_2-CaO, ZrO_2-Y_2O_3).

Coulometric titration using a stabilized zirconia electrolyte is especially suited for measuring the oxygen activities and the phase boundaries in nonstoichiometric oxides (M_aO_b). This technique

J. Nowotny and W. Weppner (eds.),
Non-Stoichiometric Compounds Surfaces, Grain Boundaries and Structural Defects, 11–26.

allows the composition of the nonstoichiometric oxide to be changed by known amounts in situ at high temperatures. Oxygen is added to or removed from one electrode compartment of the galvanic cell isothermally by applying a non-equilibrium voltage through the cell. The electromotive force (EMF) of the cell can be determined for each oxygen addition or removal[2-4]. The coulometric titration method has been applied to measure the equilibrium oxygen partial pressure, i.e., the relative partial molar free energy of oxygen for nonstoichiometric oxides such as UO_{2+x} [5-7], $Ni_{1-y}O$ [8] and $Fe_{1-y}O$ [2,9-12].

The vanadium-oxygen system is one of the more complex and interesting metal-oxygen system. In the V_2O_5-V_2O_4 system, the compounds formulated as $V_{2n}O_{5n-2}$ by Wadsley [13] are expected to exist and the existence of V_2O_5, V_3O_7, V_4O_9, V_6O_{13} and V_2O_4 has been reported [13-29]. In spite of numerous studies, there remain discrepancies in the phase relations. Although a few studies have also been made on the equilibrium oxygen partial pressures over the V_2O_5-V_2O_4 system [31-35], the conclusive result has not been established yet and there is no reliable data on the nonstoichiometry in V_2O_{5-x}.

Much interest has been focused on the perovskite-type compound $BaBiO_3$ because it is the end member of the superconducting material $BaPb_{1-y}Bi_yO_3$ [36]. The superconducting behavior of $BaPb_{1-y}Bi_yO_3$ is affected by the valence state of the bismuth cation and the nonstoichiometry of oxygen [37], which are closely related to each other. The nonstoichiometry in $BaBiO_3$ has been studied by Cox and Sleight [38]. Chaillout and Remika [39] have reported that there exist two distinct phases of $BaBiO_{2.97}$ and $BaBiO_{2.77}$ and that oxygen vacancies give rise to a superstructure in both phases. Beyerlein et al. [40] have determined the phase relation in $BaBiO_{3-x}$ ($0<x<0.5$) using thermogravimetric (TG) measurements. They have observed three nonstoichiometric phases of I, II and III. The present authors have measured the nonstoichiometry by TG measurements to confirm the phase relation [41]. On the whole, the results are similar to those reported by Beyerlein et al. [40], except that the isobars and the phase boundaries are inconsistent with each other.

The phase relations in PrO_x ($1.500<x<1.714$) also still remain in the ambiguity, though praseodymium oxide has been thoroughly investigated by many authors [42-55]. In their studies, a tensimetric measurement has been mostly used, in which the composition of the oxide is determined by the mass change as a function of the temperature and the oxygen partial pressure.

In the present paper, a review has been made on the results of the phase relations and nonstoichiometry in V_2O_{5-x} [56], $BaBiO_{3-x}$ [57] and PrO_x [58] investigated by the present authors using coulometric titration.

2. Experimental

The schematic diagram of the cell used for coulometric titration is shown in Fig.1 [56]. The 8 mol% yttria-stabilized zirconia (YSZ) crucible (inner diameter; 8 mm, outer diameter; 10 mm, height; 30 mm) was used as a solid electrolyte. Two pairs of platinum electrodes, ① - ② and ③ - ④, were attached to the outside and the inside of the crucible, respectively. One pair was used for addition and removal of oxygen, while the other pair was used for the EMF measurements. A Pt-13·Rh wire ⑤ was welded to the electrode ④ and used as a thermocouple.

About 200 mg of the oxide sample was wrapped with a platinum foil and placed in the YSZ crucible. A thin alumina disc (thickness; about 0.5 mm) was fixed to the crucible by an alumina cement (Sumiceram S-208B, Sumitomo Chemical Co., Ltd., Hyperrandom V-903HP, Showa Denko Co., Ltd., etc.). A glassy material such as boron oxide or lead glass was used to seal the enclosed electrode crucible from the surrounding gas. A large variety of glass compositions and softening temperatures are available, and the type of the glassy material chosen depends on the intended temperature range for the measurements. The powdered glass was placed on the alumina disc. The crucible was heated and the glassy material is melted to seal the crucible. The inner lead wires were brought out of the cell through the glass seal.

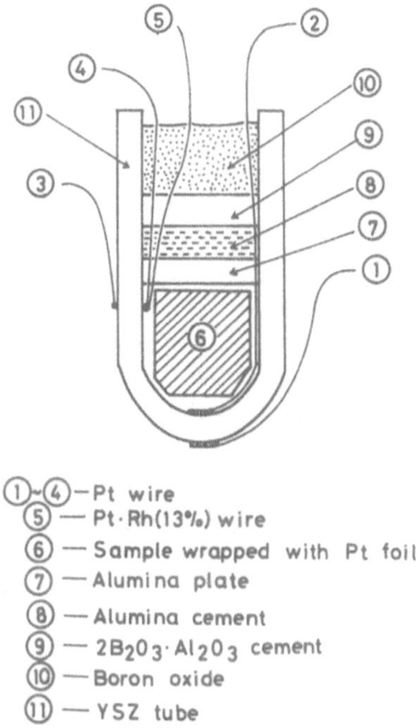

Fig.1 Schematic diagram of the cell used for coulometric titration.

The cell constructed for coulometric titration is expressed as follows:

$$Pt\ ,O_2(M_aO_b)\ |\ YSZ\ |\ O_2(air),\ Pt$$

where air is used as a reference. Air is the most commonly used reference electrode maintaining a fixed oxygen potential. Oxygen is added to or removed from the sample electrode by passing a given amount of current across the cell. The constant dc supplied from the electric source (Potentiostat / Galvanostat HA-501, Hokuto Denko Ltd.) was passed through YSZ between electrode ① and ② . The amount of oxygen transferred through the electrolyte is directly proportional to the amount of electricity, i.e., $n = It/2F$, where n denotes the number of gram-atoms of oxygen, I is the current, t is the time and F is the Faraday constant.

After the titration, a sufficient time is allowed for the electrode to equilibrate internally. The new cell voltage for each oxygen transfer gives the oxygen activity corresponding to the new composition of the oxide. The EMF between the electrodes ③ and ④ was monitored and recorded at a properly chosen time interval. The equilibrium EMF was determined when a constant EMF lasted at least 20 ks. The equilibrium oxygen partial pressure (P_{O_2}) over the sample oxide can be calculated from the Nernst equation as:

$$E = \frac{RT}{4F} \ln \frac{Po_2(\text{air})}{Po_2} \quad (1)$$

where E is the EMF of the cell and R is the gas constant. The atomic ratio of oxygen to metal (O/M) in the oxide sample was determined by assuming that the amount of oxygen in the dead volume of the electrolyte crucible (less than 2×10^{-7} m^3) was negligibly small.

3. Results and Discussion

3.1 V_2O_{5-x}

The equilibrium oxygen partial pressures over V_2O_{5-x} are shown in Fig.2 as a function of x at 880, 900, 920 and 936K. At 920K, reproducible data are obtained and a small hysteresis is observed between both runs of oxygen addition and removal. Mean values of log Po_2 are adopted as equilibrium oxygen partial pressures, and the solid line expresses the variation of the oxygen partial pressure. The value of log Po_2 decreases continuously with an increase in x below 0.055 and it becomes constant to at least O/V=2.255 (x=0.49). For the clarity of the figure, the data points at 880, 900 and 936K are omitted. The X-ray diffraction was conducted on the sample with x=0.035 after cooling from the equilibrium state at 920K to room temperature. The diffraction pattern was identical to that of the starting material and the change in spacings was negligibly small. These results indicate that V_2O_{5-x} the nonstoichiometry up to 0.055 at 920K.

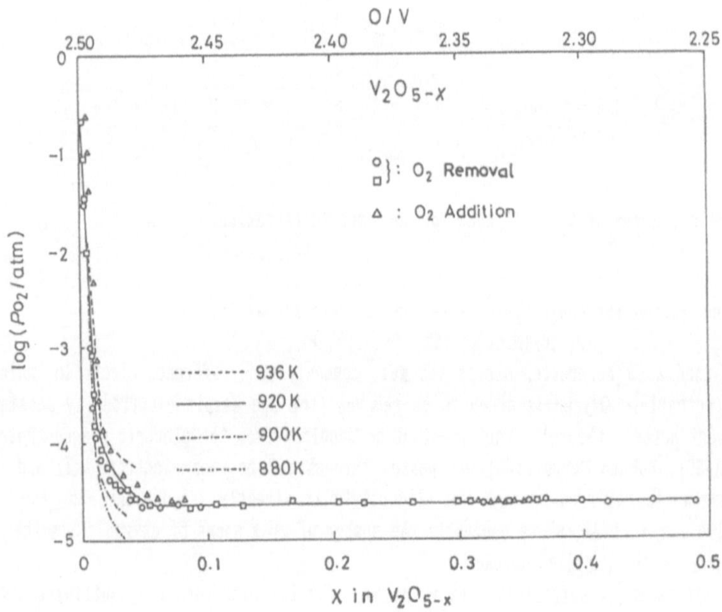

Fig.2 Logarithm of the equiliburium oxygen partial pressure (log Po_2) of V_2O_{5-x} as a function of composition (x).

The nonstoichiometry in V_2O_{5-x} has been reported to be $x=0.02$ at 873K by Dziembaj and Piwowarczyk [35]. However, the accuracy of their data is insufficient to discuss the defect structure in V_2O_5. The present data is precise enough to permit discussion. The relations between log x in V_2O_{5-x} and log Po_2 at 880, 900, 920 and 936K are shown in Fig.3.

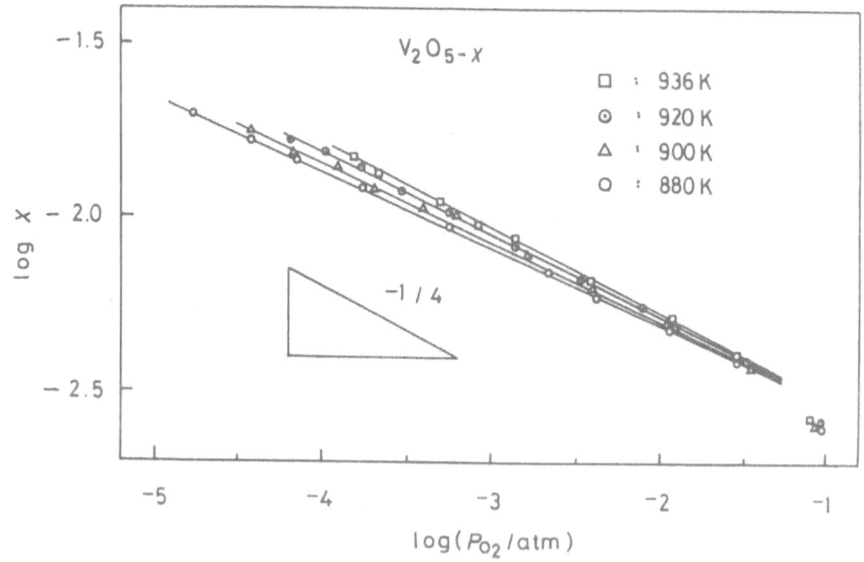

Fig.3 Relation between log x and log Po_2 in V_2O_{5-x}.

The straight lines with the slope of about -1/4 (-1/4.35 at 880K, -1/4.26 at 900K, -1/4.08 at 920K, -1/3.88 at 936K) can be drawn in the region of $x<0.017$. In the reducing process of V_2O_{5-x}, the formation of one oxygen vacancy ($\square_O^{\cdot\cdot}$) by extraction of one oxygen ion (O_O^X) at the lattice site accompanies the reduction of two V^{5+} ions (V_V^X) to V^{4+} ions (V_V') for the conservation of electrical neutrality:

$$2V_V^X + O_O^X \rightarrow 2V_V' + \square_O^{\cdot\cdot} + \frac{1}{2} O_2 \quad (2)$$

Under the assumption of the dilute defect concentration, the mass action law gives the equilibrium constant (K_2) for the reaction (2) as

$$K_2 = \frac{[V_V']^2[\square_O^{\cdot\cdot}]Po_2^{1/2}}{[V_V^X]^2[O_O^X]} = \frac{(2x)^2(x)Po_2^{1/2}}{(2-2x)^2(5-x)} \quad (3)$$

and requires the relation of $x \propto Po_2^{-1/6}$, which relation does not predict the observed data. The electron paramagnetic resonance studies of Gillis and Boesman [59] have confirmed the existence of a V^{4+} ion associated with an oxygen vacancy$(V_V' + \square_O^{\cdot\cdot})^{\cdot}$. Furthermore, they proposed the defect structure of $\overset{5+}{V}_{2-3x/2}\overset{4+}{V}_x\overset{3+}{V}_{x/2}O_{5-x}$ in the nonstoichiometric V_2O_5 assuming the formation of V^{3+} ions. This model requires the following reaction for the defect formation:

$$3V_V^{\times} + 2O_O^{\times} \rightarrow 2(V_V' + \square_O^{\cdot\cdot})^{\cdot} + V_V' + O_2 \tag{4}$$

The equilibrium constant for the above reaction (K_4) is given as follows:

$$K_4 = \frac{[(V_V' + \square_O^{\cdot\cdot})^{\cdot}]^2[V_V']Po_2}{[V_V^{\times}]^3[O_O^{\times}]^2} = \frac{(x)^2(x/2)Po_2}{(2-3x/2)^3(5-x)^2} \tag{5}$$

Equation (5) provides the relation of $x \propto Po_2^{-1/3}$ when x is small. This expected value for the oxygen pressure dependence differs from the observed value of -1/4.

The present authors introduce the following assumption. One of the two V^{4+} ions generated in the formation of an oxygen vacancy makes the defect associate $(V_V' + \square_O^{\cdot\cdot})^{\cdot}$, and the other remains free. The reaction is expressed as

$$2V_V^{\times} + O_O^{\times} \rightarrow V_V' + (V_V' + \square_O^{\cdot\cdot})^{\cdot} + \frac{1}{2}O_2 \tag{6}$$

The equilibrium constant for the reaction (6) is given as follows:

$$K_6 = \frac{[V_V'][(V_V' + \square_O^{\cdot\cdot})^{\cdot}]\,Po_2^{1/2}}{[V_V^{\times}]^2[O_O^{\times}]} = \frac{(x)(x)Po_2^{1/2}}{(2-2x)^2(5-x)} \tag{7}$$

When x is small, Eq.(7) provides the relation of $x \propto Po_2^{-1/4}$.

Regardless of the change in the ratio of O/V, the fixed value of equilibrium oxygen partial pressure indicates the two-phase co-existence as can be seen in Fig.2. The detailed measurements in the region from O/V=2.340 (x=0.32) to O/V=2.320 (x=0.36) confirm the constancy in the equiliburium oxygen partial pressure. This result implies that the V_3O_7 does not exist at 920K. For the determination of the coexisting phase with V_2O_5, X-ray diffraction was conducted on the sample with O/V=2.325 (x=0.35). The X-ray diffraction patterns the sample contains V_2O_5 and V_4O_9. Therefore, the value of log(Po_2/atm)=-4.6 is the equiliburium oxygen partial pressure over the coexisting mixture of V_2O_5 and V_4O_9 at 920K. The V_4O_9 crystal was prepared by decomposing V_3O_7 in supercritical water at 600K and 2kb for 90h in a sealed gold tube and by subsequent slow cooling [27]. On the other hand, another V_4O_9 phase was found by Taniguchi et al. [23-25] in the reducing process from V_2O_5 using SO_2 gas between 683 and 703K. The V_4O_9 with the same X-ray diffraction pattern as the Taniguchi's was also synthesized by using the hydrothermal method at 300K by Theobald et al.[26].

3.2 $BaBiO_{3-x}$

Figure 4 shows typical EMF's as a function of temperature at fixed values of x of $BaBiO_{3-x}$. For the clarity of the figure, all the EMF's are not presented. On the whole, the EMF varies lineally with temperature, and no hysteresis is observed. Identical EMF values are obtained for the different values of x (i.e. at x=0.129 and 0.167, and at x=0.372 and 0.414 around 1000K). This fact indicates the existence of two-phase region.

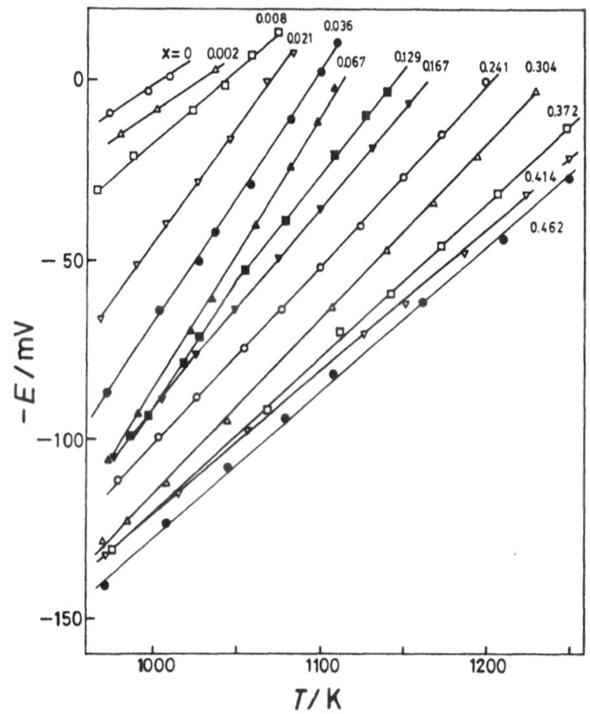

Fig.4 Temperature dependence of the EMF of the cell : Pt, $O_2(BaBiO_{3-x})$ | YSZ | O_2(air), Pt

The relative partial molar free energy of oxygen ($\Delta\bar{G}_{O_2}$) is given by the following relation:

$$\Delta\bar{G}_{O_2} = RT \ln(P_{O_2}) = -4FE + RT \ln[P_{O_2}(\text{air})] \quad (8)$$

The values of $\Delta\bar{G}_{O_2}$ calculated from measured EMF's are plotted as a function of nonstoichiometry in Fig.5. The symbols ○,● and △ represent the results obtained from three different cells. The isotherms between 1125K and 1250K give only one horizontal portion at each temperature indicating the existence of a two-phase region. At the temperatures below 1100K, another horizontal portion is observed in each isotherm. These results indicate the presence of three distinct phases of I, II and

III in $0<x<0.5$, which is consistent with the previous studies [40,41]. However, the phase boundaries have become clearer in the present study. The partial molar free energy of oxygen obtained at 1000K and 975K around $x=0.2$ are found to be dependent on the cell as shown in Fig.5. This result implies the complexity in the phase relation around $x=0.2$ where two distinct phases of $x=0.19$ and $x=0.23$ have been proposed by the different investigators [39,40].

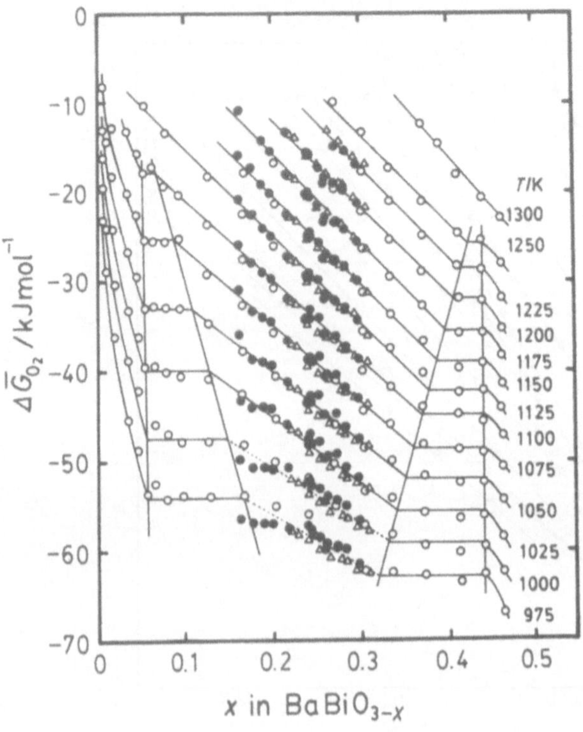

Fig.5 Relative partial molar free energy of oxygen in $BaBiO_{3-x}$ at temperatures between 975 and 1300K as a function of composition. Symbols ○, ● and △ show the results obtained from these different cells.

Figure 6 shows the isobars for $BaBiO_{3-x}$ in the temperature range of 975-1300K, which are obtained from the results shown in Fig.5. The boundary of phase I exists at x=0.05. This result is consistent with that of the previous study [41], though the two-phase region of I and II is somewhat smaller. For instance, at 1050K the two-phase region lies in the range of x=0.05-0.11 contrary to x=0.05-0.12 as reported previously [41]. The other two-phase region of II and III is found at higher temperatures and larger nonstoichiometry (i.e., x=0.37-0.44 contrary to x=0.33-0.42 [41] at 1100K).

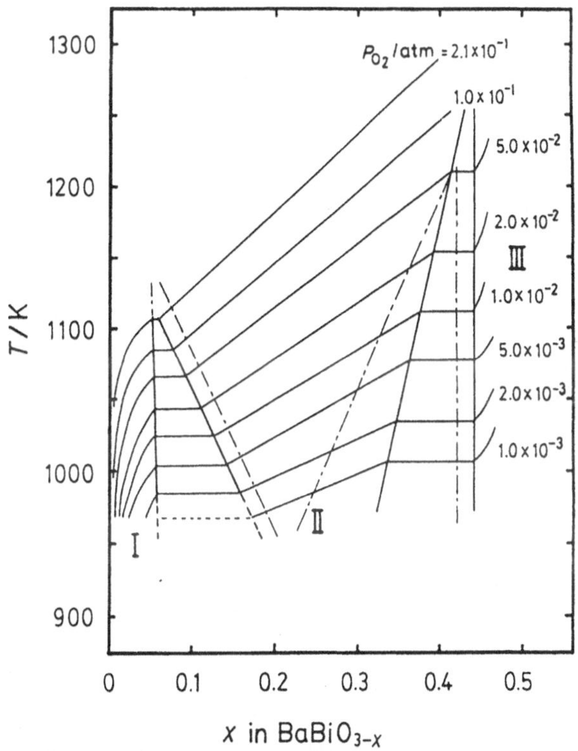

Fig.6 Relation between temperature (T) and composition (x) in $BaBiO_{3-x}$ at the oxygen partial pressures indicated. The dashed and chained lines indicate two-phase regions and phase boundaries, respectively, obtained by thrmogravimetric measurements [41].

The relative partial molar enthalpy ($\Delta\bar{H}_{O_2}$) and entropy ($\Delta\bar{S}_{O_2}$) of oxygen are respectively expressed by

$$\Delta\bar{H}_{O_2} = - \left[\frac{d(\Delta\bar{G}_{O_2}/T)\Delta\bar{H}_{O_2}}{d(1/T)} \right]_{x,p} \tag{9}$$

and

$$\Delta\bar{S}_{O_2} = - \left[\frac{d\Delta\bar{G}_{O_2}}{dT} \right]_{x,p} \tag{10}$$

The relation between $\Delta\bar{G}_{O_2}/T$ and $1/T$ above 1050K is almost linear. Using the least square method, the values of $\Delta\bar{H}_{O_2}$ are calculated from $\Delta\bar{G}_{O_2}$ above 1050K in Fig.5, and are shown in Fig.7 as a function of nonstoichiometry. The relative partial molar enthalpy of oxygen decreases in phase I and increases in phase II with an increase in x.

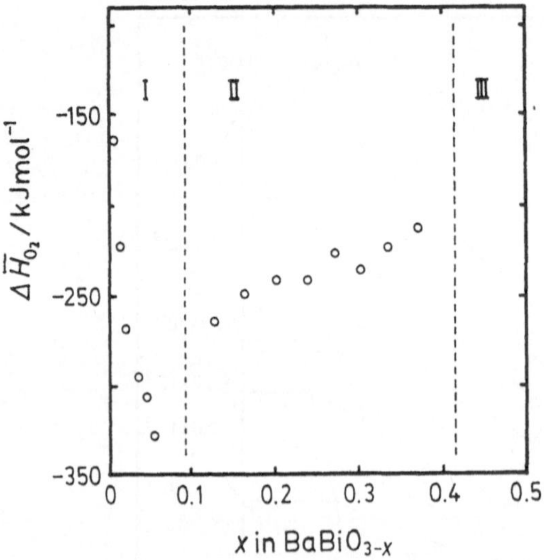

Fig.7 Relative partial molar enthalpy of oxygen in $BaBiO_{3-x}$ as a function of composition

The relative partial molar entropy of oxygen is calculated using Eq.(10) and shown in Fig.8. The oxygen deficiency in $BaBiO_{3-x}$ can perhaps be explained in terms of the oxygen vacancy as a predominant point defect. For the charge neutrality in $BaBiO_3$, it has been considered reasonable that the bismuth ion takes two valence states of Bi^{3+} and Bi^{5+}. Therefore, the dissolution of oxygen into $BaBiO_{3-x}$ (phase II) can be written as follows:

$$\frac{1}{2}O_2 + V_O^{\bullet\bullet} + Bi_{Bi}^{X} = O_O^{X} + Bi_{Bi}^{\bullet\bullet} \qquad (11)$$

where Bi_{Bi}^{X} and $Bi_{Bi}^{\bullet\bullet}$ represent bismuth ion of Bi^{3+} and Bi^{5+}. Assuming that the defects are randomly distributed and non-interacting, the configurational entropy ($\bar{S}_{conf}$) is given as follow:

$$\bar{S}_{conf} = k\left[\frac{N!}{N!} + \ln\frac{N!}{(N/2+n)!(N/2-n)!} + \ln\frac{(3N)!}{(3N-n)!n!}\right] \qquad (12)$$

where N is the number of the lattice site of Ba and Bi, $3N$ the number of the lattice site of oxygen in $BaBiO_3$, n the number of oxygen vacancy, and k the Boltzmann constant. Three terms in the square brackets are the configurational entropies for Ba, Bi and O sites, respectively. Using Stirling's formula, $\bar{S}_{conf}$ can be calculated from Eq.(11) as:

$$\bar{S}_{conf} = -kN[(0.5 + \frac{n}{N})\ln(0.5 + \frac{n}{N}) + (0.5 - \frac{n}{N})\ln(0.5 - \frac{n}{N}) + (3 - \frac{n}{N})\ln(3 - \frac{n}{N}) + \frac{n}{N}\ln(\frac{n}{N}) - 3\ln 3] \quad (13)$$

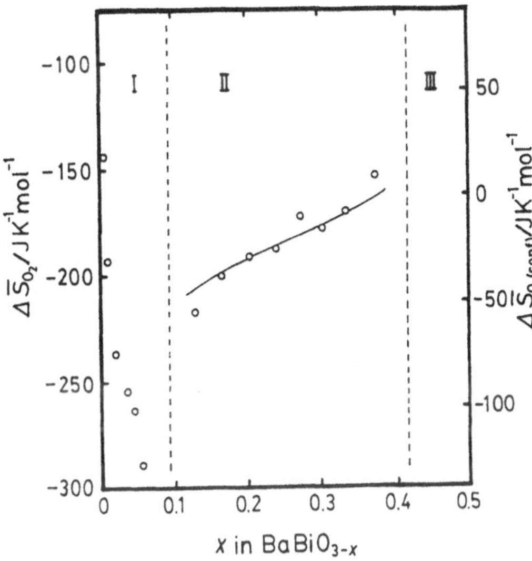

Fig.8 Relative partial molar entropy of oxygen in $BaBiO_{3-x}$ as a function of composition. The caluculated value of $\Delta\bar{S}_{O_2}$(conf) is drawn as a solid line.

Since the concentration of the oxygen vacancy is

$$[V_O^{\bullet\bullet}] = \frac{n}{N} = x \quad (14)$$

and the gas constant is

$$R = kN \quad (15)$$

Eq.(12) can be rewritten as follows

$$\bar{S}_{conf} = -R[(0.5 + x)\ln(0.5 + x) + (0.5 - x)\ln(0.5 - x) + (3 - x)\ln(3 - x) + x\ln x - 3\mathrm{in}3] \quad (16)$$

Assuming that the vibrational entropy is independent of composition, the configurational where dependence of the relative partial molar entropy of oxygen, $\Delta\bar{S}_{O_2}$(conf), can be calculated from Eq.(16) as follows:

$$\Delta\bar{S}_{O_2}(\mathrm{conf}) = -2\frac{\mathrm{d}}{\mathrm{d}x}\bar{S}_{conf} = -2R\ln\frac{(3-x)(0.5-x)}{x(0.5+x)} \quad (17)$$

The variation of $\Delta\bar{S}_{O_2}$(conf) in the phase II is also shown in Fig.8 by a solid line. The variation of the relative partial molar entropy of oxygen with composition is in good agreement with that of $\Delta\bar{S}_{O_2}$(conf) in phase II.

3.3 PrO_x

The relation between E and x in PrO_x at each temperature from 1200 to 1350K is shown in Fig.9. No hysteresis is observed between both runs of oxygen addition and removal.

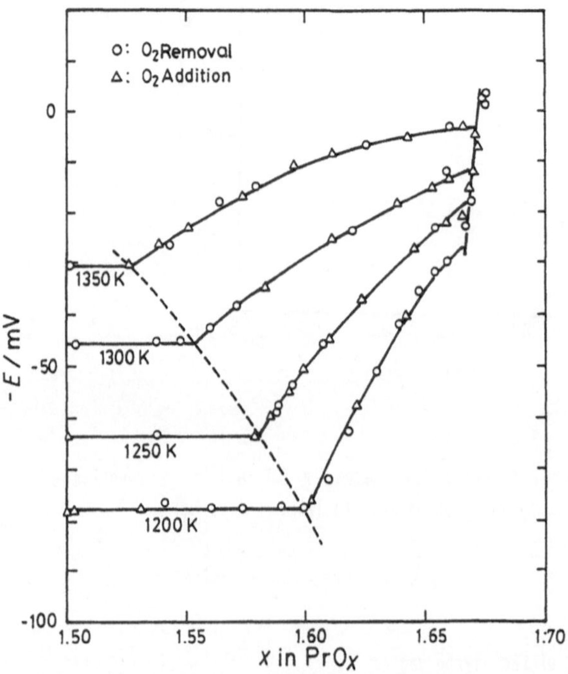

Fig.9 Temperature dependence of the EMF of the cell : Pt, $O_2(PrO_x)$ | YSZ | O_2(air), Pt.

A distinct line phase is observed at x=1.667 which corresponds to the composition of n=6 in the homologous series of Pr_nO_{2n-2}. This phase is newly discovered in the present study.

As x decreases, there exists a nonstoichiometric phase of rather wide range which may correspond to the sigma-phase designated by Hyde et al.[47]. Moreover, a diphasic region is observed between the sigma-phase and the stable sesquioxide of $PrO_{1.500}$. This diphasic region has not been informed yet. The width of the diphasic region becomes narrower as the temperature increases.

In Fig.10, log P_{O_2} is plotted against x and the results obtained by Endo et al. [60] are also plotted in dotted lines for comparison. It is clearly shown that the stability region of the sigma-phase given by the present study mostly includes their isotherms, although the gradients of both isoterms are different from each other.

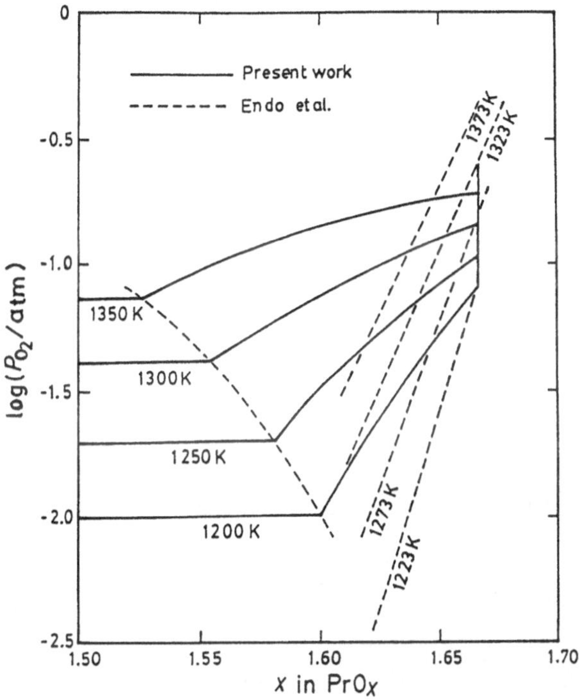

Fig.10 Logarithm of the equilibrium oxygen partial pressure (log P_{O_2}) of PrO_x as a function of composition (x).

Figure 11 illustrates the superposition of the stability region of the sigma-phase obtained in the present study on the phase diagram of Pr-O system given by Hyde et al.[47]. The difference in the stability region of the sigma-phase is clearly shown and the existence of the line phase, $PrO_{1.667}$, is conclusively indicated.

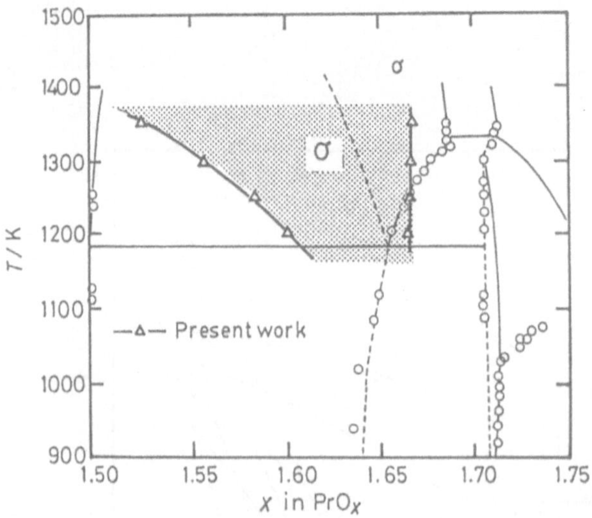

Fig.11 Stability region of the sigma-phase superposed onthe phase diagram of Pr-O system given by Hyde et al [47].

4. Conclusion

The phase relation and the nonstoichiometry in V_2O_{5-x}, $BaBiO_{3-x}$ and PrO_x have been measured by coulometric titration using yttria-stabilized zirconia electrolyte. The maximum oxygen deficiency in V_2O_{5-x} is determined to be 0.055 at 920K. The $BaBiO_{3-x}$ has three distant phases in $0<x<0.5$. The existence of $PrO_{1.667}$ phase is suggested in PrO_x. The results prove that coulometric titration is one of the powerful tool to study the phase relation particularly in the oxides with the small deviation from the stoichiometric composition.

Acknowledgment

The authors wish to express their hearty thanks to Mr. Y. Ito, Mr. A. Yamanaka and Mr. Y. Watanabe for performing the collaboration in the present study.

References

[1] K. Kiukkola and C. Wagner, *J. Electrochem. Soc.* **104** (1957) 379.
[2] R.A. Gidding and R.S. Gordon, *J. Electrochem. Soc.* **121** (1974) 793.
[3] I. Riess and D.S. Tannhauser, 'Transport in Nonstoichiometric Compounds', Elseviev Sci. Publ., Amsterdam, pp. 503-517 (1982).
[4] R.A. Rapp and D.A. Shores, 'Physicochemical Measurements in Metals Research ', Part 2, ed. by R.A. Rapp, Wiley-Interscience, New York, pp. 123-192 (1970).
[5] T.L. Markin and R.J. Bones, *Atomic Energy Research Establishment Report*, AERE-4178 (1962).
[6] K. Kiukkola, *Acta Chem. Scand.* **16** (1962) 327.
[7] A. Nakamura and T. Fujino, *J. Nucl. Mater.* **149** (1987) 80.
[8] Y.D. Tretyakov and R.A. Rapp, *Trans. Met. Soc. AIME* **242** (1969) 1235.
[9] H.G. Sockel and H. Schmalzried, *Ber. Bunsenges. Phys. Chem.* **72** (1968) 745.
[10] F.E. Rizzo and J.V. Smith, *J. Phys. Chem.* **72** (1968) 485.
[11] H.F. Rizzo, R.S. Gordon and J.B. Cutler, *J. Electrochem. Soc.* **116** (1969) 267.
[12] B.E.F. Fender and F.D. Riley, *J. Phys. Chem. Solids* **30** (1969) 793.
[13] A.D. Wadsley, *Acta Cryst.* **10** (1957) 261.
[14] F. Aebi, *Helv. Chim. Acta* **31** (1948) 8.
[15] G. Anderson, *Acta Chem. Scand.* **8** (1954) 1559.
[16] F. Theobald, R. Cabala and J. Bernard, *Compt. Rend.* **C266** (1968) 1534.
[17] K. Wilhelmi, K. Waltersson and L. Kihelborg, *Acta Chem. Scand.* **25** (1971) 2675.
[18] T. Tudo and G. Tridot, *Compt. Rend.* **C261** (1965) 2911.
[19] G. Tridot and J. Tudo, *Compt. Rend.* **C263** (1966) 421.
[20] K. Waltersson, B. Forslund and K. Wilhelmi, *Acta Cryst.* **B30** (1974) 2644.
[21] k. kosuge, *J. Phys. Chem. Solids* **28** (1967) 1613.
[22] S. Kachi, K. Kosuge, M. Shiotani and N. Nakanishi, *Shokubai* [*Catalysis*] **10** (1968) 103.
[23] M. Taniguchi, A. Miyazaki and H. Yokomizo, *Shokubai* **10** (1968) 99.
[24] M. Taniguchi, S. Kato and T. Nanao, *Shokubai* **14** (1972) 53.
[25] M. Taniguchi, in Proc. 4th Intern. Conf. Thermal Analysis, **Vol. 1** (1974) 727.
[26] F. Theobald, R. Cabala and J. Bernard Compt. Rend. **C269** (1969) 1209.
[27] K. Wilhelmi and K. Waltersson, *Acta Chem. Scand.* **24** (1970) 3409.
[28] T. Sata, E. Komada and Y. Ito, *Kogyo Kagaku Zasshi* **71** (1968) 643.
[29] T. Sata and Y. Ito, *Kogyo Kagaku Zasshi* **71** (1968) 647.
[30] T. Sata and Y. Ito, Bull. *Tokyo Inst. Technol.* **98** (1970) 1.
[31] K. Iwase and N. Nasu, Sci. *Rept. Tohoku Imp. Univ., Honda Aniversary* (1936) 476.
[32] H. Flood and J. Kleppa, *J. Am. Chem. Soc.* **69** (1947) 998.
[33] A.A. Fotiev and V.L. Volkov, *Zh. Fiz. Khim.* **45** (1971) 2610.
[34] H. Endo, M. Wakihara and M. Taniguchi, *Chem. Letters* (*Chem. Soc. Japan*) (1974) 905.
[35] R. Dziembaj and J. Piwowarczyk, *J. Solid State Chem.* **21** (1977) 387.
[36] D.E. Cox, J.L. Gillson and P.E. Bierstedt, *Solid State Commun.* **17** (1975) 27.

[37] M. Suzuki and T. Murakami, *Solid State Commun.* **53** (1985) 691.
[38] D.E. Cox and A.W. Sleight, *Acta Crystallogr.* **B35** (1979) 1.
[39] C. Chaillout and J.P. Remeika, *Solid State Commun.* **56** (1985) 833.
[40] R.A. Beyerlein, A.J. Jacobson and L.N. Yacullo, *Mat. Res. Bull.* **20** (1985) 877.
[41] Y. Saito, T. Maruyama and A. Yamanaka, *Thermochim. Acta* **115** (1987) 199.
[42] J.D. McCullough, *J. Am. Chem. Soc.* **72** (1950) 1386.
[43] R.E. Ferguson, E.D. Guth and L. Eyring, *J. Am. Chem. Soc.* **76** (1954) 3890.
[44] E.D. Guth, J.R. Holden, N.C. Baenziger and L. Eyring, *J. Am. Chem. Soc.* **76** (1954) 5239.
[45] E.D. Guth and L. Eyring, *J. Am. Chem. Soc.* **76** (1954) 5245.
[46] C.T. Stubblefield, H. Eick and L. Eyring, *J. Am. Chem.Soc.* **78** (1956) 3018.
[47] B.G. Hyde, D.J.M. Bevan and L. Eyring, *Phil. Trans. Roy. Soc.* London **259A** (1966) 583.
[48] J. Kordis and L. Eyring, *J. Phys. Chem.* **72** (1968) 2030.
[49] J. Kordis and L. Eyring, *J. Phys. Chem.* **72** (1968) 2044.
[50] D.A. Burnham and L. Eyring, *J. Phys. Chem.* **72** (1968) 4415.
[51] R.P. Turcott, J.M. Warmkessel, R.J.D. Tilley and L. Eyring, *J. Solid State Chem.* **3** (1971) 265.
[52] R.P. Turcott, M.S. Jenkins and L. Eyring, *J. Solid State Chem.* **7** (1973) 454.
[53] A.T. Lowe, K.H. Lau and L. Eyring, *J. Solid State Chem.* **15** (1975) 9.
[54] E. Summerville, R.T. Tunge and L. Eyring, *J. Solid State Chem.* **24** (1978) 21.
[55] R.T. Tunge and L. Eyring, *J. Solid State Chem.* **29** (1979) 165.
[56] Y. Ito, T. Maruyama and Y. Saito, *Solid State Ionics* **25** (1987), 199.
[57] T. Maruyama, A. Yamanaka and Y. Saito, Unpublished
[58] Y. Watanabe, T. Maruyama and Y. Saito, Unpublished
[59] E. Gills and E. Boesman, *Phys. Status Solidi* **14** (1966), 337.
[60] K. Endo, S. Yamauchi, K. Fueki and T. Mukaibo, *Bull. Chem. Soc. Japan* **49** (1976) 2379.

DEFECT STRUCTURE AND THE RELATED PROPERTIES OF UO_2 AND DOPED UO_2

Keiji Naito, Toshihide Tsuji and Tsuneo Matsui
Department of Nuclear Engineering, Faculty of Engineering,
Nagoya University, Furo-cho, Chikusa-ku, Nagoya 464-01,
Japan

ABSTRACT. The oxygen potentials and the electrical conductivities of UO_2 doped with various cations (La, Gd, Pu, Th, Nb, Cr, Ti) measured by the authors were reviewed in comparison with the previous results on UO_2 doped with other cations (Mg, Y, Gd, Th, Zr) by other investigators. The defect structures of UO_2 doped with these various cations were discussed based on these data. Diffusion coefficients of uranium in UO_2 doped with various cations (La, Y, Ti, Nb) previously measured by some researchers together with our recent results on uranium diffusion in UO_2 doped with Ti ion were also reviewed in relation to the defect structures. It was generally concluded that (1) the oxygen potential and the electrical conductivity of UO_{2+x} are increased (or decreased) by doping with lower (or higher) valent cations, and (2) the diffusion coefficients of uranium in UO_{2+x} are decreased (or increased) by doping with lower (or higher) valent cations. These facts are explained from the valence control rule, assuming that these metals are present as the substitutionals for uranium ions. However our recent results on the oxygen potential, the electrical conductivity and the diffusion coefficient of uranium in UO_{2+x} doped with Ti ion can not be explained by the general rule mentioned above, suggesting the presence of titanium interstitials instead of titanium substitutionals for uranium ions. The effects of dopants with various valences (Nb, Cr, Ti, Gd, Y) on the creep rate, the fission gas release and the heat capacity were also discussed.

1. INTRODUCTION

It has been shown that the doping of aliovalent cations to the UO_2 fuel changes the properties such as electrical and thermal conductivities, oxygen potential, creep property, release of fission gas, etc. [1-3]. Since the changes of those properties are thought to be based upon the introduction of defect structure in UO_2 by doping, the knowledge of the defect structures of UO_2 doped with these cations is important to understand the changes of these properties. In the present paper, the oxygen potentials and the electrical conductivities of UO_2 doped with various cations (La[4], Gd[5], Pu[6], Th[7], Nb[8], Cr[9], Ti[10])

J. Nowotny and W. Weppner (eds.),
Non-Stoichiometric Compounds Surfaces, Grain Boundaries and Structural Defects, 27–44.

reported by the authors are reviewed in comparison with the previous results on UO_2 doped with other cations (Mg[11], Y[12], Gd[13], Th[14], Zr[15]) by other investigators. Based on these data, the defect structures of UO_2 doped with these various cations are discussed. The diffusion coefficients of uranium in UO_2 doped with various cations (La[16], Y[16], Nb[16], Ti[17,18]) previously measured by Matzke together with our recent results on uranium diffusion in UO_2 doped with Ti ion [19] are also reviewed in relation to the defect structures. The creep rates of UO_2 doped with Nb^{5+}[2] and Gd^{3+}[20], and the fission gas release in UO_2 doped with Cr[21], Ti[22], Nb[16,22] and Y[23] are reviewed in relation to the diffusion coefficient of uranium. Finally, the origin of the excess heat capacity of UO_2 doped with Gd_2O_3[24,25] at high temperature observed recently by the present authors is discussed.

2. OXYGEN POTENTIAL AND ELECTRICAL CONDUCTIVITY

The oxygen potentials ($\Delta\bar{G}o$ =RT $\ln Po_2$) at about 1282K of UO_{2+x} doped with various cations are shown as a function of O/M ratio in Fig. 1. It is seen from the figure that (1) $\Delta\bar{G}o_2$ values of UO_{2+x} are increased by doping with cations of lower valency (La^{3+}[4], Mg^{2+}[11], Gd^{3+}[13], Pu^{3+}[26]) than the uranium ions and are decreased by doping with cations of higher valency (Nb^{5+}[8], Cr^{6+}[9]). (2) In the cases of UO_{2+x} doped with homovalent (tetravalent) cations such as Th^{4+}[14] and Zr^{4+}[15], the effect of the dopants is not clear at near-stoichiometric compositions owing to the lack of data, although in the region where x is large, the $\Delta\bar{G}o_2$ values for $(U,Th)O_{2+x}$ and $(U,Zr)O_{2+x}$ are nearly equal to and lower than that of UO_{2+x}, respectively. These facts indicate that the increase (or decrease) in $\Delta\bar{G}o_2$ can be estimated, to a first approximation, from the increase (or decrease) in the oxidation state of the remaining uranium ions after substitution of dopant ions for uranium ions. However the lower $\Delta\bar{G}o_2$ value of UO_{2+x} doped with Ti ion (tetravalency) than that of UO_{2+x} recently observed by the present authors [10] can not be interpreted by the substitutional model, but may be explained by assuming the Ti^{4+} interstitials as the predominant defect. Similarly the lower $\Delta\bar{G}o_2$ values observed in the cases of UO_{2+x} doped with Cr ion may have the possibility to be interpreted by assuming the Cr^{3+} interstitials instead of the substitutional for uranium ions.

The $\Delta\bar{G}o_2$ values of UO_{2+x} (x=0.01 and 0.04) doped with various cations at about 1282K are shown as a function of dopant concentration in Fig. 2. It is seen from the figure that the change in the $\Delta\bar{G}o_2$ values of doped UO_{2+x} at smaller x (x=0.01) is mainly controlled by the valency of the dopants as discussed above, but the effect of the dopants becomes less significant at larger x (x=0.04), since the dopant content is not so large in comparison with the oxygen nonstoichiometry x. The similar dependences of $\Delta\bar{G}o_2$ values upon O/M ratio and upon dopant concentration were observed at temperatures 1173 and 1373K, too.

The oxygen partial pressure (Po_2) dependences of the compositional deviation (x) from stoichiometry for pure and doped UO_{2+x} at about 1282

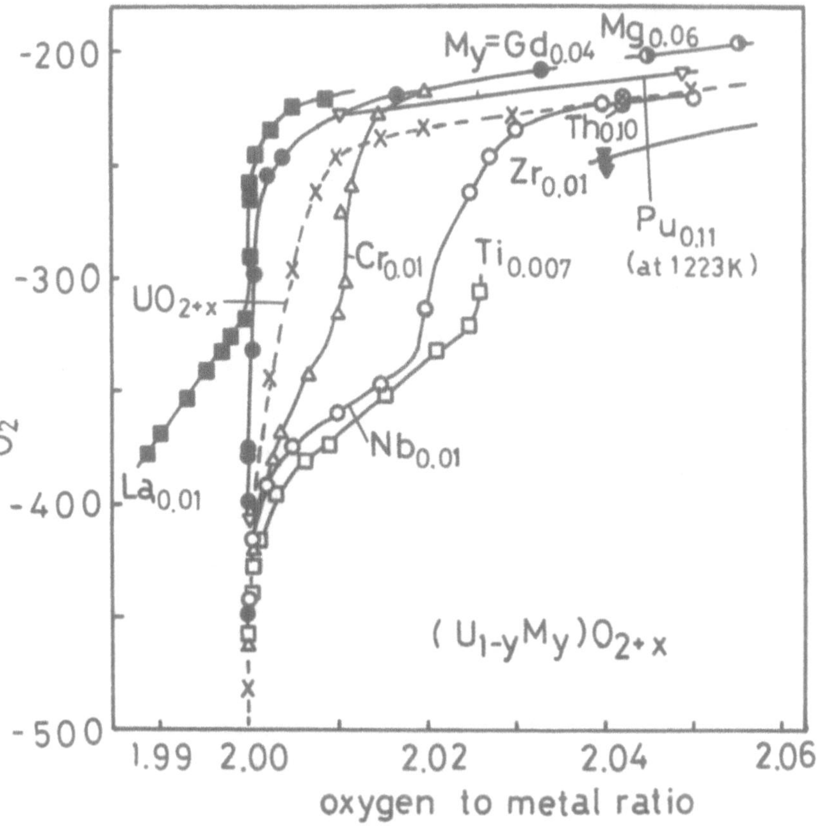

Figure 1. Oxygen potential ($\Delta\bar{G}o_2$) vs. oxygen-to-metal ratio for $(U_{1-y}M_y)O_{2+x}$ at about 1282 K.

K are shown in Fig. 3. In the figure, three different dependences of x upon the oxygen partial pressure are seen for pure UO_{2+x}. The values of n for $x \propto Po_2^{1/n}$ for pure UO_{2+x} are observed as 2, 12 and 2 in turn from the low oxygen partial pressure for three oxygen partial pressure regions. The oxygen partial pressure dependences of x for UO_{2+x} doped with various cations [4,6,8,10,11,13-15,26] are similar to those of pure UO_{2+x}, indicating the presence of the similar defect structures in the doped UO_{2+x} to those of pure UO_{2+x}.

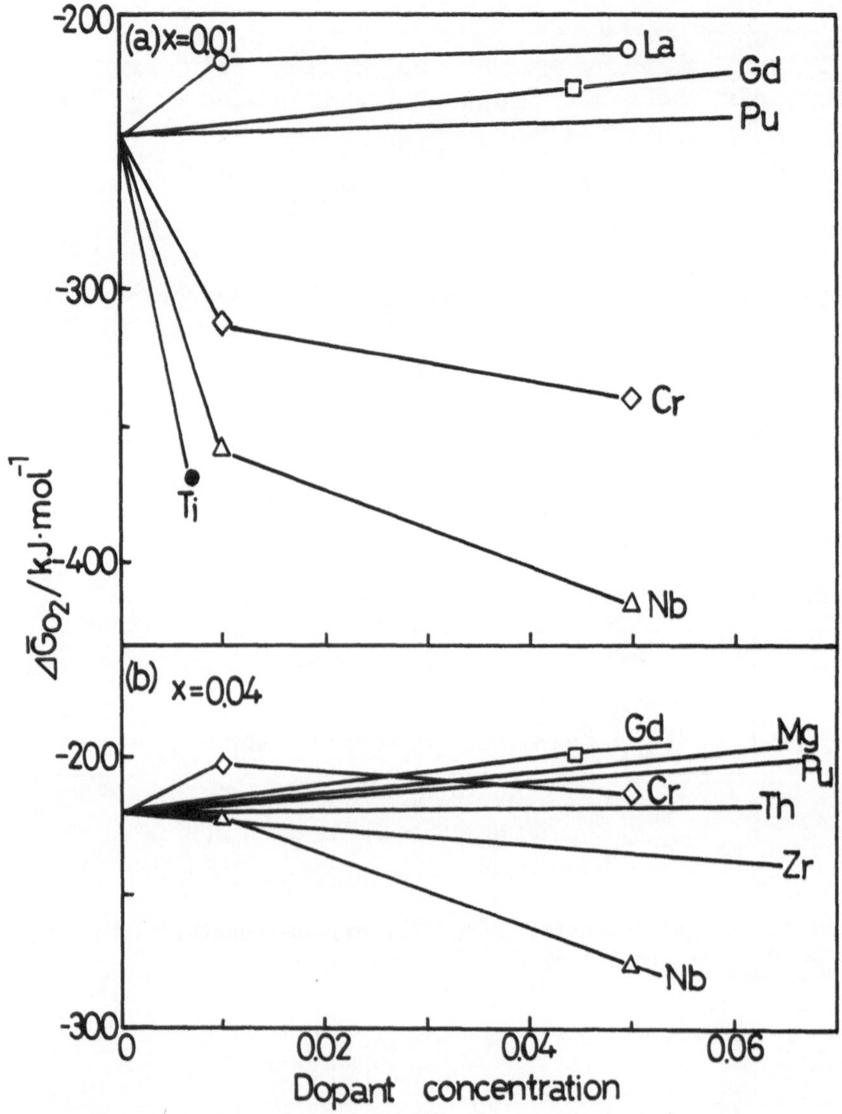

Figure 2. Oxygen potential ($\Delta\bar{G}o_2$) vs. dopant concentration (y) for $(U_{1-y}M_y)O_{2+x}$ ((a) x=0.01 and (b) x=0.04).

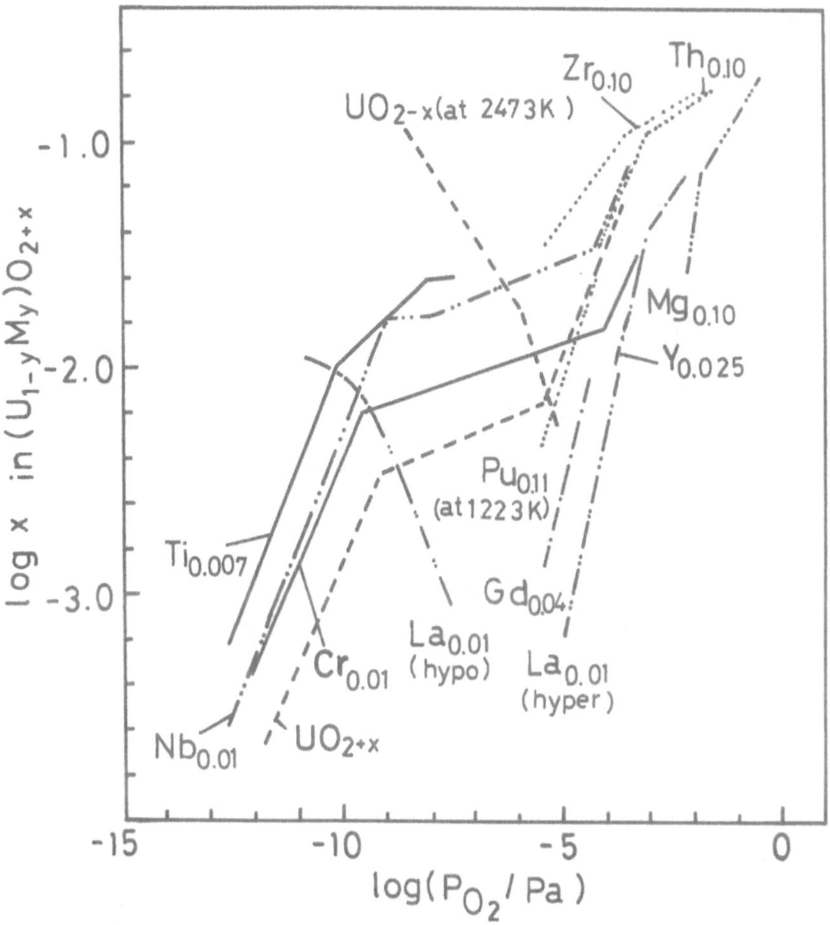

Figure 3. Dependence of x on Po_2 for $U_{1-y}M_yO_{2+x}$ at about 1282 K.

In Fig. 4, the oxygen partial pressure dependences of the electrical conductivity (σ) for pure and doped UO_{2+x} at about 1282 K are also shown. It is seen from the figure that three different dependences of σ upon Po_2 for pure UO_{2+x} exist. The values of n' for $\sigma \propto Po^{1/n'}$ for pure UO_{2+x} are seen from the figure as ∞ , 12 and 2 in turn from the low oxygen partial pressure, which corresponds to the relation between x and Po_2. The oxygen partial pressure dependences of σ for UO_{2+x} doped with various cations are similar to those for pure UO_{2+x}. Both relationships between x-Po_2 and σ-Po_2 will be discussed in the following section.

It is seen from the figure that (1) the electrical conductivity of UO_{2+x} is increased by doping with cations of lower valency (Gd^{3+}[5], Pu^{3+}[6], Y^{3+}[12]) than uranium ions, since the lower valent cations substituted for uranium ions can act effectively as hole donors. In the case of UO_{2+x} doped with La^{3+}[4], the effect of dopant seems to be small in the concentration of 1 mol%, but the electrical conductivity of UO_{2+x} doped with 5 mol% La [4] showed nearly the same value as that of $(U_{0.927}Gd_{0.073})\ O_{2+x}$ [5] in Fig. 4 as expected from the valence control rule. (2) Conversely, the electrical conductivity of UO_{2+x} is decreased by doping with cations of higher valency (Nb^{5+} [8] and Cr^{6+} [9]) which act effectively as electron donors. (3) The electrical conductivity of $U_{0.99}Th_{0.01}O_{2+x}$ [7] is nearly equal to that of pure UO_{2+x}, because Th is homovalent cations.

However the lower electrical conductivity of UO_{2+x} doped with Ti ion (tetravalency [10]) than that of UO_{2+x} can not be interpreted by the substitutional model, but may be explained by assuming the Ti interstitials as well as the case of oxygen potential. Similarly, the lower electrical conductivity observed for UO_{2+x} doped with Cr ions may have the possibility of the Cr^{3+} interstitials [9] instead of the substitutionals for uranium ions. In the case of $(U,Nb)O_{2+x}$ [8], a minimum in σ corresponding to the transition between n- and p-type conduction is seen. The band-gap energy of $(U,Nb)O_{2+x}$ was calculated

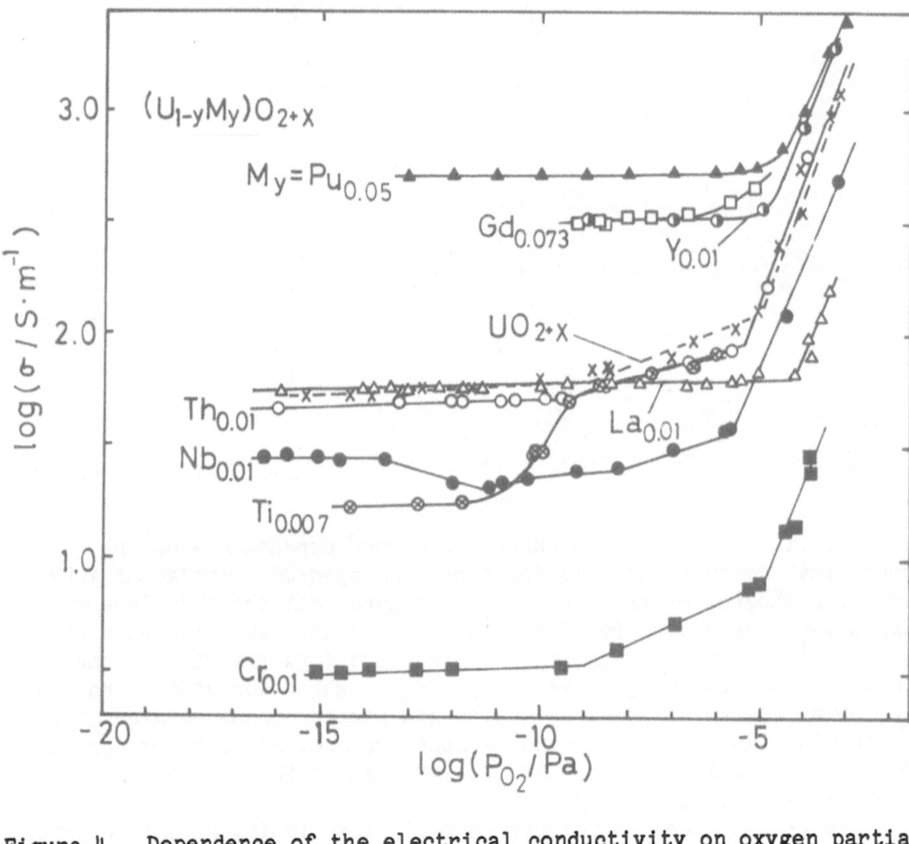

Figure 4. Dependence of the electrical conductivity on oxygen partial pressure for $(U_{1-y}M_y)O_{2\pm x}$ at about 1282 K.

from the temperature dependence of the minimum of the electrical conductivity and found to be 248±12 $kJ \cdot mol^{-1}$, which is nearly equal to that of UO_{2+x} [8].

3. DEFECT STRUCTURE

The oxygen partial pressure (Po_2) dependences of the departure from stoichiometric composition (x) and the electrical conductivity (σ) for $MO_{2\pm x}$ are usually expressed as $x \propto Po_2^{1/n}$ and $\sigma \propto Po^{1/n'}$, respectively. The defect structure can be discussed from the values of n or n'.

Based on the values of n and n' obtained from Figs. 3 and 4, the defect structures in these oxides are discussed by considering the complex defect model consisting of the oxygen vacancies (Vo) at the normal lattice sites and two kinds of interstitial oxygens (O_i^a and O_i^b) and also some electroneutrality conditions between the concentrations of the dopant cations substituted for uranium ions and those of the complex defect. The results are summarized in Table I.

Using the Kröger-Vink notation [27], the formation of the complex defect $\{2(O_i^a O_i^b Vo)\}^{m'}$ for pure UO_{2+x} [7,28] and $(U,Th)O_{2+x}$ [7] is represented by

$$2V_i^a + 2V_i^b + 2O_o + O_2(g) = \{2(O_i^a O_i^b Vo)\}^{m'} + mh^{\cdot} , \qquad (1)$$

where $h^{\cdot}$ is a hole and m' is the charge of the complex defect. The dependences of n=2 and n'=2 observed from x-Po_2 and σ-Po_2 relations, respectively, in the relatively higher oxygen partial pressure region can be interpreted by eq. (1) taking the value m=1. As for the defect structure in the composition near stoichiometry for pure UO_{2+x} and $(U,Th)O_{2+x}$, where n=n'=12 or n=6, the formation of the complex defects $(2O_i^a O_i^b 2Vo)$ is assumed:

$$2V_i^a + V_i^b + 2O_o + 1/2\ O_2(g) = \{2O_i^a O_i^b 2Vo\}^{m'} + mh^{\cdot} . \qquad (2)$$

The dependences of x and σ upon Po_2, n=n'=12 or n=6 can be interpreted by eq.(2) taking m=5 or m=2, respectively. In table I, a model for UO_{2+x} recently proposed by Nakamura and Fujino [28] is also shown for comparison. The diffence is caused from the interpretation of the values of n and n' for the intermediate oxygen partial pressure region, n=n'=12 and 6, but the n' value of 6 instead of 12 for the intermediate region is difficult to determine from the log σ vs. log Po_2 relation.

In the case of $(U,La)O_{2+x}$ [4] and $(U,Nb))O_{2+x}$ [8], the dependences of x and σ upon Po_2 can be explained by using electroneutrality conditions shown in Table I and by using either eqs. (1) or (2). For $(U,La)O_{2+x}$, the following reactions are assumed:

$$2O_i^a + O_i^b + 2V_o = \{2V_o^a V_o^b 2O_i\}^{\cdot} + e' + 1/2\ O_2(g) , \qquad (3)$$

and

$$2O_i^a + 2O_i^b + 2V_o = \{2(V_o^a V_o^b O_i)\}^{\cdot} + e' + O_2(g) , \qquad (4)$$

where e' is an electron.

TABLE I
Defect structures for UO_{2+x} and $(U, M)O_{2+x}$ (M=metal)

phase	Po_2 region	$x \propto Po_2^{1/n}$ n value	$\sigma \propto Po_2^{1/n'}$ n' value	defect model	neutrality conditions
UO_{2+x} [7]	high	2	2	$\{2(O_i^a O_i^b Vo)\}'$	$[h^\cdot]=[\{2(O_i^a O_i^b Vo)\}']$
and	intermediate	12	12	$\{2O_i^a O_i^b 2Vo\}^{5'}$	$[h^\cdot]=5[\{2O_i^a O_i^b 2Vo\}^{5'}]$
$(U,Th)O_2+x$[7]	low	2	very large	$\{2O_i^a O_i^b 2Vo\}^x$	$[h^\cdot]=[e']$
UO_{2+x}[28]	high	2	2	$\{2(O_i^a O_i^b Vo)\}'$	$[h^\cdot]=[\{2(O_i^a O_i^b Vo)\}']$
	intermediate	6	6	$\{2O_i^a O_i^b 2Vo\}''$	$[h^\cdot]=2[\{2O_i^a O_i^b 2Vo\}'']$
	low	2	very large	$\{2O_i^a O_i^b 2Vo\}''$	$[h^\cdot]=[e']$
$(U,La)O_{2+x}$[4]	high	2	2	$\{2(O_i^a O_i^b Vo)\}'$	$[h^\cdot]=[\{2(O_i^a O_i^b Vo)\}']$
	intermediate	1	very large	$\{2(O_i^a O_i^b Vo)\}'$	$[h^\cdot]=[La_U']$
$(U,La)O_{2-x}$[4]	low	-1 ~ -2	very large	$\{2(V_o^a V_o^b O_i)\}^\cdot$ or $\{2V_o^a V_o^b 2O_i\}^\cdot$	$[h^\cdot]=[La_U']$

TABLE I
Continued

phase	Po_2 region	$x \propto Po_2^{1/n}$ n value	$\sigma \propto Po_2^{1/n'}$ n' value	defect model	neutrality conditions
	high	2	2	$\{2(O_i^a O_i^b Vo)\}'$	$[h^\cdot]=[\{2(O_i^a O_i^b Vo)\}']$
	intermediate	12	12	$\{2O_i^a O_i^b 2Vo\}^{5'}$	$[h^\cdot]=5[\{2O_i^a O_i^b 2Vo\}^{5'}]$
$(U,Nb)O_{2+x}$[8]	low	very large	±10	$\{2O_i^a O_i^b 2Vo\}^{5'}$	$[Nb_U^\cdot]=5[\{2O_i^a O_i^b 2Vo\}^{5'}]$
	very low	2	very large	neutral defect or $\{2O_i^a O_i^b 2Vo\}^{5'}$	$[Nb_U^\cdot]=[e']$
	high	very large	very large	$[\{2O_i^a O_i^b 2Vo)^{5'} Ti_i^{4\cdot}\}']$	$[h^\cdot]=[\{(2O_i^a O_i^b 2Vo)^{5'} Ti_i^{4\cdot}\}']$ $=Ti_i^x$
$(U,Ti)O_{2+x}$[10]	intermediate	4	4	$[\{2O_i^a O_i^b 2Vo)^{5'} Ti_i^{4\cdot}\}']$	$[h^\cdot]=[\{(2O_i^a O_i^b 2Vo)^{5'} Ti_i^{4\cdot}\}']$
	low	2	very large	$[\{2O_i^a O_i^b 2Vo)^{4'} Ti_i^{4\cdot}\}^x]$	$[h^\cdot]=[e']$

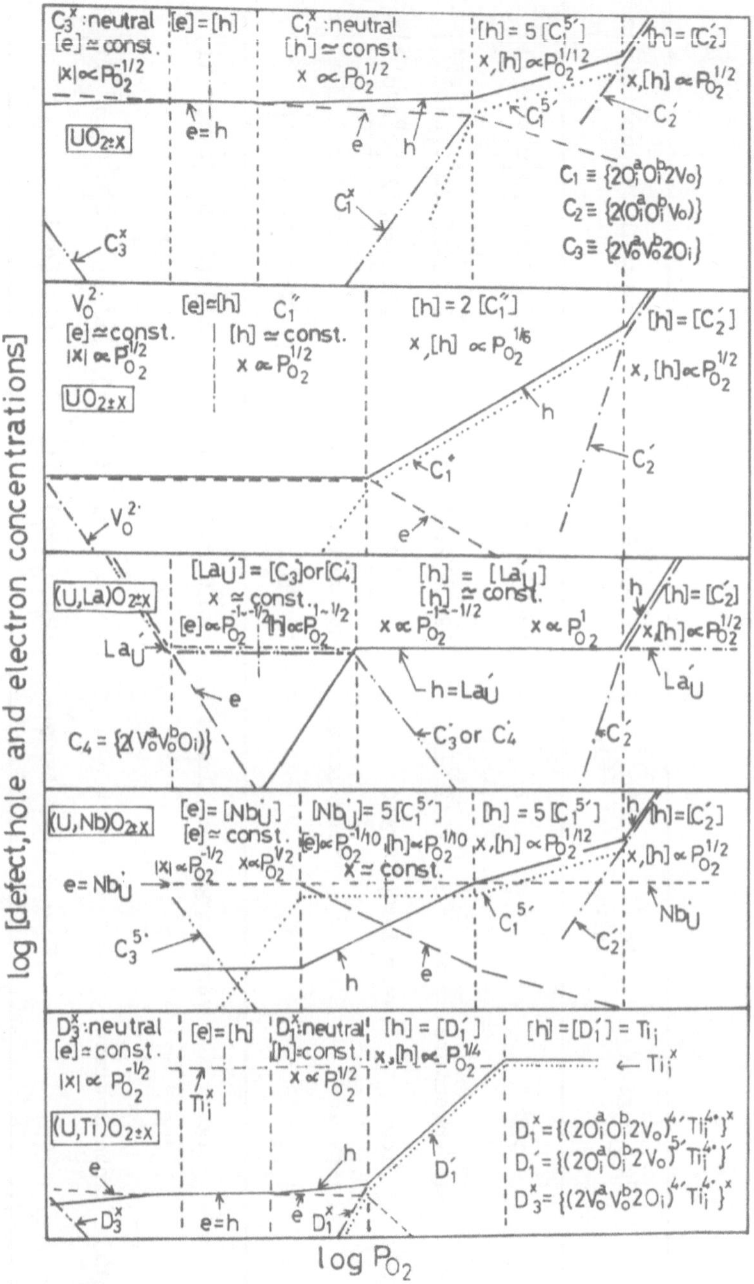

Figure 5. Schematic representation of the relationship between the relative concentration of defects and Po_2 for $UO_{2\pm x}$ and $(U,M)O_{2\pm x}$. The equations indicate the conditions for electroneutrality and the dependences of σ and x on Po_2.

In the case of $(U,Ti)O_{2+x}$ [10], interstitial model rather than substitutional one seems to be appropriate as mentioned already, and the following reaction can be assumed:

$$Ti_i^x + 1/2\ O_2(g) + 2V_i^a + V_i^b + 2O_o$$
$$= \{(2O_i^a O_i^b 2Vo)^{m'}\ Ti^{4^\cdot}\}^{q'} + qh^\cdot\ , \qquad (5)$$

where Ti_i^x is the neutral titanium in interstitial position produced by doping and q' is the charge of the complex defect $\{(2O_i^a O_i^b 2Vo)^{m'}\ Ti_i^{4\cdot}\}^{q'}$ which consists of $(2O_i^a\ O_i^b\ 2Vo)^{m'}$ and Ti_i^4 . The dependences of n=n'=∞ and n=n'=4 in the high and the intermediate oxygen partial pressure regions, respectively, can be interpreted by eq. (5) taking the value m=5 and q=1 and by using electroneutrality conditions as shown in Table I. The values of n=2 and n'=∞ can be explained by eq. (5) taking the value m=4 and q=0.

The relationships between the relative concentration of the predominant defect and log Po_2 are shown schematically in Fig. 5, where that proposed by Nakamura and Fujino [28] is also shown for the sake of comparison.

4. DIFFUSION COEFFICIENTS OF URANIUM

The ratios of the self-diffusion coefficient of uranium (D_U) in the doped UO_2 to that in pure UO_2 are shown in Table II. It is seen from the table that the self-diffusion coefficient of uranium in UO_{2+x} decreases by doping with cations of lower valency (La^{3+} [16] and Y^{3+} [16]) than the uranium ions, and increases with higher valent cations (Nb^{5+} [16]). This fact was explained by Matzke [16] according to the following valence control rule: Lower (or higher) valent cations, substituting for the U^{4+} ions in the UO_2 lattice, impart an effective negative (positive) charge to the lattice. This leads to the increase (or decrease) of the concentration of oxygen vacancies and then to the decrease (or increase) of the concentration of oxygen interstitials through Frenkel defect equilibrium, thereby decreasing (or increasing) the concentration of cation vacancies through Schottky defect equilibrium. The decrease (or increase) of the concentrations of cation vacancies is expected to cause the decrease (or increase) of the self-diffusion coefficients of uranium. However, higher self-diffusion coefficients of uranium in UO_{2+x} doped with Ti ions (tetravalency) than that of UO_{2+x} reported by Matzke [17,18] and recently observed by the present authors [19] can not be explained by the substitutional model, but by the interstitial model, that was already suggested by Matzke in his paper [16]. As was discussed in the defect structure, the Ti^{4+} interstitials should increase the concentration of oxygen interstitials and decrease the concentration of oxygen vacancies through Frenkel defect equilibrium, therby increasing the concentration of the cation vacancies through Schottky defect equilibrium. The increase of the concentrations of cation vacancies is expected to cause the increase of self-diffusion coefficients of the uranium in Ti-doped UO_{2+x}.

TABLE II

The ratios of the self-diffusion coefficient of uranium in the doped UO_2 to that in pure UO_2

Doped content	D_U(doped UO_2)/D_U(pure UO_2)		
	Temp./K, 1473	1673	1823
0.1 mol% La_2O_3[16]	——	≃0.02	0.085
0.1 mol% Y_2O_3[16]	——	≃0.02	0.11
0.1 mol% Nb_2O_5[16]	——	225	1
0.1 mol% TiO_2[17,18]	160	250	13
0.7 mol% TiO_2[19]	——	≃ 6*	——

* $Po_2 = 10^{-5.1}$Pa

Bell [28] has also shown that the self-diffusion coefficients of uranium in $U_{0.9}Y_{0.1}O_{1.98}$ nearly equal to those of pure stoichiometric UO_2 at 1733 and 1933 K. Although the decrease of the diffusion coefficient of uranium in Y-doped UO_{2+x} would be expected from the valence control rule, the effect of doping might be disturbed by the effect of nonstoichiometry of oxygen (x), due to the difficulty of controlling the same oxygen potentials in near stoichiometric composition.

5. OTHER RELATED PROPERTIES

5.1. Creep rate

It is expected from the valence control rule that the creep rate of UO_2 under the condition of the constant grain size, applied stress and temperature is increased (or decreased) by doping with cations of higher (or lower) valency than the uranium ions, since creep rate is proportional to the product of the concentration of uranium vacancy and the diffusion coefficient of uranium.

The creep rates of UO_2 doped with various cations such as Ca [30], Y [30], Zr [30], Si [30], Nb [3,31] and Gd [20] have been reported so far. The creep rates of UO_2 doped with Nb^{5+} ion (grain size 27-40 μm, applied stress 10 MPa) [2] and Gd^{3+} ion (grain size 10 μm, applied stress 12 MPa) [20] have been recently reported by Sawbridge et al. and

Hirai et al., respectively. Their data are shown in Fig. 6 as a function of the reciprocal temperatures. It is seen in the figure that (1) the creep rates of UO doped with 0.25-0.80 mol% Nb_2O_5 are two-to-three orders of magnitude higher than that of pure UO_2, although the clear dependence of the creep rates upon Nb dopant content is not observed, and (2) the creep rates of UO_2 doped with Gd^{3+} ion are decreased in proportion to the Gd content. These facts indicate that the variation of the creep rates of doped UO_2 can also be interpreted from the valence control rule. However the change in the creep rate of

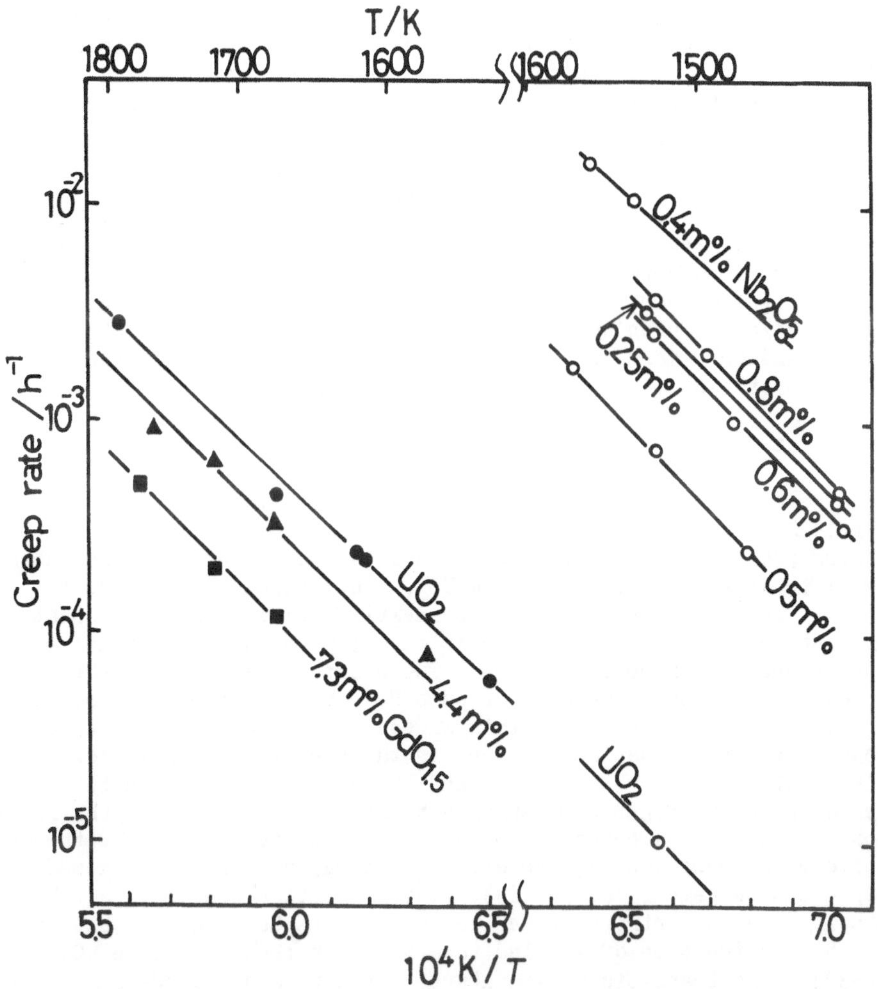

Figure 6. Temperature dependence of the creep rates of UO_2 doped with Nb and Gd
○ Sawbridge et al. [2], ●, ▲, ■ Hirai et al. [20].

UO_2 by doping with Nb^{5+} (or Gd^{3+}) ion is difficult to understand quantitatively, comparing the change in the diffusion coefficient of uranium in UO_2 by doping with Nb^{5+} (or La^{3+}, Y^{3+}) ion reported by Matzke [16], since the creep is a complex process.

5.2. Fission Gas Release

Since fission gas atoms are considered to diffuse via vacancy clusters of one-uranium vacancy and two-oxygen vacancies [3], the diffusion coefficients of fission gases calculated from the fission gas release rate may be related to the diffusion coefficients of uranium. The diffusion coefficient of fission gas is, however, a very pronounced function of damage concentration and diffusion via tri-vacancies is only effective at low gas (damage) concentrations. All the diffusion coefficients of fission gases in UO_2 doped with Cr [21], Ti [22], Nb [16,22] and Y [23] reported so far have not been influenced simply by the dopant content and the valency of the dopant. However the experimental results at high gas and damage concentration that the diffusion coefficients of ^{133}Xe in UO_2 doped with Nb_2O_5 and TiO_2 were enhanced [22], and that the diffusion coefficient of ^{85}Kr in UO_2 doped with Cr was also enhanced [21] can be interpreted from the enhancement of the diffusion coefficient of uranium due to the increase of the concentration of uranium vacancy on the assumption of the substitution of Nb^{5+} and Cr^{6+} ions for uranium ions and the presence of the Ti^{4+} or Cr^{3+} interstitials.

5.3. Heat Capacity

An increase in the heat capacity of UO_2 doped with $GdO_{1.5}$ (4.4 mol%-14.2 mol%) at high temperature has been observed by the present authors [24,25]. The heat capacities of UO_2 doped with Gd are shown in Fig. 7 together with that of pure UO_2. The increase in the heat capacity and the decrease in the transition temperature (the onset temperature of the increase in heat capacity) with the increase of Gd content are seen from the figure. The onset temperature of the increase in the heat capacity of pure UO_2 has been reported around 1500 K [32,33], and the origin of this excess heat capacity has been a subject of the discussion: the one is the formation of electron-hole pair by thermal activation [36], and the other is the formation of Frenkel pair of oxygen by thermal activation [33,35]. From the results shown in Fig. 7 the enthalpy and entropy of activation of the thermally activated process causing the excess heat capacity for each doped sample can be calculated [24,25]. The transition temperature, the enthalpy and the entropy of activation for $U_{1-y}Gd_yO_2$ are plotted in Fig. 8 as a function of y value together with those for pure UO_2 [32-39]. The transition temperature, enthalpy and entropy of activation for pure UO_2 were estimated by extrapolating those for doped UO_2 to zero-doped UO_2. The transition temperature thus obtained was in good agreement with that observed for pure UO_2 [32,33]. The enthalpy value thus obtained was higher than that for the formation of an electrical-hole pair estimated by the theoretical calculation [36], but lower than that for the formation of Frenkel pairs of oxygen [38,39]. The entropy value thus obtained was very high compared with that for the formation of electron-hole pairs [37], as seen from Fig. 8. It may be concluded that the contribution due to the formation of electron-hole pairs to the excess heat capacity is rather small and that due to the formation of Frenkel pairs of oxygen may be predominant. Furthermore, the electrical conductivity measured by the present authors [5] for UO_2 doped with 7.3 mol% $GdO_{1.5}$ showed no anomaly in the electrical conductivity curve around the transition temperature

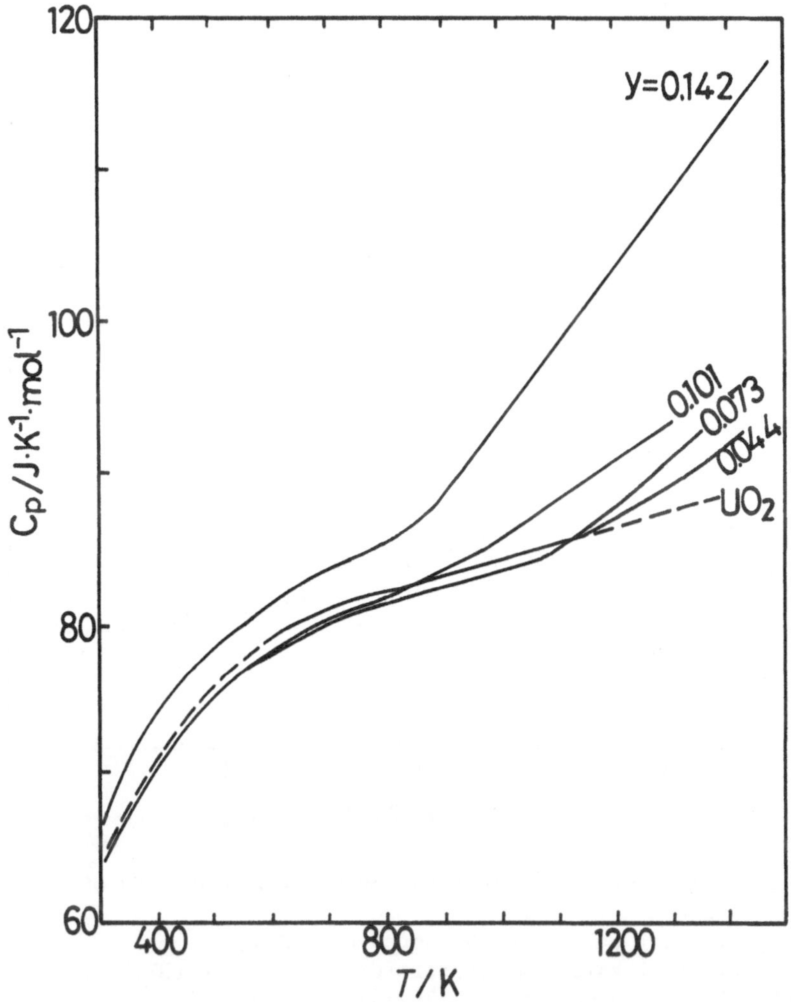

Figure 7. Heat capacity of $U_{1-y}Gd_yO_2$ [23, 24]

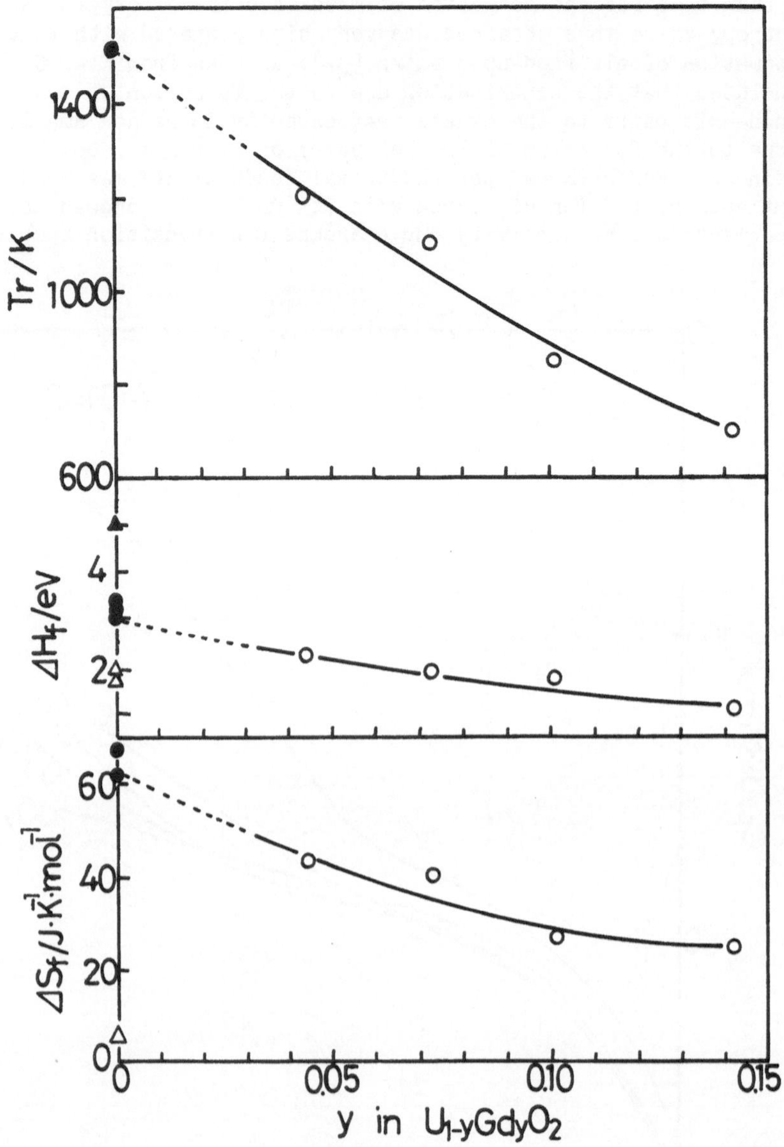

Figure 8. Transition temperature, enthalpy and entropy of activation vs. Gd content

○ experimental values for Gd doped UO_2 [23,23]

● experimental values of transition temperature [32,33], enthalpy [33,35,36] and entropy [33,35,36] for pure UO_2

Δ theoretical values for the formation of an electron-hole pair [37]

▲ theoretical values for the formation of a Frenkel pair [38,39]

found in the heat capacity curve. This may support the assumption that the predominant contribution to the heat capacity anomaly is due to the Frenkel pair formation. As seen in Fig. 8, the enthalpy and entropy of activation, ΔH_f and ΔS_f, decrease as the Gd content of $U_{1-y}Gd_yO_2$ increases. This may be explained as follows. By the introduction of Gd^{3+} ions in UO_2, oxygen sublattice around the Gd^{3+} ions tends to form an oxygen vacancy, which causes to form an oxygen interstitial in the neighborhood in order to keep nearly stiochiometric composition $U_{1-y}Gd_yO_2$. This series of event would result in the increase in the total number of oxygen clusters and the decrease in the formation enthalpy of the clusters.

ACNOWLEGEMENTS

We express our gratitude to Dr. Hj. Matzke at the European Institute for Transuranium Elements for valuable comments on this manuscript.

REFERENCES

[1] T. Matsui and K. Naito, J. Less. Common. Met. 121 (1986) 279.
[2] P.T. Sawbridge, G.L. Reynolds and B. Burton, J. Nucl. Mater. 97 (1981) 300.
[3] Hj. Matzke, Radiation Effects. 53 (1980) 219.
[4] T. Matsui and K. Naito, J. Nucl. Mater. 138 (1986) 19.
[5] T. Matsui and K. Naito, J. Nucl. Mater. 151 (1987) 86.
[6] K. Naito, T. Tsuji, M. Abe, T. Yamamoto, M. Sato, T. Fujino, K. Ohuchi and T. Yamashita, presented at the 26th Annual Meeting of the Atomic Energy Society of Japan L27 (1988) p.255.
[7] T. Matsui and K. Naito, J. Nucl. Mater. 132 (1985) 212.
[8] T. Matsui and K. Naito, J. Nucl. Mater. 136 (1985) 59.
[9] T. Matsui and K. Naito, J. Nucl. Mater. 137 (1986) 212.
[10] T. Tsuji, T. Matsui, M. Abe and K. Naito, J. Nucl. Mater. to be submitted.
[11] T. Fujino, J. Tateno and H. Tagawa, J. Solid State Chem. 24 (1978) 11.
[12] N. J. Dudney, R. L. Coble and H. L. Tuller, J. Am. Ceram. Soc. 64 (1978) 11.
[13] K. Une and M. Oguma, J. Nucl. Mater. 131 (1985) 88; 110 (1982) 215.
[14] S. Aronson adn H. C. Clayton, J. Chem. Phys. 32 (1960) 749.
[15] S. Aronson and J. C. Clayton, J. Chem. Phys. 35 (1961) 1055.
[16] Hj. Matzke, Nucl. Appl. 2 (1966) 131.
[17] Hj. Matzke, AECL-2585 (1966).
[18] Hj. Matzke, J. Nucl. Mater. 20 (1966) 328.
[19] T. Tsuji, M. Abe and K. Naito, unpublished work.
[20] M. Hirai, H. Masuda and M. Oguma, presented at the 26th Annual Meeting of the Atomic Energy Society of Japan L34 (1988) p.262.
[21] J. C. Killen, J. Nucl. Mater. 88 (1980) 177.
[22] K. Une, I. Tanabe and M. Oguma, J. Nucl. Mater. 150 (1987) 93.

[23] G. Long, W. P. Stanaway adn D. Davis, AERE-M1251 (1964).
[24] H. Inaba, K. Naito, M. Oguma and H. Masuda, J. Nucl. Mater. 137 (1986) 176.
[25] H. Inaba, K. Naito and M. Oguma, J. Nucl. Mater. 149 (1987) 341,
[26] T. L. Markin and E. J. McIver, Plutonium 1965 and Other Actinides p.845 (1965).
[27] F. A. Kroger and H. J. Vink, Solid St. Phys. 3 (1956) 588.
[28] A. Nakamura and T. Fujino, J. Nucl. Mater. 140 (1986) 113.
[29] J. Belle, WAPD-MRP-85 p.52 (1962).
[30] W. M. Armstrong and W. R. Irvine, J. Nucl. Mater. 12 (1964) 261.
[31] H. Assmann, W. Dorr, G. Gradel G. Maier and M. Peehs, J. Nucl. Mater. 98 (1981) 216.
[32] R. A. Hein, P. N. Flagella and J. B. Conway, J. Am. Ceram. Soc. 51 (1968) 291.
[33] J. F. Kerrisk and D. G. Clifton, Nucl. Technol. 16 (1972) 531.
[34] J. Ralph and C. J. Hyland, J. Nucl. Mater. 132 (1985) 76.
[35] R. Szwarc, J. Phys. Chem. Solids 30 (1969) 705.
[36] P. Browning, J. Nucl. Mater. 98 (1981) 345.
[37] G. Hyland and J. Ralph, High Temp.- High Press. 15 (1983) 179.
[38] C. R. A. Catlow, Proc. R. Soc. London A353 (1977)533.
[39] J. H. Harding, P. Masri and A. M. Stoneham, J. Nucl. Mater. 92 (1980) 73.

DEFECT INTERACTIONS, EXTENDED DEFECTS AND NON-STOICHIOMETRY IN CERAMIC OXIDES

A.N. Cormack
New York State College of Ceramics
Alfred University
Alfred NY 14802 USA

ABSTRACT. Non-stoichiometry in ceramic oxides is usually accommodated structurally through the incorporation of lattice defects. Except in regions very close to stoichiometry, these point defects tend to aggregate into clusters, or extended defects. In this presentation we discuss the energetics of the interactions giving rise to these extended defects from the point of view of atomistic computer-based simulations.

1. INTRODUCTION

Non-stoichiometry in solids can be discussed from a number of viewpoints. There is the thermodynamic approach, centring on the behaviour of the chemical potential of a phase. In a system showing bivariant behaviour, the chemical potential is a continuous function of composition across the existence range of the non-stoichiometric phase. This approach, however, says nothing about how the variable composition might be accommodated structurally, so one must also consider the crystal chemical aspects. In a simple binary oxide system (one with only one cationic species), bivariant, or non-stoichiometric, behaviour requires the metal species to be multivalent, in order to satisfy the electroneutrality requirement with a variable cation to anion ratio. This situation is sometimes called 'true' non-stoichiometry, whereas the other possible way of changing the overall cation to anion ratio, by adding a different aliovalent cation, is sometimes called simulated non-stoichiometry.

In fact, the thermodynamic definition of non-stoichiometry allows for both methods of altering the cation to anion ratio if one is in a true solid solution phase field. In this case, as well as in the binary oxide case, can one find the bivariant behaviour characteristic of thermodynamic non-stoichiometry.

Our interest is in how one can reconcile the thermodynamic treatment with a crystal chemical approach, since translational periodicity is clearly incompatible with compositional variations. Apparently, the answer must be sought in the behaviour of atomistic structural defects: point, or lattice defects, and their complexes. In this presentation, we discuss the interactions between point defects in some systems where it is known that defect interactions play an

J. Nowotny and W. Weppner (eds.),
Non-Stoichiometric Compounds Surfaces, Grain Boundaries and Structural Defects, 45–52.

important role in both their overall defect chemistry and their nonstoichiometric behaviour. First, however, we wish to make some broad remarks about interactions between defects.

2. DEFECT INTERACTIONS

Generally speaking, interactions between individual point defects may be either attractive, or repulsive. There have very few, if any, studies on the interaction potential between point defects as a function of defect separation. This is probably because the interaction distance is usually assumed to be fairly short-range. Thus one has been mainly interested in those interactions leading directly to cluster formation. Specifically, one has been concerned with the stability of those defect clusters, as usually measured by the binding energy of the cluster with respect to 'free' point defects (1). On the other hand there have been a number of studies of the long-range interactions between extended, or planar defects, such as crystallographic shear planes (2-4). In these cases, the attractive interaction leads to formation of regular arrays or superlattices that are in fact themselves stoichiometric.

Clearly, if defect complexes are to form, the interaction must be attractive; repulsive interactions can only lead to continued disorder. Non-stoichiometric behaviour in a system must be closely related to the degree of disorder of the (point) defect population, since if the interaction between defect complexes is also attractive, then ordering of the defects into extended defects will occur. Continued attractive behaviour will lead to superlattice formation and the creation of new stoichiometric phases. On the other hand, repulsive interactions only promote nonstoichiometry and disorder.

An important point to consider is that of the range of the interaction. Very short range attractive behaviour is unlikely to cause superlattice ordering, except at high defect concentrations. Very long range repulsive behaviour may lead to nearly stoichiometric behaviour through precipitation of a second phase at the crystal boundary, say surface or grain boundary. In most cases, though, this must be accompanied by a change of chemistry at the crystal boundary.

In this paper, we consider three systems, each of which exhibits an apparently different non-stoichiometric behaviour. Our objective is to rationalise their non-stoichiometric behaviour in terms of differences in the nature of the defect interactions, particularly extended defects, in these systems.

3. EXAMPLES OF NON-STOICHIOMETRIC SYSTEMS

3.1 Calcia stabilised zirconia

In this system, the addition of calcia is accompanied by the generation of oxygen ion vacancies to preserve charge neutrality. The Ca dopant and V_0 interact attractively to form the defect pair that we consider to be the extended defect building block. Our calculations (5) show that the oxygen vacancy prefers a second nearest neighbour site with respect to the Ca substitutional ion; this is similar to our earlier result for yttrium doped zirconia (6). However, in the present case, the situation is more complicated since we find an attractive binding energy, albeit smaller, for the nearest neighbour configuration as well.

Our calculations on larger defect aggregates strongly support the idea that the vacancy will prefer the next nearest neighbour position, however. In all cases with two Ca_{Zr}, if the vacancy was placed initially next to the dopants, it relaxed away towards a next nearest neighbour position. In this respect, our results concur with those of Allpress and Rossell (7), who have long argued for a next nearest neighbour position for the oxygen vacancy. Their argument is based on the observation of microdomains seen in high resolution electron lattice images. By analogy with the doped hafnia system, they suggest that the composition of these microdomains is $CaZr_4O_9$; in the hafnia counterpart, $CaHf_4O_9$, they have determined that the Ca dopant and the oxygen vacancy are ordered in such a way that the vacancy occupied a next nearest neighbour site, leaving the dopant in [8] coordination. At the time we did these calculations, there was no reported structure determination of $CaZr_4O_9$; indeed, its inclusion in the equilibrium phase diagram had not been completely established. Recent careful work by Hellman and Stubican (8) has confirmed its existence, noting that the equilibration times required to map that part of the phase diagram were extraordinarily long.

In order to examine the energetics of the interaction between the $Ca_{Zr}.V_O$ structures, we calculated the lattice energy of the $CaZr_4O_9$ phase (9). Comparison of the energies of Ca incorporation in terms of both point defects and superlattice formation ($CaZr_4O_9$) gives us an indication of the nature of the interaction between the extended [$Ca_{Zr}.V_O$] defects. Since there has been no structure determination reported in the literature, we used as our starting point an ideal model with superlattice unit cell vectors taken from the hafnia analogue. To obtain our equilibrium lattice energy, the ideal atomic coordinates were relaxed along with the lattice vectors. Using this energy, the energy of the supercell formation reaction :

$$CaO + 4ZrO_2 \text{ -------> } CaZr_4O_9$$

may be found. A very small, but negative energy is obtained. This suggests that although the reaction will proceed (and hence microdomains will form) it will do so very slowly, because of the small (thermodynamic) driving force. This is consistent with the observation of long annealing times needed to produce the phase. In contrast, the point defect mode gives an energy of 0.88 eV, reduced to 0.20 eV, including the binding energy of the extended defect pair. Clearly, there is an attractive interaction favouring the ordering of the defect pairs into a stoichiometric superlattice.

In summary, in this system the thermodynamics suggest that the 'non-stoichiometry' is due to a strong interaction between two point defects, Ca_{Zr} and V_O, to form the embryo of a more extended structure, but which then only very weakly interact so that microdomain formation is limited and establishment of the true phase equilibrium even more so.

3.2 Sr excess $SrTiO_3$

$SrTiO_3$ can accommodate only a small amount of excess Sr through point defect generation (10). The extra Sr ions occupy [6] B sites in the perovskite structure, normally occupied by titanium. The addition of greater amounts of SrO to $SrTiO_3$ results structurally in the generation of planar defects often called Ruddlesden-Popper layers after the people who first described them. Ruddlesden and Popper (11) reported the formation of Sr rich phases Sr_2TiO_4 and $Sr_3Ti_2O_7$ and showed that their structure could be discussed in terms of alternating slabs of perovskite and rock-salt structure, the variation in composition coming from the difference in the perovskite slab thickness; the rock-salt or SrO layer is only one unit cell thick, and the slabs are two-dimensional in extent.

The generalisation of these structural observations results in a homologous series of superlattice type phases with the formula : $SrO.nSrTiO_3$, with n = 1,2,3 and so on; n = ∞ corresponds to the perovskite $SrTiO_3$ end-member. Only a few of the members of this homologous series have been found phase pure, however. In an electron microscopy study, Tilley (12) found that attempts to prepare single phases with stoichiometries corresponding to n greater than 3 always resulted in the formation of intergrowths with larger n in a host matrix of the n = 2, $Sr_3Ti_2O_7$, structure.

As part of a study on alkaline earth excess alkaline earth titanates (13), we examined the energetics of Ruddlesden-Popper layer formation in this system. We found that the energy of reaction to form Sr_2TiO_4:

$$SrO + SrTiO_3 \text{ ---------> } Sr_2TiO_4$$

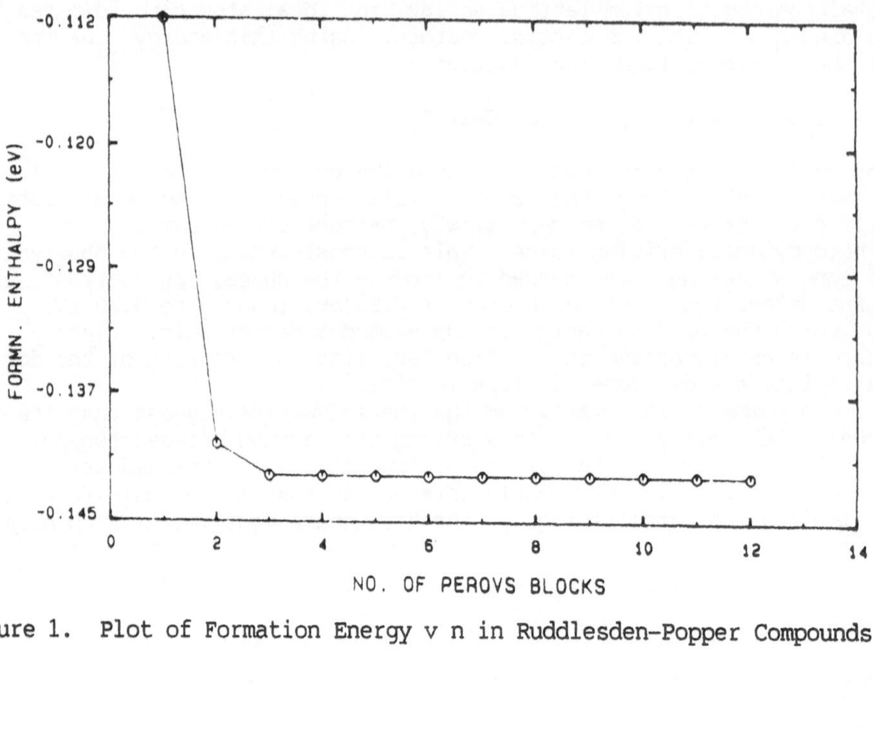

Figure 1. Plot of Formation Energy v n in Ruddlesden-Popper Compounds

was negative, showing that this phase should indeed form, in agreement with the findings of Ruddlesden and Popper. We thus set out to calculate (14) the lattice energy of others members in the homologous series, to see if there were any interaction between the SrO layers. Since only a few of the structures have been reported, we generated a series of ideal configurations and equilibrated these using energy minimisation techniques.

Our results are plotted in Figure (1). From this it can be seen that there is no thermodynamic driving force for ordering the SrO layers into regular arrays for n greater than 3. This provides an energetic basis for the results of Tilley and helps to explain why $SrTiO_3$ can accommodate excess SrO to the extent that it does. The absence of any interaction across a perovskite slab larger than 3 unit cells suggests that the effective chemical potential is essentially constant across a wide range of stoichiometry.

3.3 Fe doped $CaTiO_3$

Interest in this system arises from the observation that large amounts of Fe_2O_3 can be added to $CaTiO_3$ without changing the basic perovskite structure. Grenier et al (15) found that upto 50% of the Ti could be substituted with Fe before any superlattice ordering was detected. The end-member $Ca_2Fe_2O_5$ has the brownmillerite type of structure in which layers of octahedral Fe alternate with Fe in tetrahedral coordination.

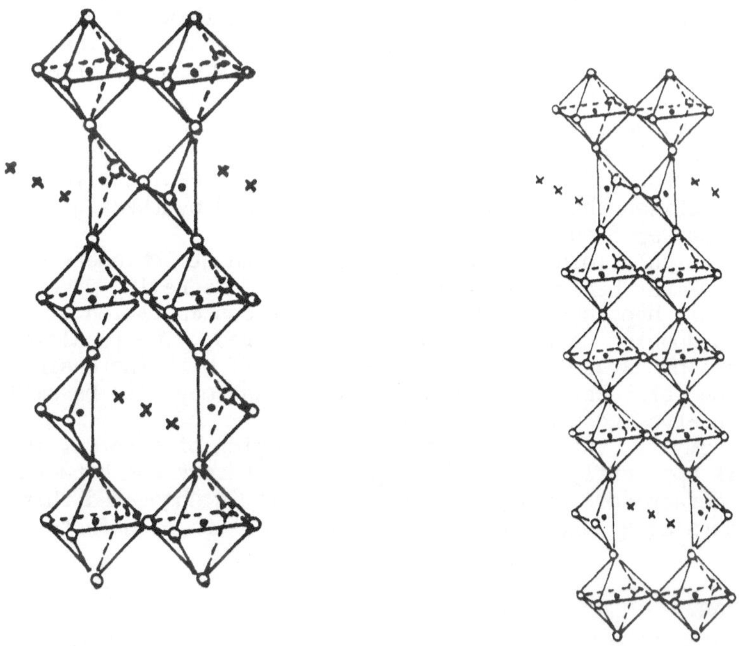

Figure 2. (a) Part of Brownmillerite Structure showing planes of [4] cations due to ordering of oxygen vacancies (indicated with an 'x') along the perovskite [110] direction, (b) Part of structure of $Ca_2FeTiO_{5.5}$ showing 3 layer octahedral slab between tetrahedral layers.

The [4] coordination comes about from the ordering of the anion vacancies introduced to compensate for the replacement of Ti by Fe. With only half of the titanium replaced, just half of the iron is in tetrahedral sites which are separated by three layers of octahedra, as shown in figure (2). Another way to view these structures is in terms of strings of oxygen vacancies aligned along [110] of the perovskite structure: ordering of these strings of vacancies leads to the superlattice formation observed. Note that the strings of vacancies adopt a zig-zag stacking pattern which leads to a doubling of the unit cell parameter along c.

By using atomistic simulation methods, our objective is to rationalise these observations on the basis of defect energetics. We record here some preliminary results (16). Firstly, we calculated the interaction energy between Fe_{Ti} and V_O to see if [4] Fe could be predicted. Our conclusion here is that there should be, based on the energetics of the defect reaction:

$$[Fe_{Ti}.2V_O] + Fe_{Ti} \text{ -----> } 2[Fe_{Ti}.V_O]$$

This reaction has a positive energy, suggesting a balance towards the left, notwithstanding the fact that we find the defect $[Fe_{Ti}.V_O]$ to be strongly bound. Secondly, we found that small strings of three V_O would tend to dissociate and thirdly that pairs of $Fe_{Ti}.V_O$ will tend to repel each other.

From these results we may infer that the Fe_{Ti} and the V_O will be disordered within the perovskite structure, at least at modest levels of Fe substitution. At higher levels, some interaction is inevitable, but since we calculate the interaction to be repulsive, any ordering that occurs will do so to minimise the repulsive energy rather than maximise an attractive energy. This may be the reason why so much Fe can be doped into $CaTiO_3$ before ordering takes place. We suggest that this is also the reason that a doubling of the unit cell is seen: the interaction distance between the strings of V_O is increased by the zig-zag pattern adopted, as depicted in figure (2).

Our calculations on perfect lattices with stoichiometries $Ca_2Fe_2O_5$ (the end member brownmillerite structure) and $Ca_2FeTiO_{5.5}$ seem to support this hypothesis. With the greatest concentration of [4] Fe planes, a doubled unit cell is clearly favoured. The repulsion between the planes of oxygen vacancy strings falls off with increasing separation (at least in our limited study) and may also be ameliorated by the intervening planes of Ti ions.

In summary, the large compositional region of disorder in this system is apparently due to the repulsive interactions between the various defect species in the structure. It is suggested that ordering only occurs as a result of the high concentration of defects.

4. SUMMARY

We have discussed the nonstoichiometric behaviour in three different systems from the point of view of defect interactions and extended defect structures. We suggest that attractive interactions will lead to superlattice formation and creation of other stoichiometric phases, whereas disorder and non-stoichiometry are promoted by repulsive interactions, or the absence of any interaction.

5. ACKNOWLEDGEMENTS

The work that formed the basis for this paper has been distilled from the theses of several of my students, K.R. Udayakumar, A. Dwivedi and R. Ward. We are grateful to Alfred University and the New York State College of Ceramics for the provision of computing facilities.

Some of the calculations discussed in this paper were performed at the Cornell National Supercomputer Facility, which is funded in part by IBM, NSF and the State of New York; we thank these institutions for their support.

6. REFERENCES

1. S.M. Tomlinson, C.R.A. Catlow and J.H. Harding, in: "Transport in Non-Stoichiometric Compounds", eds. G. Simkovich and V.S. Stubican, pp 539-550, Plenum Press NY (1985).

2. A.M. Stoneham and P.J. Durham, J. Phys. Chem. Solids 34, 2127 (1973).

3. A.N. Cormack, C.R.A. Catlow and P.W.Tasker, Rad. Effects, 74, 237 (1983).

4. A.N. Cormack, C.M. Freeman, R.L. Royle and C.R.A. Catlow, in: "Non-Stoichiometric Compounds", eds. C.R.A. Catlow and W.C. Mackrodt, Adv. in Ceramics, 23, 307-329 (1987).

5. A. Dwivedi and A.N. Cormack, in preparation.

6. C.R.A. Catlow, A.V. Chadwick, A.N. Cormack, G.N. Greaves, M. Leslie and M.L. Moroney, in: "Defect Properties and Processing of High-Technology Ceramics", MRS Symposia Proc. 60, 173-178 (1986).

7. J.G. Allpress and J.H. Rossell, J. Solid State Chem. 15, 68-78 (1975).

8. J.R. Hellman and V.S. Stubican, J. Amer. Ceram. Soc. 66, 260-264 (1983).

9. A. Dwivedi and A.N. Cormack, in preparation.

10. Y.H. Han, M.P. Harmer, Y.H. Hu and D.M. Smyth, in: "Transport in Non-Stoichiometric Compounds", eds. G.Simkovich and V.S. Stubican, pp 73-85, Plenum Press, NY (1985).

11. S.N. Ruddlesden and P. Popper, (a) Acta Crystallogr. 10, 538-539 (1957); (b) Acta Crystallogr. 11, 54-55 (1958).

12. R.J.D. Tilley, J. Solid State Chem. 21, 293-301 (1977).

13. K.R. Udayakumar and A.N. Cormack, J. Phys. Chem. Solids, in press (1988).

14. K.R. Udayakumar and A.N. Cormack, J. Amer. Ceram. Soc. Communications, in press (1988).

15. J-C. Grenier, F. Menil, M. Pouchard and P. Hagenmuller, (a) J. Solid State Chem. 20, 365-379 (1977); (b) Mat. Res. Bull. 13, 329-337 (1978).

16. R. Ward and A. N. Cormack, unpublished results.

Computer Simulation Studies of $Fe_{1-x}O$ and $Mn_{1-x}O$

S.M. Tomlinson and C.R.A. Catlow

Department of Chemistry,
University of Keele,
Staffordshire, ST5 5BG, UK.

Abstract

We show how a combination of static simulation techniques and a mass-action treatment of defect equilibria may be used to study the defect structure of $Mn_{1-x}O$, and to calculate the variation of x with $p(O_2)$. We find a defect model including 4:1 clusters with a variety of charge states may reproduce the observed behaviour. For $Fe_{1-x}O$ we update our last survey of defect cluster stabilities, and find the 12:4 cluster remains the favoured large defect aggregate. We also show that inter-defect interactions will favour the formation of defect clusters. Lastly, we use the mass-action method to show how the slightly higher binding energies of clusters in $Fe_{1-x}O$, may account for the difference in behaviour of the two oxides.

1. Introduction

In this paper we will report the recent advances we have made in understanding the complex defect structures of $Mn_{1-x}O$ and $Fe_{1-x}O$, through the use of computer modelling techniques. We will describe how the mass-action treatment of defect equilibria, presented in the proceedings of the Pennsylvania State, Non-Stoichiometry meeting four years ago[1], has been developed to enable direct calculation of the variation of non-stoichiometry with $p(O_2)$, which is, of course, directly measurable by the technique of thermogravimetry. We are now, therefore, in a position to test our predictions concerning microscopic defect models against a macroscopic, measurable quantity. We demonstrate our procedure for the

J. Nowotny and W. Weppner (eds.),
Non-Stoichiometric Compounds Surfaces, Grain Boundaries and Structural Defects, 53–75.

case of $Mn_{1-x}O$, which will enable us to make important predictions concerning defect clustering in this material.

For $Fe_{1-x}O$, several direct structural observations of clustering have been made. However, the precise form of these clusters is not clear. Attempts to clarify their nature by various techniques have continued since we last reported our predictions concerning the nature of clustering in this oxide. Here we will report calculations of the stabilities of a variety of defect clusters proposed in the last four years, and compare them with established models. We also consider the interaction of defect clusters in $Fe_{1-x}O$, and consider whether such a term in the formation energy would favour or oppose cluster formation.

In the final section of this paper we will consider the question of whether our calculations can determine the origin of the difference in behaviour of $Fe_{1-x}O$ and the oxides $Mn_{1-x}O$, $Co_{1-x}O$ and $Ni_{1-x}O$. The latter three may all be prepared with stoichiometric composition, yet $Fe_{1-x}O$ may not. We will use our extended mass-action approach to calculate the free energy of oxidation of $Mn_{1-x}O$ and $Fe_{1-x}O$ at compositions around the lower stability limit of the latter.

2. Methodology

2.1 Calculation of Defect Formation Energies

We use the techniques of static lattice simulation to calculate the energy and entropy of defect formation. The procedures are well documented elsewhere[2,3] so we will provide only a brief summary here.

The calculations are based on a Born model of the crystal. Ions in the crystal are treated by the shell model[4] in order to model ionic polarisability. This involves each ion being represented by two point ions whose total charge equals the formal charge of the ion. These two point charges - the core and shell of the ion - are coupled by an harmonic spring. The shells of the ions interact with the shells of surrounding ions via a parameterised short-range potential energy function. The coulombic interaction between different ions is calculated explicitly using an Ewald summation. We thus have a model for the crystal in which the total potential energy resulting from ionic interaction is expressed in terms of ionic co-ordinates, and ionic polarisation and short-range inter-ionic forces are effectively coupled.

The defect formation energy is calculated by a Mott–Littleton method, in which the crystal is formally partitioned into an inner region containing the defect, and an outer region. The ions in the inner region are allowed to relax explicitly to their equilibrium positions during an iterative energy minimisation process, whilst the ions in the outer region are treated by the pseudo–continuum approximation of Mott and Littleton(5). The above procedure is incorporated in the computer code HADES III(6,7).

2.2 Calculation of Defect Formation Entropies

The vibrational entropy change associated with defect formation may also be routinely calculated, using the computer code SHEOL(8,9). The crystal is again partitioned into two regions. The inner region is then treated as a "large crystallite". The vibrational entropy change of the crystallite is given by the logarithm of the ratio of the determinant of the dynamical matrix after defect formation, to the determinant of the perfect lattice dynamical matrix, plus a correction term in the case of charged defects. The outer region contributes only via the diagonal elements of the dynamical matrix.

2.3 Mass–action treatment

The Law of Mass–Action is used to determine the relative concentrations of defect species in the non–stoichiometric oxides. Equilibrium constants for a series of defect reactions are obtained directly from the defect energies and entropies calculated by the methods outlined above. Defect clusters, and different charge states of the cation vacancy, are all treated as being in equilibrium with the doubly charged vacancy and hole. We treat the hole state as a small polaron i.e. as a trivalent cation. Different charge states of the cation vacancy are modelled by placing M^{3+} ions at next–nearest neighbour or nearest neighbour cation sites to the vacancy. The same is true of different charge states of defect clusters. The mass–action equations, and the electroneutrality and non–stoichiometry equations, are then solved numerically for a range of compositions. This yields the variation of all defect concentrations with non–stoichiometry.

We have recently extended the method to include the calculation of oxygen partial pressure. The procedure for doing this is as follows. First, the partial molar enthalpy and entropy of the systems are calculated. These terms contain a constant contribution due to the reaction of the oxide with ambient oxygen, plus terms due to the association and interaction of defect species. The latter term is calculated using the results of our mass-action treatment which gives the variation of defect concentration with oxide composition. It is this term which determines the slope and shape of the $p(O_2)$ versus x curve. The constant term will shift the curve rigidly up and down the $p(O_2)$ axis of such a plot. The partial molar enthalpy term is, therefore:

$$\Delta\bar{H}(O_2) = H_o + 2 \sum_{i=1}^{T_D} h_i \, (dC_i/dx) \qquad (1)$$

where

$$\sum C_i = 1 \qquad (2)$$

and h_i is the enthalpy of formation of the ith defect, C_i its concentration at deviation from stoichiometry x, and the sum is over all defect types in the proposed set of defect equilibria.

The term H_o contains:

(i) the dissociation energy of O_2

(ii) twice the first and second electron affinities of oxygen

(iii) four times the third ionization energy of the cation

(iv) twice the lattice energy of the oxide.

The second electron affinity of oxygen, and the ionisation potential of the cation in the lattice are terms which involve some uncertainty. We therefore add a correction term at a later stage to compensate for this, and for other limitations of the model, which we will discuss later.

The enthalpy of formation of the defect, h_i, is calculated directly from the formation energies obtained from the Mott-Littleton calculations. Strictly, we are approximating a constant pressure by a constant volume term. However, previous studies(10,11) have shown that error introduced by this approximation will be small compared to other limitations of our model (see section 2.4). The derivative term in equation (1) is obtained by differencing, using the defect concentrations calculated at closely spaced consecutive values of x.

The partial molar entropy of oxidation is obtained from the expression

$$\Delta\bar{S}(O_2) = S_o - 2R \left\{ \sum_{i=1}^{T_D} (dC_i/dx)[s_i^{VIB} + \ln(C_i/1-C_i)] \right\} \qquad (3)$$

where s_i^{VIB} is the vibrational entropy change of formation of the ith defect, C_i its concentration at deviation from stoichiometry x and the sum is over all defect types. S_o is the constant entropy term. It has only one component; the entropy of gaseous oxygen at the temperature being considered. Equation (3) also contains the configurational entropy contribution to $\Delta S(O_2)$.

We note here how both the partial molar entropy and enthalpy of oxidation depend on the derivative term (dC_i/dx). This is an important point, as we shall show later when we apply the procedure to calculate the $p(O_2)$ dependence of composition for $Mn_{1-x}O$. If the concentration of a particular defect type begins to change rapidly with increasing x, then it will strongly influence $\Delta H(O_2)$ and $\Delta S(O_2)$, even though its concentration may still be low.

Once the partial molar entropies and energies have been calculated, the partial molar Gibbs free energy may be obtained from

$$\Delta\bar{G}(O_2) = \Delta\bar{H}(O_2) - T\Delta\bar{S}(O_2) \qquad (4)$$

and hence we arrive at the oxygen partial pressure using

$$p(O_2) = \exp(\Delta\bar{G}(O_2)/RT) \qquad (5)$$

A computer code has been developed to automate the above procedure[12,13]. A key feature of the program is the encoding of the mass-action equations as character strings, which are then read and interpreted by the code. This allows the analysis of a wide range of defect models, for different systems and temperatures, to be performed in a straightforward manner. The operation of the program is shown schematically in figure (1).

2.4 Limitations of the models

In order to develop our models to the point where useful predictions can be made concerning the atomistic processes underlying the macroscopic behaviour of materials, we are of course obliged to make simplifying assumptions. We summarise these below.

1. For transition metal ions there are crystal field terms which are not directly included in the parameterised short-range energy functions. This is relevant in the case of the formation of defect clusters, where there is a change of co-ordination of the interstitial transition metal ion. However, we believe this term will have a minor effect only on the formation energy of defect clusters, the major components being the Coulomb and elastic terms.
2. We are assuming the validity of the ideal solution model, i.e. we assign unit activity coefficients for all defect species; we are therefore neglecting defect interactions. This limitation will vanish as the deviation from stoichiometry tends to zero, and the model is thus most accurate at low deviations from stoichiometry.
3. We treat the hole state as an M^{3+} ion. However, the key quantity in the calculations are the energies and entropies of formation of defect clusters (including V_m' and V_m species) from cation vacancies and holes. These are obtained from the difference in the defect energy or entropy of the cluster, and the sum of these for their components. Thus inaccuracies arising from the treatment of the hole state as an M^{3+} ion will tend to cancel out in calculation of binding energies and entropies. This treatment will also have an effect on the constant term H_o in equation (1). However, inaccuracies here will lead only to a constant displacement of the $p(O_2)$ vs. x curve; the slope will not be effected.
4. As we noted earlier, we approximate the constant pressure defect parameters s_p and h_p by the constant volume terms s_v and u_v. The effect of this is discussed in more detail by Harding[9]. However, we believe that the effects of this approximation will be small compared with the limitations of the model described above.

3. <u>$Mn_{1-x}O$</u>

Tomlinson et al[1] reported a defect model for $Mn_{1-x}O$ which comprised doubly charged cation vacancies and 4:1 defect clusters (shown in figure 2). In the light of our calculations of $p(O_2)$ as a function of x, we have modified this defect model to incorporate different charge states of the 4:1 cluster. We found this was necessary in order to reproduce the form of the experimental curve, obtained by Keller and Dieckmann[14] using thermogravimetry techniques.

Figure 1. Schematic illustration of the mass-action code

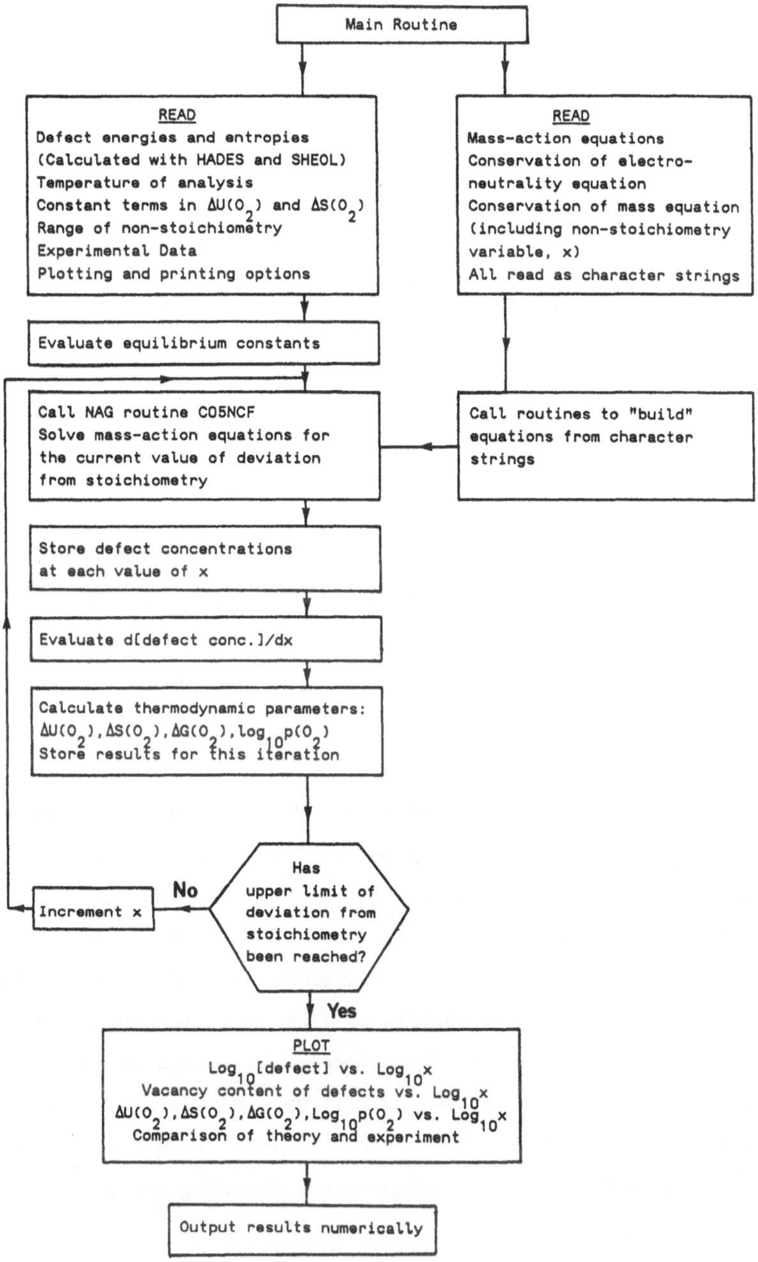

Figure 2. Basic 4:1 defect cluster

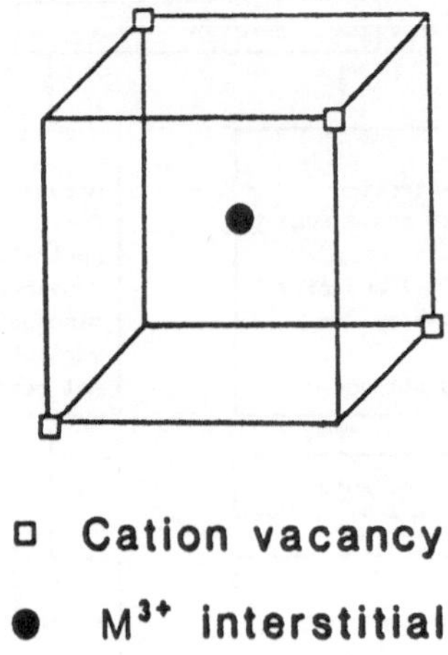

In table (1) we report the defect energies and entropies associated with the various defect reactions included in our model. We also report the binding energies, per net vacancy, of the defect clusters. Our favoured defect model is shown in figure (3), in the form of a plot of the percentage fraction of total cation vacancy population present in each defect species, as a function of $\log_{10}x$. The plot refers to a temperature of 1200°C. Such a plot gives a clear representation of the relative importance of each defect. It is apparent from figure (3) that we predict the 4:1 cluster to exist across the entire range of deviation from stoichiometry at 1200°C, by forming first in high charge states and subsequently binding holes as they become increasingly available as the material is progressively oxidised. This point deserves emphasis: cluster formation is not constrained by the electroneutrality condition to occur only at high deviations from stoichiometry, when oxidation has proceeded far enough to generate sufficient charge compensating holes. Vacancy stabilisation by clustering around a tetrahedral interstitial cation can now occur across the entire composition range, with the total cluster concentration (of all charge states) increasing gradually with deviation from stoichiometry.

We should note that at low to intermediate deviations from stoichiometry, it is insufficient simply to compare cluster binding energies to deduce the favoured cluster type. From figure (3) it is clear that the dominant defect cluster is the $4{:}1^{5-}$, which we note from table (1) has the lowest binding energy per net vacancy. It is favoured by the Law of Mass-Action, because it may be formed when hole concentrations are low. It is the merit of this type of mass-action treatment that the formation energies of clusters are directly coupled through the equilibrium constants to the concentrations of the component point defects.

The calculated variation of $p(O_2)$ with x for the defect model shown in figure (3), is plotted in figure (4). We also show the experimental values of Keller and Dieckmann[14] determined by thermogravimetry at 1200°C. The agreement between calculated and experimental curve is the best obtained for all defect models tested[15] which included the classical model of vacancies in different charge states and holes (see for example Keller and Dieckmann[14]). Indeed the agreement is very satisfactory if we remember the following: first, the limitations of our model described in section 2.4 and secondly that no fitting procedure has been used. We were obliged to shift all the $\Delta G(O_2)$ values by 1.422 eV prior to calculating $p(O_2)$. However, this is easily accounted for by the uncertainties in the second electron affinity of oxygen and the ionisation potential of the cation. Moreover, this adjustment only affects a wholesale shift of the curve along the $p(O_2)$ axis; the shape and slope, which are determined by the equilibrium defect concentrations, are not changed.

The least satisfactory feature of the calculated curve is the slight sigmoidal shape caused by the decrease in slope at higher deviations from stoichiometry. Although we would expect worst agreement in this region due to the limitations of the mass-action model, this feature may be due to the omission from our defect model of a cluster of intermediate charge between the $4{:}1^0$ and $4{:}1^{3-}$ cluster. Nevertheless, this is the best agreement between simulation and experiment we were able to obtain with the defect energies and entropies calculated in the present work, and with the current level of sophistication of the mass-action model. It shows that the general form and magnitude of the observed non-stoichiometry of $Mn_{1-x}O$ can be explained by defect clustering, with the formation of 4:1 clusters of varying charge states. Furthermore, the simple defect model, consisting of vacancies in different charge states and holes, does not reproduce the experimental behaviour if our calculated energies and entropies are used.

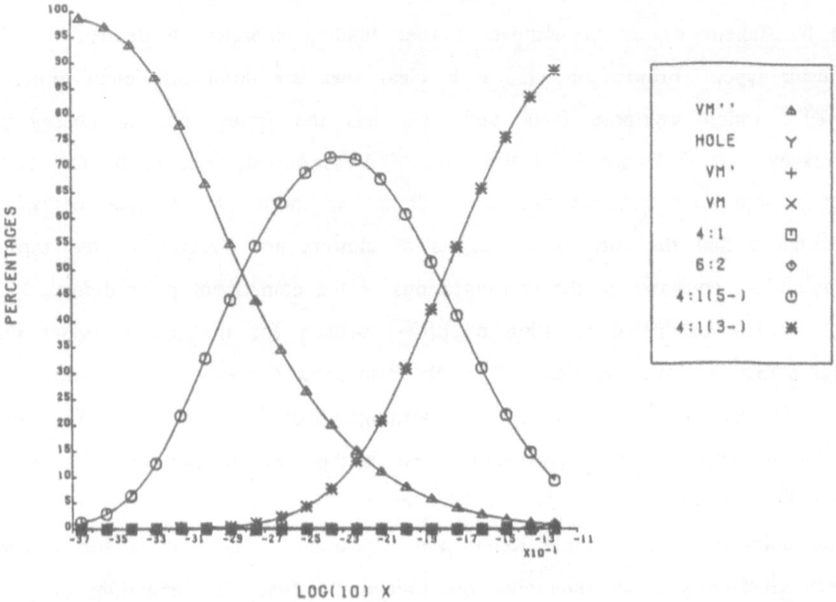

Figure 3. Proposed defect model for $Mn_{1-x}O$ at 1200°C

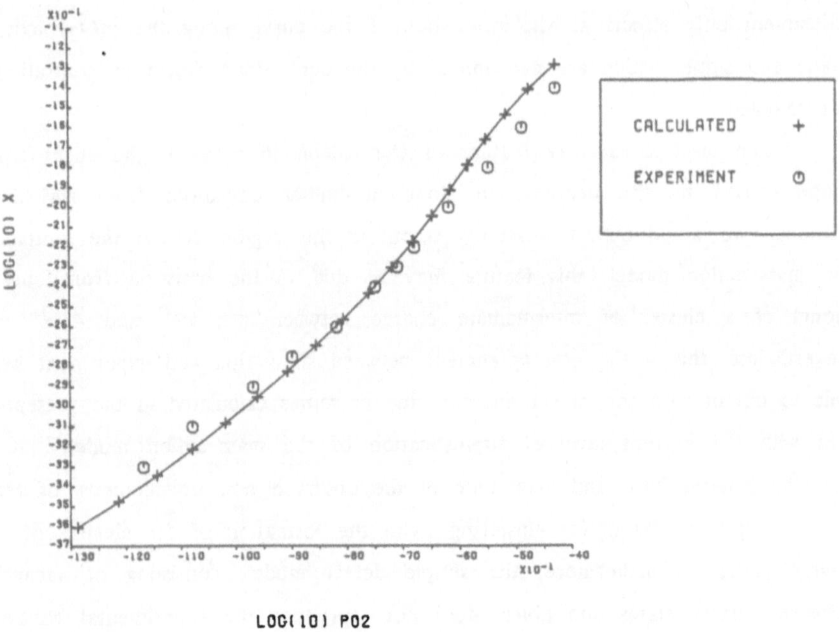

Figure 4. $Log_{10}x$ versus $log_{10}p(O_2)$ for $Mn_{1-x}O$ at 1200°C

Table 1

Defect formation energies and vibrational entropies for $Mn_{1-x}O$

Defect	u_v/eV	s_{vib}/k_B	‡Δu_v/eV/Vac
V_{Mn}''	20.937	-6.287	-
$h^{\cdot}$	-30.128	3.430	-
V_{Mn}'	-9.773	-5.153	-
V_{Mn}	-40.267	-4.970	-
$V_{Mn}'(nnn)^{*}$	-9.872	-3.127	-
$4:1^{1-}$	-92.464	-9.246	-1.545
$4:1^{3-}$	-31.618	-12.558	-1.348
$4:1^{5-}$	29.930	-18.316	-0.962
$4:1^{2-}$	-61.957	-6.553	-1.419
6:2	-165.135	-17.065	-1.965
$6:2^{2-}$	-103.470	-15.998	-1.613

* hole at next nearest-neighbour site to vacancy

‡ cluster binding energies, per net vacancy

4. $Fe_{1-x}O$

4.1 Calculations of cluster stability

A wide variety of defect clusters based on the tetrahedral 4:1 cluster have been proposed to account for the direct structural observations, by diffraction and microscopy, of some form of defect clustering or ordering in $Fe_{1-x}O$. As many reviews exist in the literature(1,16,17,18), our aim here will be to update our previous study(1).

The most significant new clusters to be proposed since our last survey are two forms of a smaller cluster consisting of five vacancies and two interstitials. This proposal is important because it represents a challenge to the widely accepted belief that the 4:1 cluster is the basic building block of larger clusters, due to the high binding energy of the tetrahedral unit. Gartstein et al(18) initially propose a particular geometry of the 5:2 cluster, depicted in figure 5, which they call the 5:2G cluster. They propose this arrangement after an analysis of their high temperature x-ray diffuse scattering data. The study is notable because it is amongst the first of this type for an oxide at high temperature. The data analysis is however very difficult, as it requires separating the diffuse intensity due to local order from that due to ion displacements, followed by Fourier inversion of the local-order intensity to obtain the short-range order parameters. Calculated short-range order parameters are then compared with the values obtained experimentally, as described above.

Gartstein et al[18] subsequently attempt to model the electrical conductivity of $Fe_{1-x}O$, which is a small polaron p-type semiconductor[19]. Their model assumes constant activation energy for the hopping process, which it is proposed occurs between nearest neighbour octahedral ions which also have two cation vacancies in nearest neighbour sites. The modelling procedure adopted is to select clusters with the required number of preferred sites, and which therefore satisfy the observed variation of the density of states of the small polaron with composition[20]. Gartstein et al[18] find that the 5:2G cluster does not fit the conduction model, and they propose an alternative geometry which they call the 5:2M cluster. We depict such a cluster in figure (6). Their final defect model consists of 5:2M clusters and 13:4 Koch-Cohen clusters[21]. However, they also suggest that various other clusters, notably the 6:2, 12:4 and 16:5, would also fit their electrical conduction model.

Our calculated defect formation energies and the binding energies per net vacancy of the 5:2G and 5:2M clusters are given in table (2). Here we also report the values for the clusters favoured by other workers for comparison. We note that the 5:2M cluster is considerably more stable than the 5:2G cluster. However, the 4:1 cluster is half an electron-volt per vacancy more stable than the 5:2M cluster. We therefore consider these clusters will probably not make a significant contribution to the non-stoichiometry of $Fe_{1-x}O$. We also recall that Gartstein et al[18] state that the electrical conductivity data could be modelled satisfactorily by 4:1, 6:2 and 12:4 clusters.

Our favoured defect model is based on the results of our calculations and the electron miroscopy work of Lebreton and Hobbs[16]. The latter authors discuss the presence of two sub-phases of $Fe_{1-x}O$, the P' and P". The former is incommensurate with the rock-salt structure, with a characteristic length of 2.6-2.7 times the lattice parameter of the FeO reference lattice. The latter is more ordered and is commensurate with the rock-salt lattice, having a repeat distance of

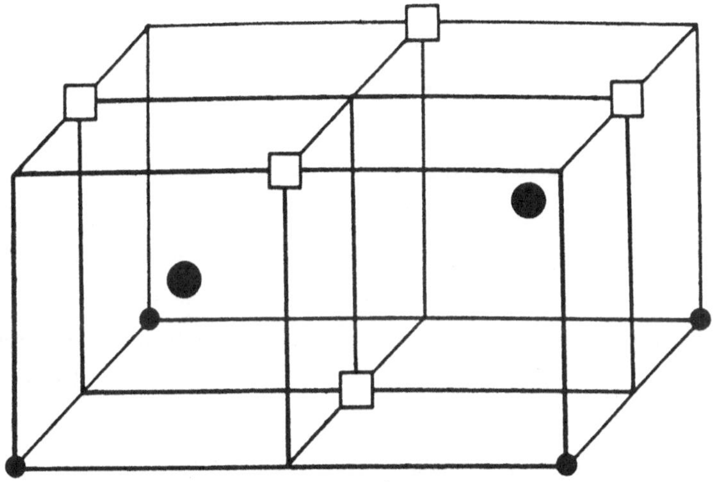

Figure 5. 5:2G defect cluster

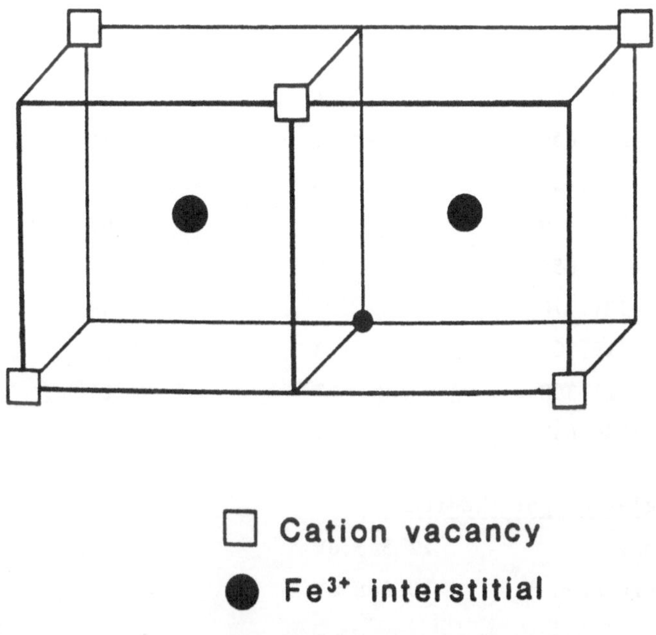

Figure 6. 5:2M defect cluster

2.5 times the reference lattice parameter. The P" phase forms extensively in the more iron deficient material subjected to slow quenching from high temperatures.

We suggest that the P' phase contains a mixture of the stable edge-shared clusters, i.e, the 6:2, 8:3 and the 12:4. These may be ordered for short periods to reduce cluster-cluster interaction. But there is sufficient disorder, for example a 6:2 cluster lying lengthways between two 12:4 clusters, to cause the observed average characteristic length in the lattice images to be incommensurate with the host lattice. Slower quenching will allow sufficient time for an ordering process to occur. The driving force is the minimisation of cluster-cluster interaction energy. The dominant cluster type will then be the 12:4, possibly ordered in the supercell arrangement as Lebreton and Hobbs(16) have suggested. Our calculations, summarised in table (2), give a high binding energy from the 12:4 cluster.

Table 2

Results of cluster stability calculations for $Fe_{1-x}O$

Cluster	Defect formation energy/eV	Net number of vacancies	Binding energy per vacancy/eV
4:1(-1)	-91.966	3	-1.75
6:2	-164.943	4	-2.04
8:3(+1)	-238.050	5	-2.24
12:4	-331.329	8	-2.22
16:5(-1)	-421.919	11	-1.97
13:4(-2)	-307.654	9	-1.85
10:4(-4)	-121.770	6	-1.68
5:2G	-119.704	3	-0.71
5:2M	-121.308	3	-1.24
12:5(-1)	-256.657	7	-1.88

Formation energies of point defects

Fe^{2+} vacancy	22.543 eV
Fe^{3+} substitution	-30.868 eV

n.b. the number in parentheses after the cluster type is residual charge on the cluster in the defect energy calculation.

4.2 Inter-cluster interaction

Our modelling techniques neglect the interaction energy between the defect species. (We are not, of course, referring to the intra defect terms arising from the formation of defect clusters and vacancy-hole complexes). In this section we describe a method by which this limitation may be removed.

We define the lattice energy of $Fe_{1-x}O$ to have three components. The first is the lattice energy of non-defective FeO, the second is the formation energy of the defect and the third is the interaction energy between defects. We will show in this section that the last term is less when defects are clustered than when unassociated.

From the above definition of the lattice energy of defective oxide, we obtain the following relationship defining the interaction energy,

$$E_{interaction} = E_{latt}(Fe_{1-x}O) - E_{latt}(FeO) - E_{defect} \qquad (6)$$

We can identify the last term, E_{defect}, with the defect formation energy obtained from the HADES calculation, i.e. the defect energy at infinite dilution. The procedure is, in essence, relatively straightforward. A supercell is set up containing the defect or defects. The structure is iteratively relaxed until the minimum energy is obtained. From this value we subtract the lattice energy of the same size block of perfect crystal, and the sum of the individual defect energies calculated using HADES. The value we obtain is positive, and represents the energy cost of creating the defects due to their interaction.

We use the METAPOCS code(22) to calculate the lattice energies of the proposed supercell. The METAPOCS code iteratively adjusts ionic co-ordinates within a specified unit cell until a minimum energy configuration is obtained. The minimisation procedure is an adapation of the method used to obtain the configuration of the explicitly relaxed region I in the HADES program. METAPOCS may also be used to adjust lattice parameter until zero bulk strain is achieved. This is often referred to as relaxation to constant pressure. We have constrained the defective supercells of $Fe_{1-x}O$ to have the observed lattice parameter of FeO, i.e. our calculations are performed at constant volume.

The method used in this section also has the advantage of defining the value of the non-stoichiometry. It could therefore be used to test models for the P" phase in FeO. Unfortunately, the large supercell proposed by Lebreton and

Hobbs contains 2541 ions. The version of METAPOCS used in the present study uses second derivatives of the energy to obtain the minimum relatively rapidly and efficiently. This requires the storage of a matrix of size $(3N)^2$, where N is the number of ions in the supercell. Consequently, even with the largest modern supercomputers, an alternative minimisation procedure would have to be used for the Lebreton and Hobbs model.

However, the present study is preliminary to the more extensive study outlined above. Here we present the results of calculating the interaction of 4:1 clusters at two values of non-stoichiometry, and comparing the result to the interaction of non-associated cation vacancies and holes. To define the value of non-stoichiometry it is necessary to adjust the size of the supercell, or change the number of vacancies in a given supercell. We used two sizes of supercell in these calculations. The first contained 64 ions (i.e. 96 species because the oxygen ions were treated by the shell model). The potential model was identical to that used in the defect formation energy calculations. The second contained 216 ions (i.e. 324 species).

In each case the 4:1 cluster was embedded in the middle of the larger cube. Five Fe^{3+} ions were located in nearest neighbour sites to the cluster to preserve charge neutrality. The value of the non-stoichiometry for the two supercells, i.e. the value of x in $Fe_{1-x}O$, may be easily determined. It is the net number of vacancies of the 4:1 cluster divided by the total number of iron ions in the supercell; i.e. for the 64 ion cell $x = 3/32 \sim 0.094$, and for the 216 ion supercell $x = 3/108 \sim 0.028$. The latter value is in fact outside the stability range of wüstite. The result from the calculation using the larger cell serves to illustrate the fall-off of the interaction energy with increasing stoichiometry.

The results of the calculations are shown in table (3). The formation term is obtained from the HADES calculation of the defect formation energy of a $4{:}1^-$ cluster, given in table (2), plus the formation energy of the additional hole. The lattice energy of the perfect cell was obtained using the METAPOCS code, as discussed earlier. For the 64 ion supercell, where $x = 0.094$, we note the interaction energy is 4.9 eV, i.e. unfavourable. This is almost as large, and opposite in sign, to the binding energy of the isolated $4{:}1^-$ cluster; −5.25 eV. However, the interaction energy will not oppose the binding energy if the interaction energy of the unassociated defects is larger. We have calculated the interaction energy of 1,2 and 4 vacancies and neutralizing holes in the 64 ion supercell. The results are given in table (4). The interaction energies are calculated using equation (6). Table (4) also contains the value for 3 vacancies in the 64 ion supercell. This was obtained by a linear interpolation of the values for 2 and 4 vacancies. We could not obtain a value for 3 vacancies directly because the calculation failed to converge. This failure is due to the lack of symmetry that results from attempting to disperse 3 vacancies in the supercell. However, we need this value in order to assess the cost of defect clustering due to their interaction.

From table (4) we note that the interaction energy for 2 vacancies (plus 4 holes), is almost as large as the interaction energy of 4:1 clusters in the same volume, which incorporate 3 vacancies. This suggests that the additional vacancy can be inserted at little extra cost in terms of interaction energy, if a 4:1 cluster forms. If a cluster were not formed, we estimate the interaction energy would increase by about 1.5 eV to 6.125 eV. This result suggests that the interaction energy term would not oppose cluster formation. Rather, the interaction energy favours cluster formation, although it is much less significant than the binding energy.

Table 3

Interaction energy of 4:1 clusters in $Fe_{1-x}O$

64 ion supercell with 4:1 cluster

Lattice energy of defective cell	=	-1382.06 eV
Lattice energy of perfect cell	=	-1264.13 eV
Formation energy of defects	=	-122.83 eV

Interaction energy (using equation 6)
= -1382.06 -(-1264.13-122.83) eV
= 4.90 eV

216 ion supercell with 4:1 cluster

Lattice energy of defective cell	=	-4388.80 eV
Lattice energy of perfect cell	=	-4266.45 eV
Formation energy of defects	=	-122.83 eV

Interaction energy (using equation 6)
= -4388.80 -(-4266.45-122.83) eV
= 0.48 eV

Table 4

Interaction energy of unassociated defects in $Fe_{1-x}O$

Number of vacancies in 64 ion supercell	Non-Stoichiometry (x in $Fe_{1-x}O$)	Interaction energy/eV
1	0.03	0.174
2	0.06	4.619
3	0.094	6.125*
4	0.125	7.631

*Estimated from results for 2 and 4 vacancies.

The results we have presented in this section allow us to conclude first that defect interaction will favour cluster formation. In addition we have also demonstrated that the supercell method is a viable and straightforward method for calculating defect interaction energies.

It would be of interest to extend the study to include the larger defect clusters and the different models for the P" sub-phase of $Fe_{1-x}O$. The interaction energy could also be added to the binding energies to allow their use as input to the mass-action code. Thus the mass-action method could be extended to study the defect structure of highly defective $Fe_{1-x}O$.

5. Comparison of $Fe_{1-x}O$ and $Mn_{1-x}O$

$Fe_{1-x}O$ cannot be prepared with stoichiometric composition. Attempting to reduce it results in phase separation and the precipitation of iron. This indicates that the free energy of oxidation has a negative second derivative with respect to composition, at around the value $x=0.05$ (the lower stability limit). Wüstite is unique amongst the series of FeO, MnO, CoO and NiO oxides in exhibiting such behaviour. The origin of this difference in behaviour is, we suggest, related to the high stability of defect clusters in $Fe_{1-x}O$.

In Table (5) we show the binding energies we calculate for 4:1 and 6:2 clusters in the different oxides. Some of the results have been presented elsewhere in this paper, but we collate them in table (5) for ease of comparison. We see that the clusters do indeed have higher binding energies in $Fe_{1-x}O$ than in the other three oxides. However, the differences appear to be small. To test whether the small differences are sufficient to account for the unique behaviour of $Fe_{1-x}O$ we have used the mass-action code to calculate $\Delta G(O_2)$ over the composition range x=0.04 to x=0.06, at 1200°C. We used as input the same defect equilibria that we found to best model the thermodynamic behaviour of $Mn_{1-x}O$; i.e. vacancies in the three possible charge states, charged 4:1 clusters and the neutral 6:2 cluster (which yields the defect model shown in figure 3).

The result of calculating $\Delta G(O_2)$ for $Fe_{1-x}O$ over this composition range is shown in figure (7). It is apparent that this is a region of instability or $Fe_{1-x}O$; the second derivative is predicted to be negative, which accords with the observation of phase separation at this composition. The origin of this behaviour is, therefore, the extra degree of defect cluster stability in $Fe_{1-x}O$. We suggest that although the differences in binding energy of clusters in the different oxides are similar, the small additional binding energy of clusters in $Fe_{1-x}O$ is responsible for its unique behaviour. The plot of $\Delta G(O_2)$ vs. x for $Mn_{1-x}O$, obtained using the same defect model as for $Fe_{1-x}O$ above, and over the same range of x, is shown in figure (8). This figure confirms the stability of $Mn_{1-x}O$ at this composition.

Table 5

Cluster binding energies for the four oxides

Oxide	Binding energy/eV	
	4:1	6:2
$Fe_{1-x}O$	-5.25	-8.16
$Mn_{1-x}O$	-4.63	-7.86
$Co_{1-x}O$	-4.61	-7.13
$Ni_{1-x}O$	-4.81	-7.18

Figure 7. $\Delta\bar{G}(O_2)$ versus $\log_{10}x$ for $Fe_{1-x}O$ at 1200°C, calculated using defect model shown in figure 3.

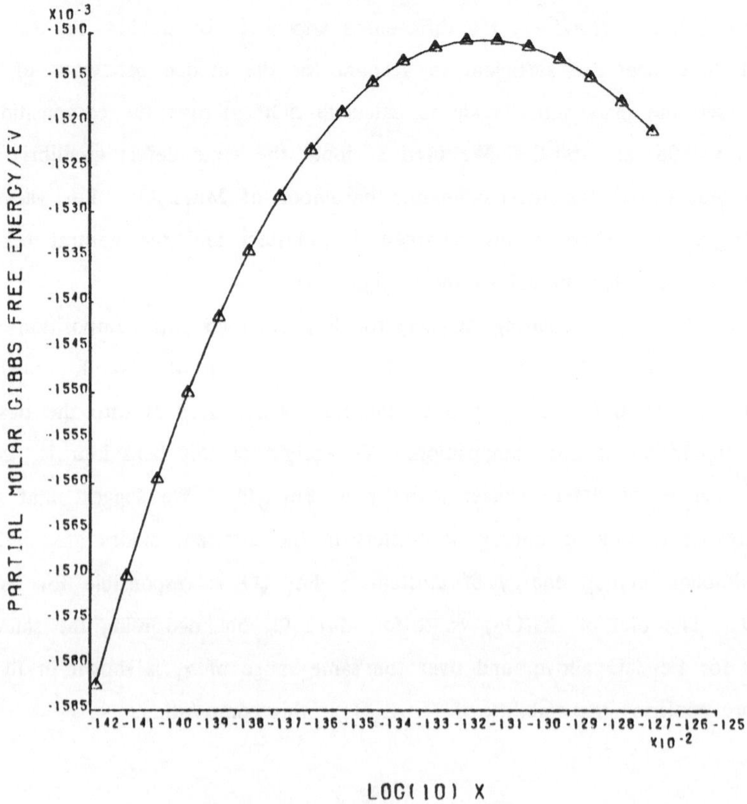

Figure 8. $\Delta\bar{G}(O_2)$ versus $\log_{10}x$ for $Mn_{1-x}O$ at 1200°C, calculated over the same range of x and using the same defect model as for $Fe_{1-x}O$ (figure 7)

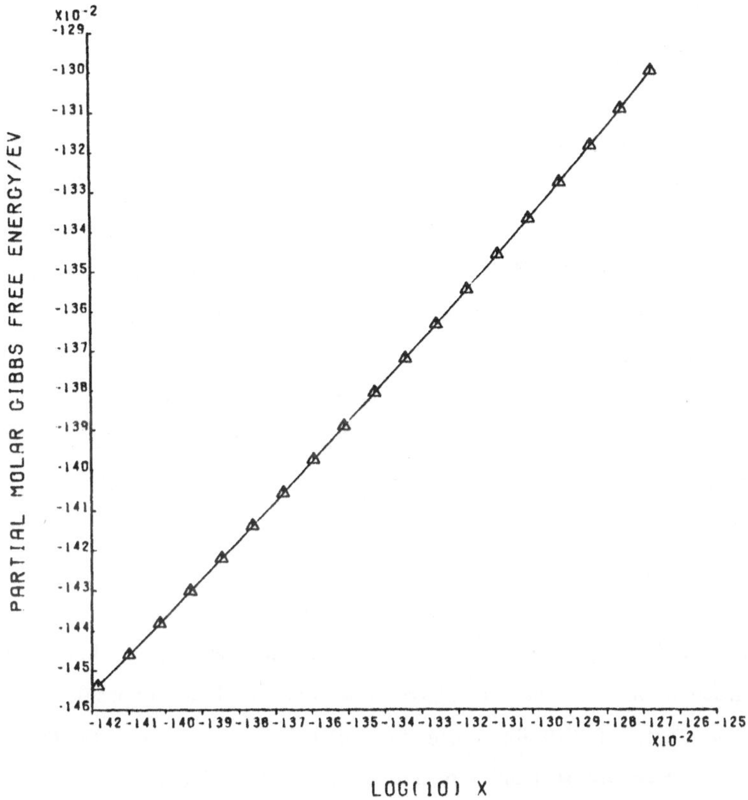

6. Conclusions

For $Mn_{1-x}O$ we have shown how a defect model containing 4:1 clusters in a variety of charge states can reproduce satisfactorily the observed variation of x with $p(O_2)$. Our analysis was dependent on the availablity of good experimental data against which we could test our predictions. We suggest that the methodology outlined in section 2 is complementary to experimental studies of highly defective oxides. Indeed the type of theoretical analysis we have presented here may be used to propose thermodynamically plausible starting points for data analysis. Our calculations on $Fe_{1-x}O$ indicate that the 12:4 cluster proposed by Lebreton and Hobbs[16] is very stable. We have proposed that the incommensurate P' sub-phase of $Fe_{1-x}O$ consists of a mixture of 12:4 clusters, the smaller edge-shared 6:2 cluster, and the 4:1 cluster. The commensurate P" sub-phase probably contains predominantly 12:4 clusters, whose formation on quenching is driven by a reduction in cluster interaction energy.

On the subject of defect interaction energy, we have presented some preliminary results for 4:1 clusters. The results indicate that the formation of the 4:1 cluster reduces the unfavourable interaction term. The supercell method we have outlined may be extended to calculate interaction energies of larger clusters. The results of such a study could be used to extend the analysis by the mass-action method into the highly defective compositions of $Fe_{1-x}O$.

Finally we have predicted, in accord with the observed behaviour, that $Fe_{1-x}O$ is unstable with respect to phase separation at the composition x=0.05. The origin, therefore, of the difference in the behaviour of $Fe_{1-x}O$ the other non-stoichiometric rock-salt structured oxides, is the higher binding energy of defect clusters.

7. Acknowledgements

We would like to thank Drs A.M. Stoneham and J.H. Harding for useful discussions.

8. References

1. Tomlinson S.M., Catlow C.R.A. and Harding J.H., in "Transport in Nonstoichiometric Compounds", eds. Simkovich and Stubican, NATO ASI Series B: Physics, Vol. 129, Plenum Press, New York (1985).
2. Catlow C.R.A., Ann. Rev. of Mat. Sci., 16, 517 (1986).
3. Catlow C.R.A. and Mackrodt W.C., "Computer Simulation of Solids", Springer-Verlag, Berlin (1982).
4. Dick B.G. and Overhauser A.W., Phys. Rev., 112, 90, (1958).
5. Mott N.F. and Littleton M.J., Trans. Farad. Soc., 34, 485 (1938).
6. Catlow C.R.A., James R., Mackrodt W.C. and Stewart R.F., Phys. Rev. B, 25, 1006 (1982).
7. Norgett M.J., AERE Harwell Report, R7650 (1974).
8. Ball R. and Harding J.H., AERE Harwell Report, M3294 (1983).
9. Harding J.H., Physica B+C, 131, 13 (1985).
10. Murray A.D., PhD Thesis, University of London (1985).
11. Jackson R.A., Murray A.D., Harding J.H. and Catlow C.R.A., Phil. Mag. A, 53, 27 (1986).
12. Thirlby P., unpublished work (1983).
13. Tomlinson S.M., PhD Thesis, University of London (1987).
14. Keller M. anad Dieckmann R., Ber. Bunsenges. Phys. Chem., 89, 883 (1985).
15. Tomlinson S.M., Catlow C.R.A. and Harding J.H., J. Phys. Chem. Solids, to be submitted (1988).
16. Lebreton C. and Hobbs L.W., Radiation Effects, 74, 227 (1983).
17. Anderson A.B., Grimes R.W. and Heuer A.H., J. Solid State Chem., 55, 353 (1984).
18. Gartstein E., Mason T.O. and Cohen J.B. in Advances in Ceramics, vol. 23 : Nonstoichiometric Compounds, eds. Catlow C.R.A. and Mackrodt W.C. (1987).
19. Gartstein E. and Mason T.O., J. Am. Ceram. Soc., 65, C24 (1982).
20. Hillegas W.J., PhD Thesis, Northwestern University, U.S.A. (1968).
21. Koch F. and Cohen J.B., Acta Cryst., B25, 275 (1969).
22. Cormack A.N. in "Computer Simulation of Solids", Springer-Verlag, Berlin (1982).

MOTT–LITTLETON AND HARTREE–FOCK CALCULATIONS ON DEFECTS IN α-QUARTZ

Fiona Sim, Department of Chemistry,
University College London, 20 Gordon Street,
London, UK, WC1H OAJ.

C.R.A. Catlow, Department of Chemistry,
University of Keele,
Staffordshire, UK, ST5 5BG.

ABSTRACT. Calculations were performed to investigate properties of defects associated with radiation damage in α-quartz, using the quantum mechanically based Hartree–Fock method, and the more classically rooted Mott–Littleton method. The results obtained using both methods were compared with those obtained by other calculations and from experiment.

1. INTRODUCTION

The theoretical study of solid-state systems can be divided into two broad classes of calculation; those based on the explicit solution of the Schrödinger equation, and those based on classical models of the solid. Both types of approaches have their strengths and weaknesses. The first approach, based on quantum mechanics, should in principle lead to exact information about any system. However, there exist many theoretical and technical problems in its application - not least of which is that the present formulation of quantum mechanics is assumed to be reversible in time, whilst clearly many processes in crystal lattices are not. The most difficult technical problem so far, is that the Schrödinger equation has proved to be exactly solvable only for one-electron systems. Thus, much effort has been expended in producing methods which give approximate solutions to this equation. These in general involve the reduction of the many-body Hamiltonian to the sum of one-electron Hamiltonians for the system. However, this leads in turn to difficulties in including electron-correlation effects in this type of calculation.

Perhaps one of the most simple models of the solid was proposed by Born in 1923 [1]. Originally proposed for ionic crystals, it is assumed in this model that ions are spherical, do not overlap and have integral charge. This picture leads to the assumption that the lattice energy of the crystal can be broken down into two main contributions; the electrostatic interaction of the ions, and a distance dependent "short-range" function which should be repulsive at small internuclear distances.

In the work presented here, calculations representing both types of approach, were performed for defects associated with radiation damage in α-quartz. This system was

J. Nowotny and W. Weppner (eds.),
Non-Stoichiometric Compounds Surfaces, Grain Boundaries and Structural Defects, 77–91.

chosen because it represents a challenge to both theoretical methods, and because it is technologically important. α-quartz is now the most commonly used material in ultrasonic devices and frequency standards. Furthermore, it is often required to operate in hostile evironments, such as in satellites, where exposure to radiation induces off-sets in its oscillation frequency. This instability on irradiation is known to be caused by point defects in the quartz crystal, the most important of which are probably those associated with a substitional aluminium ion. This ion – which may be designated Al^{3+} in the Born model – requires a charge compensating monovalent cation, such as Li^+, Na^+, H^+ or an electron-hole, when it substitutes for the tetravalent Si^{4+} at a quartz lattice site. The interaction dynamics of these compensating defects leads to changes in the elastic constants of the lattice and this causes frequency changes in the oscillator. Therefore, a detailed understanding of the behaviour of these defects is necessary if the problem of quartz oscillator instability is to be eliminated.

2. CALCULATION METHODS

2.1 Mott–Littleton Calculations

This method for the treatment of defects is based on the Born model of the solid, which provided a model for perfect crystals only. Application of the method, as employed in the computer programs HADES [2-4] and CASCADE [5], requires that a few assumptions about the solid be made. Firstly, that it is possible to separate the electronic and nuclear contributions to the energy (the adiabatic approximation). Secondly, that the displacements of the atoms in the solids form their equilibrium positions may be described as a quadratic function of their position (the harmonic approximation) and thirdly, that the quadrupole and higher order multipole contributions to the electrostatic energy may be neglected (the dipole approximation). These approximations lay the foundations for static lattice simulations of perfect crystals, the method of Mott and Littleton [6] allows this approach to be extended to systems containing defects. It involves adjusting the positions of the ions nearest the defect, using an iterative procedure like the Newton-Raphson method [7], so that the forces created by the introduction of the defect to the perfect lattice are minimised. In the volume immediately surrounding the defect, Region I, the forces on every ion are minimised. In a second larger volume, Region II, the displacements caused by the defect are assumed to be due to the macroscopic polarisation induced by the effective charge of the defect. Detailed mathematical descriptions of this method are given in Refs. [3,4,8]. For a more general description of this type of model and the quantum mechanically based methods used in the treatment of defects in solids see Ref. [9].

In the last ten years, this method has been established as a reliable tool for investigating the properties of ionic systems. Several reviews of its application have been published [10,11] and it is being used to treat an increasingly diverse range of materials including the superconducting oxide La_2BaCuO_4 [12]; the superionic conductor LaF_3 [13]; lithium intercalation in MnO_4 and Fe_3O_4 [14,15]; gas adsorption in zeolites [16,17] and silicate minerals [18,19].

Much effort is being put in to extending the method and removing some of its deficiencies, for example, the inclusion of many-body [20,21] and entropy terms [22]. One development which could lead to a significant advance in the accuracy of these calculations would be the development of a method for the production of interatomic potentials, particularly for defect interactions, from a quantum mechanical basis. Some work towards this has been carried out using the particular variant of the Hartree–Fock method, which will be outlined in the next section [23,24]. However, there are many problems concerned with the application of this method to the solid-state which must be solved, before it can be used as a standard method to produce interatomic potentials.

2.1.1 *Calculations on α-quartz* Full details of these calculations are given in Refs. [24,25]. The calculations were carried out using a Region I of 1.2 lattice units (l.u.) which contained 89 ions, using the pair-potentials given in the following References: Si...O potential [21]; O...O potential [26]; Al...O potential [27]; O...H potential [23]. Polarisation effects were included by the use of a shell-model potential for the oxygen ions [26].

The structure of right-handed α-quartz (space group $P3_121$) determined by Jorgensen [28] by X-ray diffraction was taken as the starting point in both types of calculation. Defects were introduced to the perfect lattice and the lattice was allowed to adjust the position of the ions until an energy minimum was obtained.

The calculations were carried out using the CASCADE program [5] on the CRAY-1S at the University of London Computer Centre and the FPS-164 at the SERC Daresbury Laboratory.

2.2 Self-Consistent-Field Hartree–Fock Calculations

The field of Hartree–Fock calculations on solids is relatively under-developed in comparison to both Hartree–Fock calculations on molecules and solid-state calculations involving model systems. This is partly for historical reasons (a good review of some of the aspects of this is given by Pisani *et al.* [29]) and because of the enormity of the difficulties involved in performing this type of calculation on any type of extended system.

The advent of modern supercomputers and the production of highly efficient computer programs, have gone some way towards reducing the computational problems involved (for an up to date review of the present state of the art regarding supercomputers, see Ref. [30]). However, it will not be possible in the foreseeable future to perform calculations on physically realistic portions of solids merely by increasing the number of atoms included in the calculations. Thus, it is necessary to introduce some further approximations to the basic Linear Combination of Atomic Orbital Self-Consistent-Field Hartree–Fock (LCAO SCF HF) approach which has been so successful for molecules. There exist at present many ways in which this problem has been tackled, including the use of semi-empirical SCF Hartree–Fock methods, such as CNDO and INDO on larger clusters of atoms; the use of local density functions and the development of periodic boundary Hartree–Fock methods. A description of the theory behind each of these methods and a discussion of their relative merits is far beyond the scope of this paper. However, from a present day stand point, with the rapidly increasing power of computers, it still appears that those approaches based on the *ab initio* Hartree–Fock treatment look very promising. They give reasonably accurate ground-state properties and provide a sound starting point from which to add electron-correlation effects. Furthermore, it is to be hoped that in the long term computational difficulties with this type of calculation will be ever diminishing.

2.2.1 *Calculations on α-quartz* These calculations were carried out using a basic $Si_5O_{16}^{12-}$ cluster into which various defect species were placed. Full details of these calculations, including the reasons for this choice of cluster and the 3-21G basis set [31], and a test of basis-set superposititon error are given in Ref. [32]. Especially in the treatment of electronic defects, it is necessary to include the long-range contribution to the electrostatic energy at the sites where the Hartree–Fock treated ions are placed. This can be achieved by surrounding the Hartree–Fock cluster by a series of point-ions placed at the X-ray determined lattice sites. However, it is extremely important that some thought is given to the choice of this cluster. It should be chosen in such a way that there are no spurious field effects due to the shape of the point-ion cluster, and care must be taken in assigning the charges to the point-ions, especially in the case of a semi-covalent material like α-quartz [24].

The defect clusters described here were geometry optimised using the HONDO program [33] on the LCAP systems at IBM-Kingston. Using these optimised geometries, the isotropic hyperfine coupling constants were calculated by the Unrestricted Hartree–Fock (UHF) formalism using the GAMESS program [34,35] on the CRAY-XMP at the SERC Rutherford Appleton Laboratory.

It should be noted that it was not possible to optimise the positions of the outer twelve oxygen ions, as the force on these ions due to the close proximity of the point-ion lattice was too great. This is one of the serious deficiencies of this method. In the Mott–Littleton method the lattice relaxation caused by the presence of the defect can be fully taken into account. As may be seen from a comparison of the results obtained using these two methods, this leads to the Hartree–Fock method giving larger defect binding energies in all cases.

3. RESULTS AND DISCUSSION

3.1 Defect Centres Containing Aluminium

A large number of defect centres of this type was investigated, particularly using the Mott–Littleton (ML) method. At present, the computing time required to geometry optimise the large Hartree–Fock clusters – days in some cases – make it only possible to use this method in a few selected cases. As this is primarily a comparitive study, results will be given for only one or two cases which illustrate the general trend.

Table I shows the binding energies calculated for two defects; the aluminium/hole centre, $[AlO_4]^0$ and the aluminium/proton centre, $[AlO_4H]^0$. In these defect centres, the excess charge of the Al^{3+} is compensated by an electron-hole and a proton on a neighbouring oxygen atom, respectively. These defects have been studied extensively using electron spin resonance (esr) spectroscopy in the case of the paramagnetic electron-hole defect, and infra-red (ir) spectroscopy, in the case of the proton defect. Reviews of the studies on these defect can be found in Refs. [36,37].

It is perhaps necessary at this point to explain the basic geometry of these defects in a little more detail. In perfect α-quartz, each silicon is surrounded by two pairs of oxygen atoms at distances of 1.616 Å and 1.598 Å. This slightly distorted tetrahedron forms the basic building block of the structure. Each oxygen ion is equivalent because it has one long and one short bond to the nearest silicon ions. However, the replacement of a silicon

TABLE I
Defect binding energies calculated for the aluminium/proton and aluminium/hole defects in α-quartz using the SCF-HF method and the Mott-Littleton (ML) method

Defect	Method	Binding Energy (eV)
Al/proton	SCF-HF	-5.89
Al/proton	SCF-HF (geom. opt.)	-5.26
Al/proton	ML	-1.70
Al/hole	SCF-HF	-6.36
Al/hole	SCF-HF (geom. opt.)	-5.36
Al/hole	ML	-1.82

ion by an aluminium ion breaks this equivalence, two distinct pairs of oxygen ions are created; two at the long distance from the aluminium ion and two at the short distance. It has been proposed that hopping of the electron-hole between these oxygen sites may be thermally activated or that at low temperatures, a quantum mechanical tunnelling process may occur. This electronic motion, which may also involve a flipping of the electron spin, is thought to give rise to the dielectric loss measured from this defect [38-40]. Furthermore, some debate exists about the amount of lattice distortion caused by the introduction of the electron-hole to the aluminium site. Estimates of this distortion calculated from esr data range from almost 50% [41] to 9% [42]. It is the change in distortion, in going from one compensating defect to another on irradiation, that is of fundamental importance in determining the change in oscillation frequency of the crystal.

From Table I, it can be seen that the defect binding energies calculated using the Hartree–Fock method, are much higher than those calculated using the Mott–Littleton method. It can also be seen that geometry optimisation of the inner atoms of the cluster did not reduce the binding energy by more than 1 eV. The protons in this defect centre are known to be mobile at temperatures below 77 K [43], and this would suggest that the Hartree–Fock binding energies are unreasonably high.

However, it is interesting to compare the differences in binding energies for the two defects. Subtraction of the defect binding energy of the $[AlO_4H]^0$ defect from that of the $[AlO_4]^0$ defect give –0.10 eV in the SCF–HF framework, and –0.12 eV in the ML framework. In other words, a comparison of defect binding energies within each framework will yield the same results. Therefore, considering the enormous difference in cost of the two types of computations, it may be advised that if it is changes in defect binding energies which are sought, then the cheaper Mott–Littleton method should be used.

It is not immediately obvious, that these two different types of calculations should give such similar results, especially in the case of such highly localised defects in a semi-covalent system. The fully ionic treatment of charge in the Born model, and hence in the Mott–Littleton method, has always been a major source of criticism. It has been suggested that this should be an inappropriate treatment for systems which are not completely ionic. However, the fact that the Hartree–Fock calculations, which treat the electron distribution explicitly, give such similar results suggests that the use of fully ionic charges in the ML calculation is not of primary importance in determining the results.

Of course it is not enough to simply compare the results of one theoretical method with the results of another, comparison with experiment must be made where possible. In the case of the aluminium/hole defect, good esr data is available and so it is possible to compare the calculated structures with those predicted by esr. Figures 1 and 2 show the structure calculated using the Mott–Littleton and the Hartree–Fock approaches for the $[AlO_4]^0$ defect, respectively.

In both cases, significant relaxation around the oxygen containing the electron-hole occurs (the oxygen in the upper-right hand corner of the diagram in both Figures). However, in the ML computed geometry the increase in internuclear distance of the aluminium to hole bearing oxygen is approximately 23% for both short- and long-bonded oxygen sites. Whilst in the HF computed case, the increase is only 10.8% (after geometry optimisation it was not possible to distinguish between the long- and short-bonded sites). In both calculations the distance to all other oxygens from the aluminium was reduced; by between 6% and 2% in the ML calculation and by 3% for each oxygen site in the HF case. The other main difference in computed bond lengths, was that in the ML calculation the distance from the central four oxygens to the next silicon atoms decreased by approximately 1.4%, except in the case of the oxygen bearing the electron-hole, where there was an increase in distance of almost 30%. However, in the HF computed geometry, the distance to the nearest silicon atoms increased by about 1% for the three normal oxygens and by 9% in the case of the hole bearing oxygen (labelled 4 in the diagram). Further details of these results, including changes in computed angles, are given in Ref. [24].

Although many esr experiments have been performed on this defect centre over the years, no firm conclusion can be drawn on the relaxation caused by the formation of the electron-hole at the aluminium substitutional site. This is for two main reasons; firstly, it is necessary to assume a model Hamiltonian to allow the data to be interpreted and secondly, it is necessary to use Hartree-Fock computed densities to convert the results to quantities such as bond-lengths. Of course similar problems exist in going from HF computed results to measured esr constants.

In 1975, Shirmer proposed that the Al...O^- bond length (where O^- represents the hole bearing oxygen) increased to 2.4 Å, an increase of 48% [41]. This model assumed that the electron-hole was completely localised in a non-bonded p-orbital on the oxygen ion. Nuttall and Weil [44] predicted an Al...O^- distance of 1.79 Å from their esr data to give a 11% relaxation, but this included a more or less statistical interpolation of the Al...O bond length. More recently, Adrian *et al.* have predicted a relaxation of 9% from their esr results [42]. They used a model in which the unpaired electron in the $2p_y$ orbital, is

Optimised geometry of the Aluminium hole centre

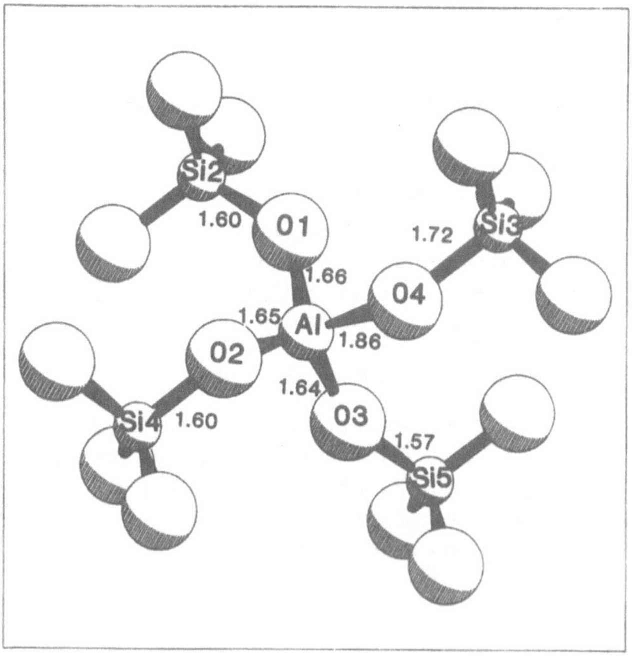

FIGURE 1. Optimised geometry of the aluminium/hole defect in α-quartz. Calculated by the SCF-HF method using an $Si_4AlO_{16}^{12-}$ molecular cluster. Internuclear distances are in Å.

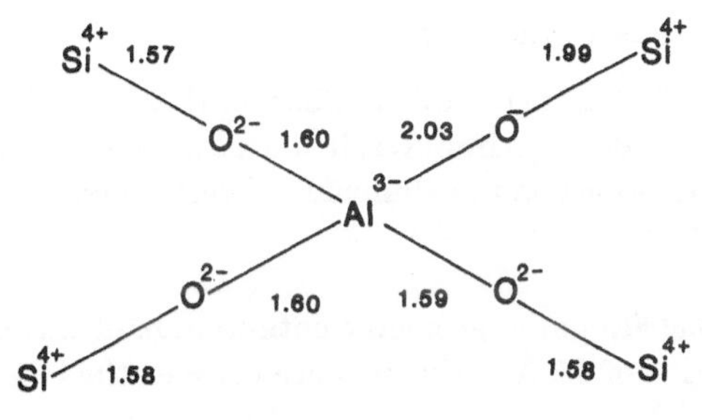

CASCADE optimised geometry of the Al /hole centre

FIGURE 2. Optimised geometry of the aluminium/hole defect in α-quartz. Calculated by the Mott-Littleton method. Internuclear distances are in Å.

able to couple, *via* an exchange mechanism, with the spin-paired electrons in the 2s and $2p_z$ orbitals. These electrons in turn effect the anisotropic hyperfine coupling constant at the aluminium nucleus, as these orbitals overlap with aluminium orbitals. Thus, inclusion of this spin polarisation in the model, reduces the estimation by almost 40% compared to that of Shirmer's, where only the classical dipole-dipole interaction from the unpaired electron to the aluminium nucleus was used to interpret the anisotropic hyperfine coupling constant.

In an SCF-HF calculation on this defect using an isolated AlO_4H_4 molecule and a minimal STO-3G basis-set, an Al...O^- bond length of 1.99 Å was calculated [45]. This represents a 17% relaxation compared to the 11% relaxation presented here for the HF method. If one takes the more recently obtained results [42,44] to be the more reliable, then it would appear that it is neccessary to use both a large model cluster and a good basis-set in the HF calculation to obtain a close agreement with experiment.

3.2 Defect Centres Containing Trapped Electrons – E′ Centres

A family of defect centres have been identified in α-quartz which are characterised by an electron trapped at a silicon site adjacent to an oxygen vacancy. For a review of these defects see Ref. [37]. These oxygen vacancies, which are formed on irradiation by neutrons, may act as a sink for the electrons displaced on the formation of the aluminium/hole centres discussed above. Three of the most important defects of this type are thought to comprise the following species:

E′ Defect Centres :

E'_1 : unpaired electron trapped at an oxygen vacancy

E'_2 : proton trapped at an oxygen vacancy plus an unpaired electron at an adjacent silicon ion

E'_4 : a hydrogen atom plus an electron trapped at an oxygen vacancy - perhaps as a hydride ion

It was not possible to perform Mott-Littleton calculations on this type of defect as there is no explicit treatment of electron density. However, it was possible to carry out Hartree-Fock calculations of the type performed on the aluminium defect centre using an $Si_5O_{15}^{12-}$ cluster and the UHF formalism.

3.2.1 *Calculated optimised geometries* Initial geometry optimisation of this basic 'oxygen-vacancy' cluster led to a reduction of 0.2 Å in the distance between the two silicon atoms, between which the vacancy had been created. The Mulliken population analyses of this cluster indicated that these two silicon atoms were 0.6e more positively charged than in the perfect lattice cluster. However, the charges on the other atoms remained largely unchanged.

Figures 3, 4 and 5 show the Hartree-Fock optimised geometries for the clusters representing the E'_1, E'_2, and E'_4 defects, respectively. Details of these calculations, and the oxygen vacancy calculation can be found in Ref. [24].

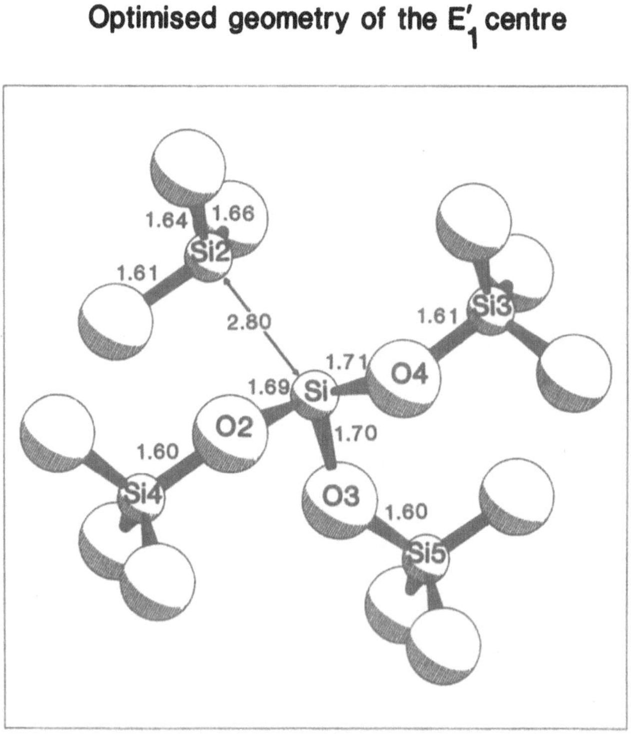

FIGURE 3. Optimised geometry of the E'_1 defect in α-quartz.
Calculated by the SCF-HF method using an $Si_5O_{15}^{13-}$ molecular cluster.
Internuclear distances are in Å.

Table II shows the binding energies calculated for these three defects. It can be seen that these energies are even larger (over 5 eV for the E'_4 defect) than those calculated for the aluminium centres. This is probably due to incomplete lattice relaxation around the vacancy site. This is a particular problem for these systems, as the creation of the oxygen vacancy effectively splits the cluster, upon which the calculation is performed, into two asymmetric clusters. In the larger cluster, around silicon 1, the three neighbouring oxygens and the next-neighbour silicon atoms can all adjust their positions to minimise the energy. However, in the smaller cluster around silicon 2, the three neighbouring oxygen atoms are held fixed at their perfect lattice positions.

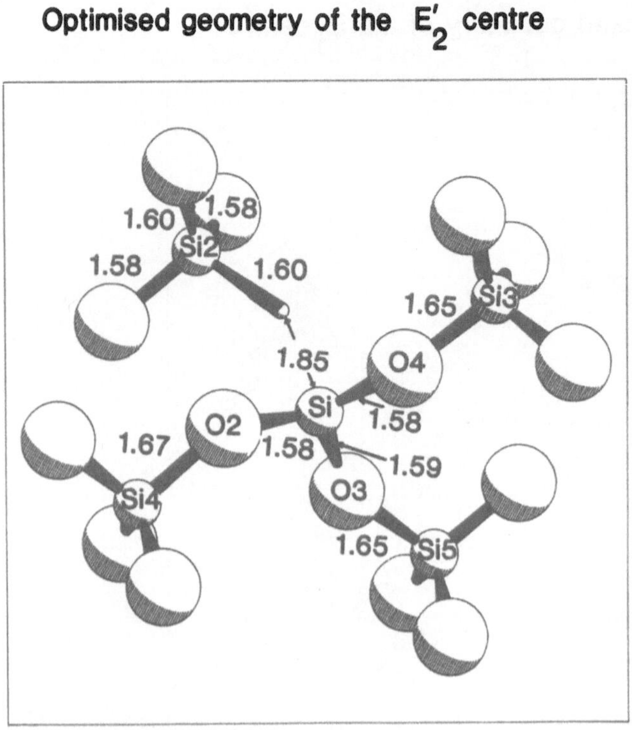

FIGURE 4. Optimised geometry of the E_2' defect in α-quartz.
Calculated by the SCF-HF method using an $Si_5O_{15}H^{12-}$ molecular cluster.
Internuclear distances are in Å.

3.2.2 *Calculated isotropic hyperfine coupling constants* Evidence that these calculations are indeed not as good as those performed on the aluminium centres, where the defect is located well within the SCF-HF treated cluster, is obtained when the calculated isotropic hyperfine coupling constants A_{iso} are compared with experimental values. These constants were calculated using the unpaired spin density at the specific nuclei and the published atomic parameters for that nucleus [46], along with the following expression for the Fermi contact potential

$$A_{iso} = \tfrac{8}{3}\pi g_e \beta_e g_N \beta_N |\psi(0)|^2 .$$

Tables III and IV show a comparison of the calculated A_{iso} constants for the E_4' defect and the aluminium/hole centre, with those obtained by experiment, respectively. In both these defect calculations the expectation value of the S^2 operator was close to the 0.75 expected of the pure doublet state; 0.753 for the aluminium/hole centre and 0.758 for the E_4' centre.

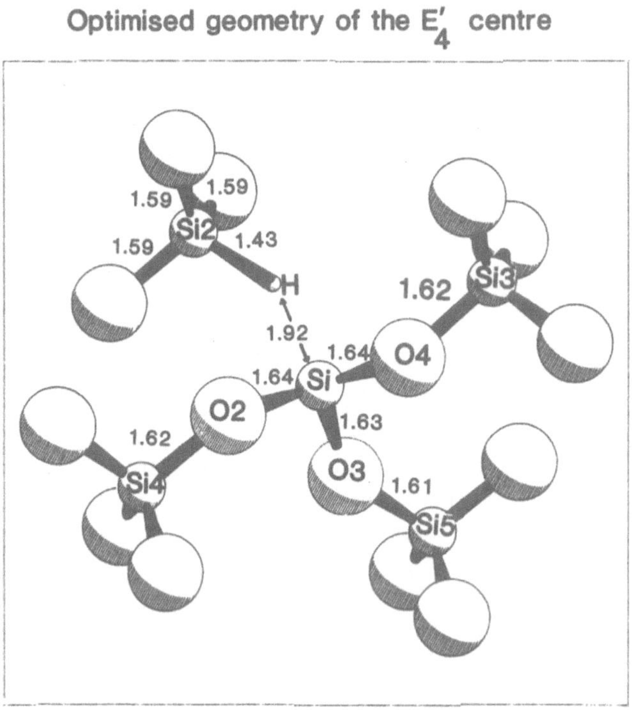

FIGURE 5. Optimised geometry of the E'_4 defect in α-quartz.
Calculated by the SCF-HF method using an $Si_5O_{15}H^{13-}$ molecular cluster.
Internuclear distances are in Å.

However, it can be seen from the Tables that only the A_{iso} calculated for the aluminium/hole defect is close to the experimentally determined value. The values for the hhc of the two silicon atoms on either side of the vacancy in the E'_4 defect are not even in good qualitative agreement with experiment. The ratio of the A_{iso} constants of Si(1) to Si(2) is approximately 6 in the experimental case and it is 22 in this calculation. However, the calculated value for the A_{iso} of the hydrogen species in this defect is much closer to the experimentally obtained value. This could be a factor of the values taken for the atomic parameters for silicon as more than one estimate is available. As mentioned previously, it is often difficult to convert directly from esr results to structural information and *vice versa*.

TABLE II
Defect binding energies calculated for the E′ centre defects using an $Si_5O_{15}^{12-}$ cluster surrounded by 956 fully ionic point-ions

Defect	Binding Energy (eV)
E'_1	-3.2834
E'_2	-2.8068
E'_4	-5.4285

TABLE III
A comparison of experimental and calculated values of the isotropic hyperfine coupling constant A_{iso} for the E'_4 defect in α-quartz.

E'_4 defect	A_{iso} Si(1) (mT)	A_{iso} Si(2) (mT)	A_{iso} H (mT)
Isoya *et al.* (1981) - Ref. [47] esr data	-44.577	-7.677	0.866
SCF-HF treated $Si_5O_{16}^{12-}$, 3-21G basis, 956 point ions at quartz lattice sites (this work)	-22.693	-1.020	1.023

TABLE IV
A comparison of some experimental and calculated values of the isotropic hyperfine coupling constant A_{iso} for the aluminium nucleus in the aluminium/hole defect in α-quartz.

Aluminium/hole defect	A_{iso} (mT)
Nuttall and Weil (1981) - esr data Ref. [44]	-0.578
Mombourquette and Weil (1985) - Ref. [48] isolated SCF-HF treated AlO_4H_4, STO-3G basis	-0.461
SCF-HF treated $Si_5O_{16}^{12-}$, 3-21G basis, 956 point ions at quartz lattice sites (this work)	-0.662

The A_{iso} calculated for the aluminium defect is in close agreement with the experimental result of Nuttall and Weil [44] and is an improvement over the previously HF calculated value of Mombourquette and Weil [48]. However, both HF calculations gave results fairly close to the experimentally obtained value and this would seem to be a good indication that, if applied with care, this method will give good quantitative results.

4. CONCLUSIONS

From the results presented here, it can be concluded that in many applications, the results of the classically based Mott-Littleton approach and the quantum mechanically based Hartree-Fock approach are in close qualitative agreement. However, in any individual systems the results must be studied with care. In the case of the relaxation around the aluminium/hole defect, the ML computed result is probably much too large due to the fact that it was based purely on an electrostatic effect, the hole bearing oxygen being treated simply as an "O^-" species. This is a place where the development of a method for the computation of interatomic potentials from first priciples would be very useful. It is not possible to obtain pair-potentials for a defect interaction of this type from experimental data and at present no clear theoretical method exists for their calculation.

For the same system, it cannot be overlooked that the HF calculation result might underestimate the relaxation around the defect, as the outer 12 oxygen ions were held fixed. This could be tested by comparing these results with those obtained using a larger cluster. However, the good agreement of the calculated isotropic hcc with that obtained experimentally, suggests that we can be hopeful that most of the relaxation effects were taken into account.

Unfortunately, the same cannot be said for the E'_4 centre. It is clear from the calculated isotropic hcc that this calculation could be improved by the use of a bigger atomic cluster. However, the reasonably good result calculated for the aluminium/hole defect, implies that a similar result could be obtained for the E'_4 centre.

The Mott-Littleton method includes most of the relaxation around the defect, it is much cheaper and therefore, can be used to calculate the energies of many defect systems. The use of the Ewald technique also means that the problem of how to truncate the defect cluster, which can present serious problems in the HF method, is completely avoided. Its theoretical limitations and the difficulties associated with obtaining defect potentials, mean that it cannot be applied to all systems – for example, the E' centres are outside its scope. However, in those systems where it can be applied, the results stand up very well when compared with Hartree-Fock computed results.

ACKNOWLEDGEMENTS

F.S. would like to thank Dr. M. Dupuis and Dr. J.D. Watts for their help with the work carried out at IBM-Kingston and also the Department of Chemistry, University College London and IBM Corporation USA for their financial support.

REFERENCES

1. Born, (1923), Atomtheorie des Festen Zustandes Springer, Berlin; Born, M. and Huang, K., (1954), Dynamical Theory of Crystal Lattices, Oxford University Press, Oxford.
2. Norgett, M.J., (1974), AERE Harwell Report R.7650.
3. Catlow, C.R.A. and Norgett, M.J., (1976), AERE Harwell Report M2936.
4. Catlow, C.R.A., James, R., Mackrodt, W.C. and Stewart, R.F., Phys. Rev. B **25**, (1982), 1006.
5. Smith, W., (1981), SERC Daresbury Laboratory Report Dl/SCI/TM25T.
6. Mott, N.F. and Littleton, M.J., Trans. Farad. Soc. **34**, (1938), 485.
7. Norgett, M.J. and Fletcher, R., J. Phys. C. **3**, (1970), 163.
8. Mackrodt, W.C., in Mass Transport in Solids, (1983), eds. Catlow, C.R.A. and Mackrodt, W.C., Plenum Press, New York.
9. Stoneham, M.A., Theory of Defects in Solids, (1975), Clarendon Press, Oxford.
10. Catlow, C.R.A. and Mackrodt, W.C., (1982), Computer Simulation of Solids, Lecture Notes in Physics Series, Vol. 166, Springer-Verlag, Berlin.
11. Mackrodt, W.C., in Transport in Non-Stoichiometric Compounds, (1984), eds. Petot-Evans, G., Matzke, H.J. and Monty, C., North-Holland, Amsterdam.
12. Islam, M.S., Leslie, M., Tomlinson, S.M. and Catlow, C.R.A., J. Phys. C **21**, (1988), L109.
13. Jordan, W.M. and Catlow, C.R.A., Proc. of the 5th Europhysical Topical Conference, Madrid, (1986) : Lattice Defects in Solids, in press.
14. Islam, M.S., Ph.D. Thesis University of London, (1988).
15. Islam, M.S. and Catlow, C.R.A., J. Phys. Chem. Solids, in press.
16. Hope, A.T.J., Ph.D. Thesis University of London, (1986).
17. Hope, A.T.J. and Catlow, C.R.A., Proc. Roy. Soc. (Lond.), in press.
18. Parker, S.C., Catlow, C.R.A. and Cormack, A.N., Acta Cryst. B **40**, (1984), 200.
19. Catlow, C.R.A., Doherty, J.M., Parker, S.C., Price, G.D. and Sanders, M.J., Materials Science Forum **7**, (1985), 118.
20. Leslie, M., Physica B **131**, (1985), 145.
21. Sanders, M.J., Leslie, M. and Catlow, C.R.A., J. Chem. Soc. Chem. Commun., (1984), 1273.
22. Jacobs, P.W.M., Nerenberg, M.A. and Govindarajan, J., in Computer Simulation of Solids, Lecture Notes in Physics Series, Vol. 166, Springer-Verlag, Berlin.
23. Saul, P., Catlow, C.R.A. and Kendrick, J., Phil. Mag. B **51**(2), (1985), 107.
24. Sim, F., Ph.D. Thesis Univeristy of London, (1988).
25. Sim, F. and Catlow, C.R.A., to be published.
26. Catlow, C.R.A., Proc. Roy. Soc. (London) **A353**, (1977), 533.
27. Lewis, G.V., Ph. D. Thesis, University of London, (1984).
28. Jorgensen, J.D., J. Appl. Phys. **53**, (1982), 477.
29. Pisani, C., Dovesi, R. and Roetti, C., (1988), Hartree-Fock Ab Initio Treatment of Crystalline Systems, Lecture Notes in Chemistry Series, Vol. 48, Springer-Verlag, Berlin.
30. Dupuis, M., (1986) Supercomputer Simulations in Chemistry

Lecture Notes in Chemistry Series, Vol. 44, Springer-Verlag, Berlin.
31. Binkley, J.S., Pople, J.A. and Hehre, W.J., J. Am. Chem. Soc. bf 102, (1980), 939; Gordon, J.S., Binkley, Pople, J.A., Pietro, W.J. and Hehre, J., J. Am. Chem. Soc. **104**, (1982), 2797.
32. Sim, F., Catlow, C.R.A., Dupuis, M., Watts, J.D. and Clementi, E. (1986), in Supercomputer Simulations in Chemistry, Lecture Notes in Chemistry Series, Vol. 44, ed. Dupuis, M., Springer-Verlag, Berlin.
33. Dupuis, M., Rhys, J. and King, H.F., J. Chem. Phys. **64**, (1976), 111.
34. Dupuis, M., Spangler D. and Wedoloski, J.J., NRCC Software Catalogue (1980), Vol. 1, Prog. No. QG01 (Manchester Regional Computing Centre).
35. Guest, M.F. and Kendrick, J., GAMESS user manual, (1985), Daresbury Technical Memorandum.
36. Weil, J.A., Rad. Eff. **26**, (1975), 261.
37. Weil, J.A., Phys. Chem. Miner. **10**, (1984), 149.
38. Taylor, A.L. and Farnell, G.W., Canad. J. Phys. **42**, (1964), 59.
39. De Vos, W.J. and Volger, J., Physica **47**, (1970), 13.
40. Schnadt, R. and Schneider, J., Phys. Kondens. Mater. **11**, (1970), 5.
41. Shirmer, O.F., Solid State Commun. **18**, (1976), 1349.
42. Adrian, F.J., Jette, A.N. and Spaeth, J.M., Phys. Rev. B **31(6)**, (1985), 3923.
43. Sibley, W.A., Martin, J.J., Wintersgill, M.C. and Brown, J.D., J. Appl. Phys. **50(8)**, (1979), 5449.
44. Nuttall, R.H.D. and Weil, J.A., Can. J. Phys. **59**, (1981), 1696.
45. Mombourquette, M.J., Weil, J.A. and Mezey, P., Can. J. Phys. **62**, (1984), 21.
46. Symons, M., (1978), Chemical and Biochemical Aspects of Electron-Spin Resonance Spectroscopy, Van Nostrand and Reinhold, Wokingham.
47. Isoya, J., Weil, J.A. and Halliburton, L.E., J. Chem. Phys. **74(10)**, (1981), 5436.
48. Mombourquette, M.J. and Weil, J.A., Canad. J. Phys. **63**, (1985), 1282.

POINT DEFECT STRUCTURE OF CHROMIUM (III) OXIDE

Ming-Yih Su* and George Simkovich
The Metals Science and Engineering Program
Materials Science and Engineering Department
The Pennsylvania State University
University Park, PA 16802

* Presently at Northwestern University, Materials Research Center, Evanston, IL 60208

ABSTRACT. Based upon studies of the electrical conductivity and Seebeck coefficient of TiO_2- and MgO-doped Cr_2O_3, the point defect structure of Cr_2O_3 was determined. It is found that the defect and transport properties in Cr_2O_3 are complicated. At high temperatures, different defects may be present depending upon the oxygen partial pressure. In general, at high P_{O_2}, Cr_2O_3 is a p-type semiconductor with electron holes and chromium vacancies as the dominant defects; at intermediate P_{O_2}, it behaves as an intrinsic semiconductor with electrons and electron holes dominant; and at low P_{O_2}, near the Cr/Cr_2O_3 equilibrium oxygen pressure, it changes to an n-type semiconductor with electrons and chromium interstitials dominant. The equilibrium constants associated with the formation of different defects are also obtained.

1. INTRODUCTION

Chromium sesquioxide (Cr_2O_3) is an oxide of extreme importance mainly because it provides a protective scale to many technological materials against high temperature oxidation and corrosion. It has been known that small additions of impurities or active elements in either the metal or oxide may have significant effects on the oxidation rate of the metal. In order to elucidate the oxidation mechanism and the effects of the additions, it is necessary to know the transport properties of Cr_2O_3 and the way in which these properties may be varied. In other words, a complete knowledge of the defect structure of this oxide is essential.

There have been extensive studies in the past 30 years with regard to the point defect structure of Cr_2O_3. Early works, summarized by Kofstad[1], indicate that the majority point defects present in Cr_2O_3 are chromium defects. In diffusion studies of Hagel and Seybolt[2], Walters and Grace[3], and Hoshino and Peterson[4], a Cr-vacancy defect model was

J. Nowotny and W. Weppner (eds.),
Non-Stoichiometric Compounds Surfaces, Grain Boundaries and Structural Defects, 93–113.

suggested. Correspondingly, Greskovich[5] studied the nonstoichiometry of Cr_2O_3 at high oxygen partial pressures, and reported a slight increase of the extent of nonstoichiometry with oxygen partial pressure. On the other hand, Kofstad and Lillerud[6] recently proposed that Cr interstitials may be the major defects under low oxygen partial pressures. This model has been supported by several investigations[7-9] including that of Matsui and Naito[10]. Also, in a study by Atkinson and Taylor[11], some indication of a changeover from Cr vacancy diffusion at high P_{O2} to Cr interstitial diffusion at low P_{O2} was observed. Apparently, the predominant defects in the high P_{O2} region and the low P_{O2} region are not the same.

The electrical properties of Cr_2O_3 have recently been reviewed by Kroger[12]. Although he was also unable to resolve the controversy between the p-type and n-type, he did point out that Cr_2O_3 is probably an intrinsic semiconductor at high temparatures (> 1523 K), and an n-type or p-type semiconductor at lower temperature caused by the presence of unavoidable impurities, or the "frozen in" defects. The studies of the electrical properties of intentionally doped Cr_2O_3[13-19] have proven that impurities play an important role in altering the defect and transport properties of Cr_2O_3. However, no conclusive evidence about the defect structure of Cr_2O_3 has been obtained.

Apparently, the point defect structure of Cr_2O_3 is very complicated, and can not be represented by a simple defect model. In order to resolve this, all the different defects need to be considered. It is generally agreed that the possible defects present in the crystal are: V_{Cr}''', $Cr_i^{\cdot\cdot\cdot}$, $h^{\cdot}$ and e'. The defect equations associated with these defects can then be written as:

$$3/2O_2(g) = 2V_{Cr}''' + 6h^{\cdot} + 3\,O_O^{x} \quad (1)$$

$$Cr_2O_3 = 2Cr_i^{\cdot\cdot\cdot} + 6e' + 3/2O_2(g) \quad (2)$$

$$\text{Null} = h^{\cdot} + e' \quad (3)$$

$$Cr_{Cr}^{x} = V_{Cr}''' + Cr_i^{\cdot\cdot\cdot} \quad (4)$$

where Eq. 1 represents the formation of a p-type, metal deficit semiconductor; Eq. 2 represents the formation of a n-type metal excess semiconductor; Eq. 3 represents the formation of the intrinsic electronic defects; and Eq. 4 represents the formation of the Frenkel defects. Applying the mass action law to the above reactions, one obtains the equilibrium constants:

$$K_1 = [V_{Cr}''']^2\,[h^{\cdot}]^6\,P_{O2}^{-3/2} \quad (5)$$

$$K_2 = [Cr_i^{\cdot\cdot\cdot}]^2\,[e']^6\,P_{O2}^{3/2} \quad (6)$$

$$K_3 = K_i = [h^\cdot]\ [e']\qquad (7)$$

$$K_4 = K_F = [V_{Cr}''']\ [Cr_i^{\cdot\cdot\cdot}]\qquad (8)$$

In addition to Eqs. 5 to 8, the electroneutrality condition gives

$$[h^\cdot] + 3[Cr_i^{\cdot\cdot\cdot}] = [e'] + 3[V_{Cr}''']\qquad (9)$$

In all these equations, the square brackets indicate the concentration of the defect involved. Apparently, in order to reveal the defect structure of Cr_2O_3, all these K's need to be solved. In this paper, studies of the electrical conductivity and Seebeck coefficient of TiO_2- and MgO-doped Cr_2O_3 have provided ways to evaluate these equilibrium constants. Accordingly, the defect structure of Cr_2O_3 is determined.

2. EXPERIMENTAL PROCEDURE

TiO_2-doped Cr_2O_3, MgO-doped Cr_2O_3 and high purity Cr_2O_3 pellets were prepared by conventional powder methods. TiO_2 and MgO powder were purchased from Alfa Products while high purity Cr_2O_3 powder was supplied by Johnson Matthey Inc. Table I lists the purity of these raw materials. In preparation of the TiO_2 and MgO doped Cr_2O_3 pellets, the Cr_2O_3 powder was first mixed with the dopant powder in certain ratios (0.1 to 0.5 mole%). The mixed powder was then put into a plastic bottle, and mixed on a mechanical shaker for 5 minutes. For the pure Cr_2O_3 pellets this step was not necessary. After the mixing, the powder was ground in a diamonite motar and pestle for two hours in order to achieve better homogeneity. The ground powder was then uniaxially pressed at 3.45×10^8 N/m^2 (50,000 psi.) into pellets of 5.1 mm. in diameter and 5.7 mm. in height without using any binder. The green density of these compacts was about 55% of the theoretical density.

Table I: Listed purity of the raw material used in this study

	Purity	Major Impurity
Cr_2O_3	99.999%	Ag,Al,Ca,Cu,Fe,Mg and Si < 1ppm
TiO_2	99.98%	---
MgO	99.999%	---

Sintering of the compacts was carried out at 1600°C in a horizontal Al_2O_3 tube furnace which was Molybdenum-wire wound and hydrogen protected. The final density of the sintered samples is strongly dependent upon the oxygen atmosphere[20]. Dense samples were obtained from low P_{O_2} sintering while porous samples were from high P_{O_2}. The oxygen atmosphere was controlled by using O_2/Ar and CO/CO_2 gas mixture with a total flow rate of 0.9 cm/sec.

After sintering, all the specimens were subsequently homogenized in air at 1300°C for three days. The dopant contents were then examined by spectrometers.

The apparatus for the electrical conductivity and Seebeck Coefficient measurements has been described elsewhere[21]. AC conductivities were performed with a GenRad 1658 RLC Digibridge at 100 and 1K Hz. Occasionally, DC conductivities were also measured for comparison, no apparent difference has been observed. Seebeck coefficient experiments were performed after the electrical conductivity measurements on the same sample. Temperature gradients were achieved by shifting the sample's position slightly away from the hot zone while the furnace temperature was controlled to maintain the sample at same average temperature. Pt-Pt10%Rh thermocouples were used to measure the temperature while the Pt leads were used for the Seebeck voltage by taking the lower temperature end as positive. The Seebeck coefficient Q was determined from the slope of the linear dependence of $\Delta V = f(\Delta T)$. Eight temperature gradients were measured in the experiments for the calculation of Q. Both electrical conductivity and Seebeck coefficient were measured as functions of dopant content, oxygen partial pressure, and temperature.

It had been of great concern about the equilibration time needed for the specimen to react with the environment atmosphere during experiments. Based on some preliminary study, the authors found that it is very difficult for a dense sample to reach equilibrium, an equilibration time of days, even weeks may be needed. In order to obtain appropriate information within a reasonable experimental time span, it was decided to use both porous and dense specimens. Porous samples, which could equilibrate to a change in oxygen atmospheres more rapidly, were used to determine the P_{O_2} dependence behavior while dense samples were used to measure the true electrical conductivities, which in turn were utilized to calculate the mobility. Throughout this study, conductivities were calculated from resistance by $\sigma = L/(A \times R)$, where L is the sample length and A is the area of the sample cross section. Also, σ will be used to denote the true conductivity, and σ_{eff} the conductivity measured from porous sample.

3. RESULTS AND DISCUSSION

3.1. TiO_2-doped Cr_2O_3

3.1.1 Results

The experimental results of the electrical conductivity are plotted in Fig. 1 as a function of oxygen partial pressure and TiO_2 content, ranging from 0.1 to 0.5 mole%, at temperatures from 1000° to 1300°C. Several interesting characteristics of these curves are described in the following.

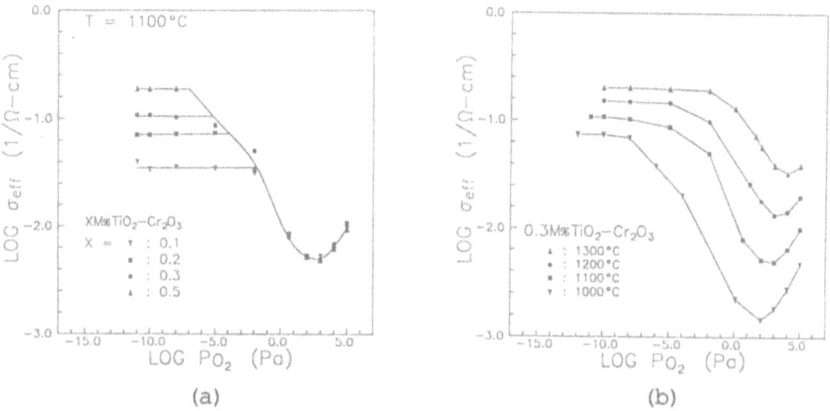

Fig. 1. Electrical conductivity of TiO_2-doped Cr_2O_3. (a) as a function of dopant content; (b) as a function of temperature.

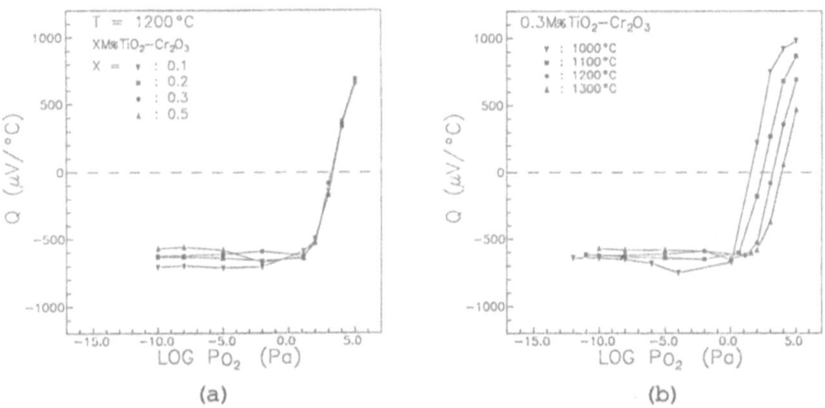

Fig. 2. Seebeck coefficient of TiO_2-doped Cr_2O_3. (a) as function of dopant content at 1200°C; (b) as a function of temperature.

A. Oxygen partial pressure dependence

(1) A conductivity minimum appears at an oxygen partial pressure $PO_2^{(min)}$. (2) At $PO_2 > PO_2^{(min)}$, the conductivity varies as $PO_2^{1/4}$, which indicates a typical p-type semiconductor behavior. (3) At $PO_2 < PO_2^{(min)}$, the conductivity varies as $PO_2^{-1/4}$, which indicates a n-type semiconductor behavior. (4) At even lower PO_2, the slope of the conductivity curve becomes smaller and tends to reach zero after a certain $PO_2^{(d)}$. Apparently, when PO_2 is decreased below this inflection point, the

electrical conductivity is governed by the dopant content. This inflection point is denoted as $P_{O_2}^{(d)}$.

B. Composition dependence

(1) In the high P_{O_2} region, the electrical conductivity does not vary with the TiO_2 content. (2) In the low P_{O_2} region, the electrical conductivity appears to be proportional to the dopant content.

C. Temperature dependence

As shown in Fig. 1(b), it appears that both the conductivity minimum $P_{O_2}^{(min)}$ and the inflection point $P_{O_2}^{(d)}$ shift to higher P_{O_2} at higher temperatures.

The corresponding Seebeck coefficients are shown in Fig. 2. The results appear in excellent agreement with the electrical conductivity. At high P_{O_2}, the Seebeck coefficient Q is positive while it is negative at low P_{O_2}. The occurrence of a reversal in the sign of Q near $P_{O_2}^{(min)}$ indicates a change of the transport mechanism from p-type to n-type conductivity in that vicinity. Also, constant negative value of Q's for $P_{O_2} < P_{O_2}^{(d)}$ implies a constant electron concentration. Furthermore, the P_{O_2} at which Q = 0 shifts with temperature in the same manner as the conductivity minimuum does.

3.1.2. Defect Structure of TiO_2-Doped Cr_2O_3

When a higher valent cation (Ti) is incorporated in Cr_2O_3, the substitution of Ti^{4+} into the Cr^{3+} site will generate a positive charge. In order to maintain the electroneutrality condition, compensation by the creation of negative charged defects are required. Both V_{Cr}''' and e' are possible candidates, and the processes may be represented by the following equations. In the case of V_{Cr}''':

$$3TiO_2 = 3Ti_{Cr}^{\cdot} + V_{Cr}''' + 6O_O^x \qquad (10)$$

$$K_{10} = [Ti_{Cr}^{\cdot}]^3\,[V_{Cr}'''] \qquad (11)$$

When the dopant concentration $[Ti_{Cr}^{\cdot}]$ is much higher than the intrinsic defect concentration, the Cr vacancies concentration is determined by

$$[V_{Cr}'''] = 1/3[Ti_{Cr}^{\cdot}] = 3^{-3/4}\,K_{10}^{1/4} \qquad (12)$$

and is independent of the oxygen partial pressure.

In the case of e':

$$2TiO_2 = 2Ti_{Cr}^{\cdot} + 2e' + 3O_O^x + 1/2O_2 \qquad (13)$$

$$K_{13} = [Ti_{Cr}^{\cdot}]^2\,[e']^2\,P_{O_2}^{1/2} \qquad (14)$$

When $[Ti_{Cr}^{\cdot}]$ is much greater than the intrinsic defect concentrations, then $[e']$ is controlled by $[Ti_{Cr}^{\cdot}]$ through

$$[e'] = [Ti_{Cr}^{\cdot}] = K_{13}^{1/4} P_{O_2}^{-1/8} \qquad (15)$$

The appearance of the $P_{O_2}^{-1/8}$ dependence indicates that the solubility of TiO_2 in Cr_2O_3 may vary with the oxygen partial pressure. The solubility increases as P_{O_2} decreases. When the dopant content is higher than the solubility, a second phase TiO_2 appears (Eq. 13). When both cases are considered, a new electroneutrality condition appears:

$$[Ti_{Cr}^{\cdot}] + [h^{\cdot}] = 3[V_{Cr}'''] + [e']. \qquad (16)$$

Theoretically, combining these equations with Eqs. 5 and 7, one may solve the four unknown defect concentrations in terms of the equilibrium constants and the P_{O_2}. With a priori knowledge of these equilibrium constants the relationship between the concentrations of the different defects and P_{O_2} can be obtained. However, when these K's are not available, this approach becomes too complicated to follow. According, the approximation method developed by Kroger and Vink is adopted in this study. In this method, these equations are solved in a piecewise linear fashion by sequentially choosing conditions for which only one term on each side of the electroneutrality equation need be considered. In other words, only the two defects with the highest concentration in the chosen condition are considered. For example, under heavily oxidizing conditions, the electroneutrality equation may be simplified to

$$[h^{\cdot}] = 3[V_{Cr}'''] \qquad (17)$$

Combining this with Eq. 5, one obtains

$$[h^{\cdot}] = 3[V_{Cr}'''] = 3^{1/4} K_1^{1/8} P_{O_2}^{3/16} \qquad (18)$$

and from Eqs. 7 and 14,

$$[e'] = 3^{-1/4} K_i K_1^{-1/8} P_{O_2}^{-3/16} \qquad (19)$$

$$[Ti_{Cr}^{\cdot}] = 3^{1/4} K_1^{-1/24} K_{13}^{1/4} P_{O_2}^{-1/16} \qquad (20)$$

The other defect regions can then be determined with successively decreasing P_{O_2}, which gives $[Ti_{Cr}^{\cdot}] = 3[V_{Cr}''']$ and $[Ti_{Cr}^{\cdot}] = [e']$. A diagram depicting the P_{O2} dependences of the defects over the different regions is presented in Fig. 3. It is noted that region IV represents the unsaturated region, a region where the level of the dopant content is below the solubility.

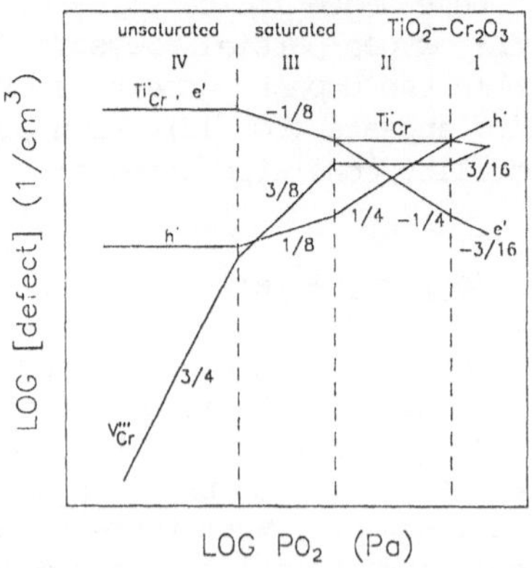

Fig. 3. A model of the defect structure of TiO_2-doped Cr_2O_3.

It is of interest at this point to compare the defect structure model with the experimental results. When the P_{O2} variations of $[h^{\cdot}]$ and $[e']$ in Fig. 3 were compared with the conductivity results, an excellent match was found. This fact suggests that the proposed model may probably represent the defect structure of TiO_2-doped Cr_2O_3. Accordingly, with this diagram and the experimental data, one may construct the real Kroger-Vink Diagram, and calculate the different defect concentrations and the equilibrium constants.

3.1.3. Determination of the Intrinsic Electron Concentration n_i

In the process of constructing the [defect] vs. P_{O2} diagram, the conductivity minimum σ_{min} and the constant conductivity in region IV are two very useful parameters. According to Becker and Frederikse's analysis(23), the electrical conductivity σ of a semiconductor containing both electrons and election holes can be expressed by

$$\sigma/\sigma_i = b^{1/2}(\alpha+1)/(b+1)\alpha^{1/2} \tag{21}$$

where $\sigma_i = n_i e(\mu_n + \mu_p)$ is the intrinsic conductivity, i.e., conductivity under the condition $n = p = n_i$, the intrinsic electron concentration; $b = (\mu_n / \mu_p)$ is the ratio of the electron and electron hole mobilities; and

$\alpha = \sigma_p/\sigma_n$ = (p/nb) is the ratio of the election hole and electron conductivities. Since σ_{min} occurs at $\sigma_p = \sigma_n$, i.e. $\alpha = 1$, from Eq. 21 one obtains

$$\sigma_{min}/\sigma_i = 2\ b^{1/2}/(b+1) \tag{22}$$

A special case, not utilized in this study, $\sigma_{min} = \sigma_i$ occurs when b = 1, i.e., $\mu_n = \mu_p$.

Eq. 22 can be written in another form

$$\sigma_{min} = 2e\ b^{1/2}\ \mu_p\ n_i \tag{23}$$

It is noted that the electrical conductivity in the unsaturated region, i.e. region IV, is controlled by the amount of dopant $[Ti_{Cr}^{\cdot}]$. The conductivity can then be expressed as

$$\sigma_{d,n} = e\ \mu_n\ n_d, \text{ where } n_d = [Ti_{Cr}^{\cdot}] \tag{24}$$

Dividing Eq. 23 by Eq. 24, one obtains

$$\sigma_{min}/\sigma_{d,n} = 2\ b^{-1/2}\ (n_i/n_d) \tag{25}$$

Thus

$$n_i = b^{1/2}\ (n_d/2)\ (\sigma_{min}/\sigma_{d,n}) \tag{26}$$

When b is known, the intrinsic electron concentration n_i can be evaluated from the amount of dopant $[Ti_{Cr}^{\cdot}] = n_d$, and the ratio $(\sigma_{min}/\sigma_{d,n})$. Since $K_i = n_i^2$, the equilibrium constant of the intrinsic ionization can in turn be calculated.

It has been suggested that in Cr_2O_3 the electron holes may have a higher mobility than electrons(15). A situation of b < 1 is expected. In order to evaluate b, the electron and electron hole mobilities need to be determined individually. The electron mobility may be calculated from the electrical conductivity in region IV, the unsaturated region, from Eq. 24. This experiment has been performed on dense samples of composition 0.2 mole% at 10^{-15} Pa of P_{O_2} and the electron mobilities were calculated. Results are shown in Fig. 4. The mobility of electron holes have been obtained from studies of the electrical conductivity on MgO-doped Cr_2O_3. A complete discussion is presented in a later section. With the evaluated μ_n, μ_p and b, the intrinsic electron concentration n_i and the equilibrium constant K_i were then calculated with Eqs. 26 and 7. In Table II, the calculated parameters are listed. Also an expression for K_i is obtained:

$$K_i = 42.19 \exp(-2.86\text{eV}/kT)$$
$$= 42.19 \exp(-66{,}000\text{cal}/RT) \qquad (27)$$

where k is the Boltzmann's constant and R the gas constant and the concentrations are given in mole fractions, i.e. N_i's.

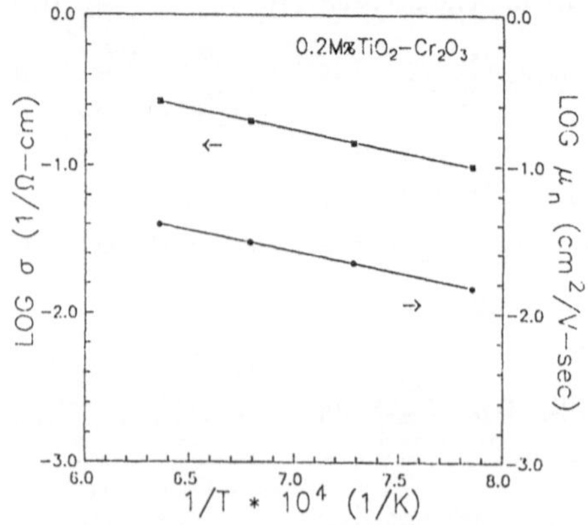

Fig. 4. Temperature dependence of the electrical conductivity of dense TiO_2-doped Cr_2O_3 sample, and the calculated electron mobility.

Table II: Calculated intrinsic parameters of Cr_2O_3.

T (^{o}C)	1000	1100	1200	1300
μ_n (cm^2/V-sec)	0.040	0.030	0.022	0.015
μ_p (cm^2/V-sec)	0.083	0.076	0.069	0.062
b	0.48	0.39	0.31	0.24
n_i (#/cm^3)	3.47×10^{18}	1.70×10^{18}	7.41×10^{17}	2.88×10^{17}
$K_i = N_{e'} \cdot N_{h^\cdot}$	2.82×10^{-8}	6.77×10^{-9}	1.29×10^{-9}	1.95×10^{-10}

In the process of constructing the [defect] vs. P_{O2} diagram, it is necessary to know the positions of (n_i, $P_{O2}^{(i)}$) and (n_d, $P_{O2}^{(d)}$). Since b $\neq$ 1 and $\sigma_{min} \neq \sigma_i$, the oxygen partial pressure $P_{O2}^{(i)}$, at which the electron concentration is equal to n_i, does not coincide with $P_{O2}^{(min)}$. However, based on Eq. 22 and the symmetric characteristics of the diagram in the vicinity of σ_{min}, the value of $P_{O2}^{(i)}$ can be easily determined from adjustment of $P_{O2}^{(min)}$. The [defect] vs. P_{O2} diagrams have been constructed for all four temperatures, and one of them is presented in Fig. 5. Based on these diagrams, the equilibrium constant K_1 associated with the formation of chromium vacancies were calculated from Eq. 5, and is expressed as:

$$K_1 = 1.13\times10^{-10}\ \exp(-5.88\text{eV}/kT)$$
$$= 1.13\times10^{-10}\ \exp(-135{,}600\text{cal}/RT) = N^2_{V_{Cr}'''} \times N^6_{h^\cdot} \times P_{O2}^{-3/2} \quad (28)$$

where concentrations are in mole fractions and P_{O2} is in atmospheres.

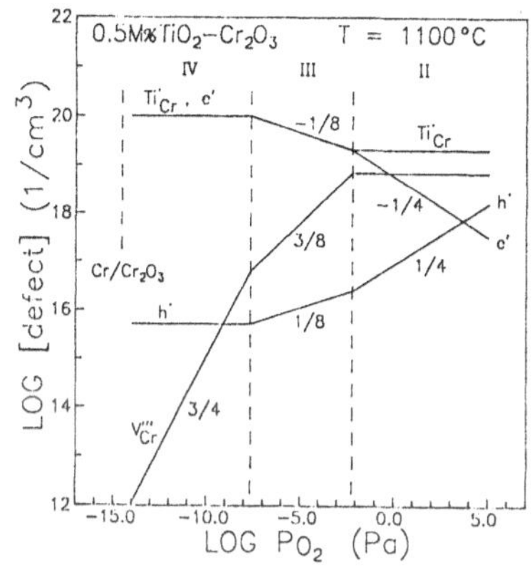

Fig. 5. The [defect] vs. P_{O2} diagram of TiO_2-doped Cr_2O_3 at 1100°C.

3.2. MgO-Doped Cr_2O_3

3.2.1. Results

The experimental results of the electrical conductivity and Seebeck coefficient of MgO-doped Cr_2O_3 are plotted as a function of oxygen partial pressure and MgO content in Figs. 6 and 7, respectively. The general characteristics of these curves are described in the following.

A. Oxygen Partial Pressure Dependence

Both the conductivity and the Seebeck coefficient Q remain relatively unchanged with little P_{O2} dependence in the high P_{O2} region. In the intermediate P_{O2} region, conductivity decreases with decreasing P_{O2} while the Seebeck coefficient increases slightly. In the low P_{O2} region, there is an indication that conductivity may vary with P_{O2} in a different manner. Also, in the corresponding Seebeck coefficients, a maximum, Q_{max}, appears at an oxygen partial pressure $P_{O2}^{(max)}$. When P_{O2} is decreased below $P_{O2}^{(max)}$, the value of the Seebeck coefficient drops relatively fast with decreasing P_{O_2}. However, negative values of Q have not been observed. This may be because the lowest P_{O_2} that can be achieved using the present experimental method is still not low enough to see a p-type to n-type transition.

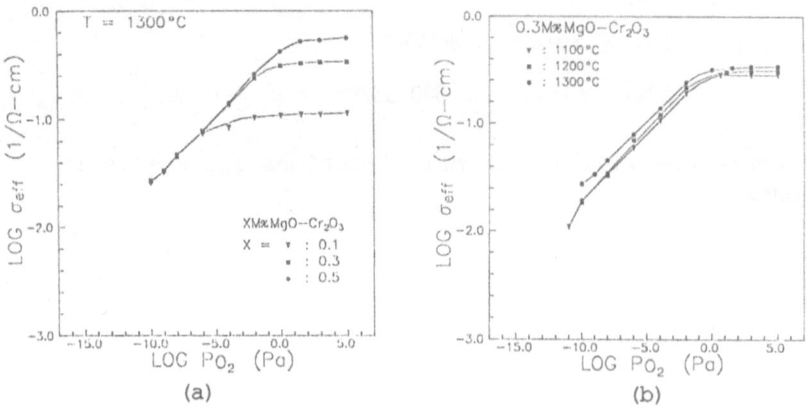

Fig. 6. The electrical conductivity of MgO-doped Cr_2O_3. (a) as a function of dopant content; (b) as a function of temperature.

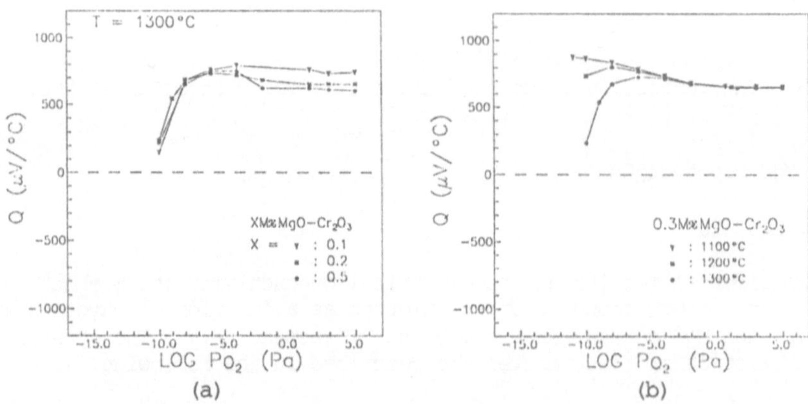

Fig. 7. Seebeck coefficient of MgO-doped Cr_2O_3. (a) as a function of dopant content; (b) as a function of temperature.

B. Composition Dependence

When the composition dependence is examined a very nice correspondence with the solubility results is observed. In the high P_{O2} region, both the conductivities and Seebeck coefficients indicate that the dopant levels used are below the solubility limit. When P_{O2} is decreased, the composition dependence diminished below a certain oxygen partial pressure $P_{O2}{}^{(d)}$. This fact indicates that at $P_{O2} < P_{O2}{}^{(d)}$ the dopant level is higher than the solubility limit. Since there is no reason for the amount of dopant to change during the experiment, apparently, the solubility limit must decrease as P_{O2} is decreased.

C. Temperature Dependence

The temperature dependence of the electrical conductivity and Seebeck coefficient are shown in Figs. 6(b) and 7(b), respectively. It appears that the activation energy of the electrical conductivity is very small. Also when temperature is increased, all the characteristics points, i.e., $P_{O2}{}^{(d)}$ and $P_{O2}{}^{(max)}$, shift to lower P_{O2}.

3.2.2. Defect Structure of MgO-doped Cr_2O_3

Since the incorporation of Mg^{2+} ions into Cr_2O_3 will generate negative charges, defects with positive charges are created in order to maintain electronic neutrality. The possibile defect reactions will then involve either $h^{\cdot}$ or $Cr_i^{\cdot\cdot\cdot}$, or both. Accordingly, the following defect equations need to be considered. For the case of $h^{\cdot}$, the relations

$$1/2O_2(g) + 2MgO = 2Mg_{Cr}' + 2h^{\cdot} + 3O_O^x \tag{29}$$

$$K_{29} = [Mg_{Cr}']^2 [h^{\cdot}]^2 P_{O2}^{-1/2} \tag{30}$$

will give the following defect concentration which are oxygen partial pressure dependent

$$[h^{\cdot}] = [Mg_{Cr}'] = K_{29}^{1/4} P_{O2}^{-1/8} \tag{31}$$

In the case of $Cr_i^{\cdot\cdot\cdot}$, the equations

$$Cr_2O_3 + 6MgO = 6Mg_{Cr}' + 2Cr_i^{\cdot\cdot\cdot} + 9O_O^x \tag{32}$$

$$K_{32} = [Mg_{Cr}']^6 [Cr_i^{\cdot\cdot\cdot}]^2 \tag{33}$$

show that the following defect concentrations are P_{O_2} independent

$$[Cr_i^{\cdot\cdot\cdot}] = 1/3[Mg_{Cr}'] = 3^{-3/4}\, K_{32}^{1/8} \qquad (34)$$

When both cases are considered, the electroneutrality condition is

$$[Mg_{Cr}'] + [e'] = [h^{\cdot}] + 3[Cr_i^{\cdot\cdot\cdot}] \qquad (35)$$

The relationships among different defect concentrations and the oxygen partial pressure can be obtained by applying the same technique discussed earlier. In Fig. 8, the theoretical prediction of the defect structure of $MgO-Cr_2O_3$ is represented. According to this diagram, four regions can be distinguished. Region I, the unsaturated region, represents the situation in which the dopant level is below the solubility limit. In this case, the concentration of electron holes is determined by the amount of dopant, i.e., $[h^{\cdot}] = [Mg_{Cr}']$. Since the solubility limit of MgO in Cr_2O_3 may decrease with decreasing P_{O_2}, in region II a second phase will be present, and the Mg concentration in Cr_2O_3 will be saturated. As a result, a $P_{O_2}^{-1/8}$ dependence of $[h^{\cdot}]$ and $[Mg_{Cr}']$ is observed. In region III, where the $Cr_i^{\cdot\cdot\cdot}$ is dominant, Eqs. 31 to 33 are applied. A situation of $[h^{\cdot}] = [e']$ occurs in this region, and a conductivity minimum is expected. Region IV represents the intrinsic behavior of pure Cr_2O_3 where chromium intersititials ($Cr_i^{\cdot\cdot\cdot}$) and electrons (e') are the majority defects.

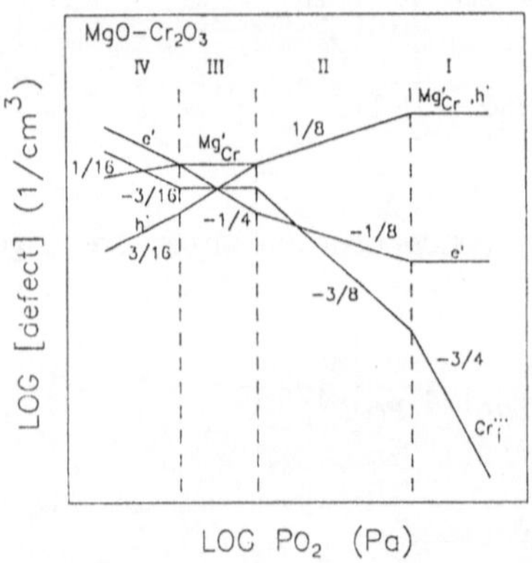

Fig. 8. A model of the defect structure of MgO-doped Cr_2O_3.

When the defect structure model in Fig. 8 is compared with the experimental results, rather good agreement is found. Although the expected conductivity minumum in the low P_{O_2} region has not been found due to the experimental limitation, this model does predict the P_{O2} variation of the conductivity and Seebeck coefficient. Based on this model, it is apparent that the mobility of electron holes may be determined by measuring the conductivity in the "unsaturated" region. This experiment has been performed on dense samples. Results are plotted in Fig. 9, and values of μ_p are listed in Table II. With this defect model and the experimental results, the actual [defect] vs. P_{O2} diagram can then be constructed. In Fig. 10, one of the diagrams is plotted. Based on Eq. 2, the equilibrium constant K_2 has been calculated, and can be expressed as

$$K_2 = 1.48\times10^{13} \exp(-20.48\mathrm{eV}/kT)$$
$$= 1.48\times10^{13} \exp(-472{,}300\mathrm{cal}/RT) = N^2_{Cr_i^{\cdots}} \times N^6_{e'} \times P^{3/2}_{O_2} \qquad (36)$$

Combining with the equilibrium constants K_1 and K_i obtained earlier, the equilibrium constant for the formation of the Frenkel defects, K_F, can be calculated, which gives

$$K_F = 4.83\times10^{-4} \exp(-4.58\mathrm{eV}/kT)$$
$$= 4.83\times10^{-4} \exp(-105{,}600\mathrm{cal}/RT) = N_{V_{Cr}'''} \times N_{Cr_i^{\cdots}} \qquad (37)$$

In equations (36) and (37) concentrations are again in mole fractions and P_{O2} is in atmospheres.

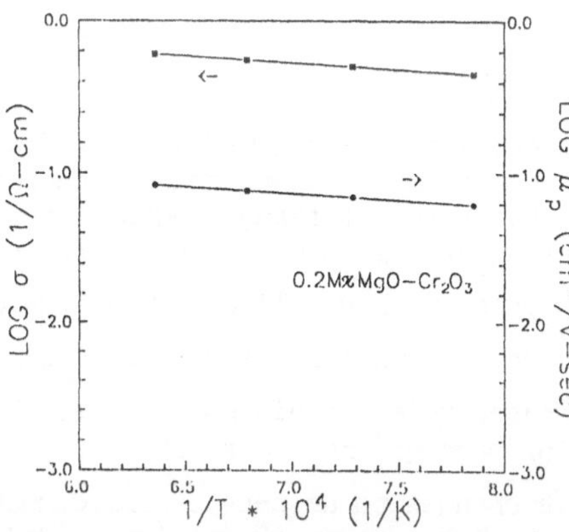

Fig. 9. Temperature dependence of the electrical conductivity of dense MgO-doped Cr_2O_3 sample, and the calculated electron hole mobility.

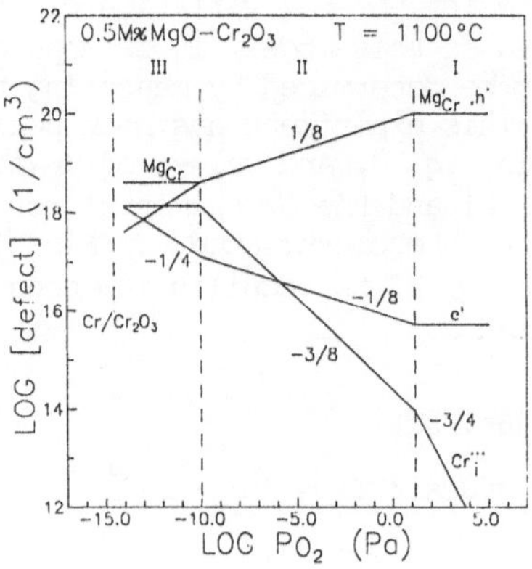

Fig. 10. The [defect] vs. PO_2 diagram of MgO-doped Cr_2O_3 at 1100°C.

3.3. Pure Cr_2O_3

3.3.1. Point Defect Structure of Cr_2O_3

Based on the results obtained from electrical conductivity and Seebeck coefficient measurements of the TiO_2- and MgO-doped Cr_2O_3, the equilibrium constants associated with the formation of different defects in Cr_2O_3, i.e., K_1, K_2, K_i and K_F, have been deduced. From these equilibrium constants and their corresponding defect equations, it is then possible to determine the point defect structure of Cr_2O_3. Accordingly, the concentrations of the different defects have been calculated as functions of both temperature and oxygen partial pressure.

In Fig. 11(a), the point defect structure of Cr_2O_3 at 1100°C is illustrated by plotting the oxygen partial pressure dependence of the defect concentrations. It appears that three different regions can be distinguished. At high PO_2's, near atmospheric oxygen pressure, Cr_2O_3 behaves as a p-type semiconductor with V_{Cr}''' and $h^{\cdot}$ as the dominant defects. As PO_2 decreases, $[V_{Cr}''']$ and $[h^{\cdot}]$ start decreasing and $[e']$ increases. When PO_2 is decreased to a certain point where $[h^{\cdot}] = [e']$, the intrinsic electronic behavior becomes important. At low PO_2's, near Cr/Cr_2O_3 equilibrium PO_2, it changes to an n-type semiconductor with $Cr_i^{\cdot\cdot\cdot}$ and e' as the dominant defects. Coincidently, a similar defect model has recently been reported by Lawrence et. al.[23] in their work of computer modeling of the defect properties of Cr_2O_3.

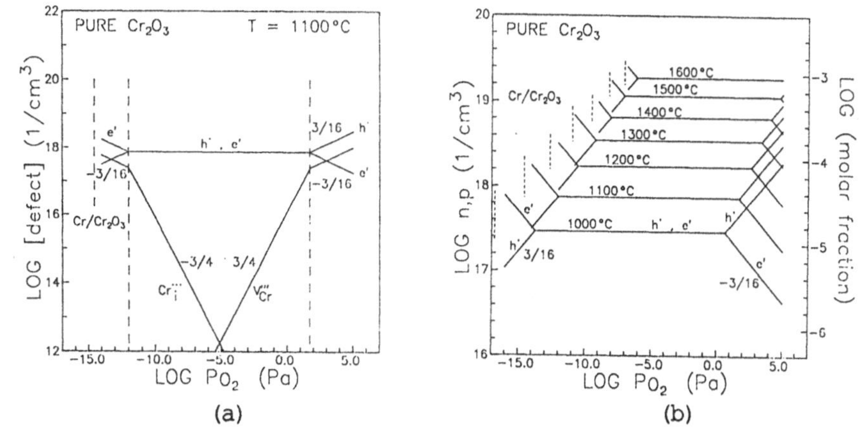

Fig. 11. A model of the point defect structure of Cr_2O_3. (a) at 1100°C (b) as a function of temperature, only electronic defects are present.

The temperature dependence of the defect structure of Cr_2O_3 is illustrated in Fig. 11(b), where only the electronic defect concentrations are displayed. When temperature is increased, a clear shift of all three regions towards high P_{O_2} is found. At temperatures above 1500°C, the intrinsic region becomes predominant even at atmospheric oxygen pressure. This kind of shift is somewhat expected since all the defect formation are thermal activated processes, and the different equilibrium constants apparently have different activation energies.

In order to verify the deduced defect structure, it is necessary to examine the different defect dependent properties of Cr_2O_3. A full explanation of these properties with the model is essential for its justification. In the following, a brief discussion on the electrical conductivity, the self-diffusion coefficient, and the parabolic growth of pure Cr_2O_3 is persented.

3.3.2. Defect Dependent Properties of Cr_2O_3

With the obtained defect structure and the electron and electron hole mobilities, the electrical conductivity of Cr_2O_3 has been calculated as functions of temperature and P_{O_2}, and plotted in Fig. 12(a). The experimental results are plotted in Fig. 12(b). It is found that these two results match very well for temperatures above 1200°C. However, at lower temperatures the experimental results do not show the same oxygen partial pressure dependence as the calculated results. This discrepancy may be due to the effect of impurities and can be easily explained by examining Fig. 11(b). Since the intrinsic electronic concentration, n_i, has a value of less than 0.005 mole/mole of Cr_2O_3 at 1100°C, the defect structure of the intrinsic region may be easily altered with an impurity level higher than this amount. When the defect structure is controlled by a small amount of impurity, a PO_2 independence of the electrical conductivity is expected.

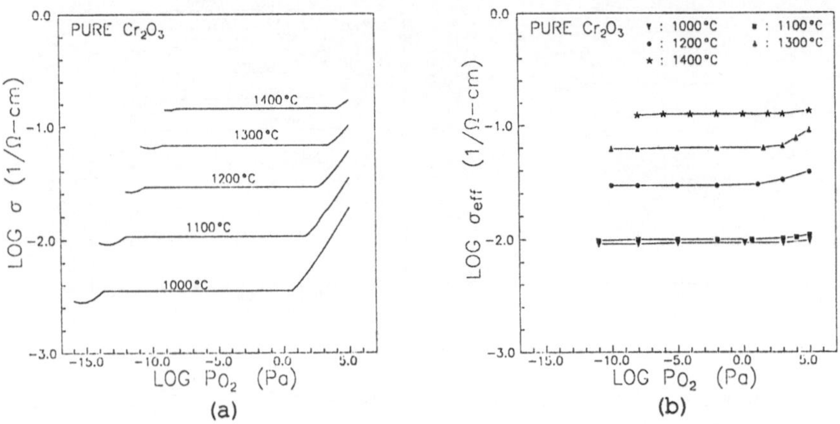

Fig. 12 The electrical conductivity vs. oxygen partial pressure diagram of pure Cr_2O_3, (a) theoretical (b) experimental results.

It has been shown[24] that the self-diffusion coefficient of the cation in an oxide can be expressed by

$$D_{Me} = \sum D_{def}\ [\text{defect}] \tag{37}$$

In the case of Cr_2O_3, the self-diffusion coefficient of Cr can then be written as

$$D_{Cr} = D_{V_{Cr}'''}\ [V_{Cr}'''] + D_{Cr_i^{\cdot\cdot\cdot}}\ [Cr_i^{\cdot\cdot\cdot}] \tag{38}$$

With information for $D_{V_{Cr}'''}$ and $D_{Cr_i^{\cdot\cdot\cdot}}$, and Fig. 10(a), the self-diffusion coefficient of Cr in Cr_2O_3 can be calculated. Also, the parabolic rate constant of the growth of Cr_2O_3 can then be evaluated. A detailed discussion will be presented in another paper. Nevertheless, from Fig. 11(a), it is apparent that D_{Cr} may be surprisingly low in the intermediate PO_2 region in comparison with the high and low PO_2 regions. If this is indeed the case, other diffusion paths, such as grain-boundaries and dislocations, may then be able to contribute to the diffusion process along with the volume diffusion. This appears to be in agreement with studies of Atkinson and Taylor[7]. They found that diffusion along dislocations actually plays a very important role in the diffusion of Cr in Cr_2O_3. This may also explain why the reported diffusion coefficients of Cr in early studies on polycrystalline Cr_2O_3 have much higher values.

4. CONCLUSION

1. It is found that the point defect structure of Cr_2O_3 is very complicated. The type of defects present are dependent upon the temperature, the oxygen partial pressure, and the amount of impurities. In general, at high temperature, Cr_2O_3 behaves as a p-type semiconductor at high P_{O_2}'s, an intrinsic semiconductor at intermidiate P_{O_2}'s, and n-type semiconductor at low P_{O_2}'s (near the Cr/Cr_2O_3 equilibrium P_{O_2}).

2. When the electronic tranport properties are considered, both electrons and electron holes appear to contribute to the process of conduction. It was found that the mobility of electron holes is greater than that of electrons.

3. Impurities have very significant effects on altering the defect structure of Cr_2O_3 and changing its transport properties. When Cr_2O_3 is doped with a higher valent cation (Ti), the electron conductivity is increased; on the other hand, when a lower valent cation is the dopant, the electron hole conductivity is increased.

4. It has also been found that the volume diffusion of Cr in Cr_2O_3 is too slow to be totally responsible for some of the diffusion dependent properties, e.g., high temperature oxidation of Cr. Other factors such as impurity effects and the short-circuit diffusion along grain boundaries, dislocations, etc., may have significant contributions.

ACKNOWLEDGEMENT

The authors would like to express their appreciation to the support of the Naval Sea System Command and the Applied Research Laboratory Exploratory and Foundational Research Program, The Pennsylvania State University.

REFERENCES:

1. P. Kofstad, pp. 203-208 in Nonstoichiometry, Diffusion, and Electrical Conductivity in Binary Metal Oxides. Wiley-Interscience, New York, 1972.
2. W. C. Hagel and A. V. Seybolt, "Cation Diffusion in Cr_2O_3," J. Electrochem. Soc., 108 [12] 1146-52 (1961).
3. L. C. Walters and R. E. Grace, "Self-Diffusion of ^{51}Cr in Single Crystals of Cr_2O_3," J. Appl. Phys., 36 [7] 2331-32 (1965).
4. K. Hoshino and N. L. Peterson, "Cation Self-Diffusion in Cr_2O_3," J. Am. Ceram. Soc., 66 [11] C-202-C-203 (1983).
5. C. Greskovich, "Deviation from Stoichiometry in Cr_2O_3 at High Oxygen Partial Pressures," J. Am. Ceram. Soc., 67 [6] C111-C112 (1984).

6. P. Kofstad and K. P. Lillerud, "On High Temperature Oxidation of Chromium, II. Properties of Cr_2O_3 and The Oxidation Mechanism of Chromium," J. Electrochem, Soc., 127 [11] 2410-19 (1980).
7. H. Hindom and D. P. Whittle, "Evidence for the Growth Mechanism of Cr_2O_3 at Low Oxygen Potentials," J. Electrochem. Soc., 130 [7] 1519-23 (1983).
8. E. W. A. Young, P. C. M. Stiphout, and J. H. W. de Wit, "N-type Behavior of Chromium(III) Oxide," J. Electrochem. Soc., 132 [4] 884-86 (1985).
9. E. W. A. Young, J. H. Gerretsen, and J. H. W. de Wit, "The Oxygen Partial Pressure Dependence of the Defect Structure of Chromium (III) Oxide,"" J. Electrochem. Soc., 134 [9] 2275-60 (1987).
10. T. Matsui and K. Naito, "Existence of Hypostoichiometric Chromium Sesquioxide at Low Oxygen Partial Pressures", J. Nuc. Mat., 136, 78-82 (1985).
11. A. Atkinson and R. I. Taylor, "Diffusion of ^{51}Cr Tracer in Cr_2O_3 and the Growth of Cr_2O_3 films," pp.285-95 in Transport in Nonstoichiometric Compounds, Edited by G. Simkovich and V. S. Stubican, Plenum, New York, 1985.
12. F. A. Kroger, "Defects and Transport in SiO_2, Al_2O_3, Cr_2O_3," pp. 89-100, High Temperature Corrosion, Proc. NACE Conf., San Diego, Ca., March 1981.
13. F. G. Hicks, D. R. Holmes and D. B. Meadowcroft, "Defect Structure and Transport Properties of Oxide Solid Solutions Containing Cr_2O_3," pp.379-84 in The 4th International Conference of Corrosion, 1969.
14. D. B. Meadowcroft and F. G. Hicks, "Electrical Conduction Processes and Defect Structure of Chromium Oxide," Proc. Br. Ceram. Soc., 23, 33-41 (1972).
15. K. A. Hay, F. G. Hicks, and D. R. Holmes, "The Transport Porperties and Defect Structure of the Oxide $(Fe, Cr)_2O_3$ Formed on Fe-Cr Alloys," D. R. Werkst. Korros., 21, 917-24 (1970).
16. W. C. Hagel, " Eletrical Conductivity of Li-Substituted Cr_2O_3," J. Appl. Phys., 36 [8] 2586-87 (1965).
17. R. F. Huang, A. K. Agarwal and H. U. Anderson, "Oxygen Activity Dependence of the Electrical Conductivity of Li-Doped Cr_2O_3," J. Am. Ceram. Soc., 67 [2] 146-50 (1984).
18. H. Nagai, T. Fujikawa and K. Shoji, "Electrical Conductivity of Cr_2O_3 Doped with La_2O_3,Y_2O_3 and NiO," Trans. Japan Inst. Metals, 24 [8] 581-88 (1983).
19. J. S. Park and H. G. Kim, "Electrical Conductivity and Defect Models of MgO Doped Cr_2O_3," J. Am. Ceram. 71 [3] 173-76 (1988).
20. M.-Y. Su, H.-Y. Chang and G. Simkovich, " Diffusion in Cr_2O_3 Via Initial Sintering Experiments," pp.385-95 in Transport in Nonstoichiometric Compounds, Edited by G. Simkovich and V. S. Stubican, Plenum, New York, 1985.
21. M. -Y. Su, "Point Defect Structure of Chromium Sesquioxide," Ph. D. thesis, The Pennsylvania State University, August, 1987.
22. J. H. Becker and H. P. R. Frederikse, "Electrical Properties of Nonstoichiometric Semiconductors," J. Appl. Phys., 33 [1] 447-53 (1962).

23. P. J. Lawrence, S. C. Parker, and P.W. Tasker, "Computer Modelling of the Defect Properties of Chromium Oxide, Cr_2O_{3-x}," pp. 247-56, Advances in Ceramics, vol. 23: Nonstoichiometric Compounds, Edited by C. R. A. Catlow and W. C. Mackrodt, Am. Ceram. Soc., Inc., 1987.
24. F. Gesmundo and F. Viani, "Application of Wagner's Theory to the Parabolic Growth of Oxides Containing Different Kinds of Defects. I. Pure Oxides," J. Electrochem. Soc., 128 [2] 460-69 (1981).

NON-STOICHIOMETRY AND DEFECT STRUCTURE OF FeO

J. Janowski[1], J. Nowotny[2] and M. Rekas[2*]
1) Academy of Mining and Metallurgy
Institute of Metallurgy
30-059 Krakow, Poland
2) Max-Planck-Institut für Festkörperforschung
7000 Stuttgart 80, F.R.G.

ABSTRACT. The non-stoichiometry in wustite ($Fe_{1-y}O$) is considered in terms of the 4:1 cluster model. It has been shown that the ionisation degree of the cluster above 1173 K is -5. The Debye-Hückel theory is applied to determine the equilibrium constant of the cluster formation. It has been demonstrated that the cluster model is well consistent with the free partial molar enthalpy of oxygen $\Delta H(O_2)$ reported for $Fe_{1-y}O$.

1. INTRODUCTION

High concentration of defects in ferrous oxide $Fe_{1-y}O$ (wustite) may result in strong interactions between the defects leading to the formation of defect assemblies and clusters. Despite many works which have been published on $Fe_{1-y}O$ its defect structure is still the subject of controversial reports. The main reason of this controversy results from conditions of measurements of the reported structural data which in major cases correspond to a quenched state. As known, however, the properties of quenched crystals are substantially different from those being equilibrated with the gas phase. Therefore, an awarness is growing that there may be so many defect strctures of oxide crystals as many quenching procedures are applicable. Accordingly, the analysis of the defect structure should be based on experimental data obtained 'in situ' i.e. in conditions of the thermodynamic equilibrium.

Several experimental data such as the dependence between the deviation from stoichiometry y and oxygen partial pressure can be considered by assuming the formation of 4:1 defect clusters (Fig. 1) composed of four iron vacancies (V_{Fe}) and one interstitial iron (Fe_i) [1-3]. According to theoretical considerations for cubic-type oxides the 4:1 cluster is stable at higher values of y corresponding to the non-stoichiometry of $Fe_{1-y}O$ [2 - 4].

The purpose of the present work is to consider the 4:1 cluster model for $Fe_{1-y}O$ against the thermodynamic data reported for this oxide material.

---

*/On leave from the Institute of Materials Science, Academy of Mining and Metallurgy, Krakow, Poland

J. Nowotny and W. Weppner (eds.),
Non-Stoichiometric Compounds Surfaces, Grain Boundaries and Structural Defects, 115–121.

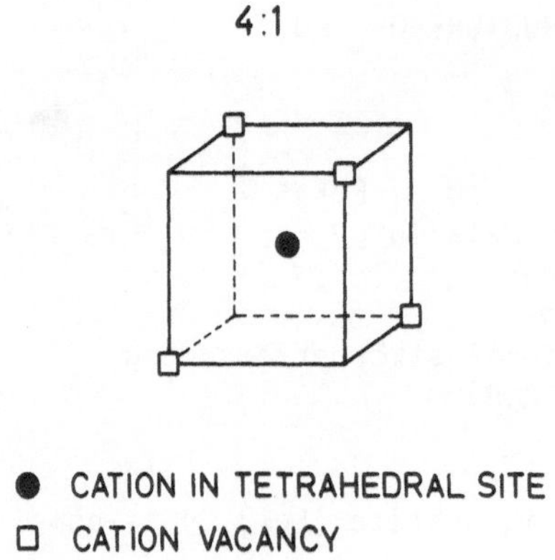

Figure 1. The 4:1 cluster

2. FORMATION OF 4:1 CLUSTERS

The following reaction can be written for the cluster formation:

$$(j+1)Fe_{Fe} + V_t + 3/2\ O_2 \rightleftarrows [(V_{Fe})_4Fe_i]^{j-} + 3O_O + jFe^{\cdot}_{Fe} \qquad (1)$$

where V_t is the interstitial (tetrahedral) site and j is the ionisation degree which may vary between 0 and -5. Assuming that the cluster with an effective charge i predominates, we obtain the following dependence:

$$n = \left(\frac{\partial \ln y}{\partial \ln p_{O_2}}\right)^{-1} = \frac{2}{3}\left[(i+1)\frac{3}{3-(i+1)y} + \frac{23y}{6-23y}\right] \qquad (2)$$

The dependence of n vs. y (Eq. 2) is shown in Fig. 2 along with experimental data[5 - 14]. As it is seen the experimental data fit best the theoretical dependence (2) for j = -5. Accordingly, Eq. (1) assumes the form:

$$6Fe_{Fe} + V_t + 3/2O_2 \rightleftarrows [(V_{Fe})_4Fe_i]^{5-} + 3O_O + 5Fe^{\cdot}_{Fe} \qquad (3)$$

Therefore, the equilibrium constant for the reaction (3) is:

$$K_c = \frac{c\{h^{\cdot}\}^5}{\{Fe_{Fe}\}^6\{V_t\}}\,p_{O_2}^{-3/2} \qquad (4)$$

where { } brakets denote activities and:

$$c = [(V_{Fe})_4Fe_i]^{5-} \qquad (5)$$

Defect activities can be considered within the Debye-Hückel theory for strong electrolytes.

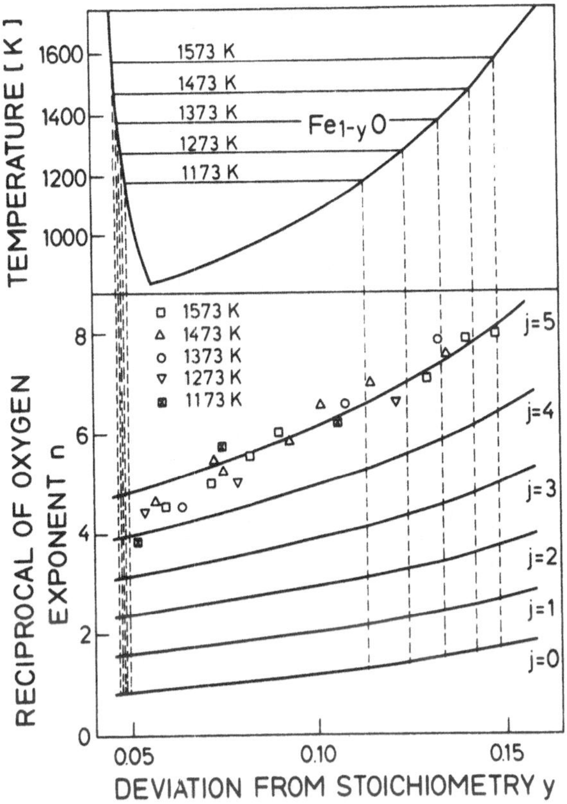

Figure 2. Fitting of the parameter n to Eq. (2) for different ionisation degree of the 4:1 complex.

Application of the Debye-Hückel theory for description of the interaction between the defects leads to the following expression for K_c^*[15]:

$$K_c^* = \frac{5^5\, y^6\, f_\pm^6}{(3-9y)^6(2-23y/3)}\, p_{O_2}^{-3/2} \qquad (6)$$

where $f\pm$ is the mean activity coefficient[15]. Its temperature dependence is shown in Fig. 3. The equilibrium constant K_c^* thus assumes the following form (Fig. 4):

$$K_c^* = 1.07\times10^{-14}\, \exp(396.7[kJ]/RT) \qquad (7)$$

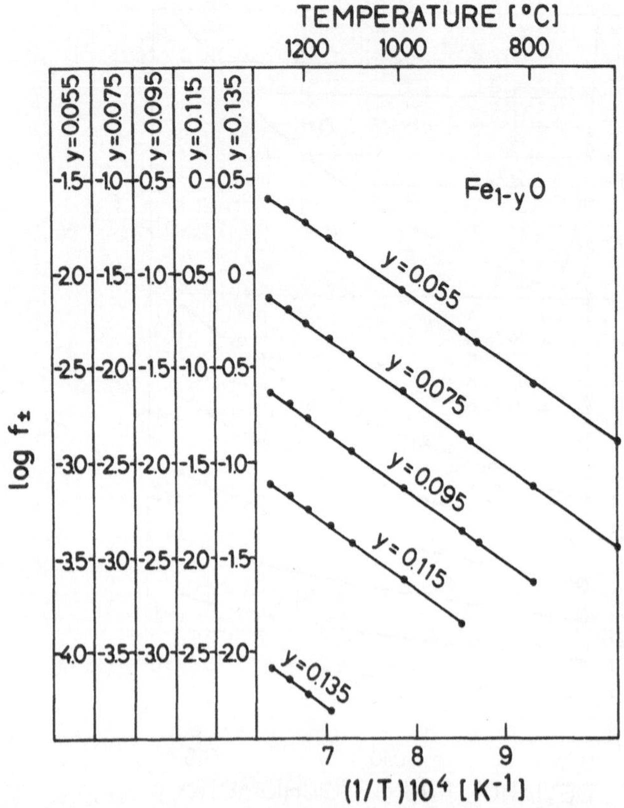

Figure 3. The plot of the activity coefficient f± vs. 1/T

3. THERMODYNAMIC PARTIAL MOLAR FUNCTIONS OF OXYGEN

The relative partial molar free enthalpy of oxygen in the wustite phase $(\Delta G(O_2))$ may be expressed as[15]:

$$\Delta G(O_2) = \Delta H(O_2) - T\Delta S(O_2) = RT \ln p_{O_2} \tag{8}$$

As can be seen from Fig. 3 the temperature dependence of f± may be expressed as follows:

$$\ln f\pm = \alpha/T + \beta \tag{9}$$

Taking into account Eqs. (6) and (9) we obtain:

$$\Delta H(O_2) = 2/3\ \Delta H_f + 4\alpha R \tag{9}$$

where ΔH_f denotes the formation enthalpy of formation of defects and α is a parameter.

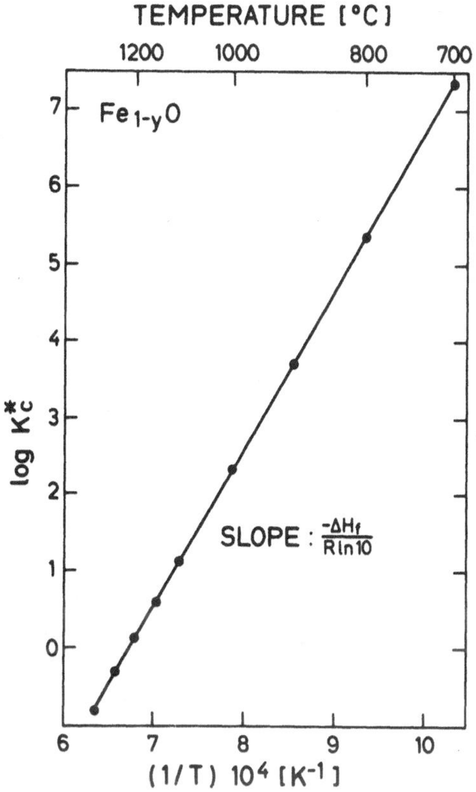

Figure 4. The Arrhenius plot of the equilibrium constant K_c^*

Fig. 5 illustrates the dependence of $\Delta H(O_2)$ as a function of wustite composition plotted along with experimental data reported in the literature[14, 16 - 19]. As it is seen the agreement between the theoretical dependence (10) and experimental results is satisfactory except the results of Marucco et al.[19]. The best agreement is observed for experimental data of Darken and Gurry[16] and Gerdanian[17].

4. CONCLUSIONS

Properties of the wustite phase can be explained by assuming the 4:1 cluster model. A good agreement between this model and experimental data of thermodynamic quantities can be attained if the Debye-Hückel theory is applicable. Thus determined equilibrium constant is valid within the entire stability range of $Fe_{1-y}O$. It has been postulated that the ionisation degree of the 4:1 cluster is -5.

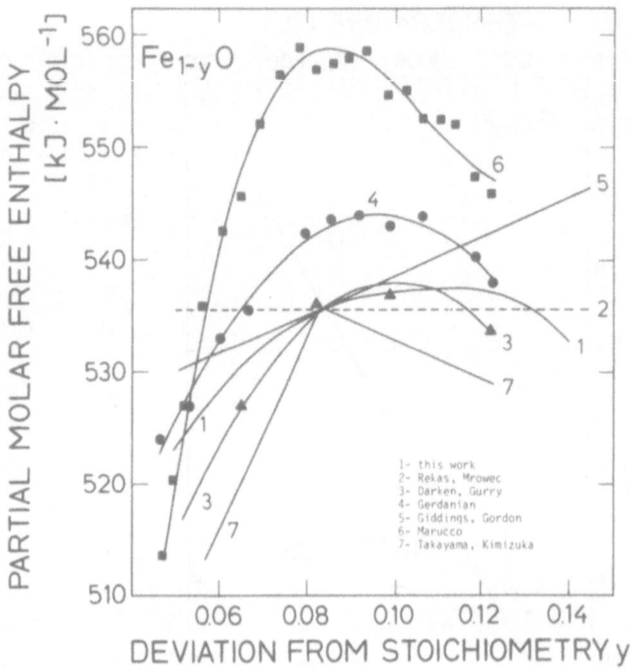

Figure 5. Plot of $\Delta H(O_2)$ vs. deviation from stoichiometry of $Fe_{1-y}O$

REFERENCES

1 M. Rekas and M. Mrowec, Solid State Ionics 22, 185 - 197 (1987)

2 M. Stoneham, Physics Today 33 [1] 34 - 42 (1980)

3 C.R.A. Catlow and B.E.F. Fender, J.Phys.C 8 [20], 3267 - 3279 (1975)

4 C.R.A. Catlow, B.E.F. Fender and D.G. Muxworthy, J.Phys.(Parsis), Colloq., 7 67 - 71 (1971)

5 L.S. Darken and R.W. Gurry, J.Am.Chem.Soc., 67 [8], 1398 - 1412 (1945)

6 K. Hauffe and H. Pfeiffer, Z.Metall., 44, 27 - 36 (1953)

7 P. Vallet and P. Raccah, Mem.Sci.Rev.Met., 62 [1], 1 - 29 (1956)

8. B. Swaroop and J.B. Wagner, Jr., Trans. AIME, 239 [8], 1215 - 1218 (1967)

9 I. Bransky and A.Z. Hed, J.Am.Ceram.Soc., 51 [4], 231 -232 (1968)

10 H.G. Sockel and H. Schmalzried, Ber.Bunsenges.Physik.Chem., 72 [7] 745 - 754 (1968)

11 B.E.F. Fender and F.D. Riley, J.Phys.Chem.Solids 30 [4], 793 - 798 (1969)

12 C. Picard and M. Dodé, Bull.Soc.Chim.France [7], 2486 - 2487 (1970)

13 W.K. Chen and N.L. Peterson, J.Phys.Chem.Solids 36, 1097 - 1103 (1975)

14 E. Takayama and N. Kimizuka, J.Electrochem.Soc., 127 [4], 970 - 976 (1980)

15 J. Nowotny and M. Rekas, J.Am.Ceram.Soc., in print

16 O.N. Salmon, J.Phys.Chem., 65 [3], 550 - 556 (1961)

17 P. Gerdanian, J.Phys.Chem.Solids, in print

18 J.F. Marucco, P. Gerdanian and M. Dodé, J.Chim.Phys., France 67 [5], 914 916 (1970)

19 J.F. Marucco, P. Gerdanian and M. Dodé, ibid., 67 [5], 966 - 973 (1970)

THERMODYNAMIC AND STRUCTURAL EVIDENCE FOR THE PRESENCE OF DEFECT CLUSTERS IN SOME NON-STOICHIOMETRIC OXIDES

O. Toft Sørensen
Risø National Laboratory, Denmark

ABSTRACT

Based on thermodynamic data for CeO_{2-x} it is first shown that the non-stoichiometric phase range for this oxide can be devided into subphase regions bounded by intermediate phases with compositions following the general formula M_nO_{2n-2}. The defect structure of these phases is then derived using a systematic packing of the basic tetrahedral defect $[V_O^{\bullet\bullet} \cdot 2Ce'_{Ce}]^x$ and it is shown that lattice parameters determined by X-ray diffraction are well reproduced by the defect structures obtained in this way. Furthermore evidence from statistical thermodynamic calculations and from structural studies (X-ray and neutron scattering) for the existence of these tetrahedral defects in non-stoichiometric fluorite oxides are presented.

1. INTRODUCTION

It is well known that many non-stoichiometric oxides exist over a considerable composition range. In the early days it was believed that the defects of this grossly non-stoichiometric phase remained randomly distributed and non-interacting throughout the whole composition range, but many recent studies have now clearly shown that this is not the case. On the contrary considerable evidence have now been accumulated for the presence of defect clusters in various non-stoichiometric oxides.

In this paper the evidence for the presence of defect clusters in general in non-stoichiometric oxides will be discussed using two approaches: (1) the thermodynamic approach by which the type of defects present is deduced from thermodynamic data measured as a function of oxygen pressure and temperature - in this part the defect structure of CeO_{2-x} will be considered, and (2) the structural approach where more direct evidence is obtained by X-ray and quasi-elastic neutron scattering measurements.

J. Nowotny and W. Weppner (eds.),
Non-Stoichiometric Compounds Surfaces, Grain Boundaries and Structural Defects, 123–136.

In this latter approach the evidence presented is based on some recent neutron scattering measurements on yttria doped ZrO_2 which is an important solid oxide electrolyte with many applications such as in oxygen sensors and fuel cells.

2. THERMODYNAMIC APPROACH - CeO_{2-x}

2.1. Defect chemistry. CeO_{2-x} is a typical grossly non-stoichiometric oxide, which at higher temperature (above 600°C) exists over a very large composition range - $0 \leq x \leq 0.25$. As the Ce^{4+} ions quite easily can be reduced to Ce^{3+} ions this non-stoichiometric oxide is therefore oxygen deficient with oxygen vacancies, $V_O^{\bullet\bullet}$ as the predominant defect as also indicated by the formula.

Formation of oxygen vacancies in CeO_2 can be expressed by:

$$O_O + 2Ce_{Ce} \rightleftharpoons V_O^{\bullet\bullet} + 2Ce'_{Ce} + 1/2\ O_2\ (gas)$$

i.e. two cations are reduced for each oxygen vacancy formed. Using the law of mass action on this equilibrium gives

$$k(V_O^{\bullet\bullet}) = [V_O^{\bullet\bullet}]\ [Ce'_{Ce}]^2 \cdot p(O_2)^{1/2}$$

which by introducing the neutrality condition

$$[Ce'_{Ce}] = 2[V_O^{\bullet\bullet}]$$

and by expressing $[V_O^{\bullet\bullet}]$ as the fraction of unoccopied sites in the oxygen lattice - x in MO_{2-x} -, i.e. $[V_O^{\bullet\bullet}] = 1/2x$ can be converted into

$$k(V_O^{\bullet\bullet}) = 4[V_O^{\bullet\bullet}]^3\ p(O_2)^{1/2} = 1/2x^3 p(O_2)^{1/2}$$

Formation of double charged oxygen vacancies, $V_O^{\bullet\bullet}$, in CeO_{2-x} thus depend on the oxygen pressure according to

$$x \propto p(O_2)^{-1/6}$$

Similar expressions can easily be derived for other types of defect - the exponent would for instance become -1/4 and -1/2 if single charged or neutral oxygen vacancies were considered - and generally

$$x \propto p(O_2)^{-1/n}$$

where n is a characteristic number depending on the type of defect considered.

The first step in identifying the predominant defect present in a non-stoichiometric oxide is therefore to determine how the composition varies with oxygen pressure (at constant temperature) - this is usually done by thermogravimetric measurement in atmospheres with controlled oxygen pressure - and from these data then to determine the value of n.

The important thermodynamic quantity $\Delta\overline{G}(O_2)$ (relative partial free energy of oxygen) depends on the composition in the following way:

$$\Delta\overline{G}(O_2) = RT \ln p(O_2) \propto -nRT \ln x$$

and the n value can thus be determined from the slope of a $\Delta\overline{G}(O_2)$ vs. log $p(O_2)$ plot, which should give a straight line if only one type of defect is predominantly formed. This plot, which is very similar to a Brauner plot is therefore very useful in defect studies.

Before considering the type of defect which can be derived from such an analysis it is worthwhile to consider the basic assumptions involved in the above treatment. First by using the simple mass-action law it is assumed that the defects are randomly distributed and non-interacting. This is probably only true for very small deviations from the stoichiometric composition but if the defects associate into still randomly distributed and non-interacting defect clusters the mass-action law can still be considered to be valid. This condition is probably also fulfilled at intermediate defect concentrations, whereas extensive ordering certainly makes the use of the mass-action law doubtful at high defect concentrations.

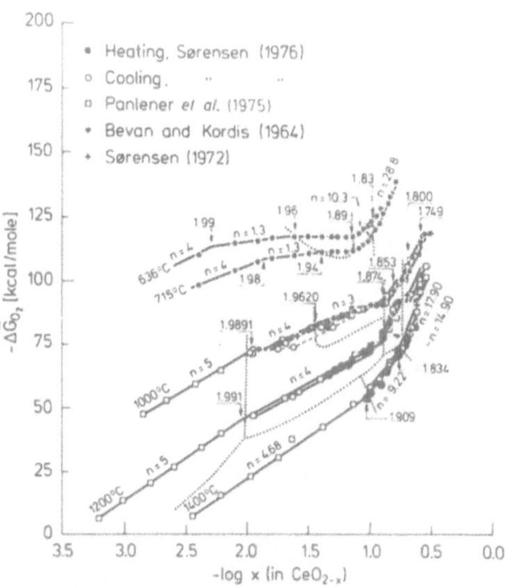

Fig. 1. Relative partial free energies of oxygen, $\Delta G(O_2)$ of CeO_{2-x} as a function of composition (log x).

<u>2.2. $\Delta G(O_2)$-log $p(O_2)$ plot for CeO_{2-x}</u>. Fig. 1 and 2 show a $\Delta\bar{G}(O_2)$-log $p(O_2)$ plot previously published by the author (1,2). From these plots, which was constructed from literature data as well as from data obtained by the author the following details about the CeO_{2-x} phase will be noticed:

(i) The straight-line relationship predicted is clearly observed, but apparently the slope changes with increasing non-stoichiometry.
(ii) According to the n-values calculated from the slope of these lines the non-stoichiometric phase range can apparently be divided into subphases each with a characteristic defect.

This subdivision into subphases can more clearly be demonstrated by plotting the composition at the break in the lines where the slope changes as a function of temperature in a normal phase diagram. The diagram obtained for the CeO_{2-x} phase is shown in Fig. 3 from which it is interesting to note that the composition of the vertical phase boundaries in all cases can be described by the general formula M_nO_{2n-2} which is the formula describing the homologous series of ordered compounds observed for these oxides.

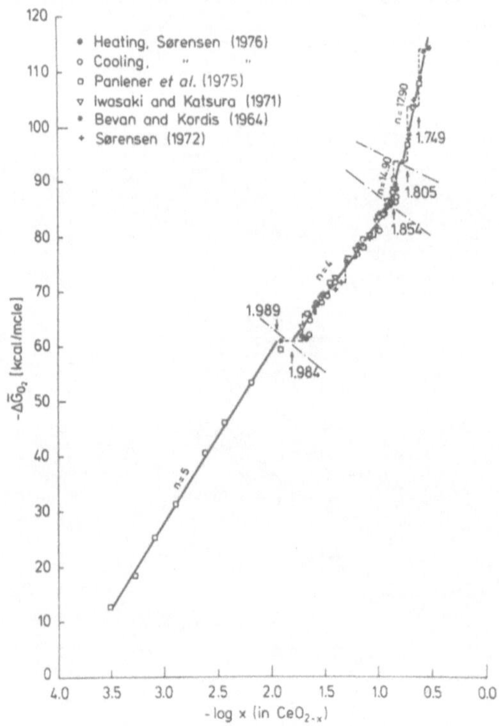

Fig. 2. Relative partial free energies of oxygen, $\Delta\bar{G}(O_2)$ at 1100°C of CeO_{2-x} as a function of composition (log x).

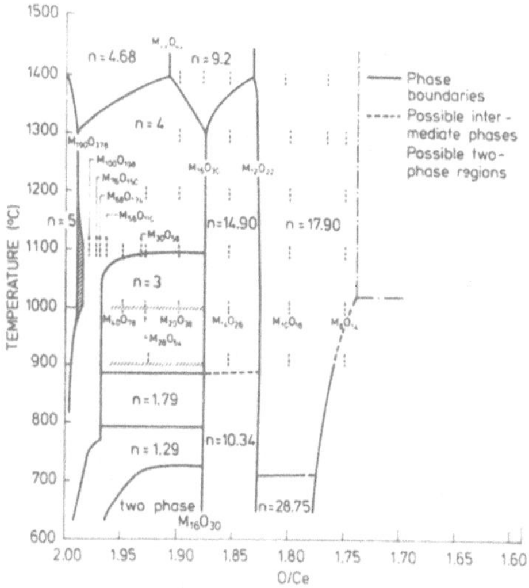

Fig. 3. Diagram of subphases with possible ordered intermediate phases for the Ce_{2-x} phase.

What can now be inferred from the n-values obtaiend by this analysis?

First it is worth noting that no matter what defect or defect cluster is considered the value of n can never exceed 6. Thus, all subphases where n≤6 can be considered to be truly non-stoichiometric with non-interacting defects or defect clusters being randomly distributed in the lattice.

For the subphases with n>6, however, substantial ordering of the defects must be expected as also indicated in Fig. 2 - data at 1100°C (from Fig. 1) drawn to a larger scale - where the steep curves actually can be considered as step curves consisting of large vertical steps (single, ordered phases) and small horizontal sections (two-phase regions).

The general conclusion from this analysis therefore is that the defect structure certainly changes with increasing deviation from the stoichiometric composition and that substantial interactions between the defects or defect clusters or at least ordering of these takes place at high defect concentrations. From this analysis it is however impossible to derive the type of defects actually present since the experimentally determined n-values in fact can be derived from many different defect types. In order to get more precise information about the actual defect structure systematic high temperature structure determination in controlled atmospheres of single crystals would be valuable, but to the author's knowledge these have not yet been performed. The only possibility we presently have to obtain a clear and consistent picture of the most probable defects in this and other non-stoichiometric oxide systems is therefore by devising models, which, however, in order to be realistic must reproduce both the thermodynamic and the structural data available. A model developed by the author and coworkers for defect formation and defect ordering in CeO_{2-x} and similar non-stoichiometric fluorite oxides which fulfil this requirement is presented in the next section.

2.3. Defect interactions in CeO_{2-x}. An important thermodynamic quantity to consider when evaluating the defect-defect interactions in a non-stoichiometric oxide is $\Delta\bar{H}(O_2)$ (relative partial enthalpy of oxygen), which easily can be calculated from $\Delta\bar{G}(O_2)$ using standard thermodynamic relationships.

First it is necessary to consider how $\Delta\bar{H}(O_2)$ depends on $\Delta H(V_O^{\cdot\cdot})$ - the enthalpy of formation of oxygen vacancies. For the formation of $V_O^{\cdot\cdot}$ it is well known that

$$\Delta G^o(V_O^{\cdot\cdot}) = \Delta H^o(V_O^{\cdot\cdot}) - T\Delta S(V_O^{\cdot\cdot})^{vibr} = -R \ln k(V_O^{\cdot\cdot})$$

or

$$k(V_O^{\cdot\cdot}) = 1/2\ x^3\ p(O_2)^{1/2} = \exp(\Delta S(V_O^{\cdot\cdot})^{vibr}/R)\ \exp(-\Delta H^o(V_O^{\cdot\cdot})/RT)$$

which by introducing $\Delta\bar{G}(O_2)$=RT ln $p(O_2)$ and rearrangement gives

$$\Delta\bar{G}(O_2)=\Delta\bar{H}(O_2)-T\Delta\bar{S}(O_2)=-2\ \Delta H(V_O^{\cdot\cdot})+T(2\Delta S(V_O^{\cdot\cdot})^{vibr})-6R\ln x+2R\ln 2)$$

Thus for randomly distributed and non-interacting defects we should expect

$$\Delta\bar{H}(O_2) = -2\Delta H(V_O^{\cdot\cdot})$$

and for a given type of defect with a fixed enthalpy of formation $\Delta\bar{H}(O_2)$ should thus be constant and independent of composition. Using now the $\Delta\bar{G}(O_2)$ data presented in Fig. 1 to calculate the value of $\Delta\bar{H}(O_2)$ and plotting these as a function of composition the plot shown in Fig. 4 is obtained - in this plot the data were calculated at 1353 K (1080°C) in order to compare these with data obtained by Campserveux and Gerdanian (3) by microcalorimetry.

First it will be noticed that the agreement between the two sets of data is excellent which gives further support to the observed subdivision of the CeO_{2-x} phase into subphases as shown in Fig. 3. But secondly it is also clear from this plot that the expected independency of composition of $\Delta\bar{H}(O_2)$ only is the case for the n=5 and n=14.9 regions whereas $\Delta\bar{H}(O_2)$ for the other phases shows a linear variation with log x with substantial slopes indicating that there must be a considerable interaction between the defects in these phases. In the next section we shall consider how the observed changes in $\Delta\bar{H}(O_2)$ for the different subphases can be interpreted.

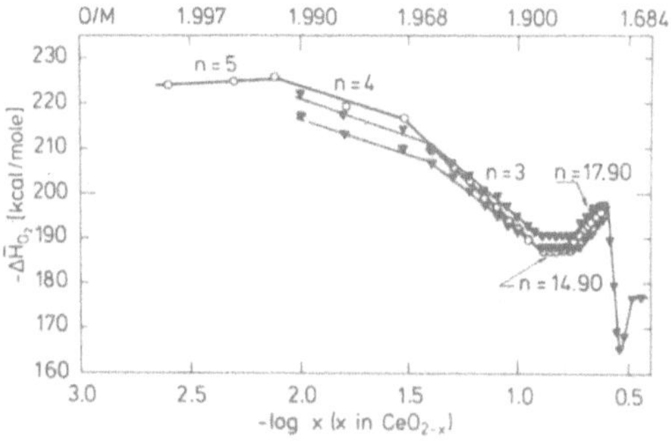

Fig. 4. Relative partial enthalpy of oxygen, $\Delta\overline{H}(O_2)$, as a function of composition (log x) for the CeO_{2-x} phase at 1080°C. Circles, data from Sørensen (1976); triangles, data reported by Campserveux and Gerdanian (1974).

<u>2.4. Defect model for CeO_{2-x}</u>. Several models have been proposed to explain the defect structure of the ordered intermediate phases in the Ce-O system. In this paper we shall only discuss the model proposed by the author and coworkers which is believed to be quite consistent as it also accounts for the observed variation of $\Delta\overline{H}(O_2)$. For a more detailed account of other models proposed for this and similar oxide systems the reader is referred to reference 2.

First the basic defect complex (cluster) considered in this model is the so-called tetrahedral defect consisting of one oxygen vacancy bonded to two adjacent reduced cations, i.e. $[2\ Me'_M \cdot V_O^{\bullet\bullet}\ ]^x$ and which naturally is formed within a cation tetrahedron of the fluorite lattice. This defect complex, which is neutral, was originally proposed by Schmitz and Marajofsky (4) for less the less ionic, oxygen-deficient oxides such as PuO_{2-x} but as will be shown below this basic defect complex is also equally useful for the description of the defect structure of the more ionic CeO_{2-x}.

Secondly it is necessary to consider how these tetrahedral defects are formed. Intuitively there must be at least two effects involved: repulsion between the positively charged oxygen vacancies which tend to keep these well separated, and strain introduced in the lattice surrounding the oxygen vacancy. If the fcc lattice is represented as cation tetrahedra containing oxygen ions, we assume that this lattice strain is inversely proportional to the distance from the central defect and that it is so large in the nearest neighbour tetrahedra that formation of additional oxygen vacancies in these positions is either excluded or at least requires considerable higher energy. This principle can be termed as the exclusion principle. As there are 22 nearest-neighbour tetrahedra to a given tetrahedron in the fluorite structure an envelope of 23 tetrahedral positions (including the central tetrahedral defect) can be identified as an extended defect - here designated as a T_{23} complex. Fig. 5 shows a photograph of such a T_{23} complex and its projection on the (111) planes. In this figure other types of defect complexes are also shown which can be derived naturally from the T_{23} complex in the following way: when the whole lattice has been filled by packing T_{23} complexes in a regular manner, the smallest increase in the lattice energy by formation of new defects will be observed when the tetrahedra under the smallest strain in the T_{23} complex - i.e. those at the largest distance from the central defect - becomes part of a neighbouring complex. A T_{19} and T_7 complex can thus be derived by removing 4 and 4+(4x3)=16 tetrahedra from the T_{23} complex as shown in Fig. 5. These complexes allow a closer packing of the defects and they are necessary to derive the defect structure at large deviations from the stoichiometric composition.

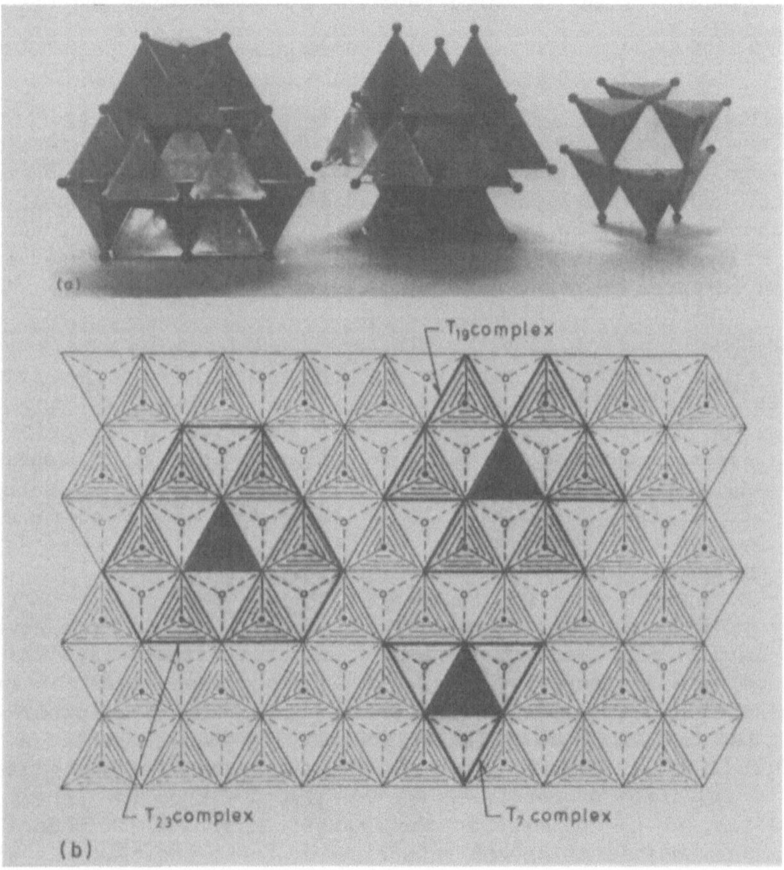

Fig. 5. (a) Photograph of models showing T_{23}, T_{19} and T_7 complexes and (b) their projection on (111).

The enthalpy changes involved in the formation of a given defect structure can be considered to be determined mainly from two opposing effects:

(i) The <u>formation</u> energy term. In subregions where only one type of defect complex is formed, a T_{23} complex for instance, this term can thus be considered to be constant and independent of composition.
(ii) The <u>interaction</u> energy term, which if coulombic in origin can be either positive or negative depending on whether there is a repulsion or an attraction between the defect complexes. The tetrahedral defect is neutral - it consists of $V_O^{\cdot\cdot}$, $2[Me_M']$, and the interactions are probably very small in these oxides. However, as explained by Manes (5), we may assume that dipoles are formed in the tetrahedral defects by charge separation, and dipole-dipole attractions must therefore also be taken into account in these systems.

Let us now consider the experimental values of $\Delta\overline{H}(O_2)$ and their dependence on log x for the CeO_{2-x} phase. From Fig. 4; it will be noted that a "lattice stabilization" (i.e. a decrease of $\Delta H(O_2)$ with increasing x) apparently takes place in some subregions whereas a "lattice destabilization" takes place in others. In terms of the two opposing effects mentioned above this must indicate that strong attractive dipole-dipole interactions operate in the former subregions, whereas the larger strain (higher lattice energy) introduced when the distance between the defect complexes becomes smaller gives the $\Delta\overline{H}(O_2)$ changes observed in the latter subregions.

Using these ideas the defect structure for different non-stoichiometric and ordered subphases observed in the CeO_{2-x} system (see Fig. 3) can now be constructed by a systematic packing of the T_{23}, T_{19} and T_7 complexes as summarized in Table I. The structure of the macrocomplexes formed by the dipole-dipole attractions consisting of either 4 T_{23} complexes (one middle and three upper) or 7 T_{23} complexes (one middle, three upper, and three lower) is shown in Fig. 6, Fig. 7a and b on the other hand, show how the $M_{12}O_{22}$ and M_7O_{12} structures can be constructed by a systematic packing of T_{19} and T_7 complexes.

Table I. Packing of defect complexes in subregions observed for CeO_{2-x}

O/M	$-\Delta\overline{H}O_2$	Packing
2.00-1.99	constant	Random distribution of T_{23}
1.99-1.96	decreasing "lattice stabilization"	Macrocomplexes of 4 T_{23} (1 middle, 3 upper)
1.96-1.92	decreasing "lattice stabilization"	Macrocomplexes of 7 T_{23} (1 middle, 3 upper, 3 lower)
1.92-1.88	decreasing "lattice stabilization"	T_{19} formed
1.88		close packing T_{19}
1.88-1.83	constant	T_7 formed
1.83 ($M_{12}O_{22}$)		T_{19} and T_7
1.83-1.82	increasing "lattice destabilization"	T_7
1.82 ($M_{11}O_{20}$)		T_{19} and T_7
1.82-1.80	increasing "lattice destabilization"	T_7 formed
1.80 ($M_{10}O_{18}$		T_{19} and T_7
1.80-1.78	increasing "lattice destabilization"	T_7 formed
1.78 (M_9O_{16}		T_{19} and T_7
1.78-1.71	decreasing "lattice destabilization"	T_7 formed
1.71 (M_7O_{12})		T_7

It is interesting to note that the lattice parameters determined (by X-ray) for many of these phases are well reproduced and from a structural point of view this model thus appears to be quite consistent.

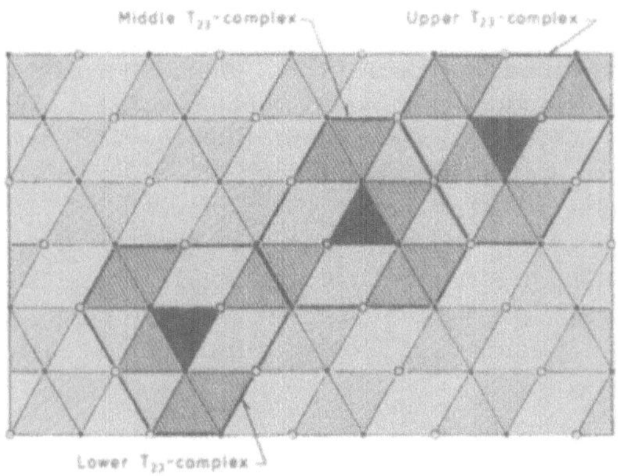

Fig. 6. Projection on (101) of 7 T_{23} complexes - 3 upper (only one is shown), 1 middle and 3 lower (only one is shown).

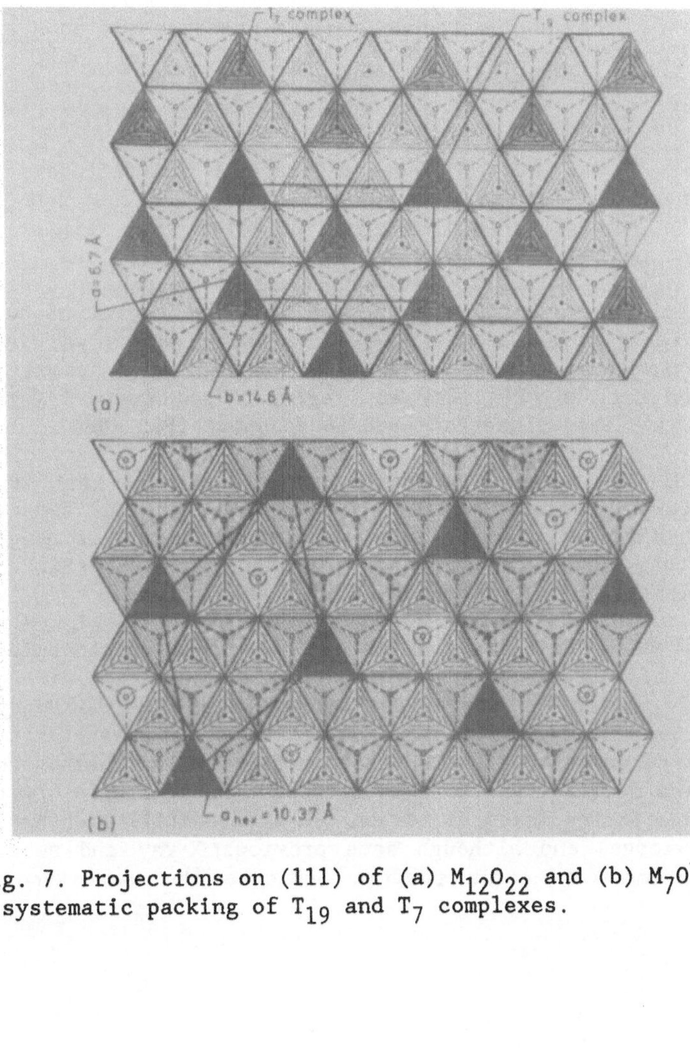

Fig. 7. Projections on (111) of (a) $M_{12}O_{22}$ and (b) M_7O_{12} constructed by a systematic packing of T_{19} and T_7 complexes.

Further evidence for the existence of this basic tetrahedral defect in these non-stoichiometric oxide systems comes from the statistical thermodynamic calculations performed by Manes (5). Based on the concepts discussed above, Manes developed a statistical thermodynamic model by which it was possible to reproduce the experimentally determined thermodynamic properties of the CeO_{2-x} phase and from a thermodynamic point of view the presence of the tetrahedral defects in these oxides thus also seem to be well established. The remaining question now is whether these defects and their packing pattern can be confirmed by direct structure determinations at high temperature. This will be discussed in the next section where the defect structure derived by some recent X-ray and neutron scattering measurements on various non-stoichiometric fluorite oxides will be considered.

3. DEFECT STRUCTURES DETERMINED BY X-RAY AND NEUTRON SCATTERING

X-ray and neutron scattering studies have been performed on powders and in some cases also on single crystals of oxides with the fluorite structure such as ZrO_2 doped with Y_2O_3, ZrO_2 doped with CaO, CeO_{2-x} and CeO_2 doped with Y_2O_3. A review of these studies has recently been presented by the author and coworkers in (6) and here we shall only consider the main conclusions about type of defect cluster proposed from these studies.

In the case of ZrO_2 doped with Y_2O_3 both diffraction and diffuse scattering measurements by X-ray and neutrons clearly indicated that the dominant feature of the disorder is oxygen vacancies with relaxations of the six nearest neighbour oxygen ions along <100> directions toward the vacancy as well as relaxations of the nearest neighbour cations along <111> directions away from the vacancy. By some recent measurements on large single crystals two specific types clusters was identified as dominating the diffuse scattering: vacancy free but tetrahedrally distorted regions (Fig. 8a) and regions with clusters formed by vacancy pairs with the relaxations mentioned above (Fig. 8b).

One drawback of diffuse neutron scattering measurements is that it is not possible by this technique to distinguish between the Zr^{4+} and the Y^{3+} ions. From ion size and charge compensation arguments it is however postulated that the central cation in the divacancy cluster (Fig. 8b) is a Zr^{4+} ion whereas 4 of the 6 relaxed cations are Y^{3+} ions.

As the Y (and vacancy) concentration increases in these doped oxides, the tetrahedrally distorted regions decreased in size and the disorder becomes completely dominated by aggregates of clusters of the vacancy pair type (Fig. 8b). This cluster thus seems to be a characteristic feature of the doped oxides but the question still remains whether this defect cluster also is important for the undoped oxides. Single crystal neutron diffuse scattering measurements on CeO_{2-x} are still lacking (to the author's knowledge) and although some previous X-ray and neutron diffraction measurement on ceria powders indicate the same type of relaxations as observed for the Y_3O_3 doped ZrO_2 the presence of the divacancy clusters in CeO_{2-x} still need to be verified.

Finally it is worth noting that it is not possible by the neutron scattering technique to distinguish between the Zr^{4+} and Y^{3+}ions. From ion size and charge compensating arguments it has been postulated that the central cation in the divacancy cluster found for Y_2O_3 doped ZrO_2 (Fig. 8b) is a Zr^{4+}ion whereas 4 of the 6 relaxed cations are Y^{3+}ions. The fundamental tetrahedral defect $[V_O^{\bullet\bullet}, 2M_M']^x$ proposed to explain the defect structure of CeO_2 and which constitutes half of the divacancy cluster thus also seem to be verified experimentally. although the arrangement of these tetrahedral defects in the lattice perhaps is different from that derived using the thermodynamic approach.

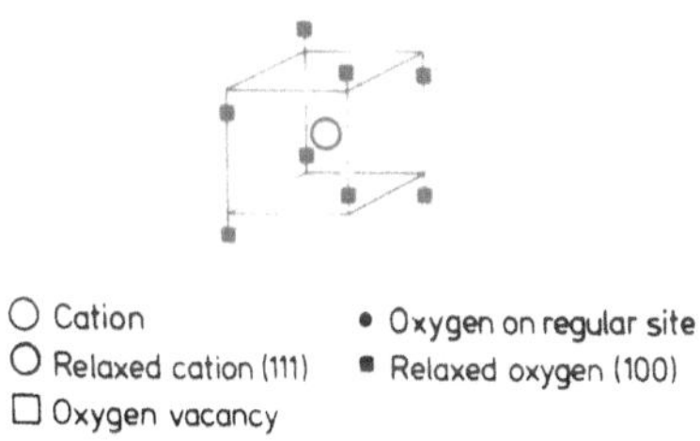

Fig. 8a. Oxygen-ion relaxation along <100> directions in the tetrahedrally distorted vacancy free regions of the fluorite structure.

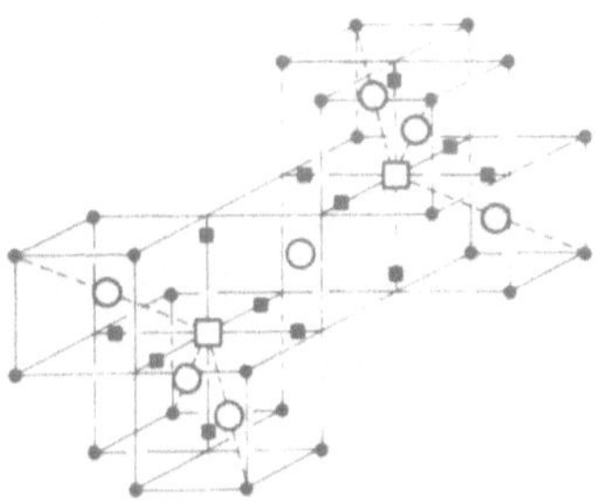

Fig. 8b. Defect cluster containing a vacancy pair, oxygen relaxations along <100> and cation relaxations along <111>.

4. SUMMARY

Evidence for the existence of defect clusters in the oxygen deficient fluorite oxides - CeO_{2-x} and Y_2O_3 doped ZrO_2 are presented using two approaches: the thermodynamic approach by which the type of defects present is deduced from thermodynamic data and a structural approach where more direct evidence is obtained from recent X-ray and neutron scattering measurements. Based on thermodynamic data for CeO_{2-x} it is first shown that the non-stoichiometric phase range for this oxide can be divided into several subphases each with a characteristic defect structure. The basic defect cluster proposed for these oxides is the so-called tetrahedral defect consisting of one oxygen vacancy bonded to two reduced cations, i.e. $[V_O^{\bullet\bullet} \cdot 2Ce'_{Ce}]$ and which naturally are formed within the cation tetrahedra of the fluorite lattice. By systematic packing of these tetrahedral defects in the lattice it is then shown that the structures of all intermediate phases can be described and it is interesting to note that the lattice parameters determined by X-ray diffraction are well reproduced by the defect structures obtained. Based on this concept a statistical thermodynamic model was developed and thermodynamic data calculated using this model showed good agreement with experimental data. From this approach the tetrahedral defect thus seem to be well established for these oxides.

The predominant defect proposed from X-ray and neutron scattering studies primarily on Y_2O_3 doped ZrO_2 is however a cluster composed of a pair oxygen vacancies connected to a common cation, in this case probably a Zr^{4+}ion, and with some relaxations of the nearest oxygen ions and cations in the surrounding lattice. The structure of this defect cluster can be considered as composed of two tetrahedral defects sharing one corner, so the presence of these defects in non-stoichiometric fluorite oxides thus also seem to be verified by these measurements but the packing pattern of these is perhaps different than that derived using the thermodynamic approach.

REFERENCES

1. Sørensen, O. Toft (1976). J. Solid State Chem. 18, 217-233.
2. Sørensen, O. Toft (1981). In: Nonstoichiometric Oxides. Edited by O. Toft Sørensen. Academic Press, New York, 1-59.
3. Campserveux, J. and Gerdanian, P. (1974). J. Chem. Thermodyn. 6, 795-800.
4. Schmitz, F. and Marajofsky, A. (1975). In: Thermodynamics of Nuclear Materials 1974, vol. 1. IAEA Vienna, 457-467.
5. Manes, L. (1981). In: Nonstoichiometric Oxides. Edited by O. Toft Sørensen. Academic Press, New York, 99-154.
6. Sørensen, O. Toft, Johannesen, Ø. and Clausen, K. (1985). In: Transport - Structure Relations in Fast Ion and Mixed Conductors. Proceedings of 6th Risø International Symposium on Metallurgy and Materials Science. Edited by F.W. Poulsen, N. Hessel Andersen, K. Clausen, S. Skårup and O. Toft Sørensen. Risø National Laboratory, Roskilde, Denmark, 93-117.

DEFECT CHEMISTRY OF $BaTiO_3$ AND $SrTiO_3$: PRACTICAL ASPECTS AND APPLICATION TO ELECTRONIC CERAMICS

Detlev Hennings and Rainer Waser
Philips GmbH Forschungslaboratorium Aachen
Weisshausstrasse
D-5100 Aachen
F. R. G.

ABSTRACT. The electrical properties of $BaTiO_3$ and $SrTiO_3$ based ceramic capacitors and nonlinear resistors are largely determined by point defects the concentrations of which are settled by selected incorporation of donor or acceptor dopes. While properties such as the bulk conductivity, grain boundary barrier layer effects, and the mobility of ferroelectric domains are widely controlled by non-mobile defects, the long term stability, i.e. the lifetime, of passive components strongly depends on the mobility of ionic defects. The defect chemistry of $BaTiO_3$ and $SrTiO_3$ has been well known for some time. This knowledge now is increasingly used for improving the processing of passive ceramic components.

1. INTRODUCTION

Alkaline earth titanate based ceramics are of considerable economic importance for the production of electronic ceramic components such as high-permittivity multilayer capacitors, PTC resistors, VDR devices, etc. An understanding of the thermodynamics and kinetics governing these materials is essential for the further development and tailoring of the components.

The first part of our paper reviews the defect structure of alkaline earth titanates in dependence on the oxygen partial pressure and on the doping by heterovalent metal ions. In addition, an introduction to recent findings concerning hydrogen defects is given. The solubility, the diffusivity, and the formation of associates of hydrogen defects is described.

The second part gives some examples of the role of non-mobile defects in determining the properties of electronic ceramics components. The effect of defect concentrations at domain walls and grain boundaries is discussed.

The third part outlines an example where mobile defects are responsible for limiting the life time of ceramic capacitors. The degradation of the insulation resistance of dielectric ceramics under prolonged temperature and dc field stress is shown to be explainable using the defect structure model as described in the first part of this paper.

J. Nowotny and W. Weppner (eds.),
Non-Stoichiometric Compounds Surfaces, Grain Boundaries and Structural Defects, 137–154.

2. NEW DEVELOPMENTS IN THE DEFECT CHEMISTRY OF PEROVSKITE TITANATES

2. 1. Model of the defect structure

Electrical conductivity measurements at high temperatures and thermogravimetric experiments as a function of the oxygen partial pressure have commonly been employed to investigate the defect structure of $BaTiO_3$ (and of $SrTiO_3$, which exhibits very similar properties in this regard) [1-7]. Undoped barium titanate was found to be dominated by a Schottky defect structure. No evidence for the presence of metal or oxygen interstitials has been reported. The Schottky equilibrium

$$n/2\ O_O + M_M \rightleftharpoons \text{"}MO_{n/2}\text{"} + n/2\ V_O^{\cdot\cdot} + V_M^{n'} \quad (1)$$

(using the defect notation of Kröger and Vink [8]) is only established at or near sintering temperatures, due to the low mobility of the metal vacancies [6]. In Eq. (1), M denotes Ba (n = 2) or Ti (n = 4). "$MO_{n/2}$" in Eq. (1) indicates an exchange with sites of repeatable growth, e.g. surfaces, grain boundaries, or second phases. Below temperatures of approx. 1200°C $V_M^{n'}$ can be regarded as virtually frozen in, so that they act as immobile acceptors. The exchange with oxygen gas of ambient atmosphere according to

$$O_O \rightleftharpoons 1/2\ O_2 + V_O^{\cdot\cdot} + 2\ e' \quad (2)$$

is activated at temperatures above approx. 600°C [6]. The mass action expression for Eq. (2) is

$$[V_O^{\cdot\cdot}]\ n^2\ P_{O_2}^{1/2} = K(T) = K_O \exp(-\Delta G/kT) \quad (3)$$

where ΔG denotes the reaction enthalpy, K(T) is the mass action constant and $n \equiv [e']$. Nominally undoped barium titanate ceramics are usually dominated by acceptor impurities (Al, Fe, etc.). Their concentration often exceeds the intrinsic acceptor concentration caused by native metal vacancies [7, 9-11].

As in case of every semiconductor, the concentrations n and p of electrons e' and holes $h^{\cdot}$ are determined by the generation/recombination equilibrium

$$0 \rightleftharpoons e' + h^{\cdot} \quad (4)$$

$$n\ p = N_V\ N_C \exp(-\Delta G_B/kT) \quad (5)$$

where ΔG_B represents the thermal band-gap energy, N_V and N_C are the effective densities of states in the valence and conduction bands, respectively.

Acceptor dopants are lower-valent metal ions Ml substitutionally accommodated on Ba or Ti sites. For example, the substitution of Al^{3+} ions on Ti sites is described by the following incorporation reaction:

$$Al_2O_3 + 2\ BaO \longrightarrow 2\ Ba_{Ba} + 2\ Al'_{Ti} + V_O^{\cdot\cdot} + 5\ O_O\ . \qquad (6)$$

The conductivity σ measured at temperatures above the onset of the equilibrium (2) shows a p-type characteristic in oxidizing atmospheres and a n-type characteristic in reducing atmospheres. Recorded as a function of pO_2, σ exhibits a minimum, the position of which depends on the temperature and the acceptor concentration. In Fig. 1 the electronic conductivity of acceptor-doped $BaTiO_3$ is shown in a numerical simulation based on Eqs. (1) to (4). The calculation is performed for the high temperature equilibrium (T = 1173 K) and for a system quenched from T = 1173 K to ambient temperature (T = 298 K). Obviously, the n-conduction as established in reducing atmospheres (i. e. below approx. 10^{-7} bar O_2 in Fig. 1) remains almost unaltered during quenching since the $V_O^{\cdot\cdot}$ donor level is very shallow, lying just below the conduction band edge. The p-conduction as established in oxidizing atmospheres decreases many orders of magnitude during quenching. This effect is essential for using titanates as insulating dielectrics in ceramic capacitor applications. It is due to the fact that acceptor levels are very deep, lying about 0.8 to 2.5 eV above the valence band.

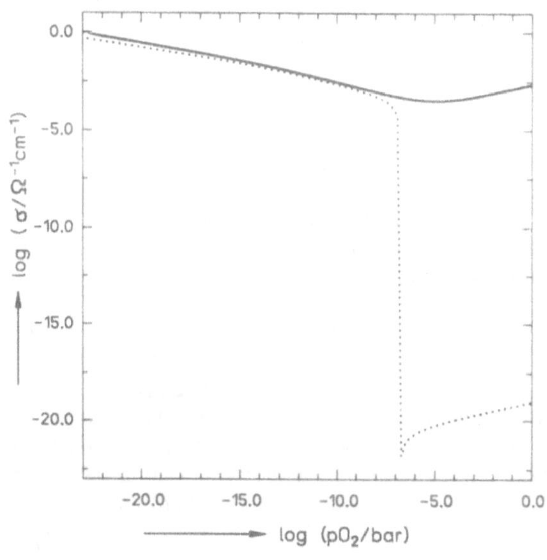

Fig. 1: Calculation of the electronic conductivity σ vs. the equilibrium oxygen partial pressure pO_2^{Eq} during annealing for 0.01 mol% acceptor-doped $BaTiO_3$. Acceptor energy level E(Ml') = 1.3 eV with respect to the valence band edge. Material constants: see Ref. [9].
(———) during annealing at T = 1173 K
(··········) after quenching from T = 1173 K to T = 298 K

Higher-valent metal ions Mh substitutionally incorporated on Ba or Ti sites act as donor dopants. For example, La^{3+} may replace the alkaline earth ion according to:

$$2\ La_2O_3 + 4\ TiO_2 \rightarrow 4\ La_{Ba}^{\bullet} + 4\ Ti_{Ti} + 12\ O_O + O_2(g) + 4\ e' \qquad (7)$$

or

$$2\ La_2O_3 + 3\ TiO_2 \rightarrow 4\ La_{Ba}^{\bullet} + 3\ Ti_{Ti} + 12\ O_O + V_{Ti}'''' \qquad (8)$$

Whether electron compensation (7) or metal vacancy compensation (8) takes place depends on the pO_2, the grain size, and the second phase [12-14]. The formation of titanium vacancies has been proved for high donor concentrations [15].

2. 2. Hydrogen Defects in Perovskites

In the last years, the thermodynamic properties and transport properties of hydrogen defects in alkaline earth titanates were investigated [16-18]. Single crystals were studied by means of IR transmission spectroscopy. Since IR spectroscopy is not applicable to the non-transparent ceramics, a thermal desorption technique with mass-spectroscopical detection was introduced in addition.

2. 2. 1. Solubility of Hydrogen Defects

As described in Ref. [16], the incorporation of hydrogen defects in the perovskite lattice of alkaline earth titantes depends strongly on the doping and the dopant type. A significant solubility is found for acceptor-doped titanates whereas no hydrogen is detected in donor-doped systems. Based on a study of the dependence of the solubility on the acceptor concentration and the water vapor pressure pH_2O, the incorporation/desorption reaction

$$H_2O + V_O^{\bullet\bullet} + O_O \rightleftharpoons 2\ (OH)_O^{\bullet} \qquad (9)$$

was deduced. For each water molecule dissolved, one oxygen vacancy vanishes and two protons are formed. The protons are bound to oxygen ions on regular sites forming hydroxide ions. Instead of being termed hydroxide ions $(OH)_O^{\bullet}$, equivalently, the hydrogen defects can be regarded as interstitial protons $H^{\bullet}$. The mass expression for Eq. (9) is:

$$\frac{[(OH)_O^{\bullet}]^2}{pH_2O \cdot [V_O^{\bullet\bullet}]} = K_H(T) = \exp\ (-\Delta G_H/kT) \qquad (10)$$

where $K_H(T)$ denotes the mass action constant and ΔG_H is the Gibbs reaction enthalpy. According to Eq. (9), hydrogen incorporation takes place as an exchange of $V_O^{\cdot\cdot}$ by $(OH)_O^{\cdot}$. This requires the presence and the mobility of $V_O^{\cdot\cdot}$. Hence, the fact that hydrogen is not detected in donor-doped titanates can be explained by the low $[V_O^{\cdot\cdot}]$ which, in turn, is a consequence of the Schottky equilibrium Eq. (1). Even in acceptor-doped titanates, hydrogen defects remain minority species, i. e. $[(OH)_O^{\cdot}] \ll [V_O^{\cdot\cdot}]$.

2. 2. 2. Diffusivity of Hydrogen Defects

The activation energy E_a and the diffusion constant D_0 of the Arrhenius equation

$$D = D_0 \cdot \exp(-E_a/kT) \qquad (11)$$

are obtained in cases of $SrTiO_3$ single crystals by determining the integrated absorption of the OH and the OD stretching band vibration during a proton/deuteron exchange at different temperatures [17]. For ceramics material, the diffusion information is extracted from the thermal desorption spectra by a numerical simulation of the desorption process and a two-dimensional fitting procedure. To interpret the diffusion data which are obtained by this technique, the ambipolar nature of the diffusion process must be considered. Since the presence and the mobility of $V_O^{\cdot\cdot}$ is essential to allow an incorporation or desorption of hydrogen defects in alkaline earth titanates, the concentration and the diffusion coefficient of $V_O^{\cdot\cdot}$ have to be taken into account as outlined in Ref. [17].

As a result hydrogen defects are found to exhibit diffusion coefficients which are about one order of magnitude above those of oxygen vacancies in alkaline earth titanates in the temperature range from 900 K to 1200 K. Activation energies of approx. 1.3 eV for Mn-doped $SrTiO_3$ single crystals and 1.5 to 1.6 eV for Ni-doped $BaTiO_3$ ceramics are determined. A thorough discussion of the diffusion data [17] reveals that hydroxide ions do not move as entities. The only mechanism which can interpret the relatively high diffusion coefficients, is a proton (or deuteron) hopping along oxygen ions on regular sites.

2. 2. 3. Formation of Defect Associates

In $SrTiO_3$, Ti sites are octahedrally surrounded by oxygen ions. If one of the oxygen ions gets protonated due to annealing of the crystal in H_2O moistened atmosphere, a hydroxide ion is formed and an absorption line of the OH stretching vibration is observed at 3495 cm^{-1}. In the case of Fe-doped $SrTiO_3$, additional absorption lines appear on the high energy side of the IR spectrum. Fe is accommodated on Ti sites. After annealing in oxidizing atmospheres most of the Fe ions are in the oxidation state +3 and, thus, act as monovalent acceptors Fe'_{Ti}. The intensity of the sideband lines grows with increasing dopant concentration [Fe]. Furthermore, the intensity distribution shifts towards higher wavelengths while the wavelength values of the absorption lines remain constant. As discussed in detail in Ref. [18], the sideband lines can be attributed to the formation of defect associates between hydroxide ions and Fe-acceptors according to a reaction

$$OH_O^{\cdot} + Fe' \rightleftharpoons \{OH,Fe\} \qquad (12)$$

It should be noted that Eq. (12) is simplified since various types of associates are formed. From the temperature dependence, a formal association enthalpy of approx. -150 kJ/mol is evaluated.

3. NON-MOBILE DEFECTS

The electrical properties of $BaTiO_3$ and $SrTiO_3$ based passive components are largely controlled by lattice defects which are produced by high donor or acceptor doping. These defects which may be considered as non-mobile at the working temperature of the passive components determine the electrical conductivity and the reversible elastic mobility of the ferroelectric domains. Defect chemical models can be used to improve the electrical properties of the materials used for the manufacture of ceramic capacitors and nonlinear ceramic resistors.

3. 1. The Aging Phenomenon in Ferroelectric Ceramics

Ceramic multilayer capacitors [19] are monolithic bodies of thin ceramic layers and Pd or Pd/Ag alloy electrodes which are cofired at temperatures of 1100°-1400°C. The use of noble metal electrodes turned out to be necessary since sintering seemed possible only in oxidizing atmospheres to avoid semiconducting ceramics (Fig. 1). Base metal electrodes (e.g. Ni electrodes) would require reducing atmospheres to protect them from oxidation. According to Eq. (2), in reducing atmospheres excessive oxygen vacancies are formed which generate conduction electrons. Calculations have revealed, however, that increasing acceptor concentrations support the trapping of these electrons and, hence, shift the transition between semiconducting and insulating ceramics (dotted line in Fig. 1) towards the reducing region on the pO_2^{Eq} scale. This has been one of the most striking conclusions drawn from the defect chemistry of $BaTiO_3$. It allows electrically insulating ceramics to be obtained by sintering in reducing atmospheres (e.g. moistened forming gas $N_2/H_2/H_2O$) which is the precondition to replace the expensive noble metal electrodes by cheaper Ni electrodes.

The acceptor dope in such Base Metal Electrode (BME) capacitors not only controls the electrical conductivity but also largely determines the dielectric properties of the ferroelectric perovskite. From several EPR measurements [20-22] it is known that acceptor ions on Ti-sites preferentially form associates with oxygen vacancies, $\{A_{Ti}^{n'},V_O^{\cdot\cdot}\}^{(n-2)'}$. The electrical dipole field of these defect clusters usually interacts with the ferroelectric polarization, thus giving rise to "internal bias" fields [23], causing asymmetric ferroelectric hysteresis loops [24] and aging effects of the dielectric properties [25].

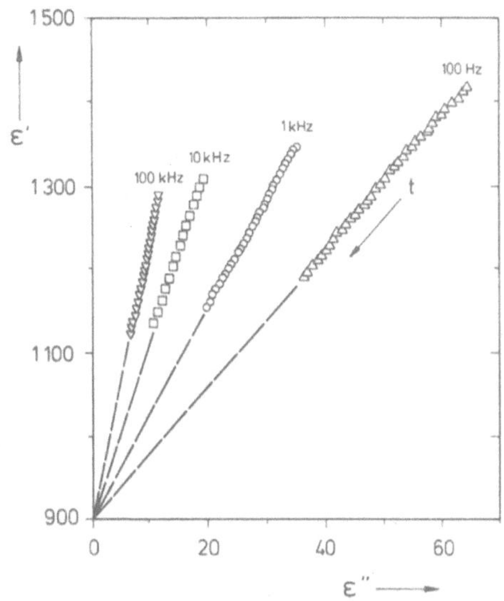

Fig. 2: Plot of ε' vs. ε'' during aging of $Ba(Ti_{0.995}Cr_{0.005})O_3$ [30]

Internal bias and dielectric aging are mainly interpreted in terms of pinning of ferroelectric domain walls by defect clusters [25, 26]. On heating or long time ac electrical treating of acceptor doped $BaTiO_3$ the domain wall pinning can be released. However, after short time these effects appear again. During the aging the elastic reversible as well as the irreversible mobility of the ferroelectric domain walls continuously decreases, following an exponential time law [27].

Since the relative dielectric constant, ε', and the losses, ε'', of ferroelectric $BaTiO_3$ are assumed to be largely dependent on the mobility of the ferroelectric domain walls [28], the continuous decrease of permittivity and losses observed during aging of acceptor doped $BaTiO_3$ is commonly attributed to the increase of a pinning of the domain walls with time. On extrapolation of the measured values of ε' in Fig. 2 for several frequencies to zero loss, ε'' = 0, which means a complete freeze of domain wall mobility, the extrapolated curves all meet in a single point at ε' ≅ 900 [25]. This extrapolated value of ε'= 900 in fact coincides with the average dielectric constant of randomly orientated, domainfree $BaTiO_3$ single crystals [29, 30].

3. 2. PTC Ceramics

One of the most interesting applications of donor doped $BaTiO_3$ is the so called PTC (Positive Temperature Characteristic) resistor. This nonlinear ceramic resistor or thermistor is a combination of ferroelectric and semiconducting effects, displaying an abrupt change in resistivity by about 5 to 6 orders of magnitude at the ferroelectric Curiepoint, Fig.3.

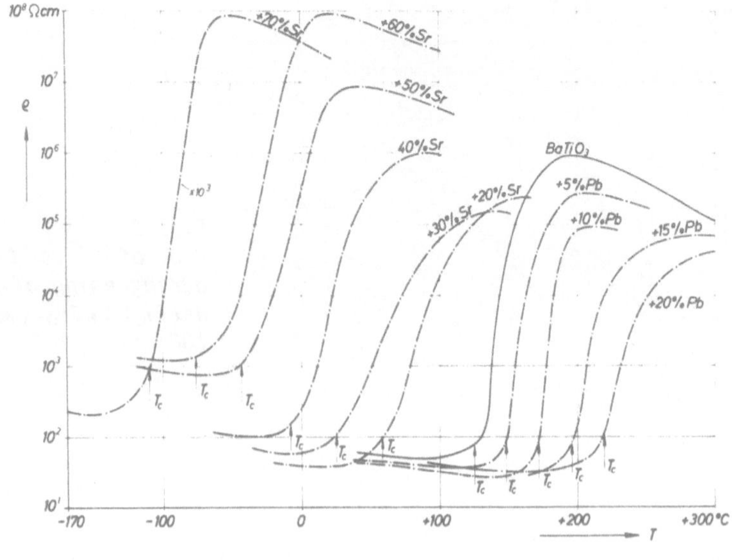

Fig. 3: Temperature characteristics of PTC resistors made from mixed crystals of $BaTiO_3$ with $SrTiO_3$ and $PbTiO_3$ [31]

Today PTC resistors have found widespread industrial application as heating elements, devices for heat sensing, and demagnetization control of colour televisions tubes etc.

The usual donor dopes in PTC resistors are Sb and Y in concentrations of about 0.3 mol%. Despite sintering in air, the donors are not completely compensated by Ba vacancies but partly by conduction electrons which give rise to a typical cold resistance of approx. 20 to 50 Ω · cm.

The PTC effect is a grain boundary barrier layer phenomenon. At temperatures above the Curiepoint thin isolating layers are formed at the grain boundaries which can be made visible by cathodoluminescence of the charge carrier density [32] or by decorating the grain boundaries with TiO_2 in hot oil suspensions under applied electrical field [31]. The PTC effect has been explained by Heywang and Jonker [33,34] in terms of a semiquantitative Schottky barrier model. According to this model the bending of the conduction band and the appearence of potential barriers at the grain boundaries are caused by the enrichment of surface acceptors at the grain boundary. In the ferroelectric state these surface acceptors are assumed to be compensated by the ferroelectric polarization charge and this results in the surface barrier layers being annihilated. However, at temperatures above the Curiepoint the surface acceptors are no longer compensated by polarization charges so that the grain boundary barrier layers again become effective.

The Heywang model is an useful working model, however, there is only rather poor knowledge about the real nature of the surface acceptors in PTC thermistors. From kinetic and diffusion experiments, Wernicke [6] has deduced that oxidation of semiconducting $BaTiO_3$ starts rather slowly at the grain boundaries. Ba vacancies which are known as strong acceptors are therefore formed at first at the grain boundaries during cooling of PTC resistors after sintering in air.

3. 3. VDR Ceramics

The nonlinearity of PTC resistors extends not only over the current-temperature but also the current-voltage characteristics. As higher voltages are applied a disproportionate increase of the current is observed. This so called "varistor" or "VDR" (Voltage Dependent Resistor) effect can also be observed in semiconducting donor doped $SrTiO_3$ ceramics which have been slightly oxidized at the grain boundaries [35].

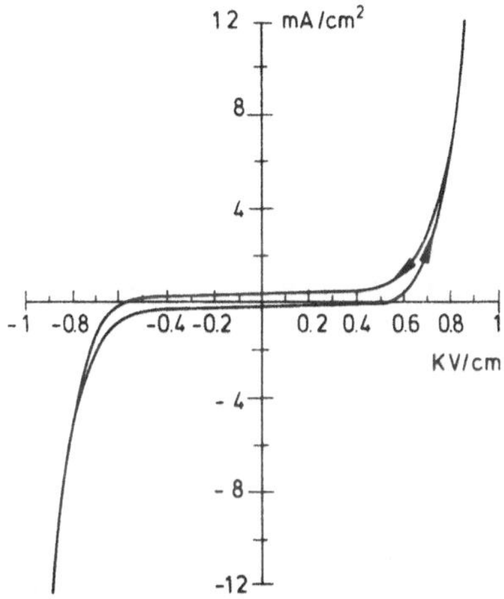

Fig. 4: Currrent-voltage characteristics of a $SrTiO_3$ varistor

Since $SrTiO_3$ is a non-ferroelectric perowskite the grain boundary barrier layers exist throughout the whole temperature scale. The nonlinear current-voltage characteristics of varistors are usually descibed by an empirical power law:

$$I = C \cdot U^{\beta} \qquad (\beta > 1) \tag{13}$$

The factor C is largely geometry dependent, whereas the "current index" β is typical for the material. The current index of $SrTiO_3$ varistors is largely determined by the reoxidation procedure. Values of β = 7-15 are usually found in $SrTiO_3$ varistors (Fig. 4) which is rather low compared with β = 30-100 of ZnO varistors [36].

Varistors are mainly applied in the field of circuit protection against overvoltage. The values of β should therefore be as high as possible because very high changes of current are then obtained at even very small changes of voltage.

4. MOBILE DEFECTS

In many high temperature devices such as the ZrO_2 oxygen sensor, mobile defects play an essential role. In this paper, we point out that defects which are mobile under operation conditions may influence the technical performance of electroceramic components. As an example, the degradation of the insulation resistance of ceramic dielectrics under temperature and dc field stress is described. According to our present understanding, the electromigration of mobile oxygen vacancies (and protons) is responsible for the process.

4. 1. Degradation under dc Stress

Ceramic capacitors may suffer a long-term degradation of their insulation resistance under moderate dc electrical field stress at elevated temperatures. The degradation may occur at dc field strengths far below the critical onset value of the dielectric breakdown and must be regarded as a typical material-inherent wear-out effect determining the life time of a capacitor.

Fig. 5 shows the typical time evolution of the insulation resistance R_i of a ceramic multilayer capacitor made from $(BaCa)(TiZr)O_3$ perovskite ceramics at a dc field of 1.5 V/μm for three different temperatures. After applying the field, at first R_i is constant or even rises slightly until the degradation starts to dominate. R_i usually decreases by many decades. The component is finally destroyed by thermal breakdown or the insulation resistance stabilizes on a new lower level.

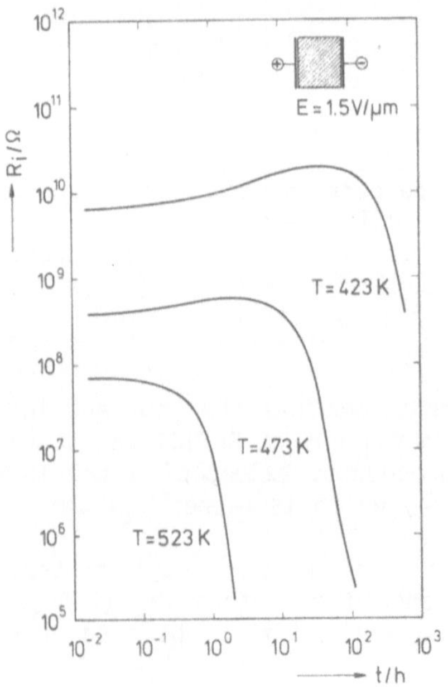

Fig. 5:
Insulation resistance R_i vs. time t for a ceramic multilayer capacitor (Z5U-type) made from $(BaCa)(TiZr)O_3$ ceramics at E = 1.5 V/μm and various temperatures.
EIA specification for Z5U:
$\Delta C/C(25^\circ C)$ = +22% ... -56% for the rated temperature interval +10°C to 85°C.

The temperature dependence of the degradation is that of a typical thermally activated process. In the case of the Z5U-type capacitors, the activation energy of the degradation is approx. 1.2 eV. For different $BaTiO_3$ and $SrTiO_3$ ceramics and single crystals activation energies in the range from 0.8 to 1.5 eV are found. In first approximation the degradation rate is determined by the field strength and not by the absolute voltage as long as an onset voltage of a few hundred mV is exceeded.

In order to study the influence of the microstructure on the degradation behaviour, undoped $SrTiO_3$ was investigated as single crystalline material, as coarse grained ceramics (mean grain size $\bar{d}_{gr} \approx$ 60 μm), and as fine grained ceramics ($\bar{d}_{gr} \approx 4\ \mu m$). Without going into the details of the character of the grain boundaries in the individual cases, Fig. 6 may qualitatively serve as a typical example illustrating that the degradation rate decreases significantly with an increasing number of grain boundaries between the electrodes.

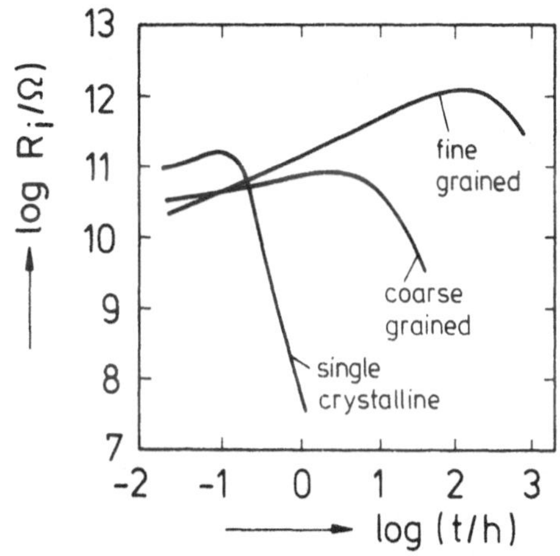

Fig. 6: Insulation resistance R_i vs. time t for undoped $SrTiO_3$ with various microstructures at T = 473 K and E = 0.5 V/μm. (Data of the ceramic samples are taken from Ref. [37])

The type of doping shows an even more striking influence on the degradation characteristics as already reported by Lehovec and Shirn [38]. While undoped and acceptor-doped alkaline earth titanates are prone to degradation, donor-doped materials (in as far as they are prepared to be electrically insulating) are found to exhibit practically no degradation at all.

Since the first comprehensive report [38], various theories have been offered to explain the mechanism of the degradation process. Some of them [39, 40] suggest that a dc field induced deterioration of the grain boundaries is responsible for the degradation phenomenon. Although these models satisfactorily describe the microstructure dependency of the degradation, they do not consistently explain the behaviour of single crystals and the influence of the type of dopant. The majority of theories assume that the relatively high mobility of oxygen vacancies is an essential part of the degradation process [38, 41-44]. Oxygen vacancies are present in significant concentrations in undoped and acceptor-doped titanates. They are positively charged with respect to the regular lattice and, thus, in a dc electrical field they suffer an electromigration towards the cathode. Although these models allow an understanding of the influence of the dopant type on the degradation, no consistent explanations about the local change of the defect state as a consequence of the electromigration of $V_O^{\cdot\cdot}$ are presented. Furthermore, the models do not take into account the electrocolouration effect.

4. 2. Electrocolouration

In order to gain a closer insight into the nature of the degradation mechanism, it is essential to obtain locally resolved information on the ceramic material. As an example, a study of the optical absorption spectrum as a function of the position between anode and cathode and as a function of time will be described.

The specific optical absorption bands of transition-metal ions used as dopants in the titanate lattice reveal information about the local valency states and, hence, about the electrochemical potential of the dopant ions. Therefore, colour changes after applying dc voltages (a phenomenon which is called electrocolouration [45, 46]) may supply

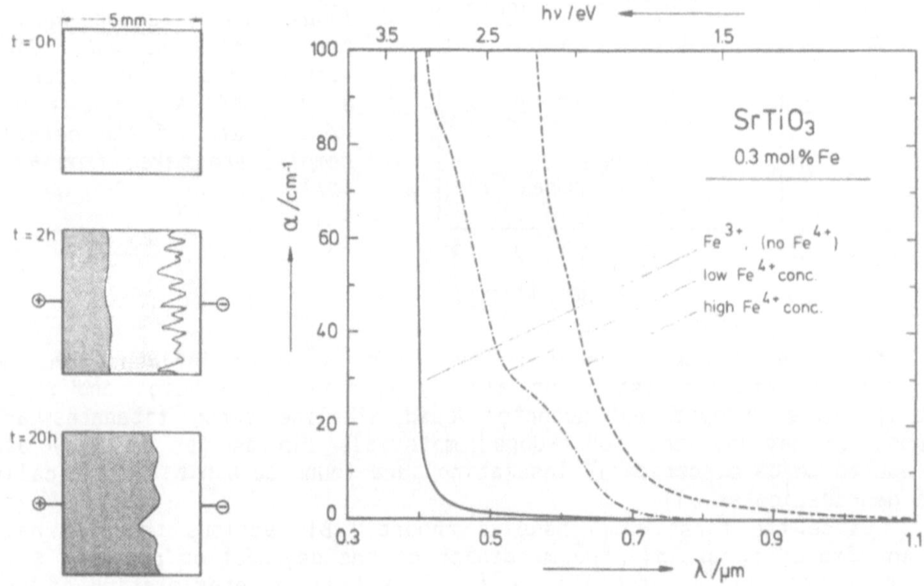

Fig. 7: Schematic sketch of the electrocolouration process of 0.3 mol% Fe-doped $SrTiO_3$ single crystal plate during degradation at T = 453 K and E = 0.125 V/μm (left hand side) and the corresponding absorption spectra (right hand side). Bright area: Fe^{3+}, (no Fe^{4+}), light hatched area: low Fe^{4+} concentration, dark hatched area: high Fe^{4+} concentration.

further information about local changes in the material during the degradation process. We have used Fe-doped $SrTiO_3$ single crystals for electrocolouration investigations since the oxidation states of Fe ions in perowskite titanates are well-known from detailed Mössbauer, EPR, optical, and thermogravimetrical studies [11, 47-49].

After annealing in oxygen, a 0.3 mol% Fe-doped $SrTiO_3$ single crystal plate of 0.5 cm thickness features a reddish-brown colour of moderate optical density. The corresponding absorption spectrum (Fig. 7) exhibits two characteristic shoulders at 0.43 and 0.59 μm which must be attributed to Fe^{4+} ions present in a certain fraction beside colourless Fe^{3+} ions. During degradation, the optical density increases at the anode and it vanishes near the cathode as sketched in Fig. 7. An electromigration of Fe ions as the cause of the electrocolouration effect can unequivocally be excluded since their mobility in the $SrTiO_3$ lattice is extremely small under the temperature and field conditions of that experiment. Hence, the electrocolouration effect during the degradation must be attributed to a change of the Fe oxidation states indicating an oxidation of the material in the anodic region and a reduction in the cathodic region.

Studies of the potential distribution between the anode and the cathode during a degradation using different techniques [50-52] reveal regions of increasing conductivity which emerge at both electrodes and which can be attributed to both, the colourless and the dark red colour region.

4. 3. Model of the Degradation Process

The findings of the electrocolouration studies and the potential distribution studies were linked to the model of the defect equilibria and defect transport properties as found in the literature. This linkage has led to a model of the degradation mechanism as sketched in Fig. 8 [53].

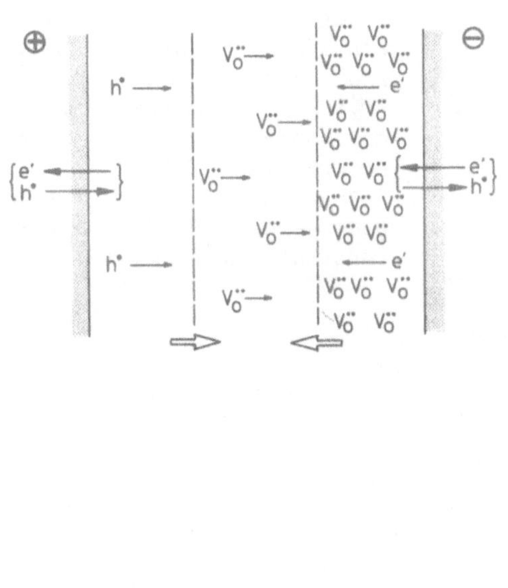

Fig. 8:
Schematic cross section of an acceptor-doped alkaline earth titanate during degradation. Only mobile defects are shown. The electromigration of $V_O^{\cdot\cdot}$ leads to a depletion in the anodic region and an enrichment near the cathode due to a (at least partial) blocking of the oxygen transfer through the metal/oxide interface. The changes of the local $V_O^{\cdot\cdot}$ concentrations result in an increase of the electronic carrier concentrations (see text).

In acceptor-doped titanates, $V_O^{\cdot\cdot}$ are presented which show a certain mobility under typical degradation conditions. An extrapolation of the diffusion coefficient of $V_O^{\cdot\cdot}$ determined at high temperatures [6, 54] down to e. g. T = 500 K and applying the Nernst-Einstein relation leads to a $V_O^{\cdot\cdot}$ mobility of approx. $3 \cdot 10^{-8}$ cm^2/Vs. Hence, a mixed ionic-electronic conduction type can be assumed. The results of the electrocolouration and potential distribution studies indicate that no stationary electromigration of $V_O^{\cdot\cdot}$ is established during degradation. Rather, the ionic transfer through the electrodes must be regarded as (at least) partially blocked. Consequently, a concentration polarization occurs as a $V_O^{\cdot\cdot}$ pile-up in the cathodic region and a depletion in the anodic region. Macroscopically, these regions of enrichment and depletion grow from the electrodes until they meet as indicated by the electrocolouration sequence in Fig. 7.

Assuming local equilibrium of the electronic defects, the electron and hole concentrations can be calculated at every position between anode and cathode from the local $V_O^{\cdot\cdot}$ concentration according to the defect model described in Sec. 2. This is indicated in Fig. 9, which shows the electronic conductivity as a function of the oxygen partial pressure pO_2^{Eq} during annealing as a result of a defect calculation for acceptor-doped $BaTiO_3$ similar to the dotted curve in Fig. 1. A fresh sample after annealing at 1 bar exhibits a conductivity as inidicated by the 'initial state' in Fig. 9. In the course of the $V_O^{\cdot\cdot}$ concentration polarization during degradation, the local equilibria depart from this point. Near the cathode, a n-conductivity develops and grows whereas in the anodic region the p-conductivity increases. A pn-junction is induced which is biased in forward direction. In contrast to Fig. 1, the calculation Fig. 9 is extended to very high oxygen partial pressures which technically are impossible to achieve. Electrochemically, however,

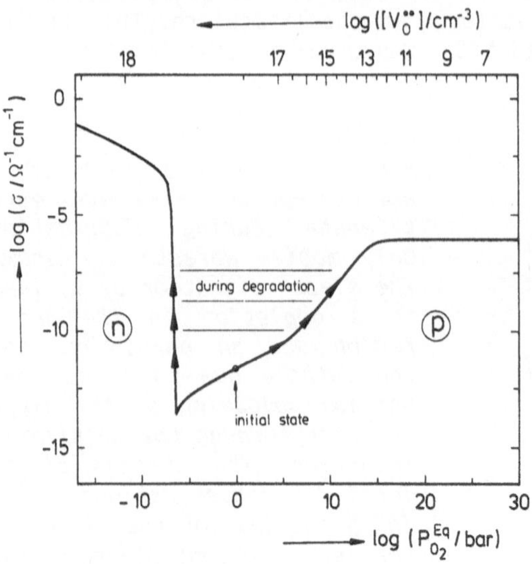

Fig. 9: Calculated electronic conductivity σ at T = 500 K for 0.01 mol% acceptor-doped $BaTiO_3$. Formal pretreatment: equilibrating at T = 1173 K and an oxygen partial pressure pO_2^{Eq} as labelled on the horizontal axis. Acceptor energy level E(Ml') = 1.3 eV with respect to the valence band edge. Further material constants: see Ref. [9].

redox potentials which formally correspond to these pO_2 values can easily be obtained by applying a few volts. Disregarding this electrochemical aspect and the possible increase in the p-conductivity, the earlier theories [38, 41-44] have tried to explain the degradation by an increase in the n-conductivity alone. According to our present understanding, the degradation phenomenon can be classified as a de-mixing process of a multicomponent phase. The de-mixing of a homogeneous phase requires the mobility of one of the components in a gradient of a generalized thermodynamic potential (which is the electrical potential in our case) and the mobile species being (at least partially) blocked from transferring across the boundary of the material. The de-mixing model of the degradation mechanism and a description of the formation of an pn-junction was first presented in Ref. [53]. It has recently been confirmed by numerical simulations [55, 56]. Details of the experimental findings and theoretical results will be published elsewhere [57].

For an increasing number of grain boundaries crossing the path between anode and cathode, a significant decrease of the degradation rate is observed (Fig. 6) although the overall characteristics of the degradation process remains the same. This effect can be explained by findings which show that the $V_O^{\cdot\cdot}$ diffusion coefficients are significantly lower for ceramics than for single crystalline titanates [6, 54, 58]. Obviously, the grain boundaries act as barriers of moderate height for a $V_O^{\cdot\cdot}$ flux independent of the driving force being a $V_O^{\cdot\cdot}$ concentration gradient or an electrical field.

REFERENCES

[1] L. C. Walter and R. E. Grace, J. Phys. Chem. Solids **28**, 239 (1967).

[2] H. Veith, Z. angew. Phys. **20**, 16 (1965).

[3] N. G. Eror and D. M. Smyth, presented at the 70th Annual Meeting, The American Society, Chicago, April 20-25, 1968 for abstract see: Amer. Ceram. Soc. Bull. **47**, 354 (1968).

[4] S. A. Long and R. N. Blumenthal, J. Am. Ceram. Soc. **54**, 515 and 577 (1971).

[5] A. M. J. H. Seuter, Philips Res. Rept. Suppl. 3, 1 (1974).

[6] R. Wernicke, PhD Thesis, Rheinisch-Westfälische Technische Hochschule Aachen, 1975.

[7] N.-H. Chan and D. M. Smyth, J. Electrochem. Soc., **123**, 1584 (1976).

[8] F. A. Kröger and H. J. Vink, p. 307 in Solid State Physics, Vol. 3. Edited by F. Seitz and D. Turnbull, Academic Press, New York, 1956.

[9] H.-J. Hagemann, PhD Thesis, Rheinisch-Westfälische Technische Hochschule Aachen, 1980.

[10] N.-H. Chan, R. K. Sharma, and D. M. Smyth, J. Am. Ceram. Soc., **64**, 556 (1981).
[11] H.-J. Hagemann and D. Hennings, J. Am. Ceram. Soc. **64**, 590 (1981).
[12] N. G. Eror and D. M. Smyth, pp. 62-74 in 'The Chemistry of Extended Defects in Non-Metallic Solids', Edited by L. Eyring and M. O'Keefe, North-Holland, Amsterdam, 1970.
[13] R. Wernicke, phys. stat. sol (a) **47**, 139 (1978).
[14] N.-H. Chan and D. M. Smyth, J. Am. Ceram. Soc., **67**, 285 (1984).
[15] N.-H. Chan, M. P. Harmer, and D. M. Smyth, J. Am. Ceram. Soc. **69**, 507 (1986).
[16] R. Waser, J. Am. Ceram. Soc. **71**, 58 (1988).
[17] R. Waser, Ber. Bunsenges. Phys. Chem. **90**, 1223 (1986).
[18] R. Waser, Z. Naturforsch. **42a**, 1357 (1987).
[19] H.-J.Hagemann, D.Hennings, and R.Wernicke, Philips Techn. Review **41**, 89 (1983/84).
[20] K. A. Müller, W. Berlinger and J. Albers, Phys. Rev. B **32**, 5837 (1985).
[21] H. Ikushima and S. Hayakawa, J. Phys. Soc. Japan **27**, 414 (1969).
[22] D. Hennings and H. Pomplun, J. Am. Ceram. Soc., **57**, 527 (1974).
[23] G. Arlt and H. Neumann, submitted to Ferroelectrics.
[24] K. Carl and K.-H. Härdtl, Ferroelectrics **17**, 729 (1978).
[25] H. Dederichs and G. Arlt, Ferroelectrics, **68**, 281 (1986).
[26] K. Okazaki, K. Sakata, Elektrotechn. Journ. Jap. **7**, 13 (1962).
[27] B. Jaffe, W. R. Cook Jr. and H. Jaffe, Piezoelectric Ceramics, ed. J.P. Roberts and P. Popper, p.28, Academic Press, London 1971.
[28] G. Arlt, D. Hennings and G. de With, J. Appl. Phys.**58**, 1619 (1985).
[29] H. Diamond, J. Appl. Phys. **12**, 909 (1961).
[30] G. Arlt and H. Peusens, Ferroelectrics **48**, 213, (1983).
[31] E. Andrich and K.-H. Härdtl, Philips Techn. Rdsch. **25**, 368 (1963/64).
[32] H. Ihrig and M. Klerk, Appl. Phys. Letters **35**, 307 (1979).
[33] W. Heywang, Solid State Electronics **3**, 51 (1961).
[34] G. H. Jonker, Solid State Electronics **7**, 895 (1964).
[35] D. Hennings, A. Schnell and H. Schreinemacher, Europ. Pat. No. 0065806, Nov. 11 (1985).
[36] L. M. Levinson and H. R. Philips, Ceramic Bulletin **66**, 639 (1986).
[37] A. Vogel, Diploma Thesis, Universität Karlsruhe, 1985.
[38] K. Lehovec and G. A. Shirn, J. Appl. Phys. **33**, 2036 (1962).
[39] E. Loh, J. Appl. Phys. **53**, 6229 (1982).
[40] H. Neumann and G. Arlt, Ferroelectrics **69**, 179 (1986).
[41] D. A. Payne, Proceedings of the Sixth Annual Reliability Physics Symposium, IEEE, California, 257 (1968).
[42] M. P. Harmer and D. M. Smyth, Proceedings of the 4th International Conference in Reliability and Maintainability, Rennes-Guirec, 132 (1984).
[43] J. Rödel and G. Tomandl, J. Mater. Sci. **19**, 3515 (1984).
[44] H. Y. Lee and L C. Burton, IEEE Trans. Components, Hybrids, Manuf. Technol. **CHMT-9**, 469 (1986).

[45] K. A. Müller, Th. von Waldkirch, W. Berlinger, and B. W. Faughnan, Solid State Commun **9**, 1097 (1971).
[46] J. Blanc and D. L. Steabler, Phys. Rev. B **4**, 3548 (1971).
[47] K. A. Müller, W. Berlinger, and R. S. Rubins, Phys. Rev. **186**, 361 (1969).
[48] B. W. Faughnan, Phys. Rev. B **4**, 3623 (1971).
[49] C. T. Luiskutty and P. J. Ouseph, Solid State Comm. **13**, 405 (1973).
[50] Y. Goto and S. Kachi, J. Phys. Chem. Solids **32**, 889 (1971).
[51] J. Freitag, Diploma Thesis, Universität Karlsruhe, 1987.
[52] A. Raith, P. Reijnen, and R. Waser, submitted to: Ber. Bunsenges. Phys. Chem.
[53] R. Waser, Ceramics Colloquium, MPI für Festkörperforschung, Stuttgart, March 19-20 (1987).
[54] A. Müller, PhD Thesis, Universität Karlsruhe, 1988.
[55] K. Wirth, Diploma Thesis, Universität Karlsruhe, 1988.
[56] T. Baiatu, PhD Thesis, Universität Karlsruhe, 1988.
[57] T. Baiatu, K. H. Härdtl, and R. Waser, to be published.
[58] C. Schaffrin, phys. stat. sol. (a) **35** (1976) 79.

STATISTICAL THERMODYNAMICS OF NON-STOICHIOMETRIC OXIDES

G.BOUREAU, M.BENZAKOUR and R.TETOT
Laboratoire des Composés non-stoechiométriques
Bâtiment 415
Université de Paris-Sud
Centre d'Orsay
91405 ORSAY Cedex
France

ABSTRACT. Some problems of statistical thermodynamics applied to non-stoichiometric oxides are examined: Two typical oxides, titanium monoxide and ceria, respectively metallic and ionic oxides, are used to discuss the role of short range and of long range interactions. A particular attention is paid to coulombic interactions which cannot be neglected in a number of ionic compounds. The validity of the use of the ideal mass action law as a tool of investigation is also discussed.

INTRODUCTION

Precise values of partial molar quantities of oxygen, free energy and enthalpy as functions of the composition are now available for a number of oxides (1,2). These rapid progress of experimental techniques justifies attempts to improve the fiability of statistical models applied to non-stoichiometric oxides. Particularly in this paper, three questions have been examined:

1-Are thermodynamic models credible? A simple example shows that the answer to this question is not obvious: For non stoichiometric ceria, at least three statistical models have been proposed and are in good agreement with experiments after their authors (3-5).

2-What informations is it possible to get from partial molar quantities?

3-Is the mass action law acceptable? This formalism has been largely used particularly since the publication of the classical book of Kofstad (6). It is clear that this approach is correct only if interactions can be neglected.

J. Nowotny and W. Weppner (eds.),
Non-Stoichiometric Compounds Surfaces, Grain Boundaries and Structural Defects, 155–161.

BASIS OF THERMODYNAMIC MODELS

Generally, the degeneracy of low energy states is small compared with those of high energy, which implies that minimum energy and maximum entropy tendancies are conficting. Therefore the Gibbs free energy which is a compromise between energy and entropy requirements is not very usefull in statistical thermodynamics and it is very common that macroscopic models are consistent with the free energy but not with enthalpy and entropy separatly (7,8). In this respect, it is necessary to use simultaneously partial molar enthalpy and entropy as basis of statistical models.

SHORT RANGE INTERACTIONS

Frequently, long range interactions between defects are small. For instance, this is probably the case in oxides with a high dielectric constant or in metallic oxides such as titanium monoxide. Therefore, the variations of partial molar quantities as function of the composition have to be understood only in term of short range interactions. An alternative possibility would be to consider a variety of different non-interacting defects, each of

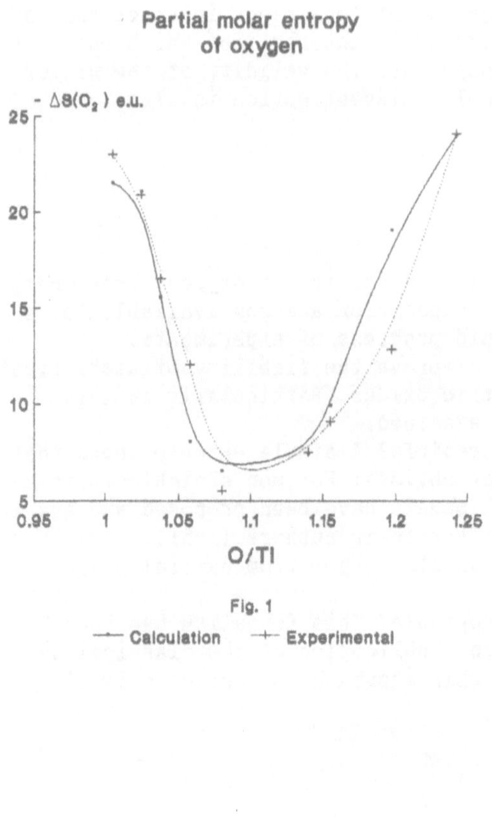

Fig. 1

them prevailing in a small range of composition, but in the absence of experimental evidence, such an approach would be an adhoc device.

In a recent paper, Boureau and Tetot (9) have exposed a new statistical approach aimed to extract informations from thermodynamic data in this case. Let it be recalled that the method consists in building the partitition function of a cube immersed in a huge reservoir. Temperature, oxygen pressure and number of sites are kept constant. Typically, we are dealing with a cube made of 64 or 96 sites. All the configurations are gathered in a file and classified ito categories according to the numbers of interactions and the numbers of defects of each type. In this way, it is possible to calculate the partition function as a function of the interaction energies. Fitting the partial molar quantities allows thus to evaluate the values of interaction energies which are supposed to be composition independent.

This method has been applied to titanium monoxide. This oxide has an interesting characteristics: At the stoechiometry O/Ti=1.00, there are 15% vacancies on each sublattice. Let the major results of our calculations be recalled: There is a strong repulsion between vacancies of the same type in first and second nearest neighbor position and also between unlike vacancies in first position. On the contrary, other interactions involving unlike vacancies are attractive. The formation of a pair of unlike vacancies is slightly exothermic (0.6 e.v.) as predicted by Huisman et al (10) from band calculations.

In figure 1, the fit of the partial molar entropy of oxygen has been compared with the experimental results of Tetot et al (11). It is noteworthy that the numbers of vacancies are not fitted. They are only consequence of calculations. Therefore the observed agreement with experiment increases the degree of confidence of the method.

COULOMBIC CONTRIBUTION

Point defects are generally charged (6). Therefore if the dielectric constant is small, coulombic interactions have to be considered. Strangely enough, this fact has been overlooked in most statistical treatments. The simplest approach would be the use of the Debye-Huckel model. Unfortunately, such a model may be used only for very dilute solutions, typically less than 10^{-4} (12). In this composition range, the role of impurities is difficult to eliminate. Therefore it is mandatory to use well established methods such as the Ewald summation method or minimum image method with a rhombic dodecahedron (13). We have used these two methods in the case of cerium dioxide and checked that they are in mutual agreement. In this oxide, from a number of independent studies (14-16), it has been concluded that the defects are doubly ionized vacancies, the electrons dissociated from the vacancies being localized on cationic sites and behaving as small polarons. It is well known (17) that fractional charges are seen in ionic compounds. In our treatment this is simply equivalent to use a larger dielectric constant. From a thermodynamic point of view, the important feature to be considered is that the departure of an oxygen atom involves the creation of on oxygen vacancy and of two polarons. Coulombic interactions are not the only interactions to be considered. It has been shown (18) that in spite of opposite claims (3) they cannot, alone, explain the partial ordering of defects. Short range interactions have simultaneously to be be considered, particularly in highly defective ceria. This part of calculations are very similar to the ones described in the above section. In this way, a rather complex model emerges, the discussion of which would be beyond the scope of this paper. Let it be only stated that, following Atlas (3), we have used a value of the dielectric constant equal to 35. As it may be seen from figure 2, this value is in qualitative agreement with the experimental variations of partial molar enthalpy in dilute solutions measured by Bevan and Kordis (19). It should not be forgotten that in this composition range, the experimental results are particularly difficult to obtain as the oxygen pressure changes dramatically as a function of the departure from stoichiometry. Moreover, in order to get a quantitative agreement, it would be necessary to take into account the role of impurities such as Ca ions (20) and short range interactions.

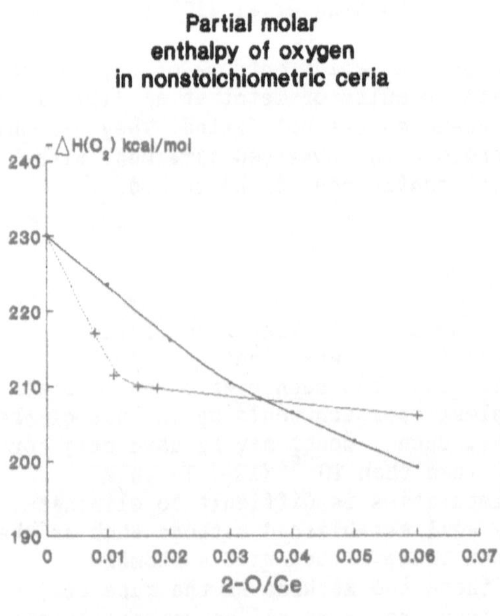

Fig. 2

DISCUSSION OF THE MASS ACTION LAW APPROACH

If the interactions between defects are taken equal to zero, then the defects are randomly distributed and the configurational contribution to the partial molar entropy may be written:

$$\Delta S(O_2)=-2\,R\,Ln\,\frac{2-x}{x}\left[\frac{1-2x}{2x}\right]^2$$

while the partial molar enthalpy may be written:

$$\Delta H(O_2)=a=constant$$

From the usual relation:

$$\Delta G(O_2) = \Delta H(O_2) - T.\Delta S(O_2)$$

it may be written:

$$RTLnpO_2=constant-6RTLnx$$

Therefore the slope $dLnpO_2/dLnx$ should be equal to -6. In the same way, it could be shown that once ionized vacancies and neutral vacancies correspond respectively to slopes equal to -4 and -2.

Let us examine how these slopes are affected in dilute solutions by coulombic interactions.

In a small composition domain, $\Delta H(O_2)$ may be written:

$$\Delta H(O_2)=a+bx$$

If the assumption of randomly distributed defects is kept, the partial molar free energy may be written:

$$RTLn\;pO_2=constant+bx-6RTLnx$$

and the slope has now to be written:

$$dLnpO_2/dLnx=b/RT.x-6$$

Let be assumed a value of b equal to (500 000 cal/mol. At the temperature T=1353 K, the following corrections have to be taken into account:

x	10^{-3}	5.10^{-3}	10^{-2}
x.b/RT	0.2	1.0	2
observed slope	-5.8	-5.0	-4

In fact a slope equal to -5 has been observed in several studies (15,21,22) and following Kevane (22) has generally been explained as caused by an equilibrium between oxygen vacancies of various states of ionization. As it may be seen from the above considerations, this analysis is not correct

So what may be kept from mass action law approach in this case? Not a lot: In very dilute solutions, it is legitimate to use slopes such as $d\log pO_2/d\log x$ to determine the nature of the defects.

SUMMARY AND CONCLUSIONS

In order to build credible statistical models, it is necessary to eliminate large errors in the quantitative treatments, which is possible with the avaibility of efficient computers. For instance, it is not acceptable to use the ideal mass action law in a nonstoichiometric oxide having a dielectric constant smaller than 50 for departures from stoichiometry larger than 0.005. Short and long range interactions can easily be dealt rigorously with Monte-Carlo methods. Thus it is possible to restrict the discussion of models to the physical basis of the assumptions. The optimisation method used seems to be a fruitful method of investigation of the non-stoichiometric compounds.

In very dilute range, the mass action law may be used to make prediction on the nature of the defects, but calculations considering a number of defects in equilibrium are highly speculative.

REFERENCES

1. C.B. Alcock,' The control of stoichiometry in oxide systems', *This conference*

2. R.Tetot, C.Picard, G.Boureau and P.Gerdanian, 'High temperature calorimetry of metal-oxygen systems', *Advances in Ceramics* ,23,455 (1987)

3. L.Atlas, 'Statistical model of partially ordered defects in a hypostoichiometric metal oxide', *J. Phys. Chem. Solids* 29, 91 (1968)

4. J.Campserveux and P.Gerdanian, 'Etude thermodynamique de l'oxyde CeO_{2-x} pour 1.5<O/Ce<2' *J. Solid St. Chem.* 23, 73 (1978)

5. L.Mannes, E.Partelli and C.M.Mari, 'A new statistical thermodynamic theory for substoichiometric fluorite structure compounds and its application', *Mater. Chem.* 6, 381,401,417 (1981)

6. P.Kofstad 'Nonstoichiometry, *diffusion and electrical conductivity in binary metal oxides*, Wiley (1972)

7. G.Boureau and J.F.Marucco, 'Equilibrium between point defects and crystallographic shear planes', *Radiat. Eff.* 74, 247 (1983)

8. G.Boureau, 'Applications of the calculation of the configurational entropy of hydrogen in metals', *Z. fur Physik. Chemie N.F.*,**143**, 89 (1985)
9. G.Boureau and R.Tetot, 'Statistical thermodynamics of nonstoichiometric oxides with high defect content. A new approach', *Cryst. Latt. Def. and Amorph. Mat* 16, 85 (1987)
10. L.M.Huisman, E.A.Carlsson, C.D.Gelatt and H.Ehrenreich,'Mechanism for energetic-vacancy stabilization: TiO and TiC' *Phys. Rev. B* 22, 991 (1980)
11. R.Tetot, C.Picard and P.Gerdanian, 'Determination of oxygen partial free energfy for non-stoichiometric TiO by e.m.f. measurements', *J. Phys. Chem. Solids,* **44**, 1059 (1983)
12. H.Schmalzried and A.Navrotsky, *Festkorperthermodynamik,* Verlag Chemie, Weinheim (1975)
13. D.J.Adams, 'Computer simulation of ionic systems: The distorting effects of the boundary conditions' *Chem. Phys. Lett.* ,329 (1979)
14. B.Touzelin, 'Etude par diffraction des rayons X à haute température en atmosphère controlée du système Ce-O', *J. Nucl. Mater.* ,101, 92 (1981)
15. H.L.Tuller and A.S.Nowick, 'Defect structure and electrical properties of nonstoichiometric CeO_2 single crystals', *J. Electrochem. Soc.,* 126, 209 (1979)
16. F.Millot and P.Gerdanian, 'The quantitative measurement of electromigration in CeO_{2-x}', *J. Phys. Chem.* **43**, 507 (1982)
17. J.B.Pendry and C.H.Hodges,' The quantisation of charge transport in ionic systems' *J. Phys. C,* 17 ,1269 (1984)
18. M.Benzakour, R.Tetot and G.Boureau, 'Statistical thermodynamics of cerium dioxide' *J. Phys. Chem. Solids,* **49**, 381 (1988)
19. D.J.Bevan and J.Kordis, 'Oxygen dissociation pressures and phase relationships in the system CeO_2-Ce_2O_3 at high temperatures' *J. Inorg. Nucl. Chem.,* **26**, 1509 (1964)
20. I.Riess, H.Janczikowski and J.Nolting, 'O_2 chemical potential of nonstoichiometric ceria, CeO_{2-x}, determined by a solid electrochemical method', *J. Appl. Phys.,* 61, 4931 (1987)
21. J.W.Dawicke and R.N.Blumenthal, 'Oxygen association pressure measurements on non-stoichiometric cerium dioxide' *J. Electrochem. Soc.* 133, 904 (1986)
22 C.J.Kevane, 'Oxygen vacancies and electrical conduction in metal oxides' ,*Phys. Rev A,* 133, 1431 (1964)

8. [illegible] Applications of the calculation of the configurational entropy of hydrogen in metals [illegible] (1963).

9. [illegible] with high [illegible] [illegible] (1963).

10. [illegible] low energy [illegible] Phys. [illegible] (1970).

11. [illegible] partial free energy [illegible] measurements [illegible] Phys. Chem. Solids, 44, 1153-1158 [illegible]

12. [illegible]

13. [illegible]

14. [illegible]

15. [illegible] properties [illegible]

16. [illegible] Phys. Chem. 42 [illegible]

17. [illegible]

18. [illegible] Phys. Chem. Solids [illegible]

19. [illegible] and phase relationships in the system [illegible] Phys. Chem. [illegible]

20. [illegible]

21. [illegible] pressure [illegible]

22. [illegible]

ELECTRICAL CONDUCTIVITY STUDY OF COBALT MOLYBDATE $CoMoO_4$

A. Steinbrunn, M. Bindo and J.C. Colson
Laboratoire de Recherches sur la Réactivité des Solides
Faculté des Sciences Mirande
B.P. 138
21004 DIJON CEDEX

ABSTRACT. Electrical conductivity of Cobalt Molybdate has been measured in the temperature range 100-600°C on pressed pellets of polycristalline sample. It has been found that $CoMoO_4$-a is a p-type semiconductor while $CoMoO_4$-b is a n-type semiconductor.

Different conduction mechanisms have been found depending on the temperature and the oxygen partial pressure. The nature of the major defects structures is discussed for both polymorphic forms of $CoMoO_4$.

Introduction

In order to improve our understanding of the preparation and functioning of hydrodesulphurization catalysts based on molybdenum, we investigated the electrical conductivity of polycrystalline samples of both varieties $CoMoO_4$-a and $CoMoO_4$-b. Despite the great deal of characterizations that have been carried out on the $Co\text{-}Mo/Al_2O_3$ catalysts, the electrical properties data are rather rare in the literature (1.2.3). The present study deals with electrical conductivity measurements of unsupported cobalt molybdate $CoMoO_4$ as a function of oxygen partial pressure and in a temperature range commonly used during the work of this kind of catalyst.

1. Experimental

1.1. Preparation and characterization of $CoMoO_4$

The $CoMoO_4$ samples used for this study were prepared by the direct solid-solid reaction from a mixture of Co_3O_4 and MoO_3 powder as described in the literature (1, 4). For this study, the Co_3O_4 and MoO_3 oxide powders were ground to fine grains and mixed to proportions so that the atomic ratio $\frac{Co}{Co + Mo}$ is equal to 0.3. The synthesis is realized under atmospheric pressure of air at 600°C during 16 hours. For these conditions, the reaction product is the pink-coloured

J. Nowotny and W. Weppner (eds.),
Non-Stoichiometric Compounds Surfaces, Grain Boundaries and Structural Defects, 163–171.

$CoMoO_4$-a variety. The green-coloured $CoMoO_4$-b variety was obtained from the previous one by mechanical crushing.

The X-Ray diffraction characterizations lead to the data of the literature (5 to 10). They show that the main reaction product is the $CoMoO_4$-a variety with residual traces of Co_3O_4 and MoO_3. The purity of our reaction products is at least equal to the source materials, i.e. :

For MoO_3 :	Cl	: max0.002 %	Co_3O_4 :
	SO_4	: max 0.01 %	
	Pb	: max0.001 %	Johnson Matthey
	Fe	: max0.0005 %	grade 1 15 ppm
	NH_4	: max0.005 %	
	P,As,Si	: max0.002 %.	

1.2. Measurements of the electrical conductivity

Electrical measurements were made by the direct current DC two probes method from polycrystalline samples obtained by pressing the oxide powders at about 10 N/cm^2 and sintering the pellet (diameter 8 mm, thickness 1.2 mm) at 600°C under vacuum for 10 hours. The electrical conductivity cell was mounted without any vacuum greased junction from a model described earlier by Juillet (11). The sample heating is performed by an annular furnace. Temperature was monitored using a Pt/Pt Rh 10 % thermocouple which was positioned as close to the powder as possible. DC conductivity measurements were carried out by using a simple circuit by application of Ohm's Law. The stabilized current was delivered by a Keithley source while the potential was continuously recorded by a Keithley 616 electrometer. The samples are maintained in a quartz tube in which gas can be streamed by an adjustable leak valve in the pressure range 10^{-3} - 10^{-5} Pa.

2. Results

The electrical conductivity (σ) of $CoMoO_4$ is found to be extremely dependent on the surrounding atmosphere, especially on oxygen.

2.1. Oxygen pressure dependence of σ

. $CoMoO_4$-a variety :

The electrical conductivity of $CoMoO_4$-a increases with increasing oxygen partial pressure at a given temperature. For temperatures lower than 450°C, the conductivity values are very low. It is only for temperatures higher than 500°C and for oxygen pressure in the 10^{-3} - 10^{-5} Pa range that they are detectable. Fig. 1 represents log σ as a function of log p_{O_2} for different temperatures. The values mentionned were obtained for increasing and decreasing p_{O_2}. The conductivity slightly increases up to 100 Torr (1 Torr = 133,3 Pa), $\sigma \simeq k_a p_{O_2}^{1/20}$. For higher oxygen pressure, the conductivity still increases as $k'_a p_{O_2}^{1/5}$.

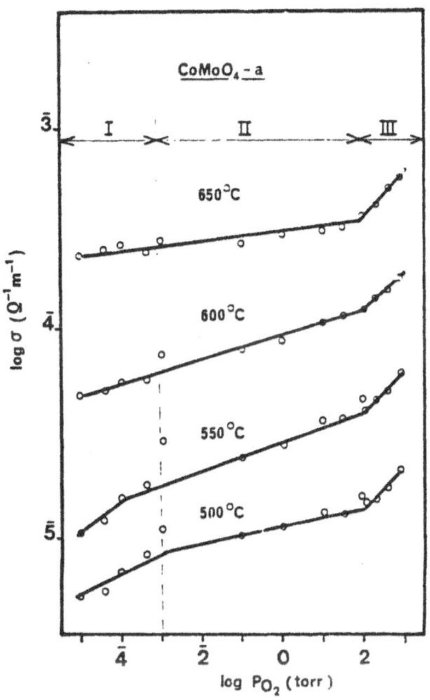

Figure 1. log σ = log P_{O_2} for $CoMoO_4$-a

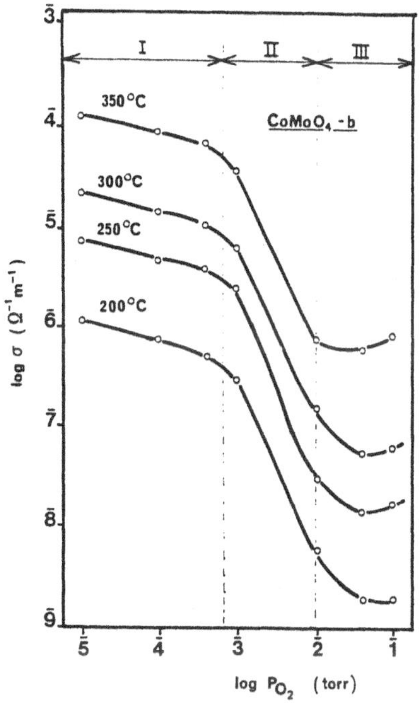

Figure 2. log σ = log P_{O_2} for $CoMoO_4$-b

. $CoMoO_4$-b variety :

Contrarily to the previous case, the electrical conductivity of $CoMoO_4$-b diminishes with increasing oxygen partial pressure. For temperatures lower than 200°C, the conductivity values are too low to be registered. Fig. 2 shows log σ as a function of log p_{O_2}. When the oxygen pressure is lower than 10^{-3} Torr, the oxygen conductivity dependence is k_b $p_{O_2}^{-1/5}$. The conductivity values are reversible for increasing and decreasing oxygen pressures.

2.2. Temperature dependence of σ

. $CoMoO_4$-a variety

As for the oxygen range, three domains have to be considered : one at intermediate pressures (10^{-3} - 100 Torr) and two others at low and high pressure ($p_{O_2} < 10^{-3}$ Torr, $p_{O_2} > 100$ Torr). In the intermediate oxygen pressure range, the experimental plot ($\ln \sigma = f(T^{-1})$) give straight lines in the 400-700 temperature range (Fig. 3). The deduced activation energy varies between 1.14 and 1.40 eV.

For the extreme pressure ranges, the plots present discontinuities at about 500°C (Fig. 4). The activation energy lies at about 0.95 eV for temperatures lower than 500°C, while their value are respectively 1.14 to 2.20 eV and 1.40 for pressures lower than 10^{-3} and higher than 100 Torr.

The σ values were registered during the increase of temperature. They do not coincide with the decreasing ones.

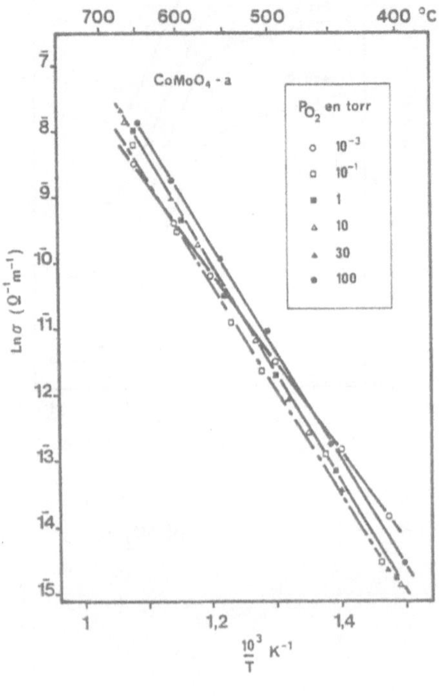

Figure 3. Ln σ = f(1/T) for $CoMoO_4$-a

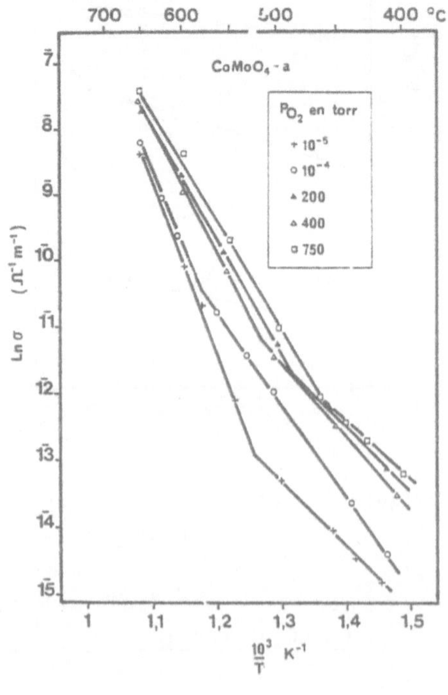

Figure 4. Ln σ = f(1/T) for $CoMoO_4$-a extreme pressure ranges.

. $CoMoO_4$-b variety :

Fig. 5 shows ln σ as a function of T^{-1} for different oxygen partial pressure ($10^{-5} < p_{O_2} < 10^{-1}$ Torr) in the 100 - 400°C temperature range. The activation energy for temperatures lower than 250°C varies in the 0.20 - 0.55 eV range as the oxygen pressure increases $p_{O_2} > 10^{-2}$ Torr. For lower oxygen pressure $p < 10^{-2}$ Torr, the activation energy is higher : 0.8 to 1.04 eV.

3. Interpretation and discussion

The comparison of the conduction type for the two molybdate varieties shows that $CoMoO_4$-b is a n-type semiconductor while $CoMoO_4$-a is a p-type semiconductor.

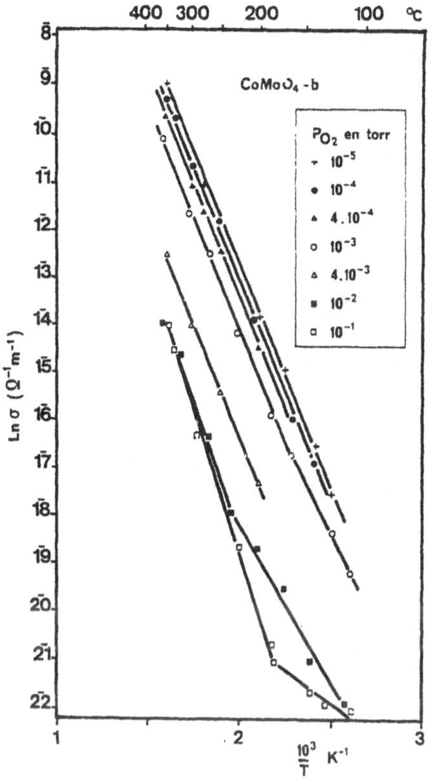

Figure 5. Ln σ = f(1/T) for $CoMoO_4$-b

3.1. Interpretation of the Partial oxygen pressure dependence results

The nature of the major point defect for both varieties $CoMoO_4$-a and -b depends on the experimental parameters T and p_{O_2}.

. $CoMoO_4$-a variety :

The plot of log σ = f log(p_{O_2}) (Fig. 1) shows three regimes where the conductivity σ is proportional to $p_{O_2}^{1/n}$ with different values for n.

- For regime I, low oxygen partial pressure ($p_{O_2} < 10^{-3}$ Torr), the n values are respectively 9 ± 0.1 and 7.8 ± 0.1 for T = 500°C and 550°C. The defects involved in that case might be more or less complex cationic defects due to the fact that Cobalt is able to occupy an interstitial position in the $CoMoO_4$-a structure. The n values do not allow us to give the ionisation degree of these defects.

- In the intermediate pressure range (regime II), the conductivity obeys a similar law with n equal respectively to 14 and 24 for T

= 500°C and 650°C. These high values of n are far from the theorical ones, 4 and 6, predicted respectively in the case of cobalt cationic vacancies singly or doubly ionized. Consequently, different possibilities have to be discussed :

* the structure defect is really complex (Interstitial Co^{2+} associated with cationic vacancy...)

* the major defect population is a mixture more or less equivalent of cationic and anionic doubly charged vacancies,

* the slope 1/n values are "apparent" if one considers that the measured electrical conductivity is due to two components, one independent of the partial pressure and the other dependent, so as $\sigma = \sigma o = f(p_{O_2})$. The variable term could be due to a pression dependence of the hole mobility (12) or to segregation phenomena taking place at the grain boundaries of the pressed pellet.

All these hypothesis have to be taken into account in order to explain the low values obtained for n.

- For regime III, i.e. the highest partial oxygen pressures, the n values are respectively equal to 4.6 ± 0.1 and 5.0 ± 0.1 for T = 650 and 500°C. These values, closer to the classical values of n, are characteristic of singly and doubly ionized cationic vacancies of cobalt.

The value n = 5 being obtained the most frequently, that allows us to think that the major defect population is constituted by a mixture of V'_{co} and V''_{co}.

. <u>$CoMoO_4$-b variety</u> :

The oxygen partial pressure dependence of σ (Fig. 2) shows that there are three regimes : $\sigma = K\, p_{O_2}{}^{-\frac{1}{n}}$ with different n values.

- For low oxygen pressures (Regime I) the n values are respectively equal to 4.5 ± 0.1 and 5 ± 0.1 at T = 200 and 300°C. These values are characteristic of oxygen vacancies $V_O^{\cdot}$ and $V_O^{\cdot\cdot}$. The proportion of the mixture cannot be deduced from our n values.

- For regime II, the n values are in the range 0.52 ± 0.02 and n = 0.60 ± 0.02. These low values permit us to think that the anionic oxygen vacancies are not the major defect any more. The nature and the ionisation degree of the vacancies are difficult to deduce from the experimental data.

- For regime III, the conductivity starts to change from n type to p type due to the increasing oxygen partial pressure.

3.2. <u>Activation energy and conduction mode</u>

. <u>$CoMoO_4$-a variety</u>

From the plots of Figures 3 and 4, the fairly good alignment obtained in the 400-700°C temperature range for oxygen partial pressure in the 10^{-3} - 100 Torr range is in favour of a band conduction mode due to ionized defects.

The discontinuities observed on Figure 4 clearly show that the conduction mode is different for $T < 500°C$. Indeed, the slight increase of electrical conductivity at low temperatures could be due to a weaker mobility. A conduction mode by small polarons (Holstein mode 13) would be :

$$\sigma = AT^{-3/2} \exp - \frac{Em}{kT}$$

with Em : activation energy of mobility.

By plotting $\log \sigma T^{3/2}$ versus T^{-1} for $T < 500°C$, we got a better alignment of the experimental points (Fig. 6).

The activation energy is between 1.00 ± 0.02 eV and 1.30 ± 0.02 - eV. The hopping mechanism of small polarons should be less probable in that temperature range. The activation energy is rather high ($\simeq$ - 1 eV) compared with the commonly retained values (0.3 eV). In that case, the $CoMoO_4$-a conduction mode for $T < 500°C$ is extrinsic conduction mode due to a small polaron band built from lattice defects. These latter ones could be due to cobalt atoms anormally positioned at tetrahedral site.

Haber and Ziolkowski (14) have shown by IR and magnetic susceptibility measurements that Co^{2+} ions can occupy tetrahedral sites in the $CoMoO_4$-a structure. Dziembaj and Ziolkowski (15) as well noticed that during the Co_3O_4-MoO_3 solid-solid reaction Co^{2+} ions randomly insert into the $CoMoO_4$ lattice and that at every stage of the reaction some Co^{2+} ions would keep their Co_3O_4 original tetrahedral coordinence and so would create structure defects.

In the intermediate pressure range (10^{-3} - 100 Torr), in the whole 400-700°C (temperature range and for $p_{O_2} < 10^{-3}$, $p_{O_2} > 100$ Torr at $T > 500°C$, the conduction mode can be considered as intrinsic. The activation energy values are in the 1.14, 2.20 range. These high values suggest a normal band conduction. The charge carriers are holes coming from the cobalt cation vacancies ionization. Molybdenum is considered without any effect in that process (highest oxidation degree).

. $CoMoO_4$-b variety :

For oxygen pressure lower than 10^{-2} Torr and for the 100-350 temperature range, the conduction mode can be considered as intrinsic with an activation energy of 0.8 eV. The charge carriers are electrons coming from the oxygen vacancies. As soon as the oxygen pressure is higher than 10^{-2} Torr, the electrical conductivity values diminish rapidly.

For $T < 250°C$, the activation energy is lower, from 0.20 to 0.55 eV. As for $CoMoO_4$-a, a change in the conduction mode occurs. It becomes a small polaron conduction mode, as shown by the Fig. 7. The deduced activation energy is 0.25 and 0.32 ± 0.02 eV. These values have the same order of magnitude as those given in the literature (0.3 eV) (16, 17, 18).

Our results are in agreement with those of Keem and Goodenough (19, 20) for a small polaron hopping mechanism.

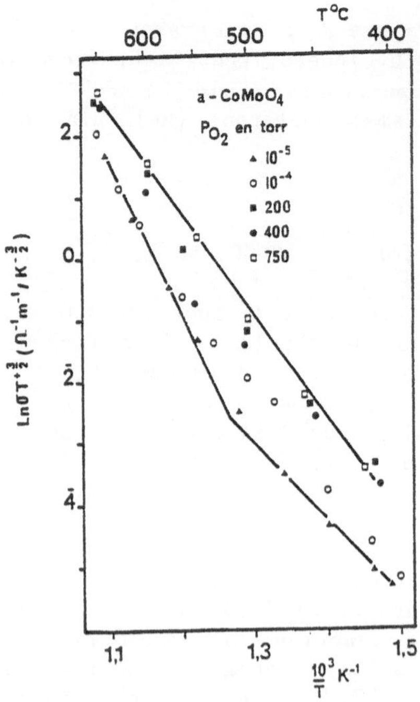

Figure 6. Ln T $+\frac{3}{2}$ = f($\frac{1}{T}$) for $CoMoO_4$-a

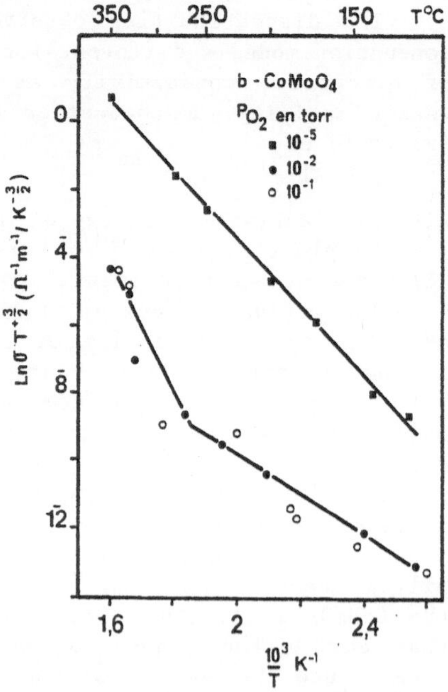

Figure 7. LnσT $+\frac{3}{2}$ = f($\frac{1}{T}$) for $CoMoO_4$-b

This conduction mode has been previously described by Sayer for MoO_3 (21) which is a n-type semiconductor as $CoMoO_4$-b.

Conclusion

The electrical properties of both varieties $CoMoO_4$-a and $CoMoO_4$-b were investigated as a function of temperature and oxygen partial pressure. The $CoMoO_4$-a variety is a p-type semiconductor while $CoMoO_4$-b is a n-type semiconductor.

. For the $CoMoO_4$-a variety, the structure of the major defects is a function of the oxygen partial pressure. For p_{O_2} < 100 Torr, the defects either would have a complex structure or may be considered as a mixture of oxygen vacancies $V_O^{\bullet\bullet}$ and cobalt vacancies V_{Co}''. On the other hand, for higher pressures (p_{O_2} = 100 - 750 Torr), the defects are constituted mainly by a mixture of cobalt cation vacancies mono and biionized.

Two conduction regimes were detected :
- an "intrinsic" one : $10^{-3} < p_{O_2} < 100$ Torr for 400°C < T < 700°C
- an "extrinsic" one : $p_{O_2} < 10^{-3}$; $p_{O_2} > 100$ Torr for T < 500°C.

. For the $CoMoO_4$-b variety, the structure of the major defects is also a function of the oxygen partial pressure.

For $p_{O_2} < 10^{-3}$ Torr, the defects might be a mixture of oxygen anionic vacancies $V_O^{\bullet}$ and $V_O^{\bullet\bullet}$.

For $p_{O_2} > 10^{-3}$ Torr, the anionic vacancies concentration would decrease in order to promote the cationic ones.

Two conduction regimes were also detected :
- an "instrinsic" one $p_{O_2} < 10^{-2}$ Torr and 100°C < T < 350°C,
- an "extrinsic" one $p_{O_2} > 10^{-2}$ Torr for T < 250°C.

At low temperature, T < 250°C, the conduction mode might be a classical mode (ionized defect band), the carrier being electrons associated to oxygen vacancies (each = 0.8 to 1 eV).

REFERENCES

1. D. INGRAIN, G. THOMAS, Ann. Chim. Fr., **6**, 1981, pp 515-523.
2. P.S. MAMYKIN, N.A. BATRAKOV, V.H. BOGATIKOVA, Tr. Ural. Petrogr Soveshch, **7**, 1975, p 45.
3. T.G. ALKHAZOV, K.YU. ADZHAMOV et A.K. KHAMNMAMEDOVA, React. Kinet. Catal-lett., **7**, 1975, p 45.
4. P. BOUTRY, J.C. DAUMAS, R. MONTARNAL, P. COURTINE, G. PANNETIER, Bull. Soc. Chim. Fr., **12**, 1968, P 4811.
5. J. HABER, J. of less Common Metals, **36**, 1974, p 277.
6. J. HABER et J. ZIOLKOWSKI, Bull. Acad. Polon. Sci., Sec. Sci. Chim. **19, n°8**, 1971, p 481.
7. J. HABER et J. ZIOLKOWSKI, in M.W. Roberts (Ed), Reactivity of Solids, Elsevier, Amsterdam, 1965, p 96.
8. G.W. SMITH, Nature, 188, 1960, p 306.
9. G.W. SMITH, Acta, Cristallogr., **15**, 1962, p 1054.
10. G.W. SMITH, et J.G. IBERS, Acta Cristallogr., **19**, 1965, p 269.
11. a) B. ARCHIROPOULOS et S.J. TEICHNER, J. Catal. **3**, 1964, p 447,
    b) F. JUILLET, La Catalyse au laboratoire et dans l'industrie, **2**, 1976, p 299, Editeurs Masson et Cie.
12. C. CLAUSS, R.J. TARENTO, C. MONTY, A. DOMINGUEZ-RODRIGUEZ, J. CASTAING and J. PHILIBERT, Transport in Non stoichiometric Compounds, Ed. G. SIMKOVICH and V. STUBICAN, Plenum Press, 1985,255.
13. T. HOLSTEIN, Ann. Phys. (NY), **8**, 1959, pp 325-343.
14. J. HABER et J. ZIOLKOWSKI in Roberts (Ed.), Reactivity of Solids, Chapman et Hall London, 1972.
15. L. DZIEMBAJ et J. ZIOLKOWSKI, Bull. Acad. Polon. Sci., Sec. Sci. Chim., **20, n°7**, 1972, p 725.
16. M.G. EROR et J.B. WAGNER, Phys. State Solid., **35**, 1969, p 641.
17. S.P. MITOFF, Chem. Phys., **35**, 1979, p 882.
18. F. FAHRI, Thèse d'Etat, Paris-Nord, 1979.
19. J.E. KEEM, J.P. HONIG, L.L. VANZANDT, Phil. Mag. B., **37, n°4**, 1978, pp 537-543.
20. J.B. GOODENOUGH, Les Oxydes de métaux de transition, Gauthier-Villards éditeurs.
21. M. SAYER, A. MANSINGH, J.B. WEBB et J. NOAD, J. Phys. C. Solid State Phys., **11**, 1978, P 315.

ELECTRONIC STRUCTURE OF TRANSITION METAL IMPURITIES AND OF SURFACE DEFECTS IN $SrTiO_3$.

M.O.SELME, G.TOUSSAINT AND P.PECHEUR
Laboratoire de physique du solide (U.A.155)
Ecole des Mines, Parc de Saurupt
54042 Nancy Cedex
France

ABSTRACT. The electronic structure of substitutional impurities of transition metal elements in $SrTiO_3$ have been investigated theoretically using a tight binding parametrisation of the band structure of $SrTiO_3$ together with the Green's functions method. Spin polarisation and crystal field effects have been included. The energy levels, spin states and the shape of the optical spectra of the defects have been calculated and compared to experiments. The localisation of the impurity wave function has also been discussed and related to the band structure. Impurity-oxygen vacancy complexes have been investigated and the corresponding binding energy estimated from the calculation. The same method has been applied to the surface defects (oxygen vacancy and Ti adatoms) in connection with photoemission experiments.

1. Introduction

Transition metal impurities substituted to titanium in $SrTiO_3$ have been widely studied experimentally both by electron paramagnetic resonance and optical absorption [1], in particular by K.A. Müller and coworkers. These studies make this material a reference substance for the behaviour of transition-metal impurities in a matrix which is both ionic and covalent, since the band structure calculation of Mattheiss [2] and the simplified model of Wolfram [3] have indeed shown that there is a large covalent coupling between the d orbitals of Ti and the p orbitals of O.

In contrast with the case of tetrahedral semiconducting compounds, there have been only a few theoretical studies of the electronic structure of these transition metal defects in $SrTiO_3$, except for some recent calculations with the X_α method [5], [6]. The purpose of the present work is to review a simple theoretical treatment of the problem, which starts from the empirical tight binding methods to describe the band structure of $SrTiO_3$ and uses the Green's function method to introduce the defects. This avoids any cluster approximation and is well suited to study band structure effects. Such methods have been widely used to treat impurities in silicon and 3-5 or 2-6 compounds [6].

The scheme of the paper is the following : § 2 gives the basis of the method. § 3 shows some examples of isolated impurities with emphasis on the band structure effects on their electronic structure ; § 4 concerns the impurity - oxygen vacancy complexes which are common in $SrTiO_3$; finally in § 5 the same method is used for the case of surface defects in relation with photoemission .

J. Nowotny and W. Weppner (eds.),
Non-Stoichiometric Compounds Surfaces, Grain Boundaries and Structural Defects, 173–186.

2. Computational Method

2.1 PERFECT CRYSTAL BAND STRUCTURE

The tight binding model includes the p orbitals of oxygen and the d orbitals of titanium, as a minimal basis. The A.P.W. calculation of Mattheiss [2] shows that the s orbital of oxygen are much lower in energy and that strontium is completely ionized. First-neighbours p-d interactions and second neighbours p-p interactions are taken into account. The latter are much weaker, buth they are necessary to obtain a correct description of the upper part of the valence bands ("non bonding" p states) which, in a first neighbours model, would appear as flat bands [3]. The diagonal elements of the Hamiltonian (E_p, $E_{d\sigma}$, $E_{d\pi}$) and the non-diagonal ones ($I_{pd\sigma}$, $I_{pd\pi}$, $I_{pp\sigma}$) are chosen to reproduce the A.P.W. results [2] and the experimental gap (E_G = 3.2 eV). They are listed in table 1.

E_p	$E_{d\sigma}$	$E_{d\pi}$	$I_{pd\sigma}$	$I_{pd\pi}$	$I_{pp\sigma}$
-0.9	5.4	3.2	-2.31	1.13	0.32

Table 1. Tight binding parameters for $SrTiO_3$ (energies are in eV).

The density of states is determined using the Green's functions method [6]. The Green's functions are calculated with the real space Lanczos-Haydock recursion procedure [7]. A cluster of 5745 atoms has been used to obtain 15 coefficient pairs (a_n, b_n) of the continuous fraction. The Green's function are reconstructed using the δ representation of Gordon [8]. This can be briefly described as follows : given a starting vector α (i.e. a p oxygen orbital, or a d titanium orbital), the Lanczos-Haydock recursion generates a new basis in which the Hamiltonian is represented by a tridiagonal N x N matrix with elements a_n and b_n. Next, this matrix is diagonalized to obtain N eigenvalues E_i and N eigenvectors $\{e_i^j\}$.Note that α is the first basis vector of the new basis. The Green's function then reads :

$$\mathrm{Re}\, G_{\alpha\alpha}(E) = \sum_i \frac{|e_i^\alpha|^2}{E - E_i} \tag{1}$$

$$\mathrm{Im}\, G_{\alpha\alpha}(E) = -\pi \sum_i |e_i^\alpha|^2\, \delta(E - E_i) \tag{2}$$

The partial density of states for the α state is then given by :

$$n_\alpha(E) = -\frac{1}{\pi}\, \mathrm{Im}\, G_{\alpha\alpha}(E) \tag{3}$$

The δ functions in $G_{\alpha\alpha}$ (E) are broadened into Gaussians of half width Γ = 0.2 eV.

The total density of states for the perfect crystal is obtained by summing up the partial densities of states for the titanium and oxygen atoms in the unit cell. It is shown in fig. 1. The valence bands are mainly built from the p states of oxygen, while the conduction bands correspond to the d states of titanium (the lower part, between 3.2 eV and 5.2 eV is due to the "π" or t_{2g} orbitals and the upper part to the "σ" or e_g orbitals).

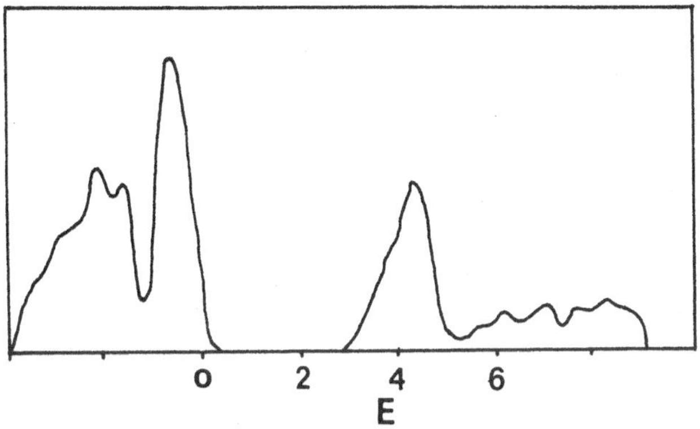

Figure 1. Perfect crystal density of states for $SrTiO_3$.

2.2 CRYSTAL-FIELD EFFECTS

The differences $\Delta_{eg} = E_{d\sigma} - E_d$ and $\Delta_{t_{2g}} = E_{d\pi} - E_d$ represent crystal field effects. In the perfect crystal $E_{d\sigma}$ and $E_{d\pi}$ are obtained directly from the A.P.W. band structure [2]. But for impurities substituted to titanium or titanium atoms with a different environment from that of the bulk (like surface atoms), the crystal field effects are different and an explicit model is needed. Crystal field effects are mainly due to covalent bonding of the metal atom with the s and p orbitals of the oxygen first neighbours. The s orbitals contribute only to Δ_{eg}. Since they are not included in the minimal basis set, this contribution can be calculated using the Löwdin folding procedure [9], with the result :

$$\Delta^{s}_{eg} = \frac{3(S_s\ E_d - I_{sd\sigma})^2}{(E_d - E_s)} \tag{4}$$

Where S_s is the sdσ overlap between s and d first neighbour orbitals. The p orbitals are included in the basis set, but S_π and S_σ overlaps (between p and d first neighbours orbitals) are not. These overlaps can be treated by first order perturbation theory [10], with the result :

$$\Delta^{p}_{d} = - \sum_{p} (S_{dp}\ H_{pd} + H_{dp}\ S_{pd}) \tag{5}$$

Where H_{pd} (S_{pd}) are the Hamiltonian (overlap) matrix elements between first neighbours orbitals and the summation over p is extended over the p orbitals of the first neighbours of orbital d. S_{dp} and H_{dp} can be expressed from the two center parameters S_σ, S_π and $I_{pd\sigma}$, $I_{pd\pi}$. The final results for the perfect crystal are :

$$\Delta^{p}_{eg} = - 6\ S_\sigma\ (I_{pd\sigma} - E_p\ S_\sigma) \tag{6}$$

$$\Delta^{p}_{t_{2g}} = - 8\ S_\pi\ (I_{pd\pi} - E_p\ S_\pi) \tag{7}$$

Using $I_{pd\sigma}$ and $I_{pd\pi}$ from Table 1 and the values given by Mattheiss [9] for other parameters, one finds Δ_{eg} = 2.2 eV and $\Delta_{t_{2g}}$ = 0.5 eV, that is $E_{d\sigma} - E_{d\pi}$ = 1.7 eV instead of $E_{d\sigma} - E_{d\pi}$ = 2.2 eV in Table 1. Although the agreement is only approximate, the method gives the right order of magnitude and has been used in defect calculations.

2.3 DEFECT CALCULATIONS

Defects calculations have been performed with the Green's functions method [6]. The Green's function G of the perturbed crystal can be obtained from those of the perfect crystal G^0, using Dyson equation :

$$G = G^0 + G^0 V G \tag{8}$$

Where V is the perturbation matrix. Alternatively, G can be calculated directly in the perturbed crystal with the recursion procedure, since this does not require periodicity and Bloch theorem. This is expecially useful for complex defects where (8) would lead to large perturbation matrices and numerical difficulties, particularly with the non diagonal elements of G^0.

For the impurity, both diagonal and non-diagonal perturbation matrix elements have been considered.

The non-diagonal matrix elements correspond to change in the transfer integrals I between the impurity and the neighbouring oxygens. They have been obtained from the values given by Mattheiss [9] for the NaCl type oxides. The differences in the nearest neighbours distances d in the NaCl oxides and $SrTiO_3$ is taken into account using the $d^{-7/2}$ variation used by Harrison [11].

The diagonal perturbation is represented by the sum of two terms :

1) a single matrix element V_d , which takes into account the changes between the diagonal elements of the impurity M and the titanium atom, together with the associated rearrangement of the electronic charge in the crystal.

2) a term which corresponds to the change in crystal field effects on the impurity, obtained from eq. (6) and (7), of the previous section.

To determine V_d , a "selfconsistency" condition has been used : the excess nuclear charge of the impurity is supposed to be compensated exactly by the change in the d orbital population on the impurity site. That is, one requires that :

$$n_d^M - n_d^{Ti} = \Delta Z \tag{9}$$

where M is the impurity and ΔZ the change in valence (ΔZ = 1 for Vanadium, ΔZ = 4 for iron), n_d^{Ti} = 1.58 for the present $SrTiO_3$ band model. In this scheme, the value of V_d depends on the occupancy of the impurity bound levels and so the effect of the impurity charge state on the level position can be obtained. The "selfconsistency" condition (9) corresponds to a short range screening of the perturbation. This scheme neglects the long-ranged screened Coulomb interaction which extends further in the crystal. Due to the high dielectric constant of $SrTiO_3$ and the high localisation of the defect bound levels, this neglect should not have a large influence on the bound levels energy.

To compare with experiments, it is necessary to introduce spin polarisation. The simplest way to do this, is to replace V_d by two different potentials $V_d\uparrow$ and $V_d\downarrow$ for majority and minority spins states. $V_d\uparrow$ and $V_d\downarrow$ are then determined using condition (9) together with the supplementary condition :

$$V_d\uparrow - V_d\downarrow = -\xi\,(n_d^{\uparrow} - n_d^{\downarrow}) \tag{10}$$

where $n_d^{\uparrow}$ and $n_d^{\downarrow}$ are the number of spin-up and spin-down d electrons and ξ is a Stoner constant, which is taken equal to 0.9 eV [12].

3. Isolated impurities

The method has been applied to the first serie of transition elements substituted to titanium [12], [13]. We give here some examples, insisting on band structure effects.

3.1 ENERGY LEVELS FOR VANADIUM AND IRON

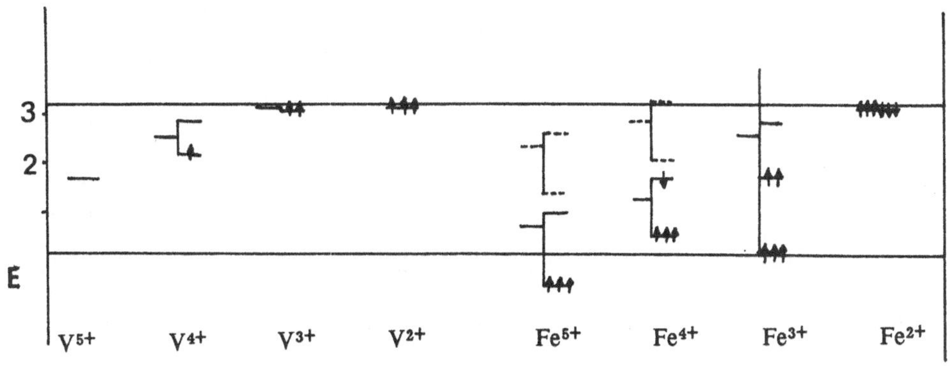

Figure 2. Impurity levels for Vanadium and Iron in the unpolarised and polarised spin models (t_{2g} levels are indicated by full lines and e_g levels by broken lines).

Charge effects are clearly seen in the unpolarised-spin calculation : the energy levels rise in the gap, when the (positive) charge state of the impurity decrease. Spin polarisation adds spin splitting, giving a more complex pattern. The band edges set limits to the number of possible charge states, since levels (or rather resonances) that are found in the valence (or conduction) bands must be completely filled (or empty). V^{4+}, V^{3+} and V^{2+} have been seen in E.P.R, as well as Fe^{5+} and Fe^{3+}.

An interesting point is the experimental result that Fe^{4+} is in a low spin state, while Fe^{3+} is in a high spin state. To investigate this further in the calculation, one has to determine which of the two spin states has the lower total energy. This can be done, using Slater transition state method. One calculates the energy levels for an occupation half-way between the high and low spin configuration, and observe which of the majority e_g level or the minority t_{2g} level has the lower energy. This indicates wether the high or low spin state is the lower energy state. The calculation correctly predicts a low spin state for Fe^{4+} and a high spin state for Fe^{3+}. These results are sensitive to the value used for ξ, the Stoner parameter in (10). If $\xi = 1.15$ eV is used, instead of 0.9 eV, the high spin state is obtained for Fe^{4+}. If $\xi = 0.7$ eV, the low spin state is the lower energy state for Fe^{3+}. Moreover, the value of 0.9 eV corresponds well with the Stoner constant calculations that have been performed for NaCl-type oxides [14].

3.2 LEVELS LOCALISATION

The localisation of the impurity levels on the impurity site and the neighbouring oxygen sites can be deduced from the calculation : they are given by the weight $|e_i^\alpha|^2$ of the corresponding δ function in the Gordon expansion (2). They can be compared with the localisations deduced from E.P.R. experiments.

The effect of the band structure is clearly seen in these localisations. When the impurity levels get close to the valence band edge the localisation on the impurity site decreases, while it increases on the first oxygen neighbours. This is a classic covalency effect : the impurity d levels hybridise with the p levels of the oxygen ligands, and when these levels get close in energy, the antibonding level (which corresponds to the impurity level in the gap) become equally represented on the impurity d orbitals and the ligand p orbitals. For Co^{4+}, for instance, the calculation [13] gives a mean localisation of 0.66 on the metal and 0.27 on the ligands, in good agreement with experiment [15].

A different situation occurs when a t_{2g} level is close to the conduction band edge (which has the same symmetry "π"). This is the case for V^{2+} in fig. 1, where the localisation on the impurity is only 0.43 and 0.02 on the oxygen ligands, so that the level extends on Ti atoms further in the crystal. This corresponds well with the E.P.R. results [16] and is related to the shape of the conduction band bottom. It can be understood most easily from the first neighbours interactions model introduced by Wolfram for $SrTiO_3$ [2]. In this model there is a discontinuity in the density of states at the bottom of the conduction band, related to the two dimensional character of π bonding in $SrTiO_3$. This gives a logarithmic singularity in the real part of the Green's function, leading to an instability of the bottom of the band : the weakest attractive perturbation is then sufficient to extract a level, as can be seen from the Dyson equation (8). The localisation of this level tends to zero with the strengh of the perturbation V_d. At the same time, its energy position tends to that of the conduction band bottom. Note that such a shallow level is not an hydrogenic level (bound by a long range screened Coulomb interaction) but rather it is due to the shape of the conduction band bottom in $SrTiO_3$.

3.3 OPTICAL SPECTRA

Many impurities substituted to titanium give rise to optical spectra. Most of these are charge transfer absorption spectra which correspond to transitions of electron from the valence band to the empty impurity levels. The intensities for such transitions can be obtained from Fermi's golden rule :

$$\mathscr{P}(h\nu) = \sum_{X,\Gamma_\alpha} |<\psi_L|\, \mathbf{A}\,.\,\mathbf{P}\, |\psi_{X,\Gamma_\alpha}>|^2\, \delta(E_L - E_{X,\Gamma_\alpha} - h\nu) \tag{11}$$

where ψ_L is the impurity localized state of energy E_L.
ψ_{X,Γ_α} the valence band state X of symmetry Γ_α and energy E_{X,Γ_α}.
A the electromagnetic potential and **P** the momentum operator.

As shown in [13], since ψ_L is localised, $\mathscr{P}(h\nu)$ can be approximated by :

$$\mathscr{P}(h\nu) = \sum_{\Gamma_{\alpha_1}} |<\phi_L|\, \mathbf{A}\,.\,\mathbf{P}\, |\phi_{\Gamma_{\alpha_1}}>|^2\, |a_L|^2\, n_{\Gamma_{\alpha_1}}(E_L - h\nu) \tag{12}$$

where $n_{\Gamma_{\alpha_1}}(E)$ are the valence band partial densities of states for the linear combinations $\phi_{\Gamma_{\alpha_1}}$ of the oxygen p orbitals on the impurity first neighbours with a symmetry

(Γ_{α_1}) of the type T_{2u} or T_{1u} (other symmetries are dipolar forbidden) and $|a_L|^2$ is the localisation of the impurity state on the impurity site d orbital ϕ_L (ϕ_L is of e_g or t_{2g} symmetry).

Both $n_{\Gamma_{\alpha_1}}(E)$ and $|a_L|^2$ are obtained directly from the Green's function calculation for the defect. Since E_L is charge dependant, a Slater transition state calculation must be performed. Moreover, for a given final charge state of the impurity, one has to consider not only the ground state of this charge state (as in fig. 2) but also possible excited state of the same charge state. Finally, the matrix element in (12) can be estimated in a way which is consistent with the rest of the calculation [12].

The results for V^{5+} and Fe^{5+} are shown in fig. 3 and 4, together with the experimental results [17]. For V^{5+}, the results are in good, quantitative agreement with experiments. The main maximum is at the right position and the calculation shows a lower energy structure which reflects a structure in the non-bonding p states at the top of valence band. This band structure effect in the optical spectra had been identified by Blazey et al. [18] and is fully confirmed by the calculation.

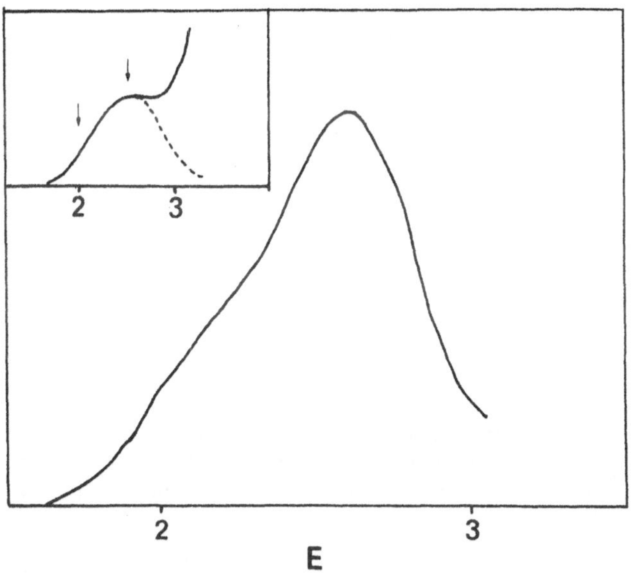

Figure 3 : Optical absorption spectrum of V^{5+}. The inset is the experimental result. The arrows indicate the experimental maxima. Energies are in eV.

For Fe^{5+}, the results also have the correct shape (showing again the low energy hump in the first peak due to the structure at the top of the valence band). However, the whole calculated curve is about 0.5 eV too low in energy. In particular, the second main peak in the calculation, at 2.8 eV, which is due to electrons originating from lower parts of the valence band, falls in fact in the intrinsic absorption range and cannot be seen experimentally. This means that the Fe^{5+} calculated energy levels, in fig. 1, are much too low in energy. In fact, the E.P.R. lines for Fe^{5+} are rather narrow and difficult to associate with the valence band resonance of fig. 1. For Fe^{5+}, the perturbation potential $V_d\uparrow$ is very large ($V_d\uparrow$ = - 5.5 eV) and the restriction of the perturbation to the impurity site is certainly too drastic. Preliminary calculations have been performed adding to V_d non diagonal perturbations of the form V_d S_{pd} between the impurity and its oxygen first neighbours. This form of the non diagonal elements does not increase the number of parameter to be determined, if one uses Mattheiss values [2] for S_{pd}. The perturbation matrix elements can still be obtained with (9) and (10). For Fe^{5+} this rises the valence band resonance just above the valence band top, giving a true bound state, as required by the E.P.R. results. For smaller value of V_d, the influence of the non diagonal perturbation is rather weak.

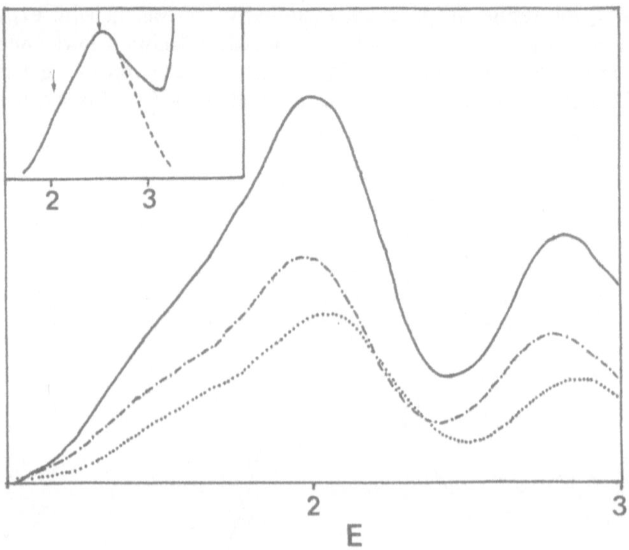

Figure 4. Optical absorption spectrum of Fe^{5+}. The full curve is the total spectrum. The dotted curve corresponds to transitions to the e_g level and the chain curve to transitions to the t_{2g} level. The inset is the experimental result. The arrows indicate the experimental maxima. Energies are in eV.

4. Associated defects : transition metal impurities - oxygen vacancy

4.1 ENERGY LEVELS

The association of a transition metal impurity and an oxygen vacancy is very commun in $SrTiO_3$. E.P.R. spectra corresponding to such associations have been reported in the case of Mn, Fe and Co [1]. For Mn^{2+}, the association is reversible at high reduction temperatures and has been followed by E.P.R. [22].

To calculate the energy levels corresponding to those associations, the same procedure has been used as for the isolated impurity. The diagonal perturbation potential on the impurity is again taken as the sum of two terms. The first one V_d is obtained from the charge condition (9). The second term, which corresponds to the difference between the impurity and the titanium crystal field effects is affected by the vacancy presence. It can be calculated as indicated in 2.2. The results for the d orbital in the axial C_{4v} symmetry are shown in Table 2.

Symmetry	Orbital	Impurity Crystal field
b_2	xy	$\Delta^M_{t_{2g}} - \Delta^{Ti}_{t_{2g}}$
e	xz, yz	$\frac{3}{4}\Delta^M_{t_{2g}} - \Delta^{Ti}_{t_{2g}}$
a_1	$3z^2 - r^2$	$\frac{2}{3}\Delta^M_{eg} - \Delta^{Ti}_{eg}$
b_1	$x^2 - y^2$	$\Delta^M_{eg} - \Delta^{Ti}_{eg}$

Table 2 : Changes in the crystal field effects for an impurity M with a first neighbour vacancy.

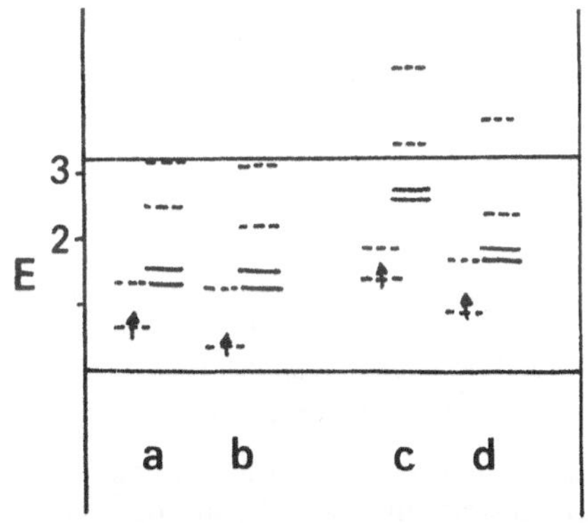

Figure 5 : Fe^{4+} - V_0 associated defect. Present calculation without iron relaxation (a) and with iron relaxation (b). X_α calculation without iron relaxation (c) and with iron relaxation (d). Full lines stand for e and b_2 levels, dashed lines stand for a_1 and b_1 levels

Spin polarisation has been included, using (10). So, in the calculation, the oxygen vacancy shows itself in two ways :

1) The lack of an oxygen neighbour modifies the coordination of the impurity atom (and also that of the titanium neighbour of the vacancy), and the corresponding Green functions are modified.

2) The crystal field term on the impurity (and on the titanium neighbour of the vacancy) is also modified, as described in Table 2.

Fig. 5 shows the results for Fe^{4+} - V_0. Isolated Fe^{4+} is in a low spin state as discussed before, while the E.P.R. experiments shows that Fe^{4+} - V_0 is in a high spin state [19]. Using the same procedure as for the isolated impurities, the difference between the total energy for the low spin state and the high spin state has been evaluated within Slater transition state approximation. The calculation shows that the high spin state is more stable for both Fe^{3+} - V_0 and Fe^{4+} - V_0. The reason is the lowering of the a_1 state (one of the e_g states of the cubic symmetry) due to the decrease of the crystal field effects for this state (see table 2). Fig. 5 also includes the result of a calculation with a 0.2 Å displacement of iron toward the vacancy. Such a displacement has been deduced from the E.P.R. spectra [20]. In the calculation, this is taken into account by changing the interactions $I_{pd\pi}$ and $I_{pd\sigma}$ of iron and its first neighbours, assuming that they vary with the distance d as $d^{-7/2}$. The tight binding results are compared in fig. 5 with those obtained with the X_α method [5]. The present calculation gives unrelaxed levels located somewhat lower in the gap. Iron relaxation has a stronger effect in the X_α calculations, so that the energy levels are similar in both calculation for the relaxed levels.

4.2 INTERACTION ENERGY

Although the calculation of the interaction energy E_B between the transition impurity and the vacancy is more complicated than that of the one-electron energy levels, an estimate is possible using the Green's function method.

The change in the density of states for the crystal when an impurity is introduced is given by [6] :

$$\delta n(E) = -\frac{1}{\pi}\frac{d}{dE}(\mathrm{Im}\,\mathrm{Log}\,\det(1 - G^0V)) \tag{13}$$

The change in the electronic energy of the crystal is then obtained by :

$$\Delta E_c = \int_{-\infty}^{E_{vb}} E\,\delta n(E)\,dE + E_L\,n_L \tag{14}$$

when the summation extends to the top of the valence band E_{vb} and the second term corresponds to the contribution of the bound states E_L, with occupation n_L.

Using (14) twice, first for a transition impurity in an otherwise perfect crystal, then for a transition impurity introduced in a site nearest neighbour of an oxygen vacancy, one obtains ΔE_c and ΔE_c^{Vac}. The binding energy is given by :

$$E_B = \Delta E_c^{Vac} - \Delta E_c \tag{15}$$

In fact, the self consistency potential contribution is counted twice in the one electron energy (14) so that one must substract this contribution once from ΔE_c and ΔE_c^{Vac}. In the present calculation this means that :

$$\Delta E_{sc} = 1/2\,[(V_d^{\uparrow}\,n_d^{\uparrow} + V_d^{\downarrow}\,n_d^{\downarrow})_{Vac} - (V_d^{\uparrow}\,n_d^{\uparrow} + V_d^{\downarrow}\,n_d^{\downarrow})_c] \tag{16}$$

has to be substracted from (15).

The results obtained in this way are given in table 3.The contribution to this binding energy due to the impurity d states alone can also be evaluated from the Green's function calculation [21]. It is given in the last column of Table 3 (E_B^d).

The calculation is admittedly crude. A comparison between E_B and E_B^d shows that only part of the binding energy originates from the d orbitals of the impurity. Most of the rest comes from the 5 oxygen first neighbours, and the present calculation does not take into account self consistency effects on these neighbours. Moreover no account has been taken of atomic relaxations. The experimental estimate [22] for M^{2+} - V_0 is 0.7 eV.

Impurity	E_B (eV)	E_B^d (eV)
$Mn^{2+} - V_0$	- 1.8	- 1.2
$Fe^{3+} - V_0$	- 2.2	- 1.6
$Fe^{4+} - V_0$	- 1.5	- 1.1
$Co^{4+} - V_0$	- 1.4	- 0.8
$Ni^{3+} - V_0$	- 1.8	- 1.0

Table 3 : Impurity-vacancy binding energy.

However, several points are in qualitative agreements with experiments. First these calculated energies are large, so that the complex formation is expected. Finally, E_B is not strongly dependant of the impurity charge state. This comes from the "selfconsistency" condition (9), which maintains the same charge on the impurity d orbitals for the different charge states, that is various occupations of the impurity bound states. This accounts for associations like $Co^{4+} - V_0$, where the impurity has the same charge state as Ti^{4+} so that a point ion model with Coulomb interactions would lead to no binding.

5. Surface defects

5.1 PHOTOEMISSION RESULTS

(100) surfaces of $SrTiO_3$ have been studied by ultraviolet photoemission spectroscopy [23] [24]. Vacuum fractured surfaces exhibit a weak photoemission in the bulk band gap with a maximum at 1.7 - 1.8 eV above the top of the valence band. When surface defects are produced by Ar ions bombardement, surface states appear in the band gap at 2.2 - 2.3 eV above the top of the valence band. Exposure to O_2 suppresses the band-gap states, showing that the lack of oxygen atoms is responsible for these gap states.

5.2 SURFACE DEFECTS

To investigate which surface defects might be responsible for the deep levels seen in photoemission, the preceeding method can be used [25]. The same tight binding model and the recursion procedure are the basis of the calculation. The recursion procedure is very helpful, because it allows to obtain the local densities of states for surface atoms and surface defects of low symmetry and because the method does not require periodicity of the system which is studied. Possible deep levels and their localisation are obtained directly from the energy and weight of the corresponding poles in the Gordon representation. Changes in the crystal field effects for surface atoms or adatoms are included as in 2.2.

The perfect (100) surfaces has been investigated first . They are of two types in $SrTiO_3$: one built from Sr and 0 atoms, the other from Ti and O atoms. Neither leads to deep surface states. The TiO_2 surface only presents a resonance just at the conduction band edge. The surface oxygen vacancy has been investigated too. Here again, no deep levels appear, only a resonance just below the conduction band bottom, in the zx orbitals of the two Ti neighbours (Fig. 6). A relaxation of these neighbours does not seem able to modify this conclusion [25].

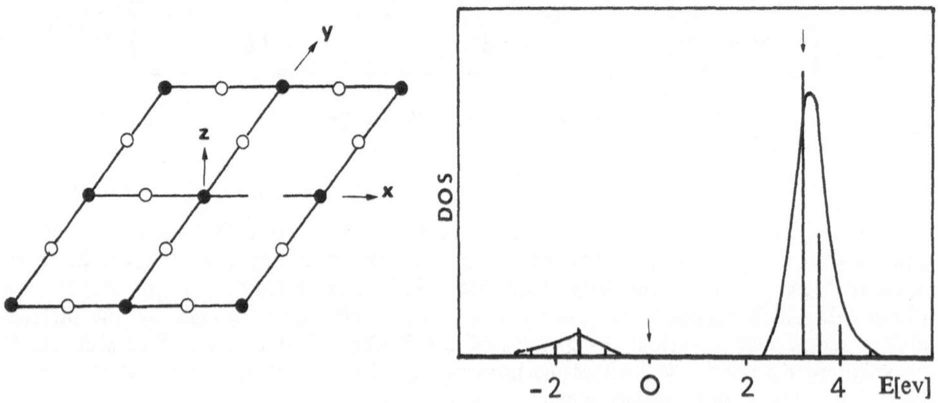

Figure 6. Surface oxygen vacancy and the corresponding partial density of states for the zx orbital of its Ti neighbours.

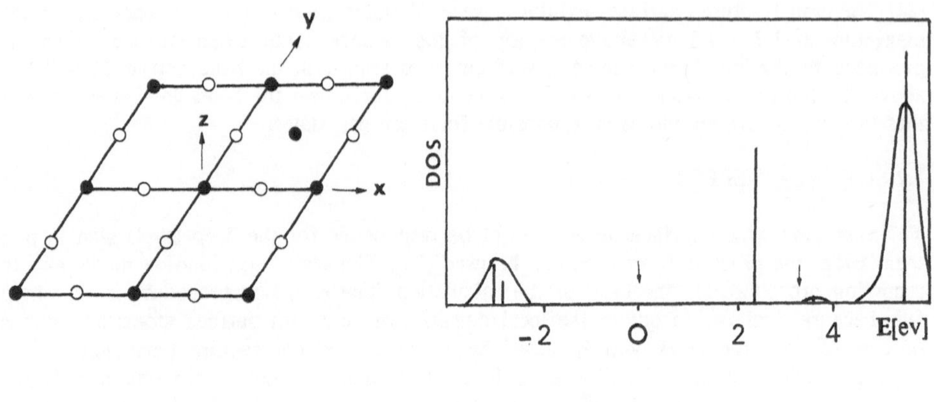

Figure 7. Ti adatom and the corresponding partial density of states for the xy orbital.

Since oxygen vacancies do not produce deep levels, other candidates have been looked for. When there is a lack of oxygen atoms, one may expect excess titanium to be adsorbed on the surface. Two such adatoms have been considered. The first one is a titanium in the (100) TiO_2 plane with four equivalent oxygen neighbours and four equivalent titanium second neighbours (fig. 7). The Ti-Ti second neighbours interactions have been obtained from those in TiO [9], with a correction for the difference in the distance d assuming a d^{-5} law. A deep level appears in the gap at 2.5 eV.

The second case that has been considered is a Ti atom on a step of a TiO_2 surface [25]. This atom has three oxygen first neighbours in the step (with the usual Ti-O distance in $SrTiO_3$) and a close Ti second neighbour (Ti-Ti interaction has been obtained as in the previous case). This time a deep level appears at about 1.3 eV. Some investigations about charge selfconsistency have been performed in [25], showing that the level might be a few tenth of eV higher in the gap.

Ti adatoms do lead to deep levels in the gap. This can be understood from simple molecular orbital considerations : the Ti adatom couples strongly with the Ti states of the $SrTiO_3$ matrix, since the distances with its Ti neighbours are much smaller than Ti-Ti distances in the perfect crystal. This leads to a bonding-antibonding structure and since E_d is at the bottom of the conduction band, the bonding level appears as a deep level in the gap.

6. Conclusion

The present approach is of a semi empirical type. It makes use of band structure parameters fitted to A.P.W. calculations [2]. It makes the assumption of a dominant local short range screening to treat the various charge states of the defects and uses a Stoner constant ξ to describe spin polarisation.

The short range screening assumption leads to the existence of three of four stable charge states in the gap, in agreement with experiments. A ξ value around 1 eV seems able to give the correct spin states for isolated and oxygen vacancy associated impurities. A higher value (1.2 eV) would lead to a high spin state for Fe^{4+}, while a lower one would predict a low spin state for Co^{4+} - V_0.

The present treatment has its own limitations. For instance, more recent self consistent band structure calculations [26] might lead to some changes in the tight binding parameters. It is also well known from the many empirical tight binding studies for defects in tetrahedral semiconductors that the uncertainty on the absolute energy level for a given impurity is several tenth of an electron volt. However, the tight binding method has the advantage of being relatively simple, devoid of any cluster approximation and easily understood in physical terms, since it remains close to the popular molecular orbital method. It could certainly be used to treat other defects of interest in $SrTiO_3$ (such Al substituted to Ti) or in other related compounds such as the new supraconductors like $YBa_2Cu_3O_7$.

REFERENCES

[1] K.A. MÜLLER, J. de Physique (Paris) *42* 551 (1981)
[2] L.F. MATTHEISS, Phys. Rev. B *6* 4718 (1972)
[3] T. WOLFRAM, Phys. Rev. Lett. *29* 1383 (1972)
[4] F.M. MICHEL CALENDINI and K.A. MÜLLER, Solid State Comm. *40* 255 (1981)
[5] F.M. MICHEL CALENDINI and P. MORETTI, Phys. Rev. B *27* (1983)
[6] S.T. PANTELIDES Rev. Mod. Phys. *50* 797 (1978)
[7] R. HAYDOCK, V. HEINE and M.J. KELLY J. Phys. C *5* 2845 (1972)
[8] R.G. GORDON, J. Math. Phys. *9* 655 (1968)

[9] L.F. MATTHEISS, Phys. Rev. B *5* 290 (1972)
[10] P. PECHEUR, G. TOUSSAINT and E. KAUFFER, Phys. Rev. B *29* 6606 (1984)
[11] W.A. HARRISON Electronic Structure and the Properties of Solids (San Francisco : Freeman) 451 (1980)
[12] M.O. SELME, P. PECHEUR and G. TOUSSAINT, J. Phys. C *19* 5995 (1986)
[13] M.O. SELME, P. PECHEUR, J. Phys. C 1779 (1988)
[14] O.K. ANDERSEN, H.L. SKRIVER and H. NOHL Pure. Appl. Chem. *52* 93 (1979)
[15] K.W. BLAZEY and K.A. MÜLLER, J. Phys. C *16* 5491 (1983)
[16] K.A. MÜLLER, M. AGUILAR, K.W. BLAZEY and W. BERLINGER, Workshop on defects in oxides, Bad Honnef (RFA) 1985
[17] K.W. BLAZEY, O.F.SCHIRMER, W. BERLINGER and K.A. MÜLLER Solid State Comm. *16* 589 (1975)
[18] K.W. BLAZEY, M. AGUILAR, J.G. BEDNORZ and K.A. MÜLLER Phys. Rev. B *27* 5836 (1983)
[19] O.F. SCHIRMER, W. BERLINGER and K.A. MÜLLER, Solid State Comm. *16* 1289 (1975)
[20] E. SIEGEL and K.A. MÜLLER, Phys. Rev. B *19* 109 (1979)
[21] M.O. SELME, P.H.D. Thesis, Université de Nancy 1 (1986)
[22] K.W. BLAZEY, J.M. CABRERA and K.A. MÜLLER, Solid State Comm. *45* 903 (1983)
[23] V.E. HENRICH, G. DRESSELHAUS, and H.J. ZEIGER, Phys. Rev. B *17* 4908 (1978)
[24] B. CORD and R. COURTHS, Surface Science *162* 34 (1985)
[25] G. TOUSSAINT, M.O. SELME and P. PECHEUR, Phys. Rev. B *36* 6135 (1987)
[26] K.H. WEYRICH and R. SIEMS, Z. Phys. B *61* 63 (1985)

CATION DISTRIBUTION AND NON-STOICHIOMETRY IN $MnCr_2O_4$-$NiCr_2O_4$ SPINEL SOLID-SOLITIONS

C. Karataş
Hacettepe University-Nuclear Engineering Department,
Beytepe, Ankara
Turkey

ABSTRACT. Cation distribution in $MnCr_2O_4$.$NiCrO_4$ spinel solid-solutions at 1300°C was calculated using non-linear disordering enthalpies obtained from the experimental cation distribution of Mn^{2+}, Mn^{2+}, Cr^{2+},Cr^{3+},Ni^{2+} and assuming random mixing on both tetrahedral and octahedral sites.The model also incorporates the electron exchange reaction:

$$Mn^{2+} + Cr^{3+} = Mn^{3+} + Cr^{2+}$$

as well as site exchanges of newly formed cations. The Gibbs free energy of mixing derived from the available experimental data obtained in our laboratory are discussed in conjuction with the values calculated from the cation distribution model as the free energy of electron exchange reaction being a parameter. The effect of size mismatch of the substituting cations was taken into account using a solution model. Observed positive deviations from ideality suggests a miscibility gat at lower temperatures.

1. INTRODUCTION

The oxide spinels comprises the major part of ternary compounds and play important role in extraction metallurgy and solid state sciences.

The vacancy and cation distribution in spinel solid solutions have been the subject of several investigations (1-5). The early models assumed the enthalpy change accompanying cation disordering as the linear function of the degree of disorder and as being equal to the difference between the site preference energies of divalent and trivalent cations. O'Neill and Navrotsky(1) (1984) modified this model adding a quadratic enhalpy term with two parameters equal in magnitude and opposite in sign. Present study uses the apparent site preference energies as calculated by O'Neill and Navrotsky(1) from the experimental cation distributions of the pertinent cations. The present model also incorporates the possible electron exchange reactions between the cations Mn^{2+} and Cr^{3+} in overall calculations.

J. Nowotny and W. Weppner (eds.),
Non-Stoichiometric Compounds Surfaces, Grain Boundaries and Structural Defects, 187–199.

Although a few study(6-9) have been performed on the end-members, no measurements of cation distribution on the solution at elevated temperatures have been reported. Recently Jacob, Iyengar and Kim(4) provided some information on spinel-corundum phase equilibrium in the system MnO-Cr_2O_3-Al_2O_3 at 1100°C. Navrotsky et al(3) performed solution calorimetry of $MgAl_2O_4$-$Al_{8/3}O_4$ defect spinels in a molten $2PbO.B_2O_3$ solvent at 975°C.

In a previous study Koç and Timuçin(6) experimentally determined the activities of $MnCr_2O_4$-$NiCr_2O_4$ solid solution at 1300°C using gas equilibration technique.

The present study was undertaken to calculate the high temperature cation distribution and free energy of mixing in $MnCr_2O_4$-$NiCr_2O_4$ solid solution from previously reported non-linear disorder enthalpies of O'Neill and Navrotsky and also seperately from site preference energies of relevant cations.

2. CALCULATION OF THE CATION DISTRIBUTION AND FREE ENERGY OF MIXING

The model incorporates site and charge exchange reactions and was applied to solid solution with and members $MnCr_2O_4$-$NiCr_2O_4$. The distribution parameters of solid solution (x,y,z,v,w) presented in Table I.

TABLE I. Distribution Parameters

Ion	Tetra	Octa	Sum
Ni^{2+}	N-x	x	N
Mn^{2+}	1-N-y	y-z	1-N-z
Mn^{3+}	v	z-v	z
Cr^{2+}	w	z-w	z
Cr^{3+}	x+y-v-w	2-x-y-z+v+w	2-z
Sum	1	2	3

where N is the mole fraction of end-member $NiCr_2O_4$ in the solution.The relevant reactions are :

$$(Ni^{2+})+\{Cr^{3+}\} = \{Ni^{2+}\}+(Cr^{3+}) \quad (1)$$

$$(Mn^{2+})+\{Cr^{3+}\} = \{Mn^{2+}\}+(Cr^{3+}) \quad (2)$$

$$\{Mn^{2+}\}+\{Cr^{3+}\} = \{Mn^{3+}\}+\{Cr^{2+}\} \quad (3)$$

$$\{Mn^{3+}\}+(Cr^{3+}) = (Mn^{3+})+\{Cr^{3+}\} \quad (4)$$

$$\{Cr^{2+}\}+(Cr^{3+}) = (Cr^{2+})+\{Cr^{3+}\} \quad (5)$$

where ( ) and { } denote the tetrahedral and octahedral sites respectively. Equations (1) to (5) except (3) are the site exchange reactions ; equation (3) represent the electron exchange between Mn^{2+} and Cr^{3+} in octahedral positions. This exchange introduces two additional cations, namely Mn^{3+} and Cr^{2+} and their distribution between two sites are given in equation (4) and (5). The pertinent equilibrium expressions are :

$$-RT\ \ell n\ \left(\frac{x(x+y-v-w)}{(N-x)(2-x-y-z+v+w)}\right)=\alpha_{(Ni^{2+}-Cr^{3+})}-T\sigma_{(Ni^{2+}-Cr^{3+})}+2\beta(x+y-v-w) \quad (6)$$

$$-RT\ \ell n\ \left(\frac{(y-z)(x+y-v-w)}{(1-N-y)(2-x-y-z+v+w)}\right)=\alpha_{(Mn^{2+}-Cr^{3+})}-T\sigma_{(Mn^{2+}-Cr^{3+})}+2\beta(x+y-v-w) \quad (7)$$

$$-RT\ \ell n\ \left(\frac{(z-v)(z-w)}{(y-z)(2-x-y-z+v+w)}\right)=\frac{\partial\Delta G_R^o}{\partial z} \quad (8)$$

$$-RT\ \ell n\ \left(\frac{v(2-x-y-z+v+w)}{(z-v)(x+y-v-w)}\right)=\alpha_{(Cr^{3+}-Mn^{3+})}\ -T\sigma_{(Cr^{3+}-Mn^{3+})}+2\beta(x+y-v-w) \quad (9)$$

$$-RT\ \ell n\ \left(\frac{w(2-x-y-z+v+w)}{(z-w)(x+y-v-w)}\right)=\alpha_{(Cr^{3+}-Cr^{2+})}-T\sigma_{(Cr^{3+}-Cr^{2+})}+2\beta(x+y-v-w) \quad (10)$$

where α's are interchange enthalpies, σ's are the electronic contributions to entropy, β is the non-linearity constant and ΔG_R^o free energy of reaction (3). O'Neill and Navrotsky assumed β as -20±5 kJ/mole for all 2-3 spinels and calculated electronic contribution seperately as in developed crystal field theory (11). They used the assumed values of β and σ to obtain the best fit values for the emprical interchange enthalpies (α_{A-B}) from the literature data on the equilibrium cation distribution at various temperatures for various spinels. Then by chosing one cation as reference, they determined the apparent site preference energies for the others. In the present study their tabulated values were used. For each combination, α_{AB} is calculated from the relationship $\alpha_{AB}=\alpha_A-\alpha_B$ where α_A and α_B are the emprical site preference energies for each cation.

By subtituting the related parameters into equations (6) to (10) we obtain :

$$\ell n\ \frac{x(x+y-v-w)}{(N-x)(2-x-y-z+v+w)} = -10.1 +3.06(x+y-v-w) \quad (11)$$

$$\ell n\ \frac{(y-z)(x+y-v-w)}{(1-N-y)(2-x-y-z+v+w)} =-14.6+3.06(x+y-v-w) \quad (12)$$

$$\ell n\ \frac{(z-v)(z)w)}{(y-z)(2-x-y-z+v+w)} =-\frac{1}{13078}\frac{\partial\Delta G_R^o}{\partial z} \quad (13)$$

$$\ln\left(\frac{v(2-x-y-z+v+w)}{(z-v)(x+y-v-w)}\right)=5.0+3.06(x+y-v-w) \tag{14}$$

$$\ln\left(\frac{w(2-x-y-z+v+w)}{(z-w)(x+y-v-w)}\right)=9.5+3.06(x+y-v-w) \tag{15}$$

The numerical solutions of equations (11) to (15) for each mole fraction of $NiCr_2O_4$ gives the extent of site and charge exchange reactions and the concentration of ions in tetrahedral and octahedral sites as the function of free energy of reaction as depicted in Table I. The free energy of cation distribution, $\Delta G^o_{cd,N}$ would be then :

$$\begin{aligned}\Delta G^o_{cd,N}=&x\{132000-20000(x+y-v-w)\}+y\{205000-20000(x+y-v-w)-9.13T\}\\&+v\{-65000-20000(x+y-v-w)\}-w\{-124000-20000(x+y-v-w)\}\\&+RT\{(N-x)\ln(N-x)+(1-N-y)\ln(1-N-y)+v\ln(v)+(x+y-v-w)\ln(x+y-v-w)\\&+w\ln(w)+x\ln(x/2)+(y-z)\ln((y-z)/2)+(z-v)\ln((z-v)/2)\\&+(2-x-z+v+w)\ln((2-x-y-z+v+w)/2)+(z-w)\ln((z-w)/2)\}.\end{aligned} \tag{16}$$

The terms inside the bracket after "RT" gives the configurational entropy. The free energy of mixing at each point, N, is then :

$$\Delta G^o_{mix.}=\Delta G^o_{cd,N}-N\Delta G^o_{cd,NiCr_2O_4}-(1-N)\Delta G^o_{cd,MnCr_2O_4} \tag{17}$$

where $\Delta G^o_{cd,NiCr_2O_4}$ and $\Delta G^o_{cd,MnCr_2O_4}$ are free energy changes on cation disordering in end-members $NiCr_2O_4$ and $MnCr_2O_4$, respectively. For the component $NiCr_2O_4$, N is unity and all parameters except x are zero, whereas for the component $MnCr_2O_4$, N and x are both zero in Table I.

Assigning these values into expressions (6) to (10), the pertinent equilibrium expressions and the resulting free energies of cation distribution for end-members are constructed. Stoichiometry for $NiCr_2O_4$ was assumed and for $MnCr_2O_4$, the electron exchange between Mn^{2+} and Cr^{3+} was accepted. The free energy of electron exchange reaction is not known, but it may be calculated from the published thermodynamic data for the anologous reaction :

$$2MnO + Cr_2O_3 = 2CrO + Mn_2O_3$$

At 1300°C, the calculated values ΔG^o_R ranges from 2 kJ/mole to 200 kJ/mole. This wide distribution is the result of different values reported on Mn_2O_3 and CrO. Elliot and Gleiser(12) report the free energy of formation for Mn_2O_3 to be -129.5 kcal/mole whereas Kubachewsky's value(13) for the same oxide is -187.2 kcal/mole. Therefore, in this study ΔG^o_R was taken as a parameter, since there is no agreement in the reported values.

3. RESULTS and DISCUSSION

System of $NiCr_2O_4$:Solution of equation (6) for 1300°C gives the degree of inversion x for end-member $NiCr_2O_4$, to be 0.0091. The corresponding free energy of disorder, $\Delta G_{cd,NiCr_2O_4}$ and enthalpy of disorder, ΔH^o_D are calculated as -239 J/mole and 1205 J/mole, respectively. Ohnishi and Terenasu (14) have determined for $NiCr_2O_4$, 10 percent deviation from normal distribution at 1200°C. This is approximately ten times larger than the value calculated using O'Neill and Navrotsky's emprical values for $NiCr_2O_4$. When Dunitz and Orgel (15)'s values were used with β=0,the degree of inversness x becomes 0.086 which is closer to the reported value of Ohnishi and Terenasu.

System of $MnCr_2O_4$:

In previous experimental study (6), the molar ratio MnO/Cr_2O_3 in manganese chromite spinels was observed as being ranged from 1.00 to 1.44. This may be possible through the formation of $CrMnO_4$ as well as $MnCr_2O_4$ in the present oxygen network with the relevant reaction :

$$MnO + MnCr_2O_4 = CrMn_2O_4 + CrO \qquad (18)$$

This reaction is not different from the reaction :

$$2MnO + Cr_2O_3 = Mn_2O_3 + 2CrO \qquad (19)$$

which may be represented in ionic terms :

$$Mn^{2+} + Cr^{3+} = Mn^{3+} = Cr^{2+} \qquad (20)$$

Moreover the behavior of manganese ion in aqeous solutions and mineral oxides suggested that it would be in different electronic states. In literature the formation of $CrMn_2O_4$ as well as $MnCr_2O_4$ was reported (8). Oxidation of manganese ion by chromium ion introduces another degree of freedom through the formation of two additional ions and creates a new equilibria which would be quite different from the one without reaction. Hence the excess MnO would be incorporated into the spinel structure with the redox reaction without precipitating as a seperate phase. The resultant distortion in oxygen network and vacancy formation due to the related non-stoichiometry could be a topic for seperate study.

Solution of equation (7) to (10) as N equals to zero, gives the exchange parameters y,z,v,w and the results are presented in Figure 1. Figure 2 depicts the concentration of ions Mn^{2+}, Mn^{3+}, Cr^{2+}, Cr^{3+} in tetrahedral and octahedral sites for pure $MnCr_2O_4$. The fraction of cations Mn^{2+} and Cr^{3+} converted to Mn^{3+} and Cr^{2+} (i.e., z) diminishes as ΔG_R^o increases.

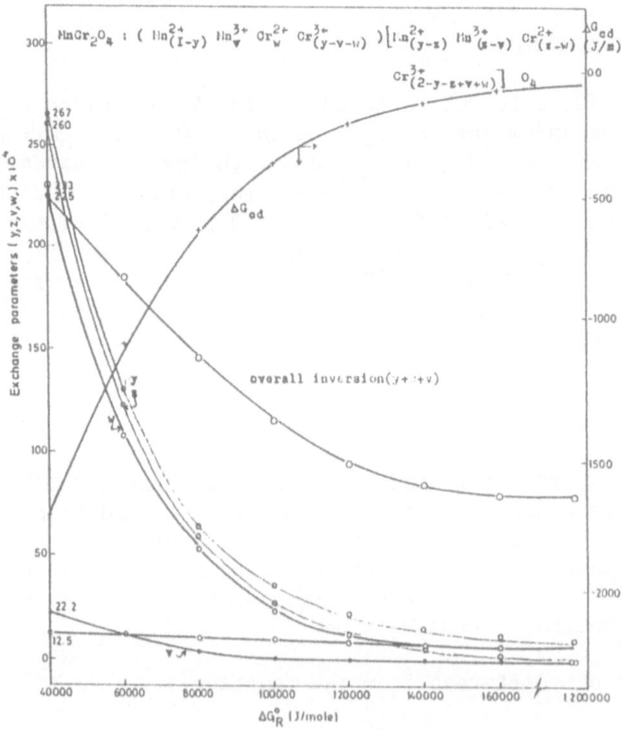

Fig.1. Calculated exchange parameters (y,z,v,w), overall inversness and free energy of disordering (ΔG_{cd}) for $MnCr_2O_4$ as a function of free energy of reduction ΔG_R^o at 1573 K.

At very large values of ΔG_R^o, z and in turn v and w become very close to zero and the system reduces to one parameter system (i.e., only y) as in the case of $NiCr_2O_4$. The overall inversness which is the total number of misplaced ions on the tetrahedral site (i.e.,y-v-w) decreases nearly two times as ΔG_R^o increases from 40kJ/mole to infinity. At ΔG_R^o=40kJ/mole,the calculated distribution is :

$$(Mn^{2+}_{0.973}Mn^{3+}_{0.0022}Cr^{2+}_{0.0232}Cr^{2+}_{0.00125})\ \{Mn^{2+}_{0.0007}Mn^{3+}_{0.0237}Cr^{2+}_{0.0028}Cr^{3+}_{1.973}\}O_4.$$

The corresponding free energy of cation distribution ranges from -1755 J/mole to -25 J/mole as ΔG_R^o approaches from 40 J/mole to infinity.

As shown, the electron exchange decreases the free energy almost 70 times.

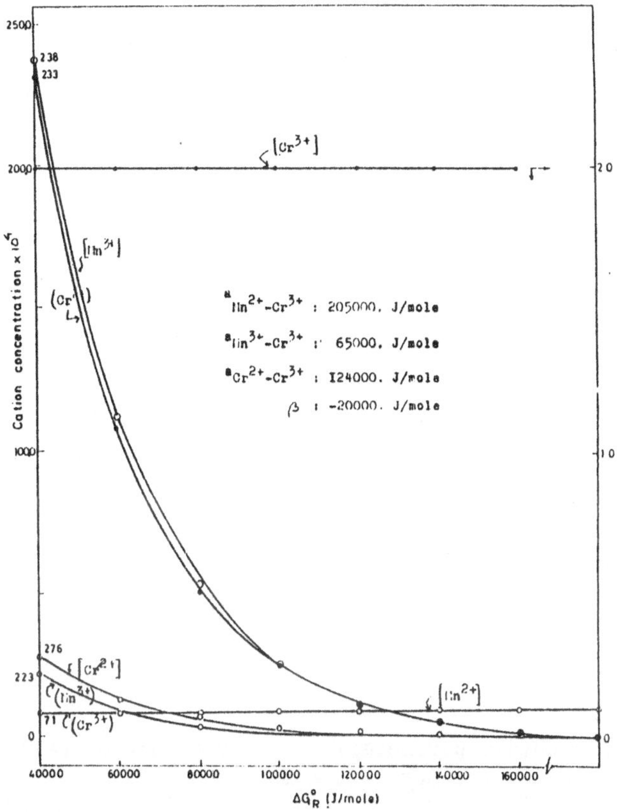

Fig.2. Calculated cation concentrations for octahedral and tetrahedral sites for $MnCr_2O_4$ as a function of free energy of reduction ΔG_R^o at 1573K.

One interesting point in Figure 1 is that the curves for parameters y,z,w are very close and almost parallel to each other. This indicates that when cation Mn^{2+} is transferred from tetrahedral to octahedral site, it is readly consumed by the existing Cr^{3+} ions forming Cr^{2+} and Mn^{3+}. Cr^{2+} out of these is then transferred back to tetrahedral site. The same can not be said for Mn^{3+}, since $\alpha_{(Cr^{2+}-Cr^{3+})}$ is approximately two orders of magnitude larger than $\alpha_{(Mn^{3+}-Cr^{3+})}$ indicating that Cr^{2+} prefers tetrahedral site more than the cation Mn^{3+} does. Figure 2 gives also the same trend. As expected the curves for each parameters do not have any common points.

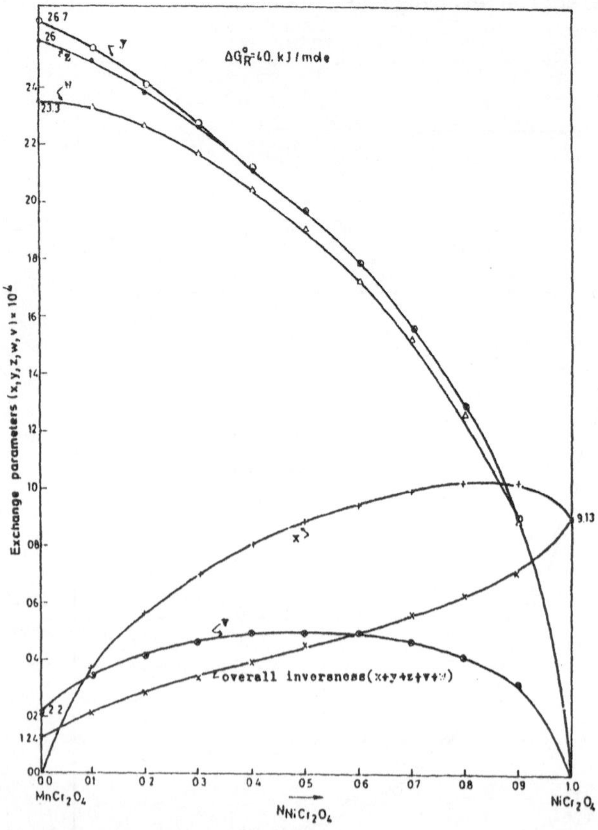

Fig.3. Calculated exchange parameters (x,y,z,v,w) and overall inversness for spinel solid-solution for ΔG_R^o=40 kJ/mole and at 1573 K as a function of composition.

System of $MnCr_2O_4$-$NiCr_2O_4$ solid solution :

Equation (11) to (15) are solved for each mole fraction N, ΔG_R^o being a parameter and the results are presented in Figure 3 and 4. Overall inversness changes from 0.001 to 0.009. The curves for parameters y,z, v,w start from the fixed values belong to the end-member $MnCr_2O_4$ and end up at zero on $NiCr_2O_4$ corner, while parameter x starts from zero and reaches to 0.009 as mole fraction gets closer to one. The concertration of ions at ΔG_R^o=40 kJ/mole is calculated from the exchange parameters as shown ih Table I and the results are summarized in Table II.

Figure 4 shows the effect to extent of charge exchange reaction on the overall transfer parameters for the solid solution containing 50 mole percent $NiCr_2O_4$ at ΔG_R^o=40 kJ/mole. Again at very high values of ΔG_R^o z,v,w approaches to y asymtotically and the system can be defined by only two parameters (x,y). In figure 5 the computed value of integral free energy of mixing of spinel solid solution at 1300°C is compared with the experimental results obtained by Koç and Timuçin (6).

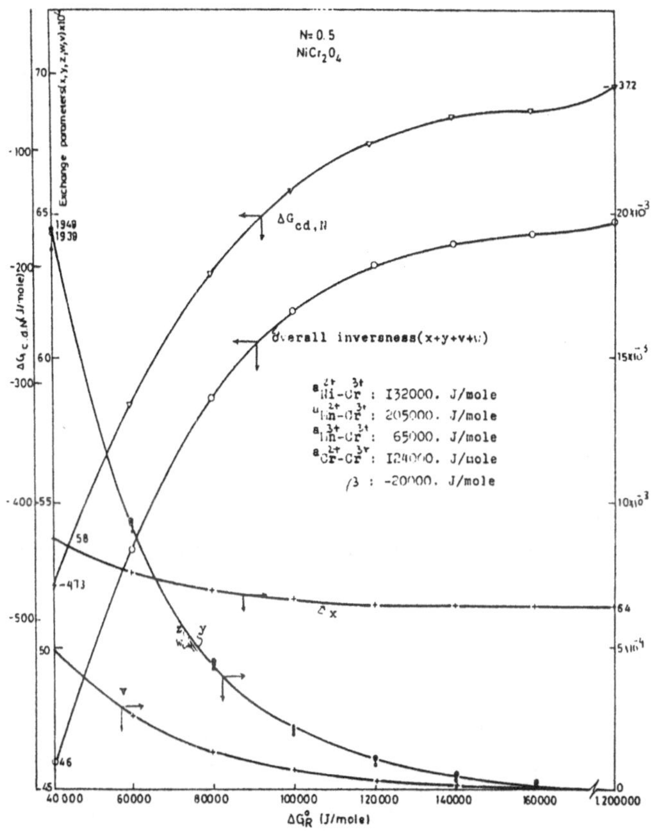

Fig.4. Calculated exchange parameters (x,y,z,v,w), overall inversness and free energy of disordering ($\Delta G_{cd.N}$) for spinel solid-solution of N=0.5 as a function of ΔG_R^o at 1573 K.

The same figure also gives the results when linear enthalpy terms are used with the site preference energies as presented by Dunitz and Orgel (14) and seperately by McClure (16). At the midcomposition the gap between the theoretical and experimental ones is 2874 J/mole for non-linear and 3810 J/mole for linear model. The gap becomes narrower as ΔG_R^o increases as expected. The most significant cause for this difference may be the strain energy arising from the distortion of the anion sublattice due to difference in cationic radii of Cr^{3+}, Cr^{2+}, Mn^{3+}, Mn^{2+}, Ni^{2+}.

TABLE II. Calculated Cation Concentrations at 1300°C Based on Emprical Interchange Enthalpies

N	(Ni^{2+}) $x10^4$	(Mn^{2+}) $x10^3$	(Mn^{3+}) $x10^4$	(Cr^{2+}) $x10^5$	(Cr^{3+}) $x10^5$	$\{Ni^{2+}\}$ $x10^5$	$\{Mn^{2+}\}$ $x10^5$	$\{Mn^{3+}\}$ $x10^4$	$\{Cr^{2+}\}$ $x10^5$	$\{Cr^{3+}\}$
0.0	0	973	22	233	124	0	71	238	276	1.9727
0.2	194	775	42	227	283	567	25	196	117	1.9733
0.4	392	578	50	205	399	814	13	163	75	1.9747
0.6	590	382	50	174	510	962	69	129	50	1.9770
0.8	790	187	42	127	637	1036	27	87	29	1.9807
1.0	991	0	0	0	913	913	0	0	0	1.9909

Following are the effective ionic radii of the pertinent cations in tetrahedral and octahedral coordination according to Shannon and Prewitt (16) scale.

Ion	radius (Å) CN-4	CN-6
Ni^{2+}	0.55	0.69
Mn^{3+}	0.58	0.65
Mn^{2+}	0.66	0.83
Cr^{3+}	-	0.62
Cr^{2+}	0.73	0.82

The lattice strain introduced because of the size mismatch has not taken into account in the thermodynamic model for computing free energies of mixing. Therefore the difference between the experimental and theoretical values would be the strain enthalpy which could be estimated from the following expression for a regular solution (18) :

$$\Delta H_{st}=\Delta G^M_{exp}-\Delta G^M_{theor.}=A_1x_1^4x_2+A_2x_1^3x_2^2+A_3x_1^2x_2^3+A_4x_1x_2^4$$

where A_1,A_2,A_3,A_4 are constants and x_1 and x_2 are the mole fractions of $NiCr_2O_4$ and $MnCr_2O_4$, respectively. Polynominal regression fit fives A_1, A_2,A_3 and A_4 to be 9491 J/mole, 19135 J/mole, 15491 J/mole and 30379 J/mole, respectively. Strickly regular solution model does not take into account the asymetry observed in strain enthalpy (see Figure 5) but subregular solution model gives a better fit through yielding different values for constants A_1,A_2,A_3 and A_4.

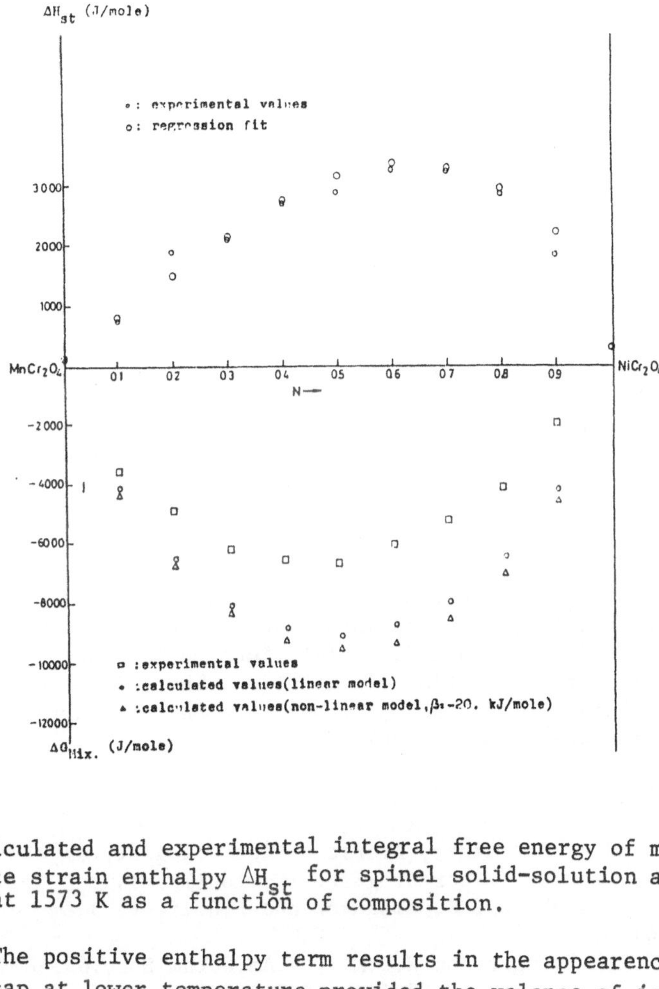

Fig.5. Calculated and experimental integral free energy of mixing $\Delta G_{Mix.}$ and lattice strain enthalpy ΔH_{st} for spinel solid-solution at ΔG_R^o=40 kJ/mole and at 1573 K as a function of composition.

The positive enthalpy term results in the appearence of a miscibility gap at lower temperature provided the valance of ions is not changed. Then ΔG^M can be expressed as :

$$\Delta G^M = \Delta H_{st} + RT(x_1 \ell n x_1 + x_2 \ell n x_2)$$

from which the critical temperature T_c and corresponding composition N_c may be calculated. T_c and N_c have been computed to be 1124°C and 0.866, respectively. No experimental evidence has been reported to date on the appearence of miscibility gap in the system $MnCr_2O_4$-$NiCr_2O_4$.

The accuracy in calculating cation distributions was limited by the accuracy of the site preference energies and the non-linearity term which are correct to, at best, ±5 kJ.

4. CONCLUDING REMARKS

1. Charge exchange reactions alters the traffic of size exchange reaction drastically. This in turn effects the configurational entropy and free energy of mixing. The effect of charge exchange of free energy of mixing becomes more profound if the system were composed of a normal and an inverse end-members. In this situation the "anchoring effect" of the tetrahedral cation for the redox in the octahedral sites would be minimal and this would render the reduction more effective in overall equilibria.

2. This model employs a quasi-chemical approach, hence treats only chemical bond contribution to the energetics of solid solution and ignores mechanical effect (strain energy) and volume effect (Madelung energy). All of these would contribute to the energetics of solution. This provides a serious limitation to the general applicability of this model. Jacob and Allcock (19) reported Madelung contribution to energetics is insignificant, whereas O'Neill and Navrotsky (2) demonstrated that electrostatic energy on disordering was very large and would swamp any crystal field stabilization energy (CFSE) term. They resulted that even at temperatures near 2000 K, Madelung energy is about two orders of magnitude larger than the configurational entropy contribution to the free energy of disordering. Glidewell(20) has shown that in many instances Madelung energy change with disordering is likely to be dominant over the CFSE term while both Dunitz and Orgel (15) and McClure (16) ignored the change in Madelung part of lattice energy cation distribution. There is no aggrement on which contribution plays the important role on the disordering and in turn on the energetics of solution.

REFERENCES

1. H. O'Neill and A. Navrotsky, American Minereologist, **69**,733(1984).
2. A.Navrotsky, American Minereologist, **71**,1160(1986).
3. A.Navrotsky, B.Wechsler, K.Geisinger and F.Seifert, J.Am.Ceram. Soc., **69**,418(1986).
4. K.T.Jacob,G.N.K.Iyengar and W.K.Kim, J.Am.Ceram.Soc.,**69**,487(1986).
5. T.O.Mason, J.Am.Ceram.Soc.,**68**,C-74(1985).
6. N.Koç, M.Timuçin, Master Thesis, Middle East Technical University, Ankara,Turkey, January(1981).
7. K.J.Jacob and K.Fitzner, J.Mat.Sci.**12**,481(1977).
8. B.Boucher, R.Buhl and M.Perring, J.Phys.Chem.Solids,**32**,1471(1971).
9. C.P.Marshall and W.A.Dollase, Am.Min.**69**,928(1984).
10. N.Renault, N.Buffier and M.Huber, J.Solid State Chem.**5**,250(1972).

11. B.J.Wood, In R.C.Newton, A.Navrotsky and B.J.Wood, Eds., Thermodynamics of Minerals (1981).
12. J.F.Elliot and M.Gleiser, Thermochemistry for Steelmaking. Addison-Wesley Co.Inc.,(1960).
13. O.Kubachewki, E.Evans, Metalurgical Thermochemical, Pergamon Press, London(1968).
14. H.Ohnishi ahd T.Terenasu, J.Phys.Soc.Japan,**16**,35(1961).
15. J.D.Dunitz,L.E.Orgel, J.Phys.Chem.Solids,**3**,318(1957).
16. D.McClure, J.Phys.Chem.Solids,**3**,311(1957).
17. R.D.Shannon and C.T. Prewitt, Acta Crystallogr.**B25**,925(1969).
18. C.Wagner,Thermodynamics of Alloy, Addison-Wesley, Co.Inc.,(1952).
19. K.T.Jacob and C.B. Alcock, J.Solid State Chem.,**20**,79(1977).
20. C.Glidewell, Inorganica Chimica Acta **19**,L45(1976).

II. Surface and Grain Boundary Phenomena

CHARACTERIZATION OF DEFECTS ON OXIDE SURFACES AND THEIR IMPACT ON SURFACE REACTIVITY AND CATALYSIS

W. Hirschwald
Freie Universität Berlin
Institut für Physikalische Chemie
Takustr. 3
D - 1000 Berlin 33

ABSTRACT: A brief survey of a few selected methods for the characterization of surface defects on oxides will be given. Reference will be made to the degree of non-stoichiometry, which can be analysed, for example, by High resolution Electron Energy Loss Spectroscopy (HREELS) and electrical conductivity measurements. Point defects like oxygen vacancies are detected by Atomic Beam Scattering (ABS) and possibly by Scanning Tunneling Microscopy (STM). Sputter induced changes are favourably characterized by Photoelectron Spectroscopy of adsorbed Xenon (PAX) and by adsorption/desorption experiments.

The influence of charged defects on the stability of oxide surfaces is illustrated for the case of zinc oxide.
Examples are given for the enhancement of adsorption and stimulation of surface reactions by defects.

The importance of these phenomena for heterogeneous catalysis is demonstrated by considering some aspects of methanol synthesis.

1. INTRODUCTION

A rather short and qualitative description of some methods to characterize defects on oxide surfaces or on metals in the early stage of oxidation will be given, supplemented by a few examples to illustrate the capability of these methods. The selection is made from a rather subjective point of view and several important techniques like

J. Nowotny and W. Weppner (eds.),
Non-Stoichiometric Compounds Surfaces, Grain Boundaries and Structural Defects, 203–219.

X-ray and ion scattering, SEXAFS, SEXELFS, LEED etc. will not be considered.

The author is, by no means, an expert in this field. He came across these problems in the course of experimental studies on oxide surfaces when annealing them in vacuo or when sputter-cleaning them.

Actually, V.E. Henrich reported in three reviews |1| on oxide surfaces with special reference to their modification by defects taking into account their characterization by ultra-violet photoelectron spectroscopy (UPS) and considering the difference in adsorption behaviour between nearly perfect and defect surfaces.

2. SOME EXPERIMENTAL METHODS

2.1 Scanning Tunneling Microscopy (STM)

In the ideal case an atomical sharp tip scans the surface at a tunneling current which is kept constant by varying the distance between surface and tip |2|.

In the following reference is made to recent results obtained by Behm et al. |3| on a close packed Al(111) surface. After sputter-cleaning and prolonged heating/annealing cycles the (111) surface exhibited a structure, characterized by extended terraces of several hundred Angström width separated by mostly monoatomic steps.

After a specific tip preparation |4| individual atoms could be resolved on this surface as shown in fig. 1. Over the entire area imaged this part of the surface is free of defects. Similar patterns were reproducibly obtained. The major part of the surface is thus characterized by a very low density of point defects. In average a corrugation of 0.3 Å was found. After adsorbing 5 % of a monolayer of oxygen on the clean (111) surface and after annealing this adsorbate at 800 K characteristic changes in the surface periodicity occured as shown in fig. 2. Six individual hollows are arranged in a regular hexagon and frequently these hexagons are grouped together in a larger ensemble. The hollows coincide with the center of three neigh-

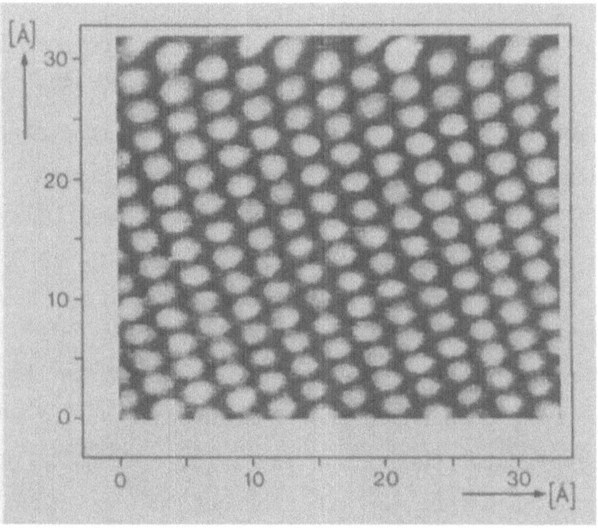

Figure 1. STM image of a clean Al(111) surface in the terrace range. From ref. |3|.

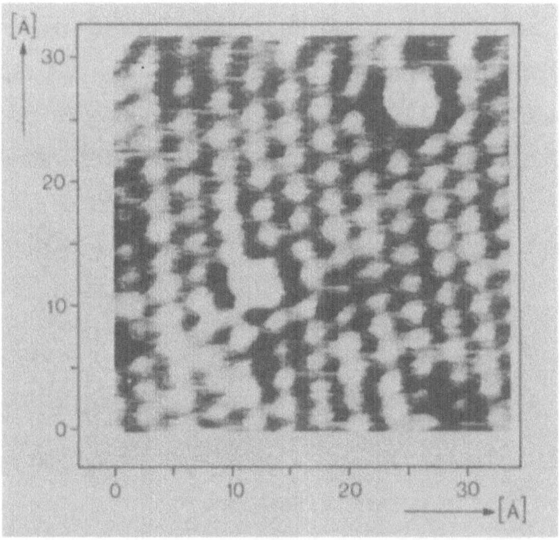

Figure 2. STM image of the surface in Fig. 1 after oxygen exposure (coverage about 5 % of a monolayer) and annealing at 800 K.From ref. |3|.

bouring Al surface atoms, which is the predicted site for adsorbed or subsurface oxygen atoms. Complementary HREELS measurements reveal that these features are most probably due to oxide nuclei |5|.

STM is obviously capable not only to observe first stages of oxidation but also to monitor defects on the atomic scale.

2.2 Photoemission of Adsorbed Xenon (PAX)

This interesting method to identify and characterize defects is based on ultra-violet photoelectron spectroscopy (UPS) and on the observation of the substrate independence of the Xe(5p) electron binding energies with respect to the vacuum level (E_B^V). Wandelt, who proposed and developed this technique |6|, pointed out that electron binding energies of adsorbed xenon atoms referred to the Fermi level (E_B^F) are sensitive to the local work function of the respective adsorption site. Consequently, separated photoemission lines of xenon atoms adsorbed on different surface sites can be resolved. One of the few oxide systems studied by this method is TiO_2(100) |7|. On this surface defects involving loss of oxygen can be easily formed either by vacuum annealing or by argon ion sputtering |8|. In fig. 3 UPS spectra of xenon adsorbed on differently pretreated TiO_2 surfaces are displayed. These spectra reveal pronounced changes due to vacuum annealing at 1123 K (curve b) and Ar^+ sputtering (curves c1) and c2). The broadened peaks of the defect surfaces can be decomposed by a best fit into three Lorentzian lines. This is shown in fig. 3 for the $5p_{1/2}$ peak of spectra b) and c). In addition to peak III, typical for the undisturbed surface two further states (I and II) exist which can be attributed to oxygen deficient sites, e.g. Ti^{+3} associated with an oxygen vacancy $V_O^{\bullet}$. Spectrum c2) with the lower Xe coverage demonstrates the prefered initial population of the defect sites.

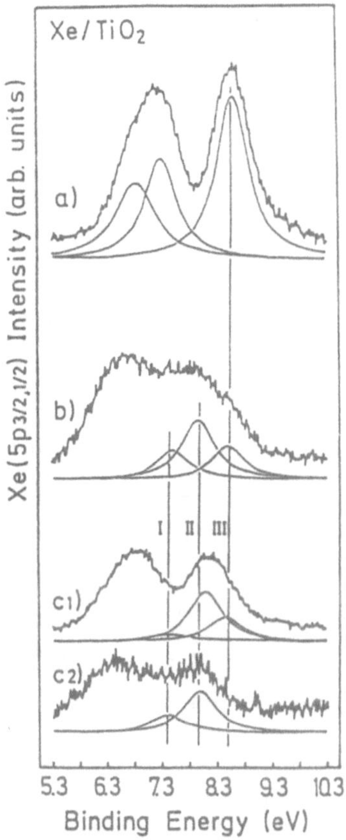

Figure 3. UV photoemission spectra of xenon adsorbed at 65 K on a) stoichiometric TiO_2 (100); b) TiO_2 annealed at 1123 K in UHV and c) sputtered TiO_2, together with peak deconvolutions. Spectra a), b) and c1) correspond to xenon coverages close to monolayer saturation, while c2) corresponds to a low coverage. From ref. |7|.

2.3 Atomic Beam Scattering and Diffraction (ABS)

For monoenergetic beams of light non-reactive gases, especially helium, scattering on ordered surfaces is predominantly elastic and if the DeBroglie wavelength is of the adequate order of magnitude, diffraction will prevail. Suitable energies are in the range of 30 to 300 meV.

He scattering has several advantages: It is strictly surface sensitive and non-destructive; it has a high sensitivity to structural disorder of the topmost layer and it is well applicable to insulators and semiconductors as no charging occurs |9|.

On scattering from a well ordered surface, nearly 15 % of the scattered helium atoms appear in the specular beam. This percentage decreases to 1 to 5 % when the surface is disordered.

Thus, measurement of the fraction of specularly scattered helium can provide information on the degree of atomic disorder in the solid surface.

To obtain structural information from diffraction studies, the measured intensities have to be correlated with the so-called "corrugation function", i.e. the scattering surface contour (following the contour of constant total electron density) which in turn has to be correlated with the actual geometric surface structure. Very often the surface corrugation provides a direct picture of the geometrical arrangement of the surface atoms. But in order to obtain precise values of bond lengths and exact locations of the atoms relative to each other, one has to relate the experimental results to ab initio surface density calculations.

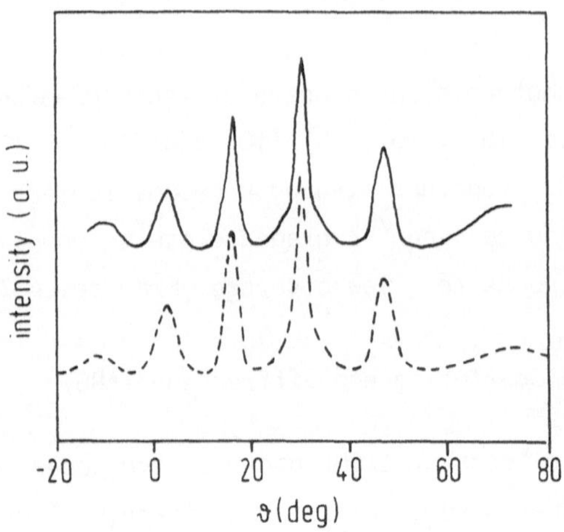

Figure 4. Diffraction patterns of He atoms scattered from a Ni(100) - c(2x2)O surface (solid line) compared to a model calculation (dashed line). From ref. |10|.

A recent thorough inspection of the diffraction patterns of He scattered from Ni(100)-c(2x2)O revealed a pronounced nearly triangular background around the specular (00) beam, fig. 4 (solid line) |10|. Measurements with different azimuth angles indicate that the total diffuse scattering contribution is cone-shaped. Best agreement with this experimental observation is attained by a model calculation considering a random distribution of 15 % empty oxygen sites ("vacancies") with a lateral relaxation of $\sim$0.25 Å of the neighbouring adatoms away from the vacancy, fig. 4 (dashed line) |10|. Evidence was given to the vacancy model by Auger electron spectroscopic (AES) determination of the oxygen surface concentration.

This example demonstrates the possibility to characterize atomic point defects by ABS and even to determine their density approximately.

2.4 High resolution electron energy loss spectroscopy (HREELS)

This method is based upon inelastic scattering of monoenergetic primary electrons with energies of some eV |11|. The excitation of

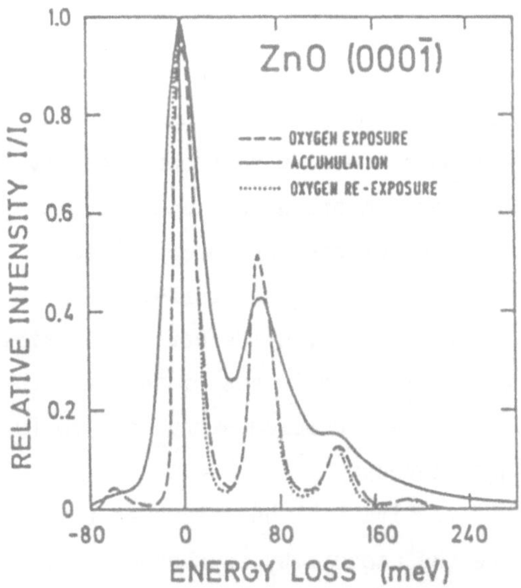

Figure 5. Specular HREELS spectra of a ZnO(000$\bar{1}$) face (E_p = 3.7 eV). The accumulation layer was produced by treatment with atomic hydrogen. From ref. |13|.

intrinsic surface phonons was studied by HREELS for a number of oxide systems compiled with references by Egdell |12|. In broad-gap semiconducting or insulating oxides multiple loss peaks occur due to sequential inelastic scattering in separate loss events. Fig. 5 shows a loss spectrum for ZnO(000$\bar{1}$) with pronounced overtones (dashed line). The strong background of intrinsic excitations can be markedly reduced by producing accumulation layers on nonconducting oxides |12, 13|. Fig. 5 (solid line) reveals the influence of an atomic hydrogen treatment on the intrinsic surface phonons |13|. A pronounced suppression is effected. Reoxidation restores the original state (dotted line). These spectra indicate that HREELS might be a good tool to study the degree of nonstoichiometry (oxygen deficiency) of oxide surfaces.

2.5 Electrical Conductivity

Collective properties of charged structural defects like vacancies and interstitials can be studied by conductivity measurements. Actually, this method is not specifically surface sensitive and it yields information about all types of shallow electronic defects being present in the sample.

This method can be rendered more surface sensitive by applying very thin substrate layers on an insulating support and by studying the dependence of conductivity on processes restricted to the surface like, for instance, ionosorption or desorption of oxygen.

Furthermore, conductivity changes correlated with changes of oxygen pressure limit the variety of defects responsible for these effects. If an unpaired electron is associated with the defect electron spin resonance (ESR) can be a complementary tool to further characterize and identify these defects.

In the following some results will be presented for zinc oxide as an example to illustrate the capability of conductivity measurements. Steady state values of the conductivity of thin ZnO layers (12 μm thick) were measured in a constant oxygen pressure of 150 mbar in dependence on temperature |14|. The results plotted in fig. 6 reveal two regimes characterized by different slopes, corresponding to energy

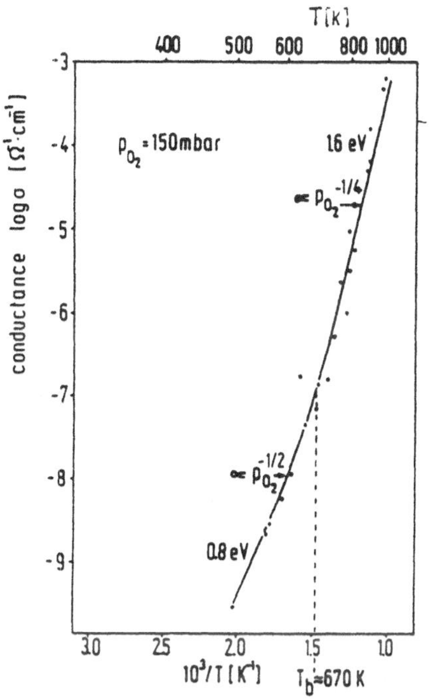

Figure 6. Temperature dependence of the conductance of a polycrystalline ZnO layer (12 μm thick) in oxygen (P_{O_2} = 150 mbar). From ref. |14|.

requirements of 0.8 eV in the low temperature range (T < 670 K) and of 1.6 eV in the high temperature range (T > 670 K). Separate measurements of the pressure dependence (P_{O_2} = 0.1 to 15 kPa) revealed a $P_{O_2}^{-1/2}$ dependence in the low temperature range and a $P_{O_2}^{-1/4}$ dependence of the conductivity in the high temperature range.

Energy considerations combined with the interpretation of the pressure dependence give evidence to a mechanism which is determined by the formation of donor defects:

$$ZnO \rightleftarrows Zn_i^+ + \frac{1}{2} O_2 + e \qquad (1)$$

One might consider oxygen vacancies ($V_O^{\bullet}$) as an alternative to interstitial zinc (Zn_i^+), but results of ESR studies are more in favour of interstitial zinc |15|. This does not exclude the existence of oxygen vacancies in the topmost layer, as proposed by Göpel |16| and by Green et al. |17|.

3. STABILIZATION OF SURFACES BY DEFECTS

If zinc oxide is annealed at high enough temperatures it will decompose to give $Zn_{(gas)}$ and $O_{2(gas)}$. At temperatures below 1000 K zinc atoms will be mainly incorporated into the ZnO lattice according to eq. (1) forming shallow donor centers there which give rise to an increase in conductivity. As no pronounced changes of the carrier mobility were observed under these conditions |18|, the increase in conductivity with temperature which extends - even in 150 mbar of oxygen - over four orders of magnitude (fig. 6) is attributed to the enrichment in electrons and hence in donors (Zn_i^+).

The enrichment of donors combined with oxygen release finally comes to saturation, depending on temperature and ambient oxygen pressure. The release of oxygen during vacuum annealing of ZnO ($10\bar{1}0$) faces is shown in fig. 7 for four different temperatures |16|. The

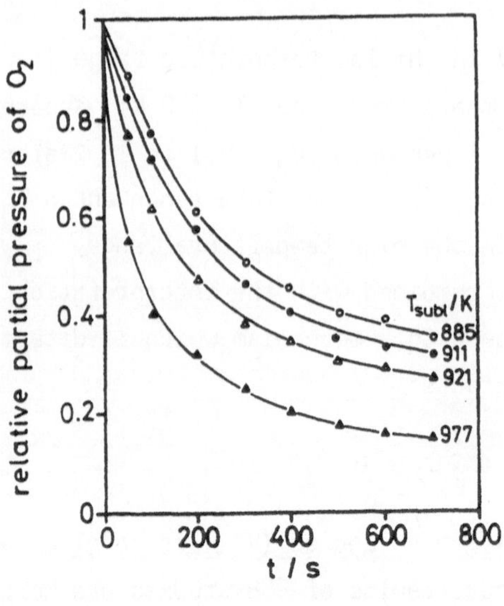

Figure 7. Oxygen release from ZnO lattice during initial heating of oxygen annealed ($10\bar{1}0$) faces. From ref. |16|.

steady-state oxygen partial pressure over these samples is markedly decreased, i.e. the surface is more stable towards decomposition. As we observed low temperature (T < 800 K) photodecomposition of oxygen pretreated ZnO by band-gap irradiation (Eg = 3.2 eV), which is obviously induced by interband transitions from bonding states (valence band) to non-bonding states (conduction band) according to

$$
\begin{array}{lll}
\text{(a)} & 2\,h\nu \longrightarrow 2\,e{\sim}h & \\
\text{(b)} & O_s^{2-} + h \longrightarrow O_s^{-} & \text{photo-decomposition} \\
\text{(c)} & O_s^{-} + h \longrightarrow \tfrac{1}{2}O_2 & \\
\text{(d)} & O_s^{-} + e \longrightarrow O_s^{2-} & \text{recombination} \\
\text{(e)} & Zn_s^{2+} + e \longrightarrow Zn_i^{+} & \text{donor formation}
\end{array}
\qquad (2)
$$

we conclude, that formation of holes in the valence band is a limiting step not only for the photochemical but also for the thermal decomposition of ZnO |19|. If this is true the increase of electron density with temperature according to fig. 6 will enhance recombination corresponding to the inverse reaction steps (a) and (d) of scheme (2) and therefore suppress further release of oxygen from the oxide lattice. Further evidence to this assumption comes from the temperature dependence of the initial quantum yield of carrier generation. Fig. 8 reveals that it passes a maximum in the temperature range where donor formation starts, indicating that recombination is enhanced by donor accumulation. In addition electron/hole separation will be reduced due to a decrease of the electrical field across the space charge layer |20, 21|.

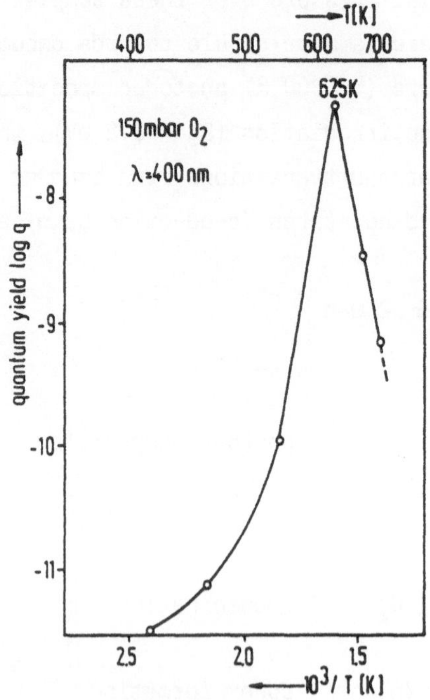

Figure 8. Temperature dependence of initial quantum yield of carrier generation in a 12 µm ZnO layer at constant oxygen pressure (P_{O_2} = 150 mbar). From ref. |20, 21|.

4. ADSORPTIVE AND CATALYTIC ACTIVITY INFLUENCED BY DEFECTS

4.1 Mechanically and chemically induced defects

A frequent pretreatment for oxide single crystals preceding surface spectroscopic investigations in ultra-high vacuum (UHV) comprises mechanical polishing and/or chemical etching. In the case of ZnO single crystals this combined pretreatment did not change the long range order as monitored by LEED but obviously leads to atom displacements or other point defects, not yet identified, which give rise to enhanced surface reactivity. This was observed in the course of CO adsorption/desorption. While from as-grown or cleaved ZnO surfaces only CO desorbed after low temperature and room temperature adsorption of CO |22| between 10 % and 100 % CO_2 desorption was observed from

etched/polished surfaces, the percentage depending obviously on details of the pretreatment |22 - 24|.
Similar observations were made by Hofmann et al. |22| after ethylene adsorption on cleaved and on etched/polished ZnO(0001) faces. Thermal desorption spectra (TDS) of C_2H_4 revealed pronounced differences for these two differently pretreated surfaces of identical crystallography.

Thermal desorption spectroscopy of simple molecules obviously constitutes a sensitive tool to probe surface disorder. Not only the bond strength between adsorbed molecule and defect site may be obtained from TDS but also information about the surface density of defects may be derived from the quantitative evaluation of the TD spectra.

4.2 Ion bombardment induced defects

It is well known from Secondary Ion Mass Spectroscopy (SIMS) and from sputter-cleaning procedures that a number of oxide systems are sputtered selectively, i.e. oxygen is removed preferentially and the oxide surface and possibly near surface regions are reduced |25|.
Removal of oxygen can be correlated with the formation of oxygen vacancies |1|. Ion bombardment is therefore another method to generate defects at oxide surfaces by intention.

Egdell et al. |26| studied water adsorption on $SrTiO_3$ before and after argon ion bombardment (1 keV Ar^+ ions, 5 μA, 10 min). These authors estimated an oxygen vacancy density of about one in every two surface unit cells after Ar^+ sputtering. While on oxygen annealed $SrTiO_3$ (100) surfaces no water adsorption could be observed at room temperature, it adsorbed readily on the defect surface, as followed by HREELS, which probes the vibrational modes of adsorbed H_2O molecules. This is in agreement with earlier observations by Henrich| 27 |. While at low temperature both the OH-stretching mode and the scissor bending mode were observed the latter vibrational mode did not occur at room temperature on the defect surface indicating restriction of this bending mode or even dissociation of the H_2O molecule.

For the technical synthesis of methanol a zinc oxide/copper based catalyst is frequently used. The feed gas is mainly composed of CO

and H_2. Addition of CO_2 to the feed gas (up to 10 %) enhances the conversion to methanol.
In the course of investigations of the mechanism of methanol synthesis we studied the interaction of CO, CO_2 and methanol with different single crystal faces of ZnO by photoelectron spectroscopy. For CO_2 we found nearly no adsorption on oxygen annealed $(000\bar{1})$ faces even at 100 K, while adsorption could be observed on this face after generation of defects by Ar^+ sputtering (3 keV Ar^+ ions, 12 μA, 10 min) |28|. The opposite is valid for CO adsorption, i.e. it adsorbs readily on the oxygen annealed surface and does not adsorb ($T \geq 100$ K) on the defect surface. Surprisingly, the XPS and UPS spectra for both CO on oxygen annealed and CO_2 on defect $ZnO(000\bar{1})$ were found to be identical. Taking this result into account we are tempted to propose for the common surface complex the following configuration (fig. 9):
Assuming the surface carbonate to be an intermediate of methanol synthesis might then explain the promoting influence of CO_2. Due to

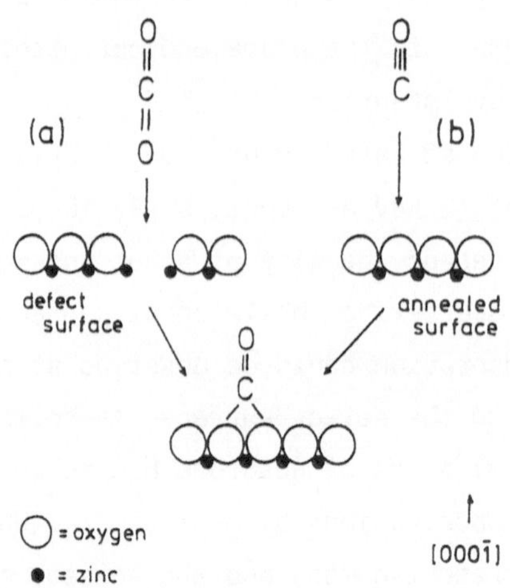

Figure 9. Spere model of a polar $ZnO(000\bar{1})$ face with (a) and without (b) defects (oxygen vacancy), interacting with CO_2 and CO, respectively. From ref. |28|.

the strong reducing potential of CO and H_2 a high density of vacancies would exist on the surface under synthesis conditions rendering further adsorption of CO highly improbable. CO_2 not only yields the required intermediate but also reoxidizes to a certain extent the surface, thereby increasing the probability for CO adsorption.
Actually, Waugh et al. |29| found that the enhancement of methanol synthesis activity is roughly proportional to the number of excess oxygen vacancies, in agreement with a mechanism proposed earlier by Kung |30|.

Finally, in a last example, it will be illustrated that defects on oxide surfaces can modify not only the activity but also the selectivity of the catalyst.
Photoelectron spectroscopic studies (XPS) and TDS indicated - at least qualitatively - a correlation between the sputter induced surface defect density and the formation of oxygen free products. XPS reveals CH_n groups on the surface (C 1s binding energy: 285 - 284 eV) |31| in correspondence to CH_4 desorption in TDS |32|.

IN SUMMARY: the capability of a few selected methods for the characterization of defects and nonstoichiometry on oxide surfaces was demonstrated. The influence of defects on the stability and catalytic reactivity of these surfaces was illustrated by some examples from current catalytic research.

REFERENCES

1. a) V.E. Henrich, Progr. Surf. Sci. 9 (1979) 143
b) V.E. Henrich, Rep. Progr. Phys. 48 (1985) 1481
c) V.E. Henrich, Chapt. 2 in Surface and Near Surface Chemistry of Oxide Materials, J. Nowotny and L.C. Dufour (Eds.), Elsevier, Amsterdam 1988, p. 23 ff.
2. G. Binnig, H. Rohrer, Ch. Gerber and E. Weibel, Physica (Utrecht) 107, B+C (1981) 1335 and Phys. Rev. Lett. 49 (1982) 57
3. J. Wintterlin, H. Brune, H. Höfer and R.J. Behm, J. Appl. Phys. A, submitted
4. J. Wintterlin, T. Gritsch, H. Höfer, J. Wiechers and R.J. Behm, Phys. Rev. B, submitted
5. R.L. Strong, B. Firey, F.D. de Wette and J.L. Erskine, J. Electron Spectrosc. Related Phenom. 29 (1983) 187
6. K. Wandelt, J. Vac. Sci. Technol. A 2 (1984) 802
7. P. Dolle, K. Markert, W. Heichler, N.R. Armstrong, K. Wandelt, R.A. Fiato and K.S. Kim, J. Vac. Sci. Technol. A 4 (1986) 1465
8. V.E. Henrich, G. Dresselhaus and H.J. Zeiger, Phys. Rev. Lett. 36 (1976) 1335
9. T. Engel and K.H. Rieder, Structural Studies of Surfaces with Atomic and Molecular Beam Diffraction, in Springer Tracts in Modern Physics 91, G. Höhler (Ed.), Springer, Berlin 1982, p. 55 ff.
10. W.A. Schlup and K.H. Rieder, Phys. Rev. Lett. 56 (1986) 73
11. H. Ibach and D.L. Mills, Electron Energy Loss Spectroscopy and Surface Vibrations, Academic Press, New York, 1982
12. R.G. Egdell, in Adsorption and Catalysis on Oxide Surfaces, M. Che and G.C. Bond (Eds.), Elsevier, Amsterdam, 1985, p. 173 ff.
13. Y. Goldstein, A. Many, I. Wagner and J.I. Gersten, Surface Sci. 98 (1980) 599

14. P. Bonasewicz, W. Hirschwald and G. Neumann, *Thin Solid Films* **142** (1986) 77; *phys. stat. sol. (a)* **97** (1986) 593 and *Appl. Surface Sci.* **28** (1987) 135
15. G. Neumann, *phys. stat. sol. (b)* **105** (1981) 605
16. W. Göpel, *Surface Sci.* **62** (1977) 165
17. M. Green and I.R. Lauks, *J. Chem. Soc. Faraday Trans. I,* **74** (1978) 2724
18. R. Littbarski, Dissertation, FU Berlin 1981
19. M. Grunze, W. Hirschwald and E. Thull, *Z. physik. Chem. Neue Folge,* **100** (1976) 201 and *Thin Solid Films,* **37** (1976) 351
20. P. Bonasewicz, W. Hirschwald and G. Neumann, *J. Electrochem. Soc.* **133** (1986) 2270
21. W. Hirschwald, *Acc. Chem. Res.* **18** (1985) 228
22. W. Hirsch, D. Hofmann and W. Hirschwald, Proc. 8th Intern.Congr. on Catalysis, Berlin 1984, (Verlag Chemie, Weinheim 1984) IV-251
23. W. Hotan, W. Göpel and R. Haul, *Surface Sci.* **83** (1979) 162
24. G. Heiland and H. Lüth, Adsorption on Oxides in *The Chemical Physics of Solid Surfaces and Heterogeneous Catalysis,* D.A. King and D.P. Woodruff (Eds.), Elsevier, Amsterdam, 1987, p. 177
25. W. Hirschwald, Selected Experimental Methods in the Characterization of Oxide Surfaces, in *Surface and Near Surface Chemistry of Oxide Materials,* J. Nowotny and L.C. Dufour (Eds.), Elsevier, Amsterdam, 1988, p. 61 ff.
26. R.G. Egdell and P.D. Naylor, *Chem. Phys. Lett.* **91** (1982) 200
27. V.E. Henrich, G. Dresselhaus and H.J. Zeiger, *Solid State Commun.* **24** (1977) 623
28. C.T. Au, W. Hirsch and W. Hirschwald, *Surface Sci.* **197** (1988) 391
29. M. Bowker, J.N.K. Hyland, H.D. Vandervell and K.C. Waugh, Proc. 8th Intern. Congress on Catalysis, Berlin 1984, (Verlag Chemie, Weinheim 1984) II-35
30. H.H. Kung, *Catal. Rev.-Sci. Eng.* **22** (1980) 235
31. C.T. Au, W. Hirsch and W. Hirschwald, *Surface Sci.*, submitted
32. W.H. Cheng, S. Akhter and H.H. Kung, *J. Catalysis* **82** (1983) 341

INTERFACE TRANSPORT IN MONOXIDES

M. Déchamps and F. Barbier
Laboratoire de Chimie des Solides, UA CNRS 446
Université Paris-Sud, Bât. 414
91405 Orsay CEDEX
France

ABSTRACT. Recent experimental data concerning bulk and grain boundary diffusion in MgO and NiO have been collected together. An attempt has been made to summarize the cross interpretation of the authors of the different works. There is a large spreading of the results which might be due to uncontrolled grain boundary chemistry and / or to artifacts. The most recent data indicate, for MgO, an enhanced cation and oxygen boundary diffusion, independent of dopes, with $\alpha D'\delta/D \approx 10^{-3}$ for chromium and oxygen diffusion. In the case of NiO, quantitative results obtained on oxide scales indicate an important enhanced cation diffusion ($\alpha D'\delta/D \approx 10^{-2}$). It was not possible to confirm these results on bicrystals, where $\alpha D'\delta/D$ was found to be less than 10^{-4}. In our opinion, these controversial results are due to oxide scale artifacts.

1. INTRODUCTION

In crystalline materials the boundaries and the intergranular phases are of great importance for properties and processes such as fracture strength, toughness, plastic deformation, creep, electrical properties, sintering, grain growth, corrosion, segregation, precipitation, or diffusion...Despite the many similarities which exist in metal and non metal grain boundaries there are some basical differences which, essentially, are due to the ionic nature of the oxides, the solute behaviour, and the intrinsic defect concentration [1].

The ionic structure leads to the formation of a space charge localized from 2 to 10 nm from the boundary, according to the dielectric constant of the oxides [1,2,3]. Yet, theoretical models [4,5] for the dependence of the interfacial space charge on temperature suggest that the effects of the space charge become negligible at high temperature, and it was shown by Yan et al [6] that, for Sr in KCl, the contribution of space charge to boundary diffusion is less than 10^3 times that of diffusion in the lattice. Thence, the phenomenon does not seem relevant to grain boundary diffusion and it appears now clearly, from recent works, that the other two differences are of prime importance to understand boundary diffusion behaviour in ceramics.

J. Nowotny and W. Weppner (eds.),
Non-Stoichiometric Compounds Surfaces, Grain Boundaries and Structural Defects, 221–236.

Table I

Some characteristics of interest for diffusion in MgO and NiO.

	MgO	$Ni_{1-y}O$
Crystallography	rocksalt a = 0.421 nm	rocksalt a = 0.417 nm
Tm / 0.6 Tm (°C)	2800 / 1680	1957 / 1065
Ionicity factor	# 0.88	0.79 - 0.84
Predominant defects	Schottky nill —> $V_{Mg}'' + V_O^{\circ\circ}$	cation vacancies $1/2\ O_2 \longrightarrow V_{Ni}^{\alpha'} + \alpha h^{\circ} + O$ $\alpha = 1$ at 0.6 Tm $\alpha = 1$ (30%) and 2 (70%) at Tm (P_{O_2} = 0.21 atm) [24]
Defect formation energy (eV)	7.5 - 7.7 [9]	≈ 1.6 for $\alpha = 1$ ≈ 2.4 for $\alpha = 2$ [25]
Defect concentration	< 10^{-6} at Tm	y ≈ { 10^{-3} at Tm, 10^{-4} at 0.6 Tm (P_{O_2} = 0.2 atm); ≈ 0 at low P_{O_2} and T [26]
Impurity level (ppm)	minimum 60 (Ca, Si)	minimum 80 (Si, Al, Fe, Ca)
Diffusion regime	always extrinsic	possibility of intrinsic diffusion

The solute behaviour is of great importance because many observations have been made in high purity metals with samples containing less than 1 ppm impurity, whereas most ceramics are difficult to purify with impurities in oxide ceramics often at the 100 ppm level (cf. table I). Moreover, they usually present a low solubility, and segregate strongly at boundaries, as shown by Kingery [7,8] in MgO, where all the dopes investigated (Al, Ca, Cr, $Fe^{(3+)}$, La, Sc, Si, Ti) were found to segregate substantially, in proportion to the difference of ionic radii of solvent and solute. It is also well established that the lower the temperature, the higher the relative segregation is, and that the fact that a boundary is special does not decrease drastically the segregation (the concentration was just observed to decrease by a factor two). Thence, it seems clear that segregation at grain boundaries in ionic oxides such as MgO is very common, if not general, which, usually, does not allow to ignore the segregation factor α in the expression of the grain boundary diffusion parameter $\alpha D'\delta$ which will be used in the following to characterize boundary diffusion.

Precipitation of second phases is yet a more crippling problem and seems to have biased some interpretations. In this respect silicon is of special importance due to its systematic presence, low solubility in MgO (≈ 400 ppm at 1800°C) and tendency to form relatively low melting compounds and/or glassy films at grain boundaries.

The defect concentration is related to the energy for defect formation, which is higher for MgO and NiO (cf. table I) than for metals [9]. This leads to an intrinsic defect concentration much lower, at least in MgO, than in metals, which, when coupled with the high impurity level of oxides and segregation at boundaries makes it difficult or impossible to control the boundary physico-chemistry and to observe intrinsic phenomena. After publications by Osenbach and Stubican [10], and by Atkinson and Taylor [11-14], respectively on grain boundary diffusion in MgO and NiO, these difficulties were somewhat downplayed. But, the recent works of Dolhert [15] and Roshko [16] on MgO, and of Barbier et al [17-19] on NiO have revived some questions and shown that the problems are far from being definitely resolved.

The lattice and grain boundary diffusion properties of monoxides, in fact NiO and MgO, will be reviewed simultaneously because the latter are dependent on the former [20-23], and it is of prime importance to obtain the lattice diffusion coefficient on the same system as that in which the boundary diffusion was studied. Some properties of MgO and NiO are listed in table I. Both oxides have a cubic rocksalt structure which give them isotropic lattice diffusion. But, from a diffusion point of view they have few other common properties. MgO is essentially a stoichiometric oxide with a very limited Schottky defect structure that leads, in practice, to a defect structure controlled by aliovalent impurities and to an extrinsic lattice diffusion behaviour. $Ni_{1-y}O$ is a metal deficient oxide with respect to stoichiometric composition, with cation vacancies (complemented by positive holes) as predominant defects. As a measure of the extent of non-stoichiometry, the value of y ranges from 10^{-5} at low T and P_{O_2} to 10^{-3} at high T and P_{O_2}. Thence, the deviation from stoichiometry is larger than the impurity content in available materials and it is possible to study lattice intrinsic properties of NiO, but specimens of high purity are required.

2. MAGNESIA

2.1. Experimental results (Fig. 1, tables II and III)

2.1.1. Cation diffusion. Wuensch and Vasilos [27] have studied grain boundary diffusion of Ni and Co in bicrystals. They observed an erratic and not reproducible preferential diffusion along boundaries. The effect was attributed to impurity segregation because impurities were present in large amounts at each of the boundaries in which enhanced boundary diffusion was noted, and no special grain boundary diffusion effect was observed when there were no impurities detectable by electron microprobe analysis.

Table II

Experimental diffusion conditions in MgO

material [reference]	major impurities (ppm)	diffusing element $\alpha D'\delta/D$	concentrat. profile determinat.	dopant (ppm)	remark
bicrystals Norton Co [27]	Ca 300-500 Si 200-300 Fe 10- 20 Al 5- 20 (e)	Ni (a) Cr (a) 10^{-2} to 1 (estimate)	electron microprobe		enhanced GB diffusion attributed to 2nd phase
tilt <100> bicrystals - Norton Co - lab. made [10]	Ca 300-500 Si 150-300 Fe 10- 30 Al 3- 25	^{51}Cr (b) up to 1	serial sectioning	Cr 0- 500	dopant increases $\alpha D'\delta/D$ by a factor 4
high purity sintered powder [16]	Si ? Ca ≈ 0- 50 Al ≈ 0- 30 P ?	^{26}Mg (c) 10^{-2} Cr (c) 10-3 ^{18}O (d) 10^{-2}	SIMS SIMS SIMS	Ca 900 Sc 0-2700 Cr 800	D'_{Cr} and D'_O unaffected by dopants
single and poly- crystals [15]	Ca 5-500 Si 15-300 Fe 3- 80 Al 10- 50 (f)	^{18}O (d) 10^{-3}	SIMS		Do and Do' unaffected by the amount of impurities (except Si)

a) Dried solution deposited at the surface.
b) Vapor deposition.
c) Electron beam evaporation.
d) Gas exchange.
e) Analysis of ref. 10 concerning Norton's bicrystals.
f) The lower range corresponds to the impurity content of a special high purity single crystal made at Oak Ridge National Laboratory.

Osenbach and Stubican [10] have studied the diffusion of ^{51}Cr in undoped and Cr doped MgO bicrystals as a function of <100> unsymmetrical tilt grain boundary misorientation (θ) :

- A large enhanced boundary diffusion was observed. For large angle boundaries $\alpha D_{Cr}'\delta / D_{Cr} \approx 1$, i.e., assuming δ = 1 nm, $\alpha D'/ D \approx 10^7$.
- The effect of the misorientation tilt angle θ on diffusitivity was studied according to two directions, // and $\perp$ to a [100] tilt axis. The ratio $\alpha D'\delta_{//}$ to $\alpha D'\delta^{\perp}$ was 10^2 for a low angle boundary (θ = 5°), and 2

in larger angle boundaries as a result of a large increase in $\alpha D'\delta^{\perp}$ with increasing θ. This decreasing anisotropy was considered consistent with Turnbull and Hoffman's model for diffusion [28] in low angle boundaries and as proof of a pipe diffusion mechanism.

- Concerning the effect of dopes, D_{Cr} was found to increase linearly with Cr concentration in the bulk, and $\alpha D_{Cr}'\delta$ was increased by a factor 4 for MgO containing 0.56 at % Cr. These results were considered to indicate that V_{Mg}'' play an important role in lattice diffusion (through the equilibrium $Cr_{Mg}^{\circ} + V_{Mg}'' \longleftrightarrow (Cr_{Mg}^{\circ} V_{Mg}'')'$ dimers) as well as in grain boundary diffusion.

Recently, Roshko and Kingery [16] have investigated cation (Mg, Cr) and oxygen diffusion in MgO. In an attempt to study controlled extrinsic diffusion effects, great precautions were taken to minimize contamination and all synthesis and processing were carried out under unusually clean conditions (class 100 clean room, high resistivity water, gloves...). The effect of homovalent (Ca) and aliovalent dopes (Cr, Sc) was studied. The dopant content and the diffusion temperatures were defined to stay within the limit of solubility of the dopes in the matrix [29,30]. The lattice diffusion results are in good agreement with previous works and are interpreted by a vacancy mechanism, as expected for diffusion of cations in MgO.

Table III

Diffusion parameters in MgO

diffusing element (dopant ppm) [ref]	bulk diffusion $D = K \exp - (Q/kT)$		boundary diffusion $\alpha D'\delta = K'\exp-(Q'/kT)$		order of magnitude of $\frac{\alpha D' \delta}{D}$
	K ($cm^2.s^{-1}$)	Q (eV)	K' ($cm^3.s^{-1}$)	Q' (eV)	
Mg [16]	5.3 10^{-2}	3.25	3.0 10^{-4}	3.21	10^{-2}
Cr [10,31]	1.0 10^{-3}	3.05		1.87	1 to 10
Cr (Ca 900) [16]	4.6 10^{-4}	2.82	2.8 10^{-10}	1.84	10^{-3}
- (Sc 800) -	3.2 10^{-3}	3.03			
- (Sc 1600) -	7.6 10^{-7}	1.89			
- (Sc 2700) -	3.6 10^{-6}	2.05			
O (Ca 900) [16]	3.2 10^{-4}	3.97	5.0 10^{-9}	2.87	10^{-1} to 10^{-2}
- (Cr 800) -					
- (Sc 2700) -					
O [15]	3.9 10^{-6}	3.45	3.2 10^{-7}	4.06	10^{-3}

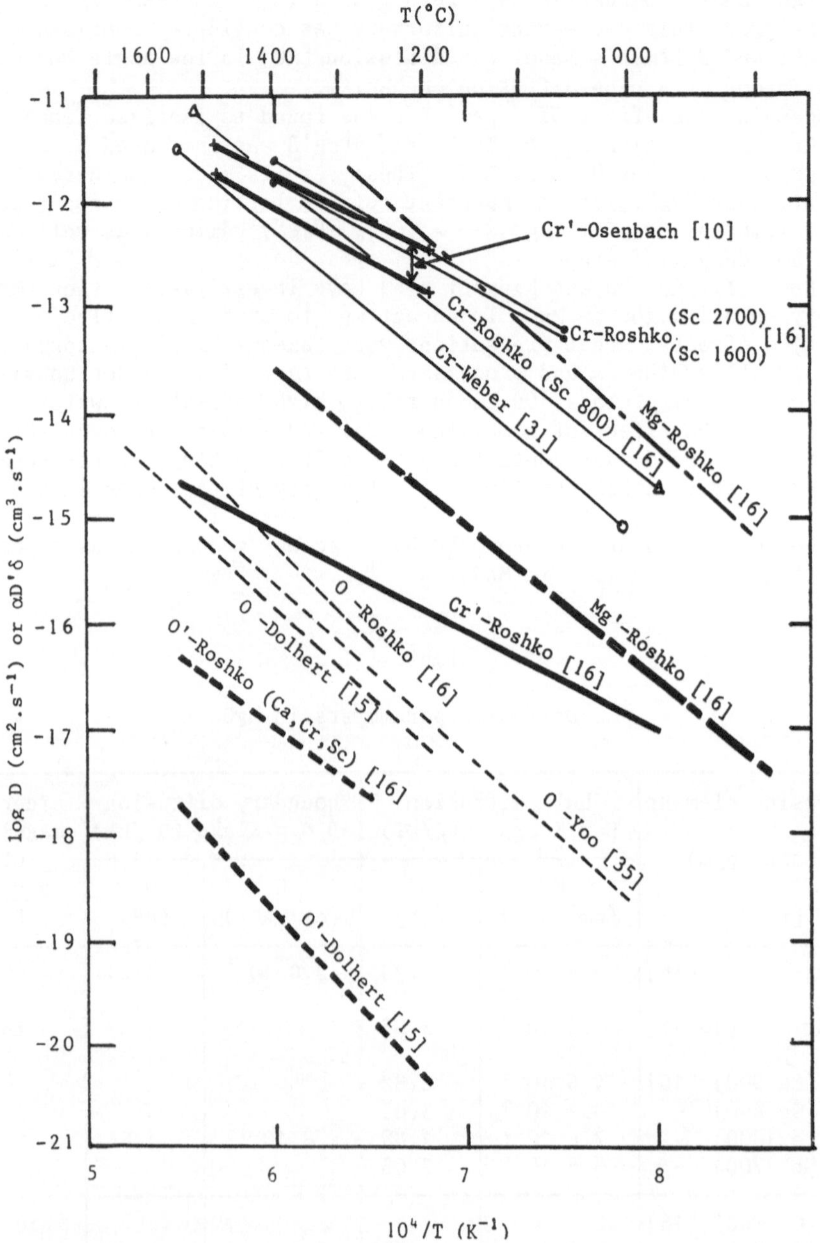

Fig. 1. Bulk diffusion coefficients (D) and grain boundary diffusion parameters ($\alpha D'\delta$) versus 1/T for MgO.
The experimental temperature ranges correspond approximately to the ends of each curve. Thick lines indicate grain boundary diffusion parameters, and thin lines lattice diffusion coefficients. Diffusing elements are indicated, with possible dopes in brackets.

The difference between the activation energy measured for bulk diffusion of Mg in *undoped MgO* (3.25 eV) and the energy for vacancy migration in MgO (2.1 eV [32]) was attributed to half the heat of solution of aliovalent impurities contained in the low purity ^{26}MgO tracer, or from the precipitates located at the grain boundaries (about 2 eV [30]).

The effect of an increasing amount of trivalent dope (Sc) was investigated on Cr diffusion. For undoped or low Sc content MgO, the diffusion mechanism is by free vacancies (activation energy = migration energy for Cr motion in MgO ≈ 3 eV [32]). But, when the concentration of vacancies is increased by adding trivalent solutes ($Sc_2O_3 \longrightarrow 2\, Sc_{Mg}^{\circ} + V_{Mg}'' + 3\, O$), the increasing concentration of associates (vacancy pairs) which results from increasing concentration of free vacancies ends up dominating lattice diffusion ($Q \approx 2$ eV).

Despite the efforts carried out to operate on "clean" MgO it was not always possible to avoid contamination by phosphorous (due to H_3PO_4 used to wash MgO) and the boundary diffusion appears to have been heavily affected,in undoped specimens, by second phases (Mg, P) present at triple junctions, and occasionally at boundaries. The authors believe that the Mg boundary diffusion parameters $\alpha D'\delta$ (about 2 orders of magnitude lower than the lattice diffusion coefficient) are actually for diffusion along second phases located at triple junctions rather than for transport along boundary cores.

The Cr boundary diffusion parameters ($\alpha D'\delta$) were determined in several doped samples.Within the experimental error they are independent on the amount and nature of dopes and 3 orders of magnitude lower than the lattice diffusion coefficient. In absence of second phase located at boundaries they are thought to be representative of the actual grain boundary diffusion.

2.1.2. Oxygen diffusion. There have been many studies of oxygen volume diffusion in MgO (essentially using gas exchange spectroscopy with crushed magnesia) but, except for the early studies of Holt and Condit [33] and McKenzie [34] carried out by the gas exchange technique and proton autoradiography, there have been only a couple of recent investigations dedicated to the oxygen diffusion in boundaries [15,16]. The result of Yoo et al [35] concerning the activation energy for bulk diffusion ($D = 1.8\ 10^{-6} \exp -3.24/kT$) is considered most accurate (the study was over a 700° range), and there is a reasonable agreement on the reported values of lattice diffusion coefficients if it is considered that the 24 results reviewed by Dolhert [15] spread over about an order of magnitude.

Dolhert [15] has investigated oxygen lattice diffusion in several undoped (and doped with Fe, Ca and Ni at the 300 ppm level) single crystals containing different amounts of impurities, and in high purity polycrystals. No sensible difference was observed among data for the different samples. The oxygen lattice diffusion measured by Roshko [16] is consistent with previous results and was also found independent of dopes [15,36]. The diffusion mechanism is difficult to ascertain because the activation energy for lattice diffusion does not fit with any simple diffusion mechanism. A vacancy pair mechanism, $(V_O^{\circ\circ}\ V_{Mg}'')^X$, was proposed by Ando et al [36] as plausible due the relatively high concentration of such associates and to the apparent impurity insensitivity of oxygen diffusion. But the activation energy for this mechanism (about 7 eV according to [16]) is uncompatible with its now well established experimental value (≈ 3.2 eV).

Oxygen grain boundary diffusion was studied by Dolhert [15] in single crystals of different origin and purity which were deformed and annealed to increase the subgrain boundary density [37] and on high purity polycrystals. The grain boundary oxygen diffusion parameter ($\alpha D'\delta$) was 3 orders of magnitude lower than the lattice diffusion coefficient and no difference was noted among data of different origin. The closeness of the values of the activation energies for bulk diffusion (3.45 eV) and for boundary diffusion (4.05 eV) suggests that the same mechanism, undefined, is valid in both cases. The boundary diffusion parameter measured by Roshko [16] is also independent of dopes, but it is higher by a factor 100. This enhanced boundary diffusion relative to the result of Dolhert is believed to be due to diffusion in second phases located at triple junctions (as for diffusion of Mg, see 2.1.1.).

2.2. Comments

Bulk and grain boundary diffusion results are collected together in figure 1. There is a general agreement for bulk diffusion as well for cation as for oxygen diffusion. Although the oxygen diffusion mechanism is still not understood, the effect of the dopes on cation diffusion confirms that the cation vacancy mechanism is valid. It is interesting to note that *solute* dopes (or impurities), which do not act on oxygen diffusion within the experimental incertitude, increase cation bulk diffusion by a limited amount. Likely, this is due to the low defect concentration and to the limited solubility of dopes in MgO. In any case, the diffusivity is increased by less than a factor 10.

With respect to grain boundary diffusion it is important to note that, except for two recent investigations [15,16], the materials under study were contaminated by uncontrolled impurities: Ca and Si essentially (but traces of phosphorous seem to have played some role in the formation of high diffusivity intergranular films in some experiments of Roshko). It is clear that it is unrealistic to interpret grain boundary diffusion without taking into account the influence of this contamination,and it is the reason invoked by Roshko to explain the important discrepancies observed with the results of Osenbach and Stubican on boundary Cr diffusion [10] which differ, by a factor 10^3 compared to Roshko's [16]. These important differences, and the high boundary diffusivity observed for Mg, and for one experiment on oxygen, have been attributed to boundary diffusion in second phases precipitated at boundaries, possibly liquid at experiment diffusion temperatures. The data obtained in the purest magnesia indicate an enhanced cation and oxygen boundary diffusion, independent of dopes, with $\alpha D'\delta/D \approx 10^{-3}$ for chromium and oxygen diffusion.

3. NICKEL OXIDE

3.1. Experimental results (Fig. 2, tables IV and V)

3.1.1. Cation diffusion. Chen and Peterson [38] have observed an enhanced grain boundary diffusion of Co and Cr in two bicrystals and large-grained (grain size > 500 μm) polycrystalline NiO grown from the melt. For both tracers, the activation energy for $\alpha D'\delta$ is approximatively the same as for the lattice diffusion. Furthermore, the authors found $\alpha D'\delta$ for cobalt diffusion to be independent of P_{O_2}.

Atkinson and Taylor [11-13] have studied the diffusion of Ni, Co, Cr and Ce along grain boundaries in fine grained (grain size ≈ 10 μm) polycrystalline scales obtained by oxidation of nickel. These diffusants cover a large range of characteristics : Ni and Co are soluble in the lattice and do not segregate appreciably to the boundaries, Ce has a negligible lattice solubility and is only present at the boundaries, Cr is intermediate between these two extremes. It was found that activation energy for intergranular diffusion is lower than that for lattice diffusion, and grain boundary diffusion coefficients decrease in the order Co, Ni, Cr, Ce, as for lattice diffusion. The authors conclude that boundary and lattice diffusion both take place by a similar mechanism, and the typical ratio $\alpha D'\delta$ / D is of the order of 10^{-2}. Last, an attempt has been made to study Ni diffusion in sintered specimens [40]. The autoradiographs have shown an almost complete absence of preferential diffusion along grain boundaries. The authors concluded that diffusion is blocked by contamination at boundaries.

The effect of aliovalent impurities on diffusion of Ni was studied in oxide scales by Chadwick and Taylor [41]. Diffusion at boundaries was not significantly affected by the presence of cerium. Moosa et al. [42] have reported similar experiments and results. They found that the presence of 0.1 wt % Y had no significant influence on lattice and grain boundary diffusion of Ni. Reversely, Atkinson [39] found that lattice diffusion of Ni was increased in Cr doped NiO scales, whereas diffusivity at grain boundaries was decreased.

Barbier et al. [17,18] have studied Ni and Co diffusion in eleven <001> and <011> symmetrical tilt bicrystals grown from the melt. No obvious enhanced grain boundary diffusion was detected despite a systematic exploration of several parameters (bicrystal misorientation, dopes, temperature and time conditions). Experiments performed on Y and Cr doped NiO bicrystals did not indicate important preferential boundary diffusion, as in undoped NiO. These results indicate that grain boundary diffusion parameters are at least two orders of magnitude lower than those previously published, and that $\alpha D'\delta$ / D is less than 10^{-4}.

Barbier et al [19] have also carried out diffusion experiments in NiO scales obtained by thermal oxidation, such as those used by Atkinson and Taylor [11-14]. It was shown that the diffusion of Ni was influenced by the specimen preparation. In particular, when the sample surface was carefully prepared, no important enhanced grain boundary diffusion was detected. Reversely, an apparent intergranular diffusion was observed when the surface was left intact after oxidation.

Table IV

Experimental diffusion conditions in NiO

material [reference]	major impurities (ppm)	diffusing element $\alpha D'\delta/D$	concentrat. profile determinat.	dopant (ppm)	remark
tilt <100> bicrystals and polycrystals [38]	?	^{60}Co (a) 10^{-2} to 0.1; ^{51}Cr (a) 10^{-1} to 7	serial sectioning		no dependence on PO_2 at 850°C
oxides scales [11-14]	Mg 10 (e) Ca 40	^{63}Ni (a), ^{57}Co (a), ^{51}Cr (a) 10^{-2} to 0.1, ^{139}Ce (a)	serial sectioning (d)		dependence on PO_2 at 700°C $D'_{Co} > D'_{Ni} > D'_{Cr} > D'_{Ce}$
		^{18}O (b) 10^{-2}	nuclear reaction		$D'_{Ni} > D'_{O}$
oxides scales [41]	?	^{63}Ni (a)	serial sectioning (d)	Ce 1000	$\alpha D'\delta$ unaffected
oxides scales [42]	Al 190, Ca 150, Fe 100, B 30, Cu 30, Mn 10	^{63}Ni (a) 10^{-2}	serial sectioning (d)	Y 1000	$\alpha D'\delta$ unaffected by dopant
oxides scales [39]	?	^{63}Ni (a) 10^{-3}	serial sectioning (d)	Cr 1000	$\alpha D'\delta$ is decreased
tilt <100> and <110> bicrystals [17,18]	Fe 3-10, Al 2-20, Si 2-30, Ca 1-10	Co (c); ^{63}Ni (a) $<10^{-4}$	electron microprobe; serial sectioning		no obvious preferential boundary diffusion
oxides scales [19]	?	^{63}Ni (a)	serial sectioning		influence of surface preparation
sintered polycrystals [40]	?	^{63}Ni (a)	autoradiograph		no obvious boundary diffusion

a) Dried solution deposited at the surface. b) Anneal in oxygen gas enriched in ^{18}O. c) Radiofrequency sputtering of CoO. d) Sectioning by sputtering. e) Major metallic impurities contained in Ni foils before oxidation (in ppm) : Na 450, Ca 300, K 300, Al 120, Mg 20.

Table V

Diffusion parameters in NiO

diffusing element (dopant ppm) [ref]	bulk diffusion $D = K \exp - (Q/kT)$		boundary diffusion $\alpha D'\delta = K'\exp-(Q'/kT)$		order of magnitude of $\frac{\alpha D' \delta}{D}$
	K ($cm^2.s^{-1}$)	Q (eV)	K' ($cm^3.s^{-1}$)	Q' (eV)	
Co [38]	$9.1\ 10^{-3}$	2.35	$3\ 10^{-3}$	2.5	
			estimate		
Co [12]	$2.5\ 10^{-2}$	2.43	$3.8\ 10^{-7}$	1.86	10^{-2} to 0.1
Ni [11]	$2.2\ 10^{-2}$	2.56	$4.3\ 10^{-8}$	1.78	
Cr [38]	$8.6\ 10^{-3}$	2.93	$5.5\ 10^{-3}$	2.93	10^{-1} to 7
			estimate		
Cr [13]	$8.6\ 10^{-3}$	2.93	$6.5\ 10^{-10}$	1.54	10^{-2} to 0.1
Ce [13]	no soluble		$D'=6.3\ 10^{-4}$	$e^{-(2/kT)}$	
Ni(Ce 100)[41]	-	-	$4.3\ 10^{-8}$	1.78	
Ni(Y 1000)[42]	$3.1\ 10^{-4}$	2.22	$8.1\ 10^{-8}$	1.83	10^{-2}
Co [17]	$3.5\ 10^{-2}$	2.52	-	-	$<10^{-4}$
O [14,44]	50	5.6	-	2.5	10^{-2}

3.1.2. <u>Oxygen diffusion.</u> Data for diffusion of O in the lattice and along grain boundaries are scarce. Oxygen lattice diffusion was first investigated by O'Keeffe and Moore [43]. More recently, Dubois et al [44] have found diffusion coefficients three or five orders of magnitude lower in NiO single crystals. Then, Atkinson et al [14] have reported intermediary results in NiO scales. The large discrepancies are attributed either to differences in technique (isotopic exchange and SIMS) or to the surface roughness of the scales. An enhanced oxygen diffusion along boundaries has also been detected in NiO scales [14].

3.2. Comments

Diffusion data are collected together in figure 2. Results for cation diffusion in NiO lattice are in good agreement, and it is established that self diffusion takes place by a vacancy mechanism [45]. Concerning lattice diffusion of O, the mechanism is not well understood and, according to Dubois et al [44], oxygen vacancies or oxygen interstitials may contribute.

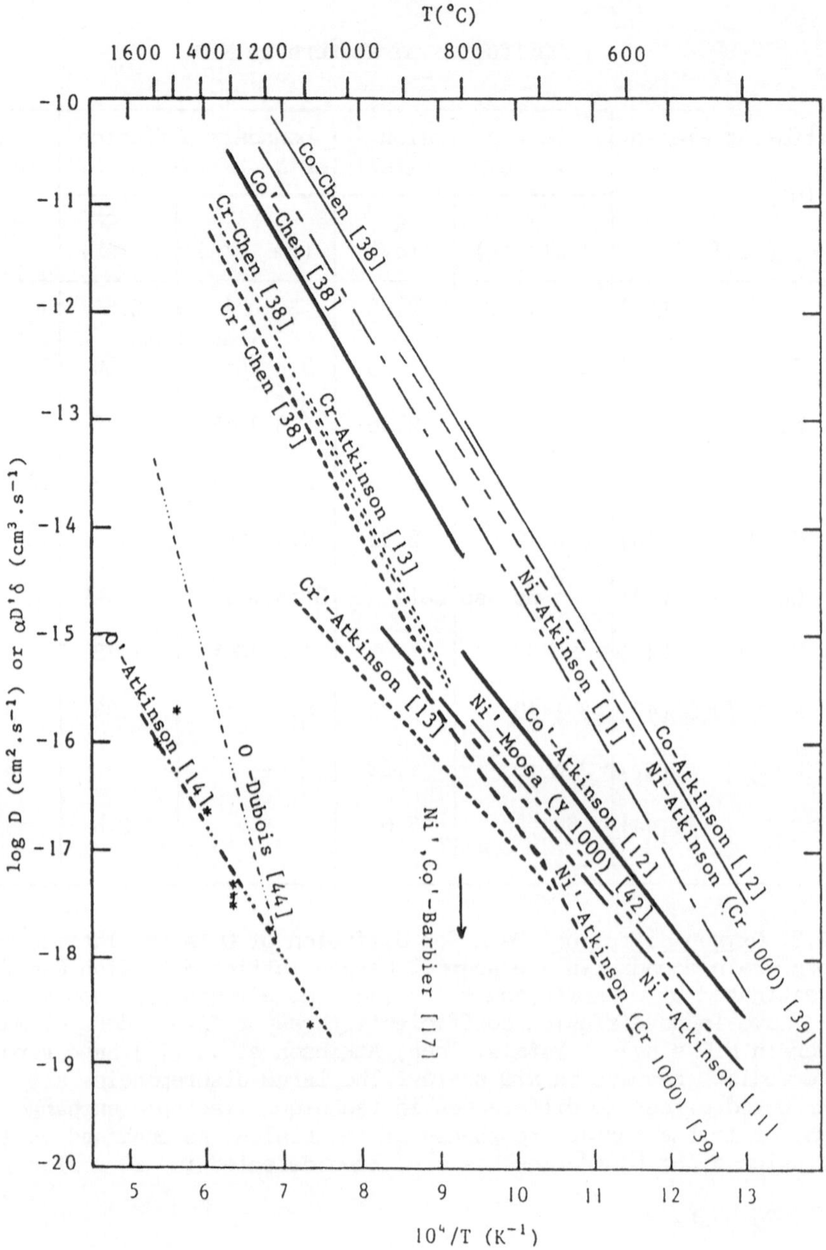

Fig. 2. Bulk diffusion coefficients (D) and grain boundary diffusion parameters (αD'δ) versus 1/T for NiO.
The experimental temperature ranges correspond approximately to the ends of each curve. Thick lines indicate grain boundary diffusion parameters, and thin lines lattice diffusion coefficients. Diffusing elements are indicated, with possible dopes in brackets.

In contrast, cation grain boundary diffusion data are not in agreement. Atkinson and Taylor's results [11-14] for Co and Cr diffusion are different of those of Chen and Peterson [38] in connection with the Co grain boundary diffusion dependance on P_{O_2}, the activation energy and the order of magnitude for $\alpha D'\delta$. In fact, there is no much information about the samples used by Chen and Peterson, but it is worth to note that their values of $\alpha D'\delta/D$ are very high and surprisingly close to those obtained by Osenbach and Stubican [10] for Cr diffusion in MgO. This exceptional diffusivity was attributed by Roshko [16] to precipitation of second phases at boundaries. In absence of other information, it is perhaps speculative, but attractive, to suggest that the origin of this high diffusivity is the same : uncontrolled segregation of impurities at boundaries and increased diffusitivity due to precipitation.

On another hand, the results of Barbier et al [17,18] indicate a rather low enhanced diffusivity in relation to Atkinson's. This discrepancy has aroused a controversy, which is two-sided:

- According to Atkinson and Taylor [13,39,40], the bicrystals are contaminated and, for some reason, impurities at the boundaries, such Ca and Si, could suppress intergranular diffusion.
- In our opinion, the enhanced boundary diffusion detected on the scales is actually due to artifacts.

Concerning the first point, it is worth to note that the influence of impurities is far from being clear and, for example, the effect of higher valency cations (Ce^{4+}, Y^{3+}, Cr^{3+}) on diffusion is controversial (Chadwick and Taylor [41], Moosa et al [42], Atkinson [39]). It can also be noticed that, usually, precipitation of second phases at boundaries increases boundary diffusivity, as shown in MgO, rather than decreases it. More specifically, the bicrystals used by Barbier et al. [46] were relatively "clean" (impurity level < 60 ppm), and STEM analyses have revealed impurities segregated at boundaries, but the level of segregation was low and independent of boundary crystallographic characteristics. Moreover, direct observations of extended and well defined intrinsic dislocation arrays in the $\Sigma = 3$ and $\Sigma = 11$ special boundaries [46,47] is absolutly incompatible with the existence of intergranular films or heavy precipitation. Thence, in our opinion, the impurities are not responsible for the relatively low diffusitivity observed at boundaries in these bicrystals.

Concerning the second point, Barbier et al [19] showed that diffusion in NiO scales grown by thermal oxidation was influenced by the sample surface. Indeed, an open porosity, which can easily generate diffusion artifacts, was observed between oxide grains in the outer scale. If such flaws are not removed before any diffusion experiment, the transport of the tracer can take place by surface diffusion rather than by intergranular diffusion. This conclusion is supported by the closeness of the nickel surface diffusion coefficients and of the nickel grain boundary diffusion coefficients determined by Atkinson and Taylor (fig. 3).

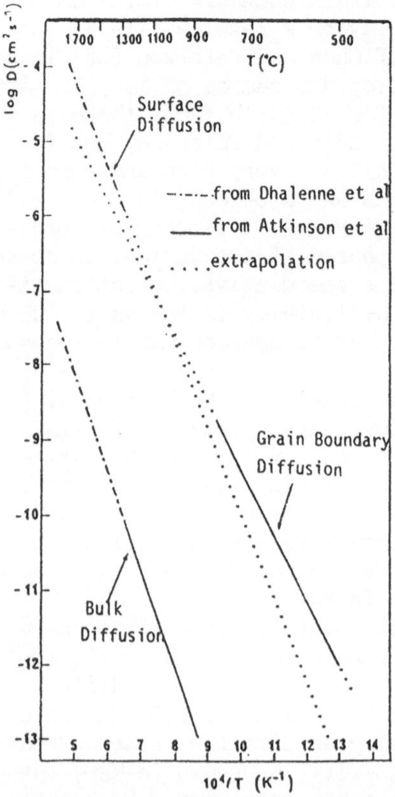

Fig. 3. Bulk, grain boundary and surface diffusion coefficients for Ni in NiO (Atkinson et al. [11], Dhalenne et al [48]).

4. CONCLUSION

With the exception of some early studies [27,49] experiments carried out on stoichiometric, or near stoichiometric oxides got the impression that grain boundary behaviour in these oxides was just a transposition of phenomena previously observed in metals. From an experimental point of view the latest results, obtained in high purity MgO or in NiO bi-crystals, cast doubt on this conclusion. In particular, intrinsic grain boundary diffusion is perhaps not so high as it has been suggested, and the influence of the impurities on this property must be seriously investigated before any definite conclusion may be drawn. Indeed, besides their "conventional" role on intrinsic and extrinsic properties, theoretically analysable in terms of point defects, or association of such defects, other features often play a role of prime importance due to the high impurity contend of theses materials and low solubility of a number of impurities : segregation, cosegregation, precipitation of second phases possibly glassy or liquid at diffusion experiment temperatures. Up to now, most of these characteristics have not yet been sufficiently appehended.

REFERENCES

1. W. D. KINGERY, J. Am. Ceram. Soc., 57 (1974) 1.
   W. D. KINGERY, J. Am. Ceram. Soc., 57 (1974) 74.
2. K. LEHOVEC, J. Chem. Phys., 21 (1953) 1123.
3. R.E. MISTLER, R.L. COBLE, J. Appl. Phys. 45 (1974) 1507.
4. R.B. POEPPEL, J.M. BLAKELEY, Surf. Sci., 15 (1969) 507.
5. J.M. BLAKELEY, S. DANYLUK, Surf. Sci. 40 (1973) 37.
6. M.F. YAN, R.M. CANNON, H.K. BOWEN, R.L. COBLE, J. Am. Ceram. Soc., 60 (1977) 120.
7. W.D. KINGERY, T. MITAMURA, J.B. VANDER SANDE, E.L. HALL, J. Mater. Sc. 14 (1979) 1766.
8. W.D. KINGERY, Pure and Appl. Chem., 56 (1984) 1703.
9. W.C. MACKRODT, in Lecture Notes in Physics 166 : Computer simulation of Solids, eds C.R.A. Catlow and W.C. Mackrodt, Springer-Verlag, Berlin, 1982, p. 175.
10. J.W. OSENBACH, W. STUBICAN, J. Am. Ceram. Soc., 66 (1983) 191.
11. A. ATKINSON, R.I. TAYLOR, Phil. Mag., 43 (1981) 979.
12. A. ATKINSON, R.I. TAYLOR, Phil. Mag., 45 (1982) 583.
13. A; ATKINSON, R.I. TAYLOR, J. Phys. Chem. Solids, 47 (1986) 315.
14. A. ATKINSON, F.C.W. PUMMERY, C. MONTY, in Transport in Nonstoichiometric Compounds, ed. by G. Simkovich and V.S. Stubican, Plenum New York, 1985, p.359.
15. L.E. DOLHERT, PhD Thesis MIT (1985), Cambridge, USA.
    L.E. DOLHERT, W.D. KINGERY, 89th Annual Meeting Am. Ceram. Soc., Pittsburgh, April 26-30 (1987).
16. A. ROSHKO, PhD Thesis MIT (1987), Cambridge, USA.
    A. ROSHKO, W.D. KINGERY, 89th Annual Meeting Am. Ceram. Soc., Pittsburgh, April 26-30 (1987).
17. F. BARBIER, C. MONTY, M. DECHAMPS, Phil. Mag. A, 58 (1988) ....
18. F. BARBIER, M. DECHAMPS, International Conference "Interface Science and Engineering'87", Lake Placid, July 1987, to be published in Journal de Physique.
19. F. BARBIER, J. BERNARDINI, F. MOYA, M. DECHAMPS in Ceramics microstructure'86 : role of interfaces, ed. by A. Pask and A.G. Evans, Materials Science Research, 21 (1987) 549.
20. J.C. FISHER, J. Appl. Phys.,22 (1951) 74.
21. R.T. WHIPPLE, Phil. Mag. A, 45 (1954) 1220.
22. T. SUZUOKA, Trans. Jap. Inst. Met. 2 (1961) 35.
23. A.D. LECLAIRE, Br. J. Appl. Phys., 14 (1963) 3
24. R. FARHI, G. PETOT-ERVAS, J. Phys. Chem. Solids, 39 (1978) 1175.
25. P. KOFSTAD, in Non-stoichiometry, Diffusion and Electrical Conductivity in Binary Metal Oxides, R.E. Krieger Publishing Co, Malabar, USA, 1983, p. 248.
26. A. DOMINGUEZ-RODRIGUEZ, J. CASTAING, Rev. de Phys. Appliq.,11 (1976) 387.
27. B.J. WUENSCH, T. VASILOS, J. Am. Ceram. Soc., 47 (1964) 63.
    B.J. WUENSCH, T. VASILOS, J. Am. Ceram. Soc., 49 (1966) 433.
28. D. TURNBULL, R. HOFFMAN, Acta Metall., 2 (1954) 419.

29. R.C. DOMAN, J.B. BARR, R.N. McNALLY, A.M. ALPER, J. Am. Ceram. Soc., 46 (1963) 313.
30. A.F. HENRIKSEN, W.D. KINGERY, Ceram. Inter. 5 (1979) 11.
31. G.W. WEBER, W.R. BITLER, V.S. STUBICAN, J. Am. Ceram. Soc., 60 (1977) 61.
32. W.C. MACKRODT, R.F. STEWART, J. Phys. C : Solid State Phys 12 (1979) 5015.
33. J.B. HOLT, R.H. CONDIT, in Materials Science Research Vol 3, eds W.W. Kriegel and H. Palmour, Plenum Press, New York, 1966, p. 13.
34. D.R. McKENZIE,A.W. SEARCY, J.B. HOLT, R.H. CONDIT, J. Am. Ceram. Soc., 54 (1971) 188.
35. H.I. YOO, B.J. WUENSCH, W.D. PETUSKEY, in Advances in Ceramics Vol 10, ed. W.G. Kingery, The Am. Ceram. Soc., Columbus, 1984, p. 394.
36. K. ANDO, Y. KUROKAWA and Y. OISHI, J. Chem. Phys., 78, (1983) 6890.
37. J. DODSWORTH, C.B. CARTER D.L. KOHLSTEDT, in Advances in Ceramics Vol 6, eds. M.F. Yan and A.H. Heuer, The Am. Ceram. Soc., Columbus, 1983, p. 73.
38. W.K. CHEN, N.L. PETERSON, J. Am. Ceram. Soc., 63 (1980) 566.
39. A. ATKINSON, AERE Harwell Report R 12404, January 1987.
40. L.B. HARRIS, R.I. TAYLOR, A. ATKINSON, J. Mat. Sci., 22 (1987) 1993.
41. A.T. CHADWICK, R.I. TAYLOR, Solid State Ionics, 12 (1984) 343.
42. A.A. MOOSA, S.J. ROTHMAN, L.J. NOWICKI, Oxidation of Metals, 24 (1985) 115.
43. O'KEEFFE, W.J. MOORE, J. Phys. Chem., 65 (1961) 1438, 2277.
44. C. DUBOIS, C. MONTY, J. PHILIBERT, Phil. Mag. A, 46 (1982) 419.
45. M.L. VOLPE, N.L. PETERSON, J. REDDY, J. Phys. Rev. B, 3 (1971) 1417.
46. F. BARBIER, M. DECHAMPS, unpublished work.
47. F. BARBIER, M. DECHAMPS, Scripta Met. 19 (1985) 1461.
48. G. DHALENNE, A. REVCOLEVSCHI, C. MONTY, in Transport in Nonstoichiometric Compounds, ed. by G. Simkovitch and V.S. Stubican, Plenum Press, New York, 1985, p. 371.
49. I. ZAPLATYNSKY, J. Am. Ceram. Soc., 46 (1963) 1358.

ZnO INTERFACE ELECTRICAL PROPERTIES-ROLE OF OXYGEN CHEMISORPTION

Mary H. Sukkar# and Harry L. Tuller
Crystal Physics & Optical Electronics Laboratory
Department of Materials Science & Engineering
Massachusetts Institute of Technology
Cambridge, MA, 02139, USA

ABSTRACT In the present studies single ZnO grain boundary or electrode interfaces are isolated and examined as a function of dopant, temperature, atmosphere, and potential. ZnO-Ag junctions were found to be rectifying but with marked sensitivity to ambient conditions (T, Po_2) and applied bias. The I(V) characteristics were examined in relation to the predictions of Sze for a metal-semiconductor junction in which both interface states and the nature of the metal influence the barrier height. Reasonable surface state densities are obtained by application of the model to our data. Difficulties associated with the derivation of thick interfacial layers are discussed. The existence of volatile surface states is confirmed by examination of atmosphere and voltage induced transients in barrier height. Manganese and manganese/praseodymium doped grain boundaries are characterized by varistor-like behavior with 3-4 volt breakdown and nonlinearity factor of three. The apparent non-activated behavior of the leakage current was traced to a strong temperature dependent barrier height characterized by $\partial\phi_B/\partial T = 1.5 \times 10^{-3}$ eV/k, a result of absorbed oxygen at the boundary.

1. INTRODUCTION

A number of ZnO interfaces that exhibit highly nonlinear current-voltage characteristics are known to be sensitive to the amount of oxygen present in the atmosphere in which they are fabricated or operated. In the processing of ZnO varistors, an oxidizing anneal is required to establish potential barriers at grain boundaries[1-3]. At

---

Present address: Ferro Corporation, Electronic Materials Division, Santa Barbara, CA

J. Nowotny and W. Weppner (eds.),
Non-Stoichiometric Compounds Surfaces, Grain Boundaries and Structural Defects, 237–263.

ZnO free surfaces, ambient oxygen is well known to give rise to acceptor states via chemisorption which, in turn, result in strong band bending and a decrease in surface conductivity[4-6]. Oxygen chemisorption is also believed to be linked to ambient-induced changes in electrical properties observed in ZnO pressed powder samples,[7] nominally undoped and In-doped sintered polycrystalline ZnO[8] and polycrystalline ZnO thin films.[9] In ZnO varistors, oxygen chemisorption/desorption processes at grain boundaries have been proposed[10] as one possible mechanism of degradation and recovery, i.e., in which prebreakdown I(V) characteristics are altered reversibly upon introducing ambient, electrical or other "stresses."[11]

In this study, the effects of atmosphere, temperature and voltage overload on the electrical properties of ZnO/Ag junctions and individual Mn- and Mn/Pr-doped ZnO grain boundaries are examined. The interfaces chosen for this study were all found to exhibit a remarkable sensitivity to ambient conditions and electrical stress. The responses of these interfaces to various stimuli, which were both rapid and pronounced, were found to be similar to "degradation" and "recovery" phenomena observed in ZnO varistors.

As we show, changes in electrical properties of these ZnO interfaces are believed to be related to the amount of oxygen chemisorbed therein. At ZnO free surfaces, the stages of oxygen incorporation are well known to proceed according to the following sequence:

$$O_2\,(gas) \rightarrow O_2\,(phys) \rightarrow O_2^-\,(chem) \rightarrow 2O^-\,(chem)$$

in which reactants to the right are favored at higher temperatures and bulk electron concentrations. The first stage of oxygen incorporation, i.e. O_2 physisorption,[12] occurs at ~100-450°K.[13,14] At 300-650°K, chemisorbed O_2^- and O^- species are formed upon reaction between surface electrons and physisorbed oxygen,[15-19] i.e.

$$O_2\,(phys) + e' \rightarrow O_2^-\,(chem) \qquad (1a)$$

$$O_2^-\,(chem) + e' \rightarrow 2O^-\,(chem) \qquad (1b)$$

Desorption of chemisorbed oxygen by hole capture can be induced upon illumination by bandgap radiation.[20,21] Similarly, electric field-enhanced oxygen chemisorption or desorption can be induced by controlling the magnitude and polarity of the applied field.[22] Unlike other ZnO interface states, chemisorbed oxygen species are relatively "volatile," i.e. their occupancy (coverage) and hence stability are not fixed but are dependent on the Fermi level at the interface.

A number of studies have been carried out on the effects of ambient atmosphere on the electrical properties of ZnO/metal junctions. In an early study, Heiland[23] found that a porous ZnO/Pt contact becomes rectifying upon exposure to oxygen and ohmic when heated at ~600°K in a vacuum. In a later study[4] on ZnO {0001} faces cleaved in Hg at room temperature, the electrical properties of these surfaces, which were ohmic initially, gradually grew more rectifying upon exposure to air and developed pronounced diode-like I(V) characteristics after ~5 minutes. Conversely, Ito[24] observed that ZnO/Pd contacts become less rectifying upon exposure to H_2-containing atmospheres.

Mead[25] has shown that ZnO/Ag junctions prepared by cleavage in a vacuum followed by the vapor deposition of a Ag electrode are rectifying with barrier heights of 0.68 eV. In this paper, we show, however, that ZnO/Ag junction barrier heights can attain values as high as 1.05 eV in oxygen (and as low as ~0.56 eV in reducing environments), which demonstrates the key role that acceptor-like chemisorbed oxygen interface states play in determining the magnitude of a large portion of the potential barrier.

In the following section, we review the appropriate theory which relates to barrier formation at various types of electrically active interfaces as a consequence of a superposition of both volatile and non-volatile sources of interface charge.

2. THEORY

2.1 Semiconductor/Metal Junctions

Electronic transport across Schottky barriers is given by

$$I = I_s[\exp(qV/nkT) - 1] \quad (2)$$

where I_s is the reverse bias saturation current and n is given by

$$n = q/kT(\partial V/\partial \ln I) \quad (3)$$

The amount that n deviates from unity provides a measure of the degree of nonideality in the junction.

Assuming thermionic emission,[26] the near-zero bias barrier height, ϕ_B, is given by

$$q\phi_B = kT\ln(AA^* T^2/I_s) \quad (4)$$

where A* is the Richardson Constant and A, the cross sectional area. In Schottky barriers, ϕ_B can be influenced by both the metal work function and the surface (interface) state density. There are two limiting cases: i) ϕ_B is determined mainly by the metal work function, ϕ_m, and semiconductor electron affinity, X, and ii) ϕ_B is pinned by surface states.

To estimate the number of chemisorbed oxygen species formed at our ZnO/metal junction at zero bias under a given set of conditions from the measured barrier height, a model developed by Sze and Cowley(26) is used. This model accounts for the contributions of both the metal work function and the surface state density to ϕ_B and is employed here to calculate Q_s, the density of interface charge. In this model, summarized below, the following assumptions are made:

1) Between the metal and the semiconductor is an interfacial layer of atomic dimensions (~3-5Å). This layer is transparent to electrons (i.e. tunneling can occur) and able to withstand potentials across it.

2) The origin and electronic structure of the surface states are properties of the semiconductor surface and are independent of the metal.

A band diagram(26) of a metal/n-type semiconductor barrier with acceptor surface states of density Q_{ss} from $q\phi_o$ to E_F is shown in Figure 1. By inspection, the semiconductor surface state charge density is given by*

$$Q_s = -qQ_{ss}(E_g - q\phi_o - q\phi_B) \qquad (5)$$

where $q\phi_o$ represents the energy at which the Fermi level coincided before Schottky barrier formation. (For surface state-pinned barrier heights, $q\phi_o$ corresponds to the energetic position of the surface state in the band gap.) At thermal equilibrium, the depletion region space charge is given by

$$Q_{sc} = [2q\varepsilon_s N_D(\phi_B - V_n)]^{1/2} \qquad (6)$$

The total surface charge density is given by the sum of Equations (5) and (6). In the absence of interfacial layer space charge effects(26), an equal and opposite charge, Q_m, develops on the metal surface to preserve interface charge neutrality,

$$Q_m = -(Q_s + Q_{sc}) \qquad (7)$$

---

* Schottky barrier lowering ($\triangle\phi$) and kT/q terms have been neglected here.

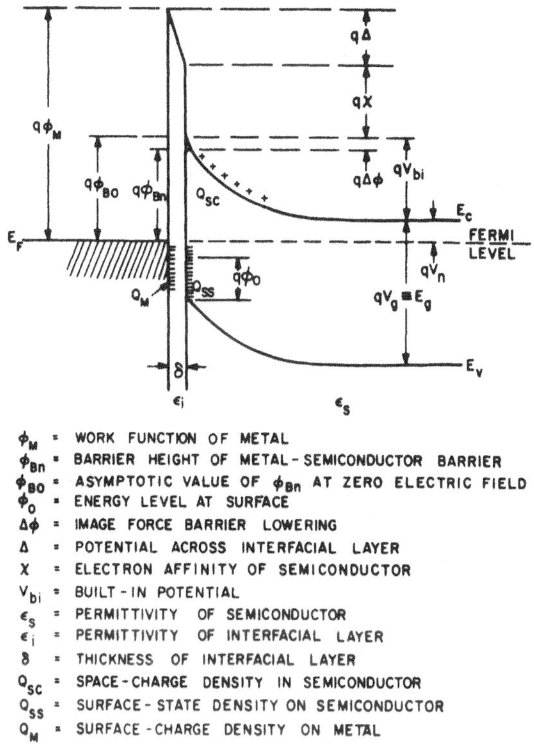

Figure 1. Detailed energy-band diagram of a metal n-type semiconductor contact with an interfacial layer of the order of atomic distance (from Ref. 26).

The potential across the interfacial layer is obtained from Gauss´ law,

$$\Delta = Q_m / \varepsilon_i \tag{8}$$

where ε_i is the interfacial layer permittivity. By inspection of Figure 1, it can be seen that Δ is also given by,

$$\Delta = \phi_m - (X + \phi_B) \tag{9}$$

Eliminating Δ from Equations (8) and (9) and defining

$$c_1 \equiv 2q\varepsilon_s N_D \delta^2/\varepsilon_i^2 \tag{10a}$$

$$c_2 \equiv \varepsilon_i/(\varepsilon_i + q^2\delta Q_{ss}) \tag{10b}$$

an expression for ϕ_B is obtained:[26]

$$\phi_B = [c_2(\phi_m - X) + (1-c_2)(E_g/q-\phi_o)] + \{c_2^2 c_1/2 - c_2^{3/2}[c_1(\phi_m - X) + (1-c_2)(E_g/q-\phi_o)c_1/c_2 - c_1/c_2(V_n) + c_2 c_1^2/4]^{1/2}\} \tag{11}$$

A similar expression for ϕ_B in terms of Q_s, employed here, was obtained by following Sze's derivation of Equation (11) and eliminating the unknown quantities, $q\phi_o$ and Q_{ss} that result from substituting Equation (5) into Equation (7). Defining $q\phi_B' = q(\phi_m - X)$ as the "ideal" value for ϕ_B in the absence of surface states, ϕ_B is given by

$$\phi_B = \phi_B' + Q_s(\delta/\varepsilon_i) + c_1/2 - (c_1(\phi_B' - V_n) + Q_s c_1(\delta/\varepsilon_i) + (c_1/2)^2)^{1/2} \tag{12}$$

For $Q_s = 0$, Equation (12) reduces to

$$\phi_B = \phi_B' + c_1/2 - \{c_1(\phi_B' - V_n) + (c_1/2)^2\}^{1/2} \tag{13}$$

which reduces to ϕ_B' for $\delta = 0$. Note that Equation (13) predicts that ϕ_B should not reach its ideal value if an interfacial layer is present even when $Q_s = 0$. Solving for Q_s in Equation (12) and defining $\Delta\phi_B \equiv \phi_B - \phi_B'$, the amount that ϕ_B deviates from its ideal value, gives

$$Q_s = -(\varepsilon_i/\delta)\Delta\phi_B + [2q\varepsilon_s N_D(\phi_B - V_n)]^{1/2}$$
$$= -(\varepsilon_i/\delta)\Delta\phi_B + Q_{sc} \tag{14}$$

In our measurements, we found that our ZnO/Ag junction exhibited very large and ambient-dependent nonideality factors ($n>2$) which indicate a large barrier height dependence on applied voltage. From Equations (2) (setting $n = 1$), (3) and (4), it can be readily shown that

$$\partial\phi_B/\partial V = (n-1)/n \tag{15}$$

which, for $n > 1$, corresponds to an increase and decrease of ϕ_B under forward and reverse biases, respectively. For highly nonideal diodes, this effect can result in both diminished forward bias characteristics (i.e. lower currents at a given voltage) and "soft" reverse bias characteristics in which the reverse bias leakage current does not approach saturation.

Assuming thermionic emission, n is given by[26]

$$n = \{1 + \partial \Delta\phi/\partial V + kT/q[\partial (\ln A^*)/\partial V]\}^{-1} \quad (16)$$

which indicates that nonideality and barrier height lowering can arise from either strong image force lowering or large variations in A* with voltage. In practice, however, these effects do not sufficiently account for very high n values (>2).[27]

Two effects that give rise to highly nonideal Schottky barrier electronic transport are known: 1) an insulating layer between the metal and semiconductor and ii) interface states. Card and Rhoderick[27] derived an expression for n in terms of both the interfacial layer width, δ, and the interface state densities at the metal and at the semiconductor:

$$n = 1 + (\delta/\varepsilon_i)(\varepsilon_s/W + qQ_{ss})/(1 + (\delta/\varepsilon_i)qQ_m) \quad (17)$$

where W is the zero bias depletion layer width and Q_{ss} and Q_m are the interface state densities in equilibrium with the semiconductor and metal, respectively. (Note that, for δ = 0, n = 1 for all values of Q_{ss} and Q_m). In most cases, $Q_m \approx 0$[26,27] and Equation (17) reduces to,

$$n \approx 1 + (\delta/\varepsilon_i)(\varepsilon_s/W + qQ_{ss}) \quad (18)$$

For $Q_m \approx 0$, Q_{ss} is a measure of the voltage dependence of Q_s[27], i.e.,

$$Q_{ss} = -n/q\ \partial Q_s/\partial V \quad (19)$$

From Equations (18) and (19), it can be readily shown that the rate of increase of $|Q_s|$ with voltage is given by

$$\partial Q_s/\partial V = -1/n\{(n-1)/(\delta/\varepsilon_i) - \varepsilon_s/W) \quad (20)$$

integration of Equation (20) gives the voltage dependence of Q_s:

$$Q_s(V) = Q_s(v{=}0) - (V/n)\{(n-1)/(\delta/\varepsilon_i) - \varepsilon_s/W\} \quad (21)$$

where Q_s (V=0) is given by Equation (14). Equation (21) shows that $|Q_s|$ increases linearly with forward bias, n and δ.

In our ZnO/Ag junction, we have found that ϕ_B and n can be modified by changes in both atmosphere and applied bias.

In later sections, we discuss experiments which we devised to examine more quantitatively the sensitivity of the junction to these parameters and the kinetics and suggest additional sources for large n values in Schottky barriers.

2.2 Grain Boundaries

Pike et al[28] have derived equations for electronic transport across grain boundary potential barriers assuming thermionic emission, i.e.

$$J = A^*T^2 \exp[-(L+\phi_B)/kT][1 - \exp(-eV/kT)]$$

$$\approx A^*TeV/k \exp[-(L+O_{Bo})/kT] \qquad (V\approx 0) \qquad (22)$$

where $L \equiv (E_C - E_f)$, A^* is the Richardson constant and ϕ_{Bo} the zero bias barrier height,

$$\phi_{Bo} = q^2 Q^2/8\varepsilon\varepsilon_o N_D \qquad (23)$$

where Q defines the interface state density, $\varepsilon\varepsilon_o$ the dielectric constant and N_D the donor density. Assuming Boltzmann statistics are valid, substitution of $L = kT\ln(n/N_C)$ where $n=N_D=F(T)$ yields,

$$J = A^*T^2(N_D/N_C) \exp(-\phi_B(V)/kT)[1-\exp(-eV/kT)]$$

$$\approx A^*TeV/k(N_D/N_C) \exp(-\phi_{Bo}/kT) \qquad (V\approx 0) \qquad (24)$$

Hence, $\ln(G_o T^{1/2})$ vs. $1/T$, where G_o is near-zero bias conductance, should be linear with a slope of $-\phi_{Bo}/k$ for temperature-independent ϕ_{Bo}.

However, the ZnO grain boundaries in this study were found to exhibit a weak temperature dependence, i.e. compared to that observed in ZnO varistors. This near temperature-independence can be accounted for by a large $\partial\phi_{Bo}/\partial T$ which decreases the effective activation energy, i.e.,

$$E_A = \phi_{Bo} - T\partial\phi_{Bo}/\partial T \qquad (25)$$

As we will show, large $\partial\phi_{Bo}/\partial T$ values can be explained by an increase in Q that results from oxygen chemisorption at the grain boundary.

3. SPECIMEN PREPARATION

The ZnO/Ag junction was fabricated from a ZnO c-axis crystal grown at Airtron ($n \approx 10^{16}$ cm^{-3}). The principle crystallographic axes were verified by interference microscopy using polarized light. To enhance sensitivity to ambient conditions, one of the crystal surfaces was polished then heavily etched in H_3PO_4 (84%) for 2 minutes. After ~15 seconds, the surface exposed to the etch grew dull and cloudy. Previous observations of the etching behavior of the ZnO polar surfaces[29] suggest a (0001) polarity for the etched surface. A layer of Ag was evaporated onto the etched surface which was masked to expose a ~2 mm^2 area. An ohmic In contact was soldered to the back side of the junction.

Single electrically active "buried" grain boundaries were fabricated by diffusing Mn or Mn + Pr into polycrystalline ZnO pellets. This was accomplished by first applying Mn and/or Pr dopants onto polished surfaces in the form of aqueous nitrate solution** followed by firing of the "sandwich" structures at 1440°C for 24 hours while buried in ZnO powder. The Mn nitrate solution was applied to a single ZnO specimen surface at a concentration of ~$8.4 \times 10^{18} cm^{-2}$ and fired in that manner. Mn and Pr nitrate solutions were both applied at concentrations of $8.4 \times 10^{17} cm^{-2}$ to a set of two samples which were then fired in a sandwich configuration. The initially undoped pellets had been fired previously in air at 1420°C for one hour and cut into 3mm squares of thickness 0.5 mm. The initial ZnO room temperature electron concentration was determined to be $1.6 \times 10^{16} cm^{-3}$ from conductivity data which increased to $1.1 \times 10^{17} cm^{-3}$ upon refiring.

4. EXPERIMENTAL APPARATUS AND PROCEDURE

Current-voltage measurements were made in air, O_2 and Ar + 4.3% H_2 atmospheres from 25-130°C in a kanthal-wound furnace. O_2 and Ar/H_2 atmosphere measurements were carried out in a Vycor tube chamber in which the pressure was varied from ~10^{-5} to one atmosphere by bleeding in gases at controlled flow rates. Sub-atmospheric pressures were obtained with a mechanical pump and monitored with a Pirani 11 gauge.

I(V) measurements were made using a Tektronix Type 575 Transistor curve tracer. For the ZnO/Ag junction, the bias polarity, i.e. forward (positive) and reverse (negative) bias, was defined by that of Ag with respect to ZnO. For all specimens, the maximum applied voltage was stepped from ±0.05 to ≥ V_B (breakdown voltage) logarithmically and the I(V) characteristics were recorded at each stage to check for possible irreversible changes therein due to specimen

---

**Fisher Scientific Co., Fair Lawn, N.J.

overheating. For the single grain boundary specimens, the third (bulk limited) regime was not observed due to deleterious effects of high current densities.

For the Mn/Pr-doped boundary, additional measurements of the sample voltage were made at 128°C at 1 = 1.5 x 10^{-4} A (from which the near-zero bias resistance was obtained) using a dc voltage supply and chart recorder to monitor the response of the boundary to bias and atmospheric stresses.

5. RESULTS AND DISCUSSION

5.1 ZnO/Ag Junction

Rectifying I(V) characteristics were obtained for ZnO/Ag junctions by equilibration in oxygen at 130°C as shown in Fig. 2a. Upon exposure to reducing atmospheres, a type of "degradation" is induced, resulting in time-dependent increases in current at a given voltage. Fig. 2b shows the I(V) characteristics upon subsequent exposure to an Ar/H_2 atmosphere at room temperature. The degradation was found to be very pronounced and essentially complete after only 5 hours. Similar behavior was induced upon exposure to considerably more oxidizing conditions, e.g. 10^{-5} atmospheres O_2 (T ≈ 130°C). In this case, changes in the I(V) characteristics occurred over a longer time period (~12-24 hours) and were not as pronounced.

This dependence of the I(V) characteristics on atmosphere indicates that the degree of ZnO/Ag junction rectification is linked to the ambient oxygen partial pressure. The changes in electrical properties, and the relatively low temperatures employed in this study, both suggest processes limited to the interface, i.e. oxygen chemisorption.

Degradation of the junction characteristics can also be electrically induced by application of high reverse bias voltages. At low voltages, the junction remains blocking but at voltages > 2V, irreversible changes in the trace occur. At sufficiently high voltages, a very marked and rapid degradation occurs.

Assuming that the degree of junction rectification is linked directly to the amount of chemisorbed oxygen, reverse bias degradation can be explained by oxygen desorption induced by holes attracted to the interface. The holes neutralized the negatively charged chemisorbed oxygen species, resulting in a higher concentration of physisorbed O_2 molecules which are unstable at higher temperatures. Degradation also occurs under forward bias but only at much higher potentials (≥ 10-20 V).

To determine more quantitatively the relation between the degree of rectification and ambient oxygen content,

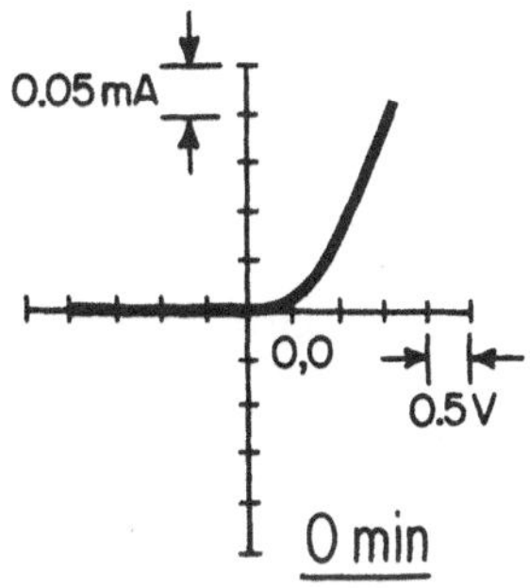

Figure 2a. I(V) characteristics of a ZnO/Ag junction annealed in air showing rectifying behavior.

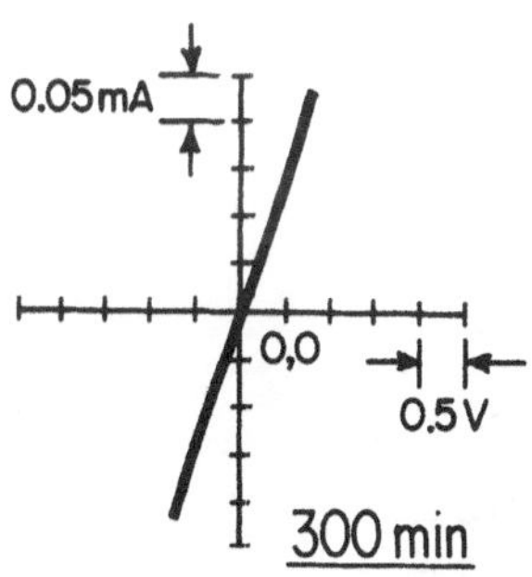

Figure 2b. Degradation of above ZnO/Ag junction in an Ar/H_2 atmosphere at room temperature.

"degradation and recovery" experiments were devised in which the ZnO/Ag interface was first "scrubbed," in situ, of chemisorbed oxygen by applying a large bias (± 20V) for 60 seconds using a curve tracer. The electrical stress was then removed and recovery was monitored by observing the forward bias I(V) characteristics as a function of time. To avoid spurious degradation effects during the measurements, the maximum voltage settings were well below those which induce forward bias degradation.

Degradation/recovery measurements were made from 88-130°C in 13-15°C intervals at 3-4 different oxygen pressures ranging from $1\text{-}10^{-5}$ atmospheres. Recovery was monitored until an approach to equilibrium in the I(V) characteristics was evident (~2-8 hours). Isotherms were obtained in ~12-24 hour intervals. Afterwards, the I(V) characteristics were allowed to re-equilibrate in O_2 for at least 12-24 hours before changing temperatures. A typical recovery transient is shown in Fig. 3.

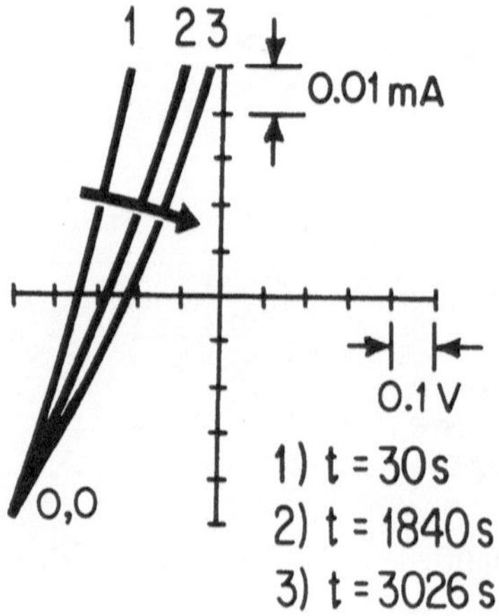

Figure 3. Recovery of the ZnO/Ag junction in oxygen at 113°C after electrical degradation, i.e. application of large forward and reverse bias (±20V) for one minute.

The I(V) curves were well fitted to Equation (2) from which ϕ_B values were obtained using Equation (4). I_s and n were obtained by finding the minimum in the sum of squares, S, with respect to I_s and n:

$$S = \sum_{(i=1)}^{N} \{I_s[\exp(qV_i/nkT) - 1] - I_i\}2 \qquad (26)$$

where N is the number of I,V data points (as obtained from the traces) and I_i and V_i are the current and voltage values for the ith data point. Corrections were made in the I(V) data for an extra contribution to the junction resistance, which is evident at high voltages. This extra contribution ranged from ~500-2000 Ω (25-130°C) and was nearly voltage-independent.

The barrier height vs. time (recovery) behavior of the junction after removal of the electrical stress is shown in Fig. 4 as a function of ambient conditions. In general, it can be seen that ϕ_B increases with both oxygen pressure and temperature, consistent with the known effects of chemisorbed oxygen on the electrical properties of ZnO free surfaces. Upon bias-induced degradation, barrier heights of ~0.68-0.75eV were obtained which recovered to values ranging from ~0.88-1.0 eV at one atmosphere to ~0.72-0.85 eV at 10^{-5} atmospheres oxygen pressure. The long term recovery data were well fitted to equations of the form,

$$\phi_B(t) = \phi_{Bi} + [\phi_{Bf} - \phi_{Bi}][1 - \exp(t/\tau)] \qquad (27)$$

where $\phi_B(t)$ represents the barrier height at time t, ϕ_{Bi} and ϕ_{Bf} are the initial and final (equilibrium) values, respectively, and τ is the "time constant" for recovery. Calculated τ values were found to increase from the order of 100s of seconds at one atmosphere to ~3000-8000 seconds at 10^{-5} atmospheres with little systematic dependence on temperature. The noticeable dependence of τ on oxygen pressure and small activation energy suggest permeation of oxygen from the surrounding atmosphere to the interface as a possible rate limiting step in recovery.[30]

Nonideality factors ranging from 2-10 were also obtained from the I(V) data. Except at short times, the n values were found to vary smoothly during recovery. Upon fitting the calculated $\phi_B(t)$ and n(t) values to Equation (27), one finds the n factors exhibit an inverse correlation with ϕ_B under a given set of ambient conditions and are consequently both characterized by similar τ values. This inverse correlation between n and ϕ_B is shown in plots of $\phi_B(t)$ vs. n(t) as a function of oxygen pressure (Fig. 5). n(t) increases with both ambient oxygen pressure and temperature for a given value of the barrier height, $\phi_B(t)$, during recovery.

From Equation (21), a large n value indicates a correspondingly large increase in Q_S with applied forward bias. Hence, the results in Fig. 5 suggests a greater tendency for the interface to acquire more chemisorbed oxygen under a given forward bias with increasing oxygen

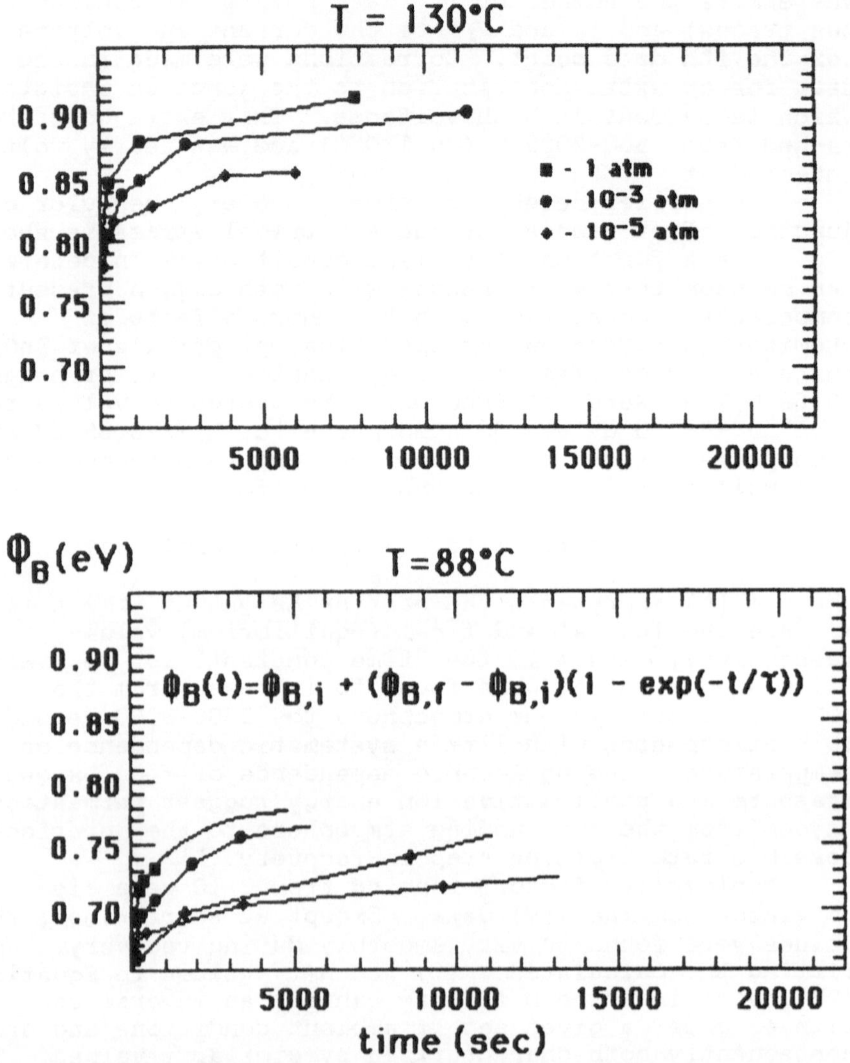

Figure 4. Plots of ϕ_B vs. time during recovery as a function of temperature and oxygen partial pressure (a)T = 130°C and (b) T = 88°C.

pressure and/or temperature. Furthermore, this tendency is greater at lower $\phi_B(t)$ values. Thus, the presence of oxygen at the ZnO/Ag interface results in an increase in both the barrier height and its dependence on applied voltage due to the formation of acceptor interface states.

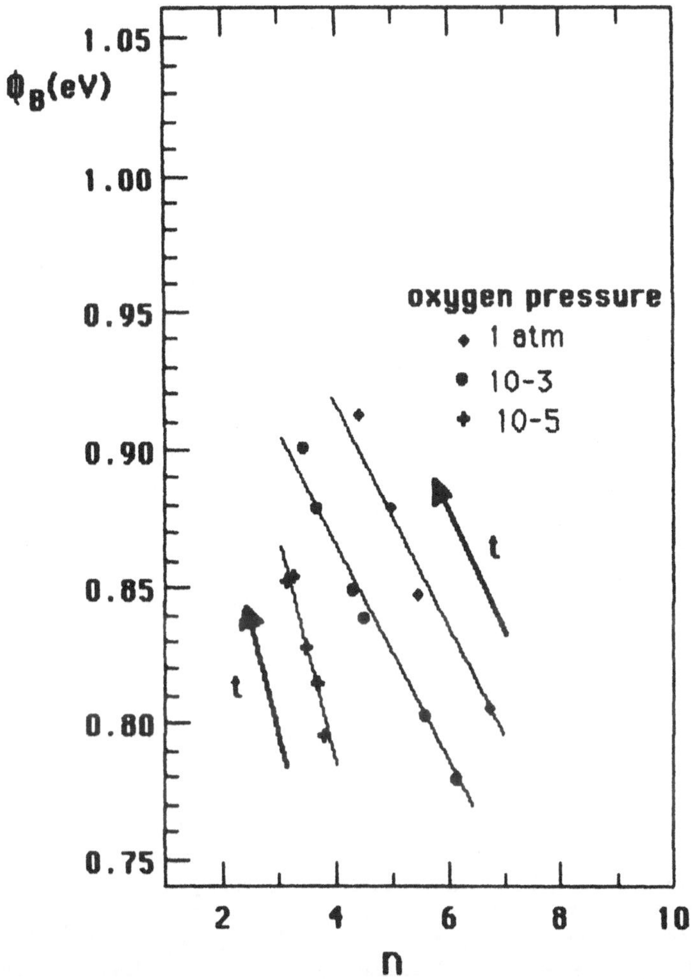

Figure 5. Plot of ϕ_B vs. n at 130°C for the ZnO/Ag junction during recovery as a function of ambient oxygen pressure. The arrows indicate schematically the direction of n and ϕ_B as a function of time during recovery.

To estimate the fixed portion of ϕ_B in the absence of surface states, I(V) measurements were made at room temperature in Ar/H_2 at pressures of 1-10^{-5} atmospheres. A nearly constant value for ϕ_B of 0.52-0.56eV was obtained under strongly reducing conditions which, we believe, corresponds to ϕ_B at Q_s = 0. This substantial deviation from the ideal value, i.e. ϕ_B (0.68 eV) suggests the presence of an interfacial layer. Upon rearranging Equation (13) and substituting the above ϕ_B and ϕ_B' values, δ was estimated at ~300 Å, which is consistent with the very large n values obtained. The presence of a thick interfacial layer in highly nonideal ZnO Schottky barriers is consistent with Ito's[24] observations of a strong deviation of the measured capacitance of highly nonideal Pd/ZnO diodes (n > 2) from the linear C^{-2} vs. V dependence expected for ideal Schottky barriers.

Calculations of the surface state density, Q_s, vs. ϕ_B from Equation (14) indicate that Q_s varies almost linearly with ϕ_B and is weakly dependent on both temperature and oxygen pressure. For $N_D = 10^{16}\,cm^{-3}$ and ϕ_B (Q_s = 0) = 0.56 eV, the calculated values for Q_s, for the barrier heights obtained in the measurements, were found to range from -2 to -4 x $10^{11}\,cm^{-2}$. Although these values are in reasonable agreement with O_2^- coverages obtained at 300°K by Göpel[13] at oxygen-annealed ZnO prismatic planes, modifications to Sze's model need to be made given that our calculated δ is much larger than 3-5Å, contrary to Sze's original assumptions.

Another, perhaps more plausible reason for the smaller barrier height measured by us under hydrogen as compared to Mead's value of 0.68eV in vacuum, which is unrelated to the existence of an interfacial layer, is the possibility that hydrogen forms donor surface states. More careful measurements over a broad Ph_2 range are necessary to confirm or deny this possibility.

5.2 Mn-Doped Single Grain Boundary

Additions of Mn to nominally undoped polycrystalline ZnO were found to result in the formation of a single electrically active grain boundary with two well-defined varistor-like regimes, a breakdown voltage of ~4V and an α value of 3 in the breakdown regime as illustrated in Fig. 6. Their symmetry suggests a correspondingly symmetrical distribution of Mn on both sides of the boundary likely due to Mn in-diffusion along the grain boundary. The fact that there is only one active boundary suggests that it lies one grain in from the surface.

The I(V) characteristics at V>1V measured in air are plotted in Fig. 7 as a function of temperature. Temperature variations in the I(V) characteristics, which are small, occur primarily in the prebreakdown regime, with an approach to nearly temperature-independent behavior at breakdown, characteristic of ZnO varistors. An activation energy of only ~0.04 eV was obtained from a fit of $\ln G_o T^{1/2}$ vs. 1/T (see Eq. 24) at 50-130°C.

ϕ_{Bo} values were obtained independently at each temperature from the conductance data near V=0 by rearranging Equation (24). The results, plotted in Fig. 8, show ϕ_{Bo} to be large in magnitude and strongly temperature-dependent. It is the latter feature which is the cause of the weak temperature dependence of G_o. For this boundary, the calculated ϕ_{Bo} values, which are much larger than might

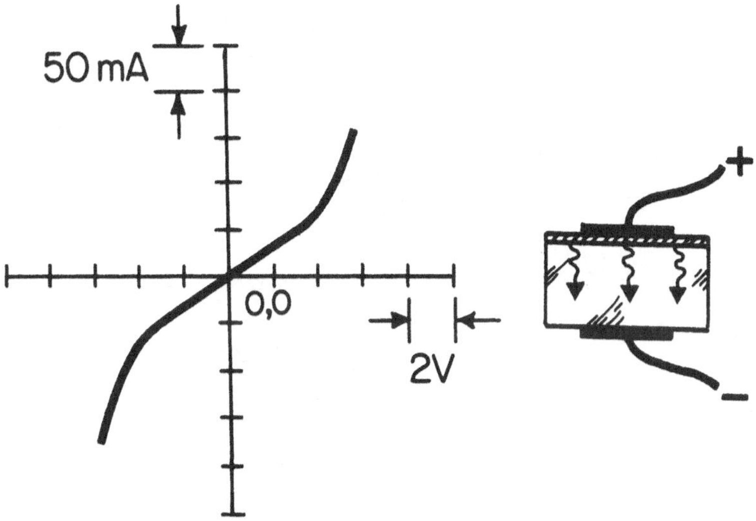

Figure 6. I(V) characteristics of undoped ZnO + Mn nitrate-doped (~5m%) interface at 50°C. (The wavy arrows indicate schematically the direction of inward diffusion of Mn into the bulk.)

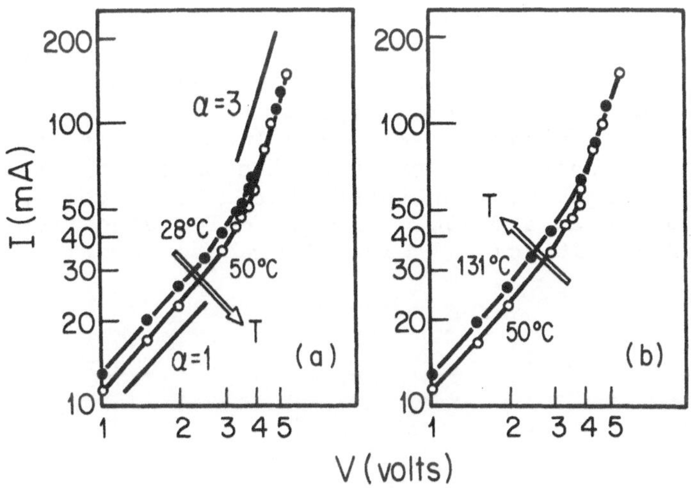

Figure 7. log I vs. log V plots for a specimen described in Fig. 6. A comparison of the I(V) characteristics at a) 28°C and 50°C and b) 50°C and 131°C.

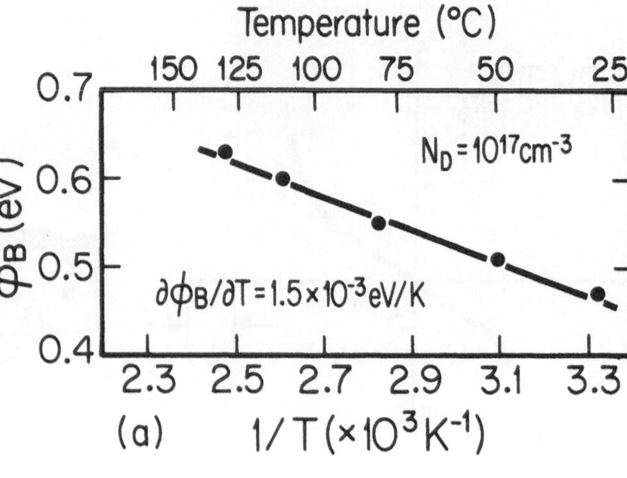

Figure 8. ϕ_B (V=0) vs. 1/T for specimen described in Fig. 6. Note strong temperature dependence.

be expected for the activation energy obtained, were found to increase linearly with temperature from ~0.47-0.63 eV at a rate of 1.5 x 10^{-3} eV/°K. This value is approximately minus 10 times that normally observed in ZnO varistors, which typically display strongly thermally activated prebreakdown behavior ($E_A \approx \phi_{Bo}$).

Large and positive $\partial\phi_{Bo}/\partial T$ values such as those obtained for this boundary suggest substantial increases in the core charge, Q, with temperature. From the ϕ_{Bo} values calculated at 25 and 130°C, ΔQ was calculated from Equation (23) at 2.2 x 10^{11} cm^{-2}. Given the above-described results for the ZnO/Ag junction, one likely source of increase in Q with temperature is an increase in the amount of chemisorbed oxygen. Similar ambient-induced changes in ZnO grain boundary electrical properties have been observed at the Mn/Pr-doped boundary as described below.

5.3 Mn/Pr-Doped Single Grain Boundary

The I(V) characteristics of this specimen are shown in Fig. 9. For this boundary, subohmic prebreakdown behavior, not observed in the "buried" Mn-doped boundary, is evident and breakdown occurs at ~3.8V. Variations in the I(V)

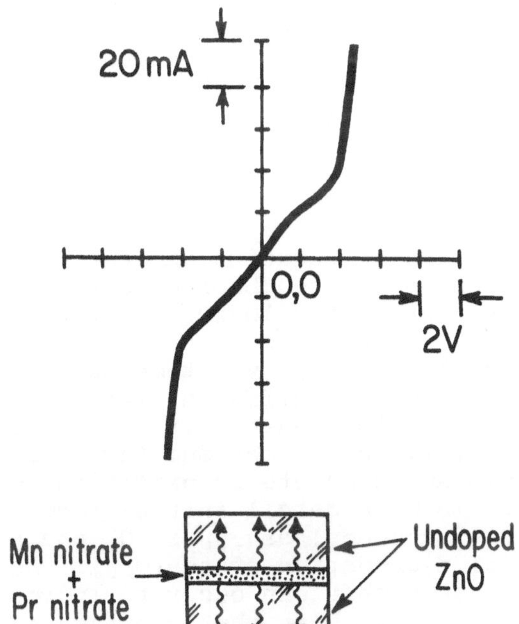

Figure 9. I(V) characteristics of an undoped ZnO + 0.5Mn/0.5 Pr nitrate-doped interface at room temperature.

characteristics with temperature as well as the calculated $\partial\phi_{Bo}/\partial T$ value (1.7 x10^{-3} eV/°K) are very similar to those of the Mn-doped boundary. The Mn/Pr-doped boundary was found to be more sensitive to voltage and ambient stresses than the Mn-doped boundary. Degradation in the I(V) characteristics was found to be induced easily upon application of biases slightly > V_B, the breakdown voltage.

Degradation/recovery experiments similar to those carried out for the ZnO/Ag junction were performed, except that, in this case, the near-zero bias resistance was monitored. To measure the rate of bias-induced degradation, a 4V pulse was applied to the boundary for various times (1-20 seconds), and the near-zero bias resistance was measured immediately thereafter. Prior to application of the pulse, the boundary was allowed to equilibrate in O_2 to attain a well-defined, initial reference state. "Recovery" of the resistance was monitored as a function of time upon removal of the pulse. To study the effects of ambient atmosphere on both bias-induced degradation and subsequent recovery, measurements were performed in O_2 and Ar/H_2 atmospheres. ϕ_{Bo} values were derived from the I(V) measurements by applications of Equation (24).

Fig. 10 shows a plot of the change in the zero bias barrier height, $\Delta\phi_B$ ($\equiv \phi_{Bf} - \phi_{Bi}$) as a function of the electrical stress pulse time in both O_2 and Ar/H_2 atmospheres. For these experiments, ϕ_{Bi} represents the maximum potential barrier height (i.e. after equilibration in oxygen) and ϕ_{Bf} represents the final barrier height after application of the pulse. The amount that ϕ_B decreases after a given pulse time appears to be significantly increased upon exposure to reducing ambients. In Ar/H_2, decreases in ϕ_B of 0.08eV were observed after only a 10 second pulse which represents ~10-20% decrease in ϕ_B for this boundary.

Fig. 11 shows plots of $\Delta\phi_B$ vs. time during recovery in O_2 and Ar/H_2 atmospheres for different initial recovery conditions (i.e. ϕ_{Bi}). In general, it can be seen that barrier height recovery to its maximum (equilibrium) value in O_2 is more rapid and complete in oxidizing ambients.

Fig. 12 shows in more detail initial recovery transients (t ≈ 0-60 seconds) of $\Delta\phi_B$ in O_2 and Ar/H_2 as a function of initial degradation state (O_{Bi}). During recovery, two processes that can occur to increase the amount of chemisorbed oxygen at the interface are possible, i.e., readsorption of O_2 from the surrounding atmosphere,

$$O_2(g) \rightarrow O_2(phys) \qquad (28)$$

and conversion of physisorbed O_2 to chemisorbed O_2^- (Equations 1a-b). If the first mechanism is rate-limiting, the recovery rate should increase with decreasing $[O_2]_{phys}$

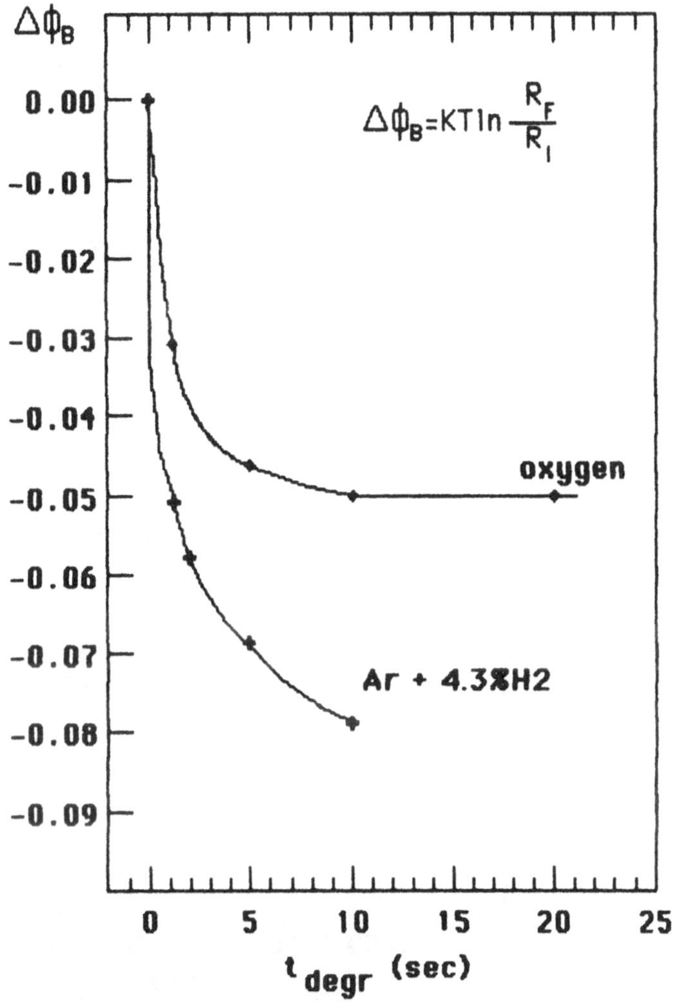

Figure 10. Degradation rate of single boundary specimen described in Fig. 9 as a function of atmosphere.

left behind after degradation due to the enhanced O_2 concentration gradient between the interface and the surroundings. If the second mechanism is rate-limiting, the recovery rate should increase with $[O_2]_{phys}$ left behind since more physisorbed O_2 is available for immediate re-conversion back to chemisorbed oxygen. In Fig. 12, it appears that the initial rate of increase in ϕ_B during recovery <u>increases</u> with ϕ_{Bi}, suggesting that re-conversion of physisorbed oxygen back to chemisorbed oxygen is a possible rate-limiting step in initial recovery. In Fig. 13, selected short term transients which were found to be well-fitted to single exponentials of the form of Equation (27) demonstrate more quantitatively that τ, the recovery time constant, does indeed increase with decreasing ϕ_{Bi}.

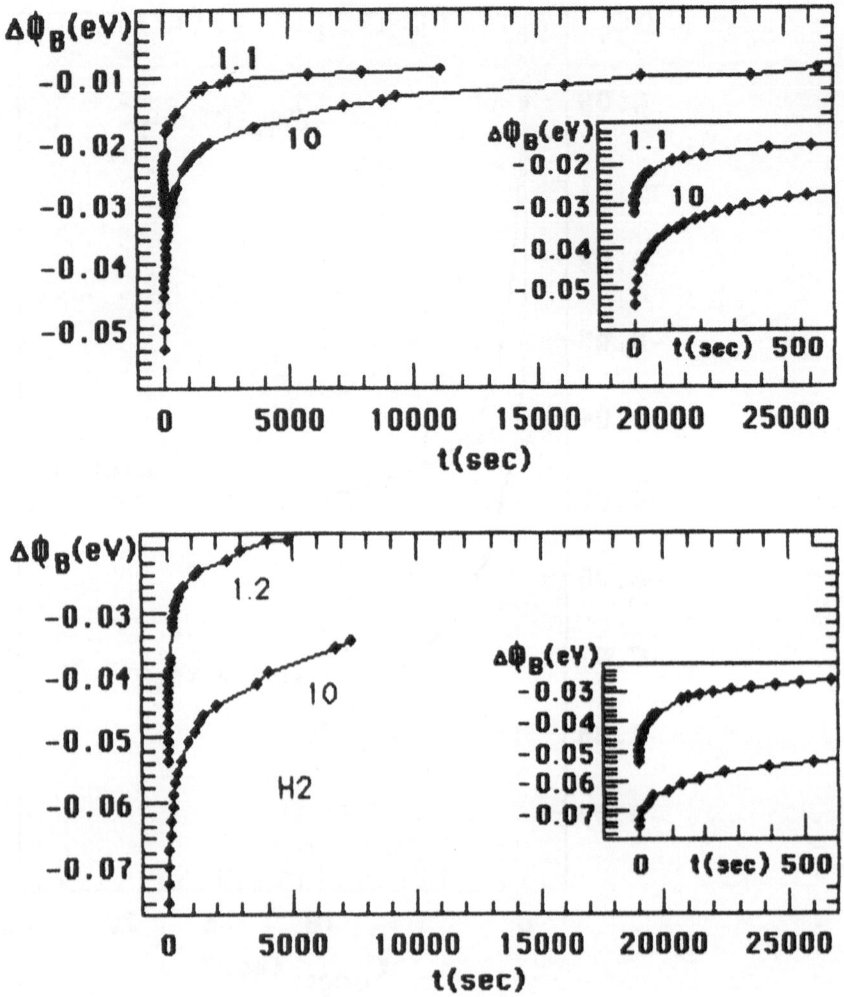

Figure 11. Overall recovery behavior of a Mn/Pr-doped single grain boundary in a) oxygen and b) Ar + 4.3% H_2 after degradation pulses of one and ten seconds. Insets show recovery transients at shorter times (≤ 600 seconds).

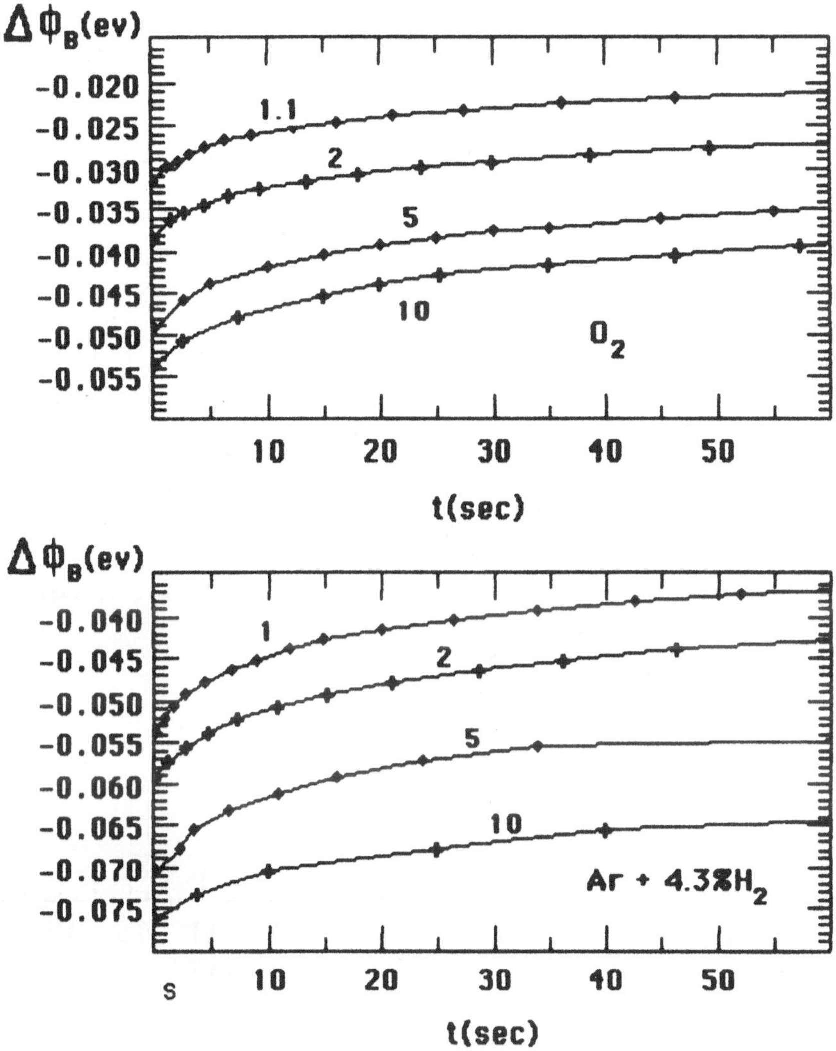

Figure 12. Recovery transients at short times (See Fig. 11) in a) oxygen and b) Ar + 4.3% H_2 for different initial conditions (ϕ_{Bi}). The numbers for each curve indicate the duration (sec) of the degradation pulse prior to recovery.

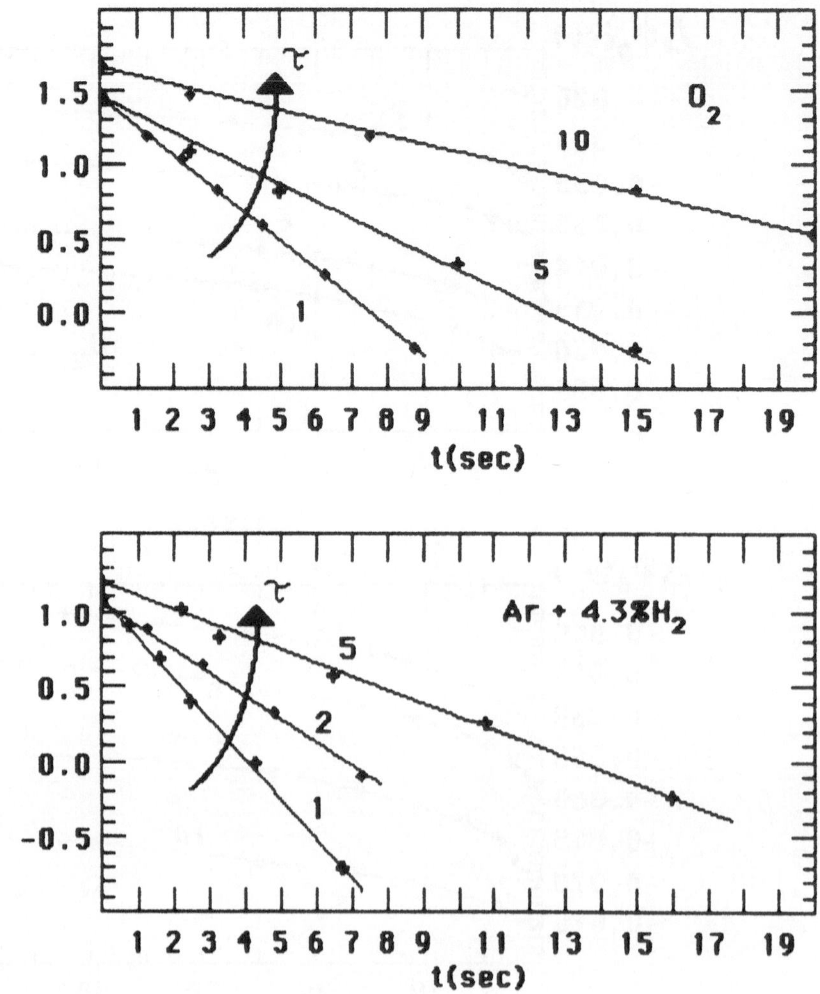

Figure 13. Plots of $\ln(V(t) - V_f)$ vs. t in a) oxygen and b) Ar + 4.3% H_2 for different initial conditions. The numbers for each curve indicate the duration (sec) of the degradation pulse prior to recovery.

6. SUMMARY AND CONCLUSIONS

At ZnO interfaces, potential barrier "degradation" and "recovery" processes of the type that occur in ZnO varistors can be induced by changing the ambient conditions and/or amount of electrical stress. We have shown that these changes in electrical properties result from oxygen chemisorption/desorption processes at the interface.

For the ZnO/Ag interface, oxygen chemisorption results in increases in both ϕ_B and $\partial\phi_B/\partial V$ in the forward bias direction. The chemisorbed oxygen acts as "volatile" acceptor-like surface states. Questions remain regarding the existence of an interfacial layer as is utilized by Sze[26] for purposes of explaining the characteristics of ZnO/Ag junctions.

For the single grain boundaries, large and positive $\partial\phi_B/\partial T$ values are also believed to result from oxygen chemisorption. Furthermore, we propose that the initial recovery kinetics after degradation are limited by the reconversion of physisorbed oxygen back to chemisorbed oxygen.

Experiments such as these provide a means for studying, in a systematic manner, the effects of changes in the amount of interface charge on the barrier height and electronic transport properties. Further techniques for the "in situ" variation in surface state density by changes in the ambient atmosphere would be of interest to understand better the influence of different types of surface states on the electronic structure and formation of ZnO interface potential barriers.

7. ACKNOWLEDGEMENTS

This research and related work on ZnO interfacial properties was funded by a National Science Foundation Grant (DMR-84-18896) and (DMR-87-20017) from the Ceramics and Electronic Materials Program in the Division of Materials Research . MHS would like to thank Dr. D. Gabbe for assistance with interference microscopy and Dr. J. Lagowski for providing ZnO single crystals.

8. REFERENCES

1. W.G. Morris, J. Vac. Sci. Technol. 13 926 (1976).

2. F.A. Selim, T.K. Gupta, P.L. Hower, and W.G. Carlson, J. Appl. Phys. 51 765 (1980).

3. R. Salmon, J.P. Bonnet, M. Graciet, M. Onillon, and P. Hagenmuller, Solid State Commun. 34 301 (1980).

4. G. Heiland, and P. Kunstmann, Surf. Sci., 13 72 (1969).

5. Y. Margoninski, Surf. Sci., 94 L167 (1980).

6. H. Watanabe, M. Wada, and T. Takahashi, Jpn. J. Appl. Phys., 4 945 (1965).

7. A. Cimino, E. Molinari, and F. Cramarossa, J. Catal., 2 315 (1963)

8. M.A. Seitz, F. Hampton, and W.C. Richmond, Advances in Ceramics, 7 60 (1983)

9. A.P. Roth and D.F. Williams, J. Appl. Phys., 52 6685 (1981).

10. K. Takahashi, T. Miyoshi, K. Maeda, and T. Yamazaki, in Grain Boundaries in Semiconductors, ed. by G.E. Pike, C.H. Seager, and H.J. Leamy, Elsevier, 1982, pp. 399-404.

11. H.R. Philipp and L.M. Levinson, Advances in Ceramics, 7 1 (1983).

12. R. Dorn and H. Luth, Surf. Sci., 68 385 (1977).

13. W. Gopel, Surf. Sci., 62 165 (1977).

14. W. Gopel, J. Vac. Sci. Technol., 15 1298 (1976).

15. M. Iwamoto, Y. Yoda, N. Yamazoe, and T. Selyama, J. Phys. Chem., 62 2564 (1978).

16. J.E. Cope and I.D. Campbell, J. Chem. Soc. Farad. Trans. 1, 69 1 (1973).

17. A.J. Tench and T. Lawson, Chem. Phys. Lett., 8 177 (1971).

18. M. Codell, J. Weisberg, H. Gisser, and R.D. Iyengar, J. Am. Chem. Soc., 91 7762 (1969)

19. G. Neumann, Phys. Stat. Sol. (b), 105 605 (1981).

20. D. Eger, Y. Goldstein, and A. Many, RCA Review, 36 508 (1975).

21. K.I. Tanaka and G. Blyholder, J. Phys. Chem., 76 3184 (1972).

22. S.A. Hoenig and J.R. Lanem, Surf. Sci., 11 163 (1968).

23. G. Heiland, Z. Phys., 148 15 (1957).

24. K. Ito, Surf. Sci., 86 345 (1979).

25. C.A. Mead, Solid State Electron, 9 1023 (1966).

26. Sze, S.M., Physics of Semiconductor Devices, 2nd ed., Ch. 5 ("Metal-Semiconductor Contacts"), John-Wiley and Sons, Inc., 1981.

27. H.C. Card and E.H. Rhoderick, J. Phys. D: Appl. Phys., 4 1589 (1971); E.H. Rhoderick, in Metal-Semiconductor Contacts, The Institute of Physics, 1974, pp. 3-19.

28. G.E. Pike and C.H. Seager, J. Appl. Phys. 50 3414 (1979).

29. Y. Mariyoshi, S. Shirasaki, H. Ooshima, T. Tsutsumi, and I. Shindo, Kristall und Tecknik, 13 1225 (1978).

30. H. Dietz, Solid State Ionics, 6 175 (1982).

NEAR-SURFACE DEFECT STRUCTURE OF CoO IN THE VICINITY OF THE CoO/Co_3O_4 PHASE BOUNDARY

J. Nowotny, and W. Weppner
Max-Planck-Institut für Festkörperforschung
7000 Stuttgart 80, FRG
M. Sloma
Institite of Metallurgy
Academy of Mining and Metallurgy
30-059 Krakow, Poland

ABSTRACT. The near-surface defect structure of CoO was studied by 'in situ' work function measurements in the temperature range 790 - 905°C at oxygen partial pressure near the equilibrium with Co_3O_4. The studies indicate that the near-surface oxide layer is enriched in cobalt interstitials which are not observed in the bulk. While CoO is oxidized and its non-stoichiometry increases, the observed phenomena in the outer surface layer may be considered within several consecutive regimes:

1. The formation of cobalt vacancies,
2. The formation of both cobalt vacancies and cobalt interstitials,
3. The formation of the Co_3O_4 superficial layer over CoO grains,
4. The phase transition $CoO \rightarrow Co_3O_4$ in the bulk phase.

Cobalt interstitials which are formed in the outer layer of CoO at higher non-stoichiometry may be considered as precursors of the spinel phase. It has been observed that the $CoO \rightarrow Co_3O_4$ phase transition within the near-surface layer takes place at p_{O_2} still within the existence of the CoO phase field in the bulk. Equilibrium p_{O_2} values corresponding to the bulk phase transition $CoO \rightarrow Co_3O_4$ are in a good agreement with literature data.

1. INTRODUCTION

Cobaltous monoxide ($Co_{1-y}O$) is known to be cation deficient. The major defects are cation vacancies and electron holes [1-3]. Depending on temperature and composition the vacancies may be neutral (V_{Co}) or singly (V'_{Co}) and doubly ionized (V''_{Co}). For long time the defect structure of CoO was considered within the framework of an ideal defect model [1, 2]. There is, however, a number of experimental data which are not in agreement with the ideal defect model, especially at higher nonstoichiometry [4-6].

J. Nowotny and W. Weppner (eds.),
Non-Stoichiometric Compounds Surfaces, Grain Boundaries and Structural Defects, 265–277.

To explain these inconsistencies it has been assumed that the defect structure of CoO is more complicated involving also defect complexes which are formed as a result of defect interactions, especially at higher deviation from stoichiometry [7, 8]. Recently, the properties of CoO were considered by assuming 4:1 complexes which are composed of four cobalt vacancies and the cobalt ions located in tetrahedral positions [3]. Also, the formation of oxygen vacancies was considered for CoO [8, 9]. So far, however, there is no direct evidence for the formation of either cobalt interstitials or oxygen vacancies.

Most reports on the defect structure of CoO are concerned with the crystalline bulk. The outer oxide layer may differ, however, substantially from the bulk [10-12]. The difference involves the non-stoichiometry and resulting defect structure. It has been shown that the near-surface region of ionic crystals is enriched in both intrinsic and extrinsic defects [11, 13]. It has been reported that the NiO surface is enriched in nickel vacancies by a factor of 5 [11]. Similar enrichment was recently determined for CoO [14]. One should, therefore, expect that the enrichment results in much stronger defect interactions in the near-surface region than in the bulk. The near-surface chemical diffusion studies performed for Cr-doped CoO have indicated that segregation of Cr may even lead to a structural transition involving the formation of a spinel-type overlayer [15]. The formation of low dimensional surface structures has been later confirmed by electron spectroscopy studies [16]. Comparison of the near-surface and bulk chemical diffusion data indicates that such structures may also be formed in the case of undoped CoO [17].

The purpose of the present work is to perform 'in situ' studies on the surface properties of CoO in order to evaluate the defect structure of the outer crystal layer at higher deviations from the ideal stoichiometry. The High Temperature Kelvin Probe is used in the present studies [18]. The method enables the measurement of the work function at elevated temperatures and under controlled gas atmospheres. So far, this is the only surface sensitive method for studying solids under these conditions which correspond closely to practical applications.

The experimental part of the work will be preceded by the descrption of fundamental relationships between work function and defect concentrations in oxide crystals.

2. WORK FUNCTION, FERMI ENERGY AND DEFECT STRUCTURE OF CoO

A very convenient method for the determination of the work function of solids at elevated temperatures is based on the measurement of the contact potential difference (CPD) between the investigated sample (s) and the reference level (ref):

$$CPD = (\phi_s - \phi_{ref})/e \qquad (1)$$

where ϕ denotes the work function (WF) and e is the elementary charge. The absolute value of the WF in the case of non-stoichiometric oxides has a very complex physical meaning and is difficult to interpret.

In the present work we will determine changes in WF accompanyning chemical processes such as oxidation and reduction of CoO. The interaction of CoO with O_2 leads to the formation of chemical defects. Using the Kröger-Vinck notation this process can be described according to the following equilibria:

$$1/2\ O_2 + Co_{Co} \rightleftarrows V_{Co}^{x} + CoO \qquad (2)$$

$$V_{Co}^{x} \rightarrow V_{Co}^{'} + h^{\cdot} \qquad (3)$$

$$V_{Co}^{'} \rightleftarrows V_{Co}^{''} + h^{\cdot} \qquad (4)$$

where V_{Co}^{x}, $V_{Co}^{'}$ and $V_{Co}^{''}$ denote neutral, singly and doubly ionized cation vacancies, respectively, and $h^{\cdot}$ are electron holes. Application of the mass action law to reactions (2) - (4) leads to the following dependence between the defect concentration and p_{O_2}:

$$[V_{Co}^{z'}] = p_{O_2}^{1/n} \exp(-\Delta G_f/RT) \qquad (5)$$

where the ionization degree z may assume the numbers 0, 1 and 2, ΔG_f is the free enthalpy of formation of defects and n is the parameter which is sensitive to the defects structure:

$$n = 2(z + 1) \qquad (6)$$

A shift of equilibrium (2) to the right hand side will lead to an increase in acceptor centers. Their ionization (equilibria 3 and 4) results in the decrease in the Fermi energy level (E_F) and corresponds to an increase in WF. Accordingly, changes in ΔE_F may be expressed as:

$$\Delta E_F = -\Delta\phi_s = \Delta\phi_{ref} + e(\Delta CPD) \qquad (7)$$

Evaluation of the component $\Delta\phi_{ref}$ in oxygen has already been described for Pt [18]. Therefore, eq. (6) may be used for the determination of ΔE_F for the investigated oxide sample. Then, assuming that $\Delta\phi = -\Delta E_F$, we may write the following dependence between the parameter n in Eq. (5) and ϕ:

$$\frac{1}{n} = \left(\frac{1}{kT}\frac{\partial\phi}{\partial \ln p_{O_2}}\right)_T \qquad (8)$$

3. EXPERIMENTAL

Nominally undoped CoO single crystals were prepared by the Verneuil method. Spectral analysis has shown the following amounts of impurities: 10^{-3} a/o Cr, 3×10^{-2} a/o Fe, 2×10^{-2} a/o Al and 10^{-3} a/o Si.

The investigated sample in the shape of a disc (ϕ 8 mm and 1 mm thick) was prepared by cleavage of the single crystal boule.

The High Temperature Kelvin Probe [18] was applied for work function measurements against the Pt reference electrode. The work function of the reference level was determined earlier [19].

The oxygen partial pressure over the oxide sample was adjusted by $Ar-O_2$ mixtures flowing through the reaction chamber. The oxygen activity was controlled by a zirconia gauge.

The experimental procedure involved measurements of the CPD while oxygen was isothermally step-wise increased (oxidation runs) or decreased (reduction runs). The studied temperature range was 790 - 905°C.

4. RESULTS

Figs. 1 - 4 illustrate the imposed changes in oxygen activity in the reaction chamber and the resulting changes in CPD as a function of time in the temperature range 790 - 905°C. The time intervals between the succesive changes in p_{O_2} (termed below as experimental runs) were about 1 - 2 h. The observed character of the CPD changes may be categorized within several regions exhibiting different behaviour. The following regions (as marked at the top of Figures 1 - 4) can be distinguished for oxidation runs:

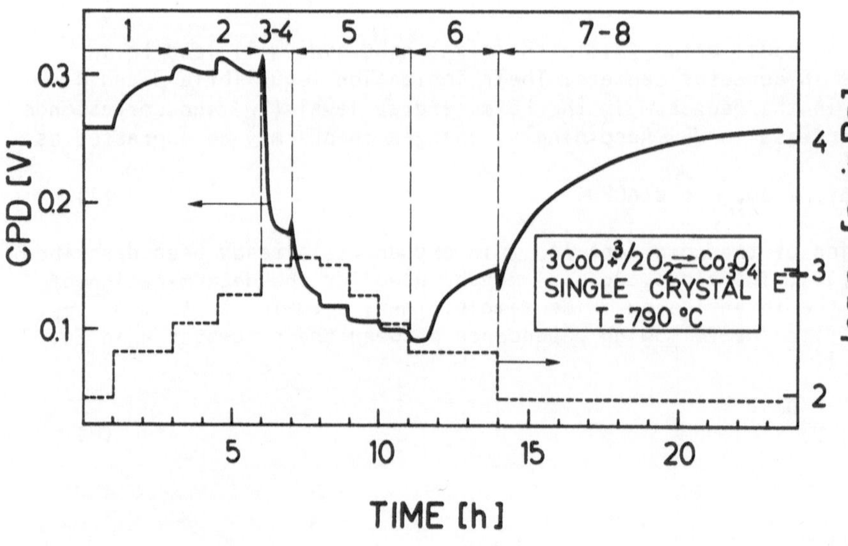

Figure 1. Plots of both CPD and log p_{O_2} at 790°C. E denote equilibrium p_{O_2} [20]

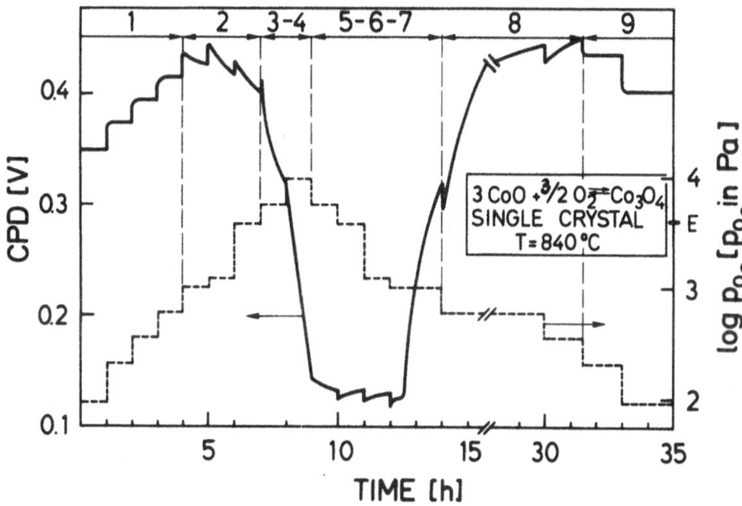

Figure 2. Plots of both CPD and log p_{O_2} at 840°C.

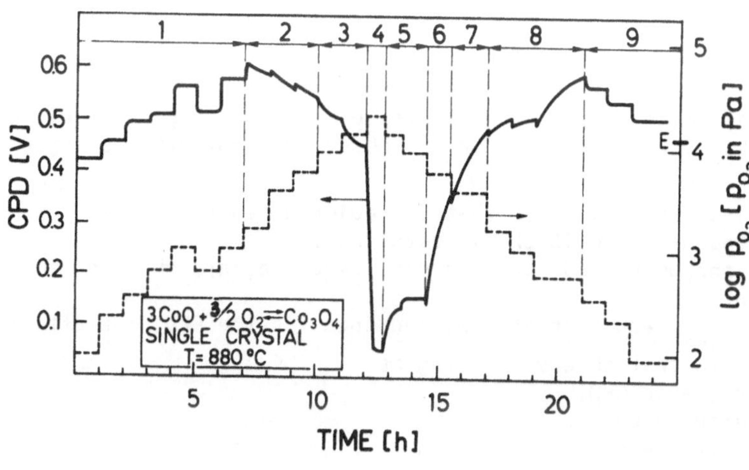

Figure 3. Plots of both CPD and log p_{O_2} at 880°C.

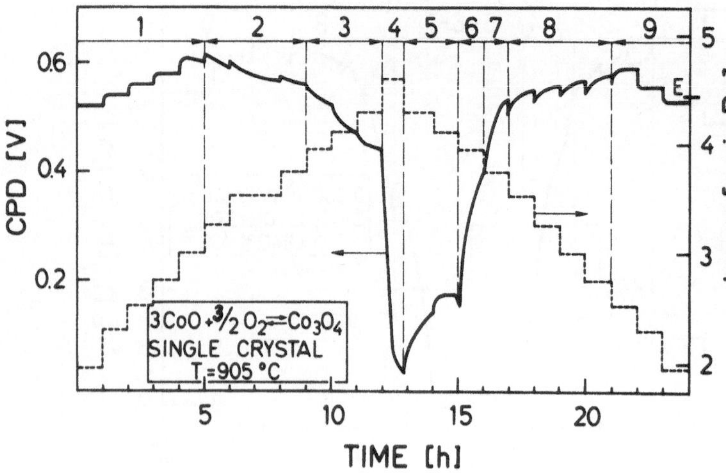

Figure 4. Plots of both CPD and log p_{O_2} at 905°C.

1. The characteristics of the CPD changes is essentially the same as the imposed changes in p_{O_2}. The CPD initially sharply increases and then assumes a stable value in time. The time required to reach a stable CPD varies between 2 h at 790°C and 5 min at 905°C.
2. After imposing a new, larger p_{O_2} the CPD initially increases rapidly, reaches a maximum and then slowly decreases. The decrease does not tend to reach a stable value even after 2 h at 905°C.
3. The initial rapid increase in CPD (as observed in region 2) disappears and a slow decrease in CPD is observed right after changing p_{O_2}.
4. The behaviour is similar to that in region 3. In this range, however, the decrease in CPD is very fast.

Similar phenomena, but in a reverse sequence, are observed for reduction runs:

5. Changes in p_{O_2} result in step-wise changes in the CPD. The effect is similar to that observed in regime 1.
6. Sharp increase in CPD.
7. Slow increase in CPD.
8. Sudden initial decrease followed by a slow increase of CPD,
9. Step-wise decrease with equilibration times comparable to those during oxidation runs.

The experimental data are well reproducible for independent experimental runs concerning both the character of the observed dependence as well as the absolute values.

5. DISCUSSION

In analysis of the experimentally observed CPD data we assume that (i) changes in WF of the reference Pt electrode in oxygen are significantly smaller than $\Delta\phi$ of the CoO sample, and (ii) changes in CPD are rate controlled by the CoO surface. As it results from ealier experiments [19] the parameter n in Eq. (7) for Pt under the conditions of the present investigation is equal to 7.1. In comparison with the magnitude of electrical effects at the CoO surface the resulting $\Delta\phi$ of Pt can be neglected in a first approximation. Accordingly, the measured changes in CPD are practically equal to the WF changes of CoO with the same sign.

The phenomena observed in region 1 indicate that the surface of CoO assumes equilibrium within 5 min (> 840°C). The increase in oxygen pressure in this region leads to an appopriate increase of the concentration of cation vacancies at the surface (acceptors). Accordingly, oxidation and reduction of CoO (regions 1 and 9, respectively) may be considered in terms of equations (2) - (4). The resulting WF changes are described by Eq. (7). At higher oxygen pressures (region 2) the initial rapid increase in WF is followed by a slow decrease which does not reach a constant value within the time of experimental runs (1 h). The observed subsequent slow decrease in WF results in the formation of donor centers. In the case of CoO both oxygen vacancies ($V_O^{\bullet\bullet}$) and cobalt interstitials ($Co_i^{\bullet}$ or $Co_i^{\bullet\bullet}$) may be considered as donors. It seems unlikely, however, that the increase in p_{O_2} might result in the formation of oxygen vacancies. On the other hand one may expect that under the conditions of high p_{O_2} and the resulting composition of CoO close to the Co_3O_4 phase boundary the formation of cobalt interstitials is energetically favourable. The interstitials may be considered as precursors of the spinel phase. Therefore, besides equilibria (2) - (4) the following defect reactions may be considered in region 2:

$$Co_{Co} + V_i \rightleftarrows Co_i + V_{Co}^{x} \qquad (8)$$

$$Co_i + h^{\bullet} \rightleftarrows Co_i^{\bullet} \qquad (9)$$

$$Co_i^{\bullet} + h \rightleftarrows Co_i^{\bullet\bullet} \qquad (10)$$

$$V_{Co}^{x} \rightleftarrows V_{Co}'' + 2h^{\bullet} \qquad (11)$$

As a consequence of the experimental data the electronic effect resulting from the formation and ionization of cobalt vacancies (equilibria 2 - 4) is much faster than the formation of the donors (equilibria 8 and 9). Moreover, the electronic effect resulting from Eq. (9) becomes predominant at higher p_{O_2}.

As p_{O_2} approaches the value CoO/Co_3O_4 phase boundary the initial increase in WF observed in region 2, dissappears and the decreasing part becomes more and more steep. Depending on the rate of the WF decrease one can clearly distinguish two regions involving the slow and the rapid WF changes (regions 3 and 4, respectively). We assume that the observed effect is characteristic for the formation of the Co_3O_4 structure. In region 3 the formation of the spinel-type structure is limited to the outer surface layer (1-2 nm) while in the bulk the CoO structure remains stable.

The oxygen pressure corresponding the CoO/Co_3O_4 equilibrium in the bulk [20] is denoted by E at the log p_{O_2} axis. When p_{O_2} surmounts this value then the WF decreases dramatically as a result of the phase transformation within the entire crystal (region 4). As it is seen from Figs. 1 - 4 the rapid dramatic change in the CPD in this range corresponds very well to the equilibrium value of p_{O_2} for the coexistance $CoO-Co_3O_4$ [20]:

$$\log p_{O_2}\ [kPa] = -(19720/T\ [K]) + 18.06 \qquad (12)$$

Reduction runs lead to the phase transition $Co_3O_4 \rightarrow CoO$ through similar regions as described above. Initially the reduction process involves the variations of the Co_3O_4 phase composition (region 5). Then the character of the resulting changes in WF are stepwise. The defect structure of Co_3O_4 is not well known but cation vacancies seem to be predominant lattice defects in this structure. When p_{O_2} decreases below the E (equilibrium) value then the bulk $Co_3O_4 \rightarrow CoO$ phase transition takes place (region 6). The provided experimental data allow to distinguish precisely between the runs involving sharp and slow CPD changes corresponding to the bulk and in the overlayer (region 7), respectively. The region 8 corresponds to defect processes described in region 2 but in the reverse direction. Similarly as in region 8 the electronic effect related to changes in cation vacancy concentration (equilibria 2 - 4) is much faster than that corresponding to shifts to the left of equilibria (8) and (9). Finally, region 9 involves again to changes in composition of CoO which are described by shifts in Eqs. (2) - (4) to the left.

6. DEFECT STRUCTURE CONSIDERATIONS

It has been shown recently that bulk properties of CoO at elevated temperatures may be well described by assuming only one type of predominant defects: doubly ionized cobalt vacancies [3, 21]. The defect structure of $Co_{1-y}O$, which can also be considered as a solid solution of cobalt vacancies in CoO, and may be expressed by the formula $Co_{1-y}(V_{Co})_yO$ (Fig. 5). The concentration of other defects as neutral and singly ionized cobalt vacancies may be assumed to be negligibly small

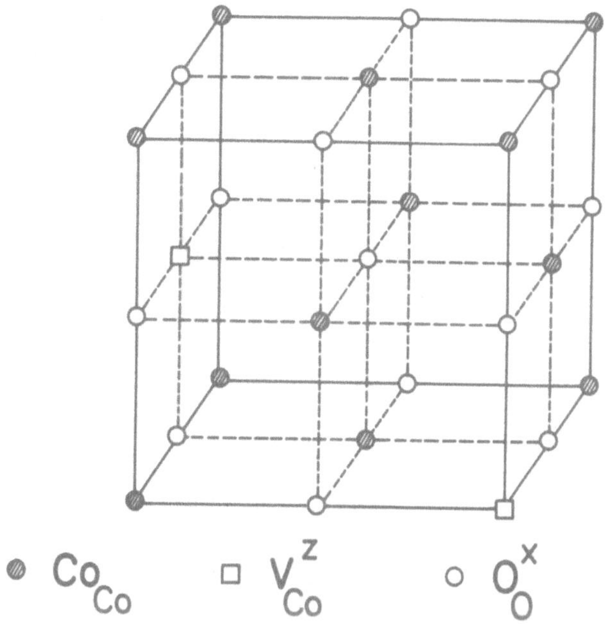

Figure 5. The unit cell of $Co_{1-y}O$ illustrated as a solid solution of cation vacancies in CoO

It has been shown that defects may segregate leading to an enrichment of the interfaces in intrinsic defects [11]. It has been reported that the near-surface region of such oxides as NiO and CoO is enriched in cation vacancies by a factor of 3-5 [11-14]. Consequently, interactions between defects at the surface are much stronger than in the bulk resulting in the formation of more complex defect assembles as, e.g., 4:1 clusters [3] or spinel-like Co_3O_4 microdefects [22]. It has been claimed by Tscherkhasin [22] that $Co_{1-y}O$ can be considered as a solid solution of Co_3O_4-type defects in CoO. One may expect that the concentration of these defects increases as the non-stoichiometry in CoO increases and assumes maximum at the Co_3O_4 phase boundary. Both cobalt vacancies and cobalt interstitials should be considered as precursors of the Co_3O_4 structure which can be described by the following formula:

$$(Co^{+2}_{tet})_8(Co^{+3}_{oct})_{16}(V_{oct})_{16}O^{-2}_{32} \qquad (13)$$

According to the Kröger-Vink notation the formation of the Co_3O_4-type microdefect can be expressed by the following reaction of O_2 with CoO:

$$V_t + 1/2O_{2(g)} + 4\ Co_{Co} \rightleftarrows Co_i^{\cdot\cdot} + 2\ Co^{\cdot}_{Co} + 2\ V''_{Co} + CoO \qquad (14)$$

The Co_3O_4 structure is illustrated in Fig. 6.

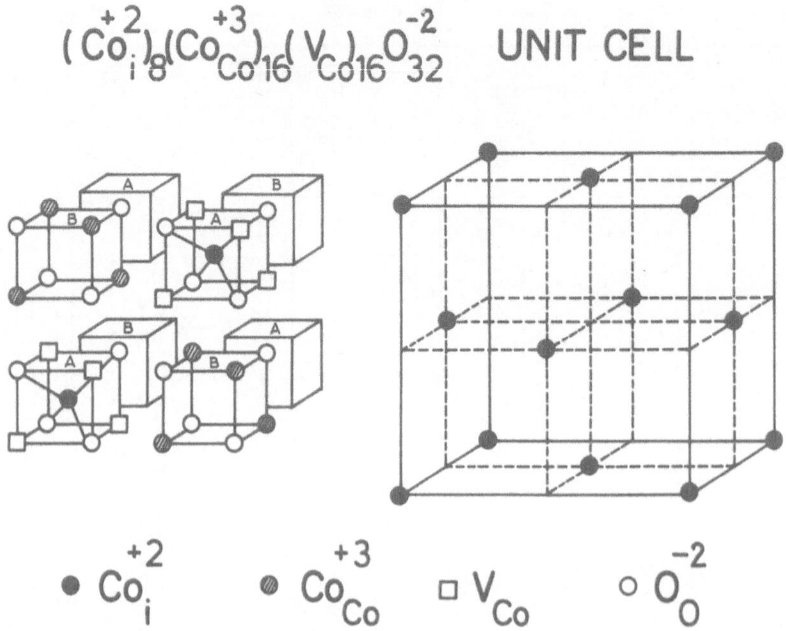

Figure 6. The unit cell of Co_3O_4. The structure consists of alternating $Co^{+2}O_4$ tetrahedra (A) and $Co_4^{+3}O_4$ cubes (B) build up into a face centered cubic lattice of 32 oxygen ions

The present studies indicate that besides cobalt vacancies, which are generally assumed to be the predominant defects in CoO, also cobalt interstitials need to be considered. Their concentration may be neglegibly low as far as the crystalline bulk is concerned. These defects, however, may reach a high concentration in the near-surface layer. Their formation has been observed for CoO in the vicinity of the Co_3O_4 phase boundary within well defined conditions of $T - p_{O_2}$. Since the outer surface regions exhibit a much higher concentration of both interstitials and cobalt vacancies one should, therefore, expect especially favourable conditions to form a bidimensional Co_3O_4 core-layer. Again, we would like to emphasize that the Co_3O_4 structure forms within the near-surface layer under $T - p_{O_2}$ conditions which correspond to the stability of the CoO phase in the bulk. Thus formed bidimensional layer remains stable only in the specific energetical conditions of the surface.

7. THE NEAR-SURFACE PHASE DIAGRAM

The present experimental data may serve to evaluate the phase diagram of the Co - O_2 system which corresponds to the outer oxide layer.

Figs. 7 and 8 illustrate the phase diagram resulting from experimental data for oxidation and reduction runs, respectively.

Fig. 7 corresponds to the CoO → Co_3O_4 phase transition. The precursor line determines the T - p_{O_2} field corresponding to the CoO → Co_3O_4 transition within the overlayer (regions 2 and 3). The boundary between the two regions can be precisely determined at 905°C.

There is a good agreement between the bulk phase boundary determined in this work (transition line) and the reported thermodynamic data [20]. The observed shift of the boundary line resulting from WF measurements for oxidation runs towards higher p_{O_2} values is within the experimental error. The shift, however, is more significant for reduction runs (Fig. 8) and is apparently produced by an oversaturation required to form the nuclei of the CoO phase on the very stable structure of Co_3O_4. The oversaturation increases with decreasing temperature.

Comparison of Figs. 7 and 8 indicates that the precursor line, corresonding to the formation of Co interstitials in reduction runs, is shifted toward lower p_{O_2}. It is also interesting to note that in both cases (oxidation and reduction) the phase field involving the formation of cobalt interstitials increases substantially with temperature.

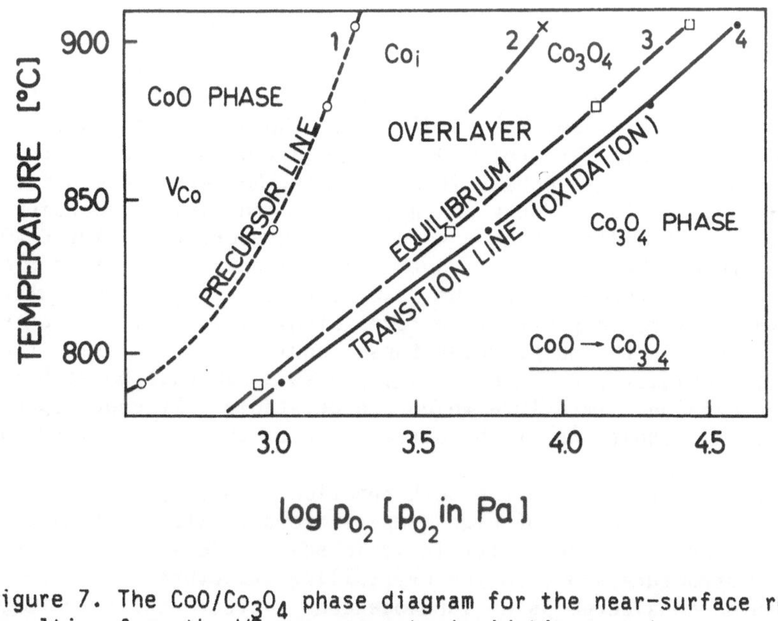

Figure 7. The CoO/Co_3O_4 phase diagram for the near-surface region resulting from the WF measurements (oxidation runs)

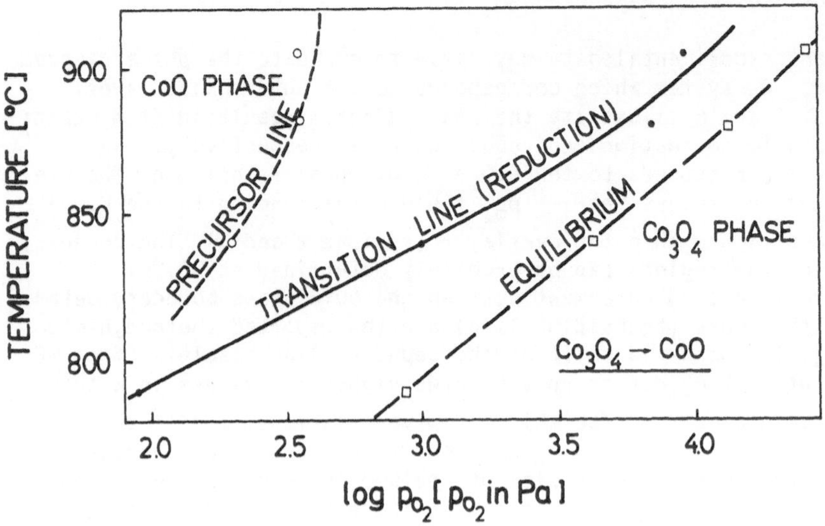

Figure 8. The CoO/Co_3O_4 phase diagram for the near-surface region resulting from the WF measurements (reduction runs)

8. CONCLUSIONS

The near-surface layer of CoO exhibits a different defect structure than the crystalline bulk. The difference results in segregation of intrinsic defects. The outer layer of CoO is enriched in cobalt vacancies by about one order of magnitute compared to the bulk within the entire stability range. The studies have shown that at higher non-stoichiometry, close to the CoO- Co_3O_4 equilibrium, the outer layer is also enriched in cobalt interstitials. So far, these defects were not identified in the bulk phase of CoO. Both cation vacancies and cobalt interstitials can be considered as precursors of the Co_3O_4 spinel structure. The presence of both these defects causes favourable conditions for the formation of the Co_3O_4 structure within the outer bidimensional layer. This structure forms within the stability range of the CoO phase in the bulk. Under these conditions thus formed the spinel-type overlayer can not be considered as a separate phase but as a local bidimensional structure which can be stable only under specific energetical conditions of the surface due to the excess of surface energy.

This work confirms that the work function method can be a very powerful tool in the hands of high temperature chemists in studies of the near-surface defect chemistry in ionic solids. Because of different defect structure and also the crystalline structure the oxide crystal can be considered as an inhomogeneous system even in thermodynamic equilibrium. The system involves the homogeneous bulk phase and the outer layer. This layer may be responsible for alterations of many oxide parameters due to changes of several physical and chemical properties. Accordingly, the evaluation of the near-surface properties is of substantial importance for controlling properties of ceramics materials.

ACKNOWLEDGEMENTS

This work was supported in part by the Institute of Materials Science, Academy of Mining and Metallurgy, Krakow, Grant No CPBR 6.6.64

REFERENCES

1. B. Fisher and D.S. Tannhauser, J.Chem.Phys., 44 (1966) 1663
2. R. Dieckmann, Z.Physik.Chem., Neue Folge, 10- (1977) 189
3. J. Nowotny and M. Rekas, J.Am.Ceram.Soc., sent for publication
4. H.C. Chen, E. Garstein and T.O. Mason, J.Phys.Chem.Solids 43 (1982) 991
5. E.M. Logothetis and J.M. Park, Solid State Comm., 43 (1982) 543
6. G. Petot-Ervas, P. Ochin and T.O. Mason, "Transport in Nonstoichiometric Compounds", V.S. Stubican and G. Simkovich, Eds., Plenum Press, London, 1985, p.61
7. A.-Z. Hed, J.Chem.Phys., 50 (1969) 2935
8. G.M. Raynaud and F. Morin, J.Phys.Chem.Solids 46 (1985) 1371
9. H.L. LeBrusq and J.P. Delmaire, Rev.Int.Haut.Temp.Refract., 10 (1973) 15
10. J. Nowotny, J.Chim.Phys., France, 75 (1978) 689
11. J. Nowotny, Materials Sci.Forum, 29, 99 (1988)
12. J. Nowotny, Solid State Ionics, in print
13. J. Nowotny and M. Sloma, "Surface and Near-Surface Chemistry of Oxide Materials", J. Nowotny and L.C. Dufour, Eds., Elsevier, Amsterdam, 1988, p. 281
14. J. Nowotny, in preparation for publication
15. J. Nowotny, I. Sikora and J.B. Wagner, Jr., J.Am.Ceram.Soc., 65 (1982) 192
16. J. Haber, J. Nowotny. I. Sikora and J. Stoch, Appl.Surf.Sci., 17 (1984) 324
17. J. Nowotny, I. Sikora and J.B.Wagner, Jr., Mater.Sci.Monographs No 10, Elsevier, Amsterdam, 1982, p. 225
18. J. Nowotny, M. Sloma and W. Weppner, Adv.Ceram., 23 (1988)159
19. J. Nowotny and M. Sloma, J.Phys.Chem., France, C1 47 (1986) C1 807
20. E.J. Grimsey and K.A. Reynolds,J.Chem.Thermodynamics 18 (1986) 473
21. J. Nowotny and M. Rekas, J.Am.Ceram.Soc., sent for publication
22. A.E. Tscherkhashin, PhD thesis, Institute of Catalysis, Siberian Branch of the Sov.Acad.Sci., Novosibirsk, 1971

GRAIN BOUNDARY PHENOMENA IN SOME ELECTRONIC AND MAGNETIC CERAMICS

R. Freer, V.A. Roberts and F. Azough
Materials Science Centre
University of Manchester/UMIST
Grosvenor Street
Manchester M1 7HS

ABSTRACT. Additives play a key role in the preparation of ceramics and the control of microstructure, but they may also affect the grain boundary chemistry and the physical properties. A brief review is given of grain boundary phenomena in some electronic ceramics. The devices considered include boundary layer capacitors, PTC thermistors, and zinc oxide varistors. The effect of segregation on the electrical losses of a ZTS dielectric is described, and the role of processing techniques and sintering atmosphere on the performance of a Mn-Zn ferrite is discussed.

1. INTRODUCTION

Monolithic ceramics are polycrystalline solids which have been prepared from assemblages of individual grains or aggregates. In many cases additives are used to control the manufacturing process or adjust the final properties. The behaviour of the additives, and indeed unwanted impurities, in the presence of the host material often results in modification of the structure or chemistry of the grain boundaries, or the properties of the grain boundaries.

Kingery [1] discussed grain boundary phenomena in electronic ceramics from seven viewpoints:

(i) grain boundary structure - point, line or planar defects may exist, and different forms of periodic structures occur to enable lattice matching. The boundaries usually act as rapid diffusion paths.

(ii) boundary compositions - cores may be disturbed for a few atoms which give rise to a new phase which separates the grains.

(iii) boundary precipitates - in many ionic systems selected impurity atoms have limited solubility, but adequate mobility to form precipitates during cooling. This is well documented in MgO, e.g,. [2,3], and important for the development of grain boundary structures in ZnO [4,5].

(iv) boundary-core chemistry - common solutes in oxide ceramics tend to be adsorbed at the boundary, and unusual oxidation states may occur. Non-stoichiometry in the boundary is equivalent to an electrically-charged layer.

J. Nowotny and W. Weppner (eds.),
Non-Stoichiometric Compounds Surfaces, Grain Boundaries and Structural Defects, 279–297.

(v) oxygen diffusion - changes in the local oxygen partial pressure will modify diffusion rates and lead to compositional changes adjacent to the boundaries.

(vi) boundary charge and space charge - in ionic systems containing impurities, there will be an electrical charge in the boundary which is a function of composition, boundary orientation and heat treatment. The resulting charge may be positive or negative.

(vii) Charge transfer processes - these are generally temperature-dependent and sensitive to many of the features outlined above.

In this paper we try to illustrate some important grain boundary phenomena in electronic ceramics, and the effect they have on structure and properties. We first consider boundary layer capacitors, PTC thermistors and ZnO varistors. We then examine a ceramic dielectric where additives can modify the bulk structure and electrical losses. Finally, we discuss Mn-Zn ferrites and demonstrate the importance of preparation techniques and processing conditions.

2. BOUNDARY LAYER CAPACITORS

Ceramics have been used as dielectrics for capacitors for over 60 years. Early developments were marked by the appearance, almost every decade, of a new class of dielectric materials exhibiting a ten fold increase in relative permittivity over the previous best value [6]. In this way porcelain was succeeded by rutile, and this in turn by barium titanate. By the mid 1950's there was no sign of new materials of higher intrinsic dielectric constants, and so efforts were directed towards investigations of new geometries and technologies, e.g. multilayer (MLC) construction, electrode barrier layers and grain boundary barrier layers, GBBL [6]. Although MLC capacitors have captured a significant fraction of the world market in terms of volume output, boundary layer capacitors are still of considerable interest because they potentially offer high capacitance density [6] and very high effective permittivity [7] in appropriate circumstances.

Most GBBL capacitors are based upon either $BaTiO_3$ or $SrTiO_3$, and the fabrication involves a series of stages such that the grain boundary has different chemistry and properties to that of the bulk. In one form [8] a porous semiconducting ceramic is prepared by sintering $BaTiO_3$ plus small amounts of Dy_2O_3 and SiO_2 in nitrogen. A slurry of an oxide such as MnO_2, CuO, TeO_2 or Bi_2O_3 etc, is then painted onto the surface of the ceramic and the assemblage is re-fired. The oxide diffuses preferentially along the grain boundaries to form an insulating layer, yielding a product with a high density and high effective relative permittivity. The frequency response of the dielectric properties and resistivity is shown schematically in Figure 1, and this can be conveniently modelled by a simple grain - grain boundary - grain structure [6,9]. Waku [8] suggests that the thickness of the insulating layer is typically of the order of 1 μm. The properties of two GBBL capacitors are summarised in Table 1.

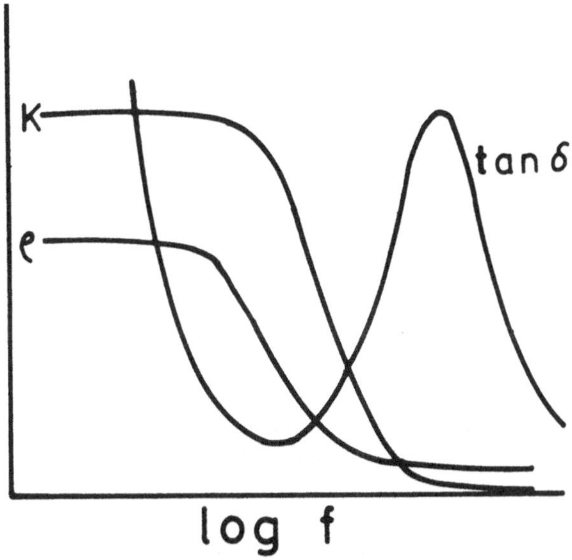

Figure 1. Schematic frequency dependence of dielectric constant, resistivity and tan δ.

TABLE 1 Properties of two GBBL capacitors (after Waku [10])

	$Ba(Ti_{0.9}Sn_{0.1})O_3$	$SrTiO_3$
ϵ_{app} (at 1 kHz)	20000	14000
tan δ	2-4	1-2
resistivity (Ωcm)	5×10^{10}	10^{12}
ϵ_r (ins. layer)	1000	250
σ (semiconducting part $\Omega^{-1}cm^{-1}$)	1	14
volume semiconducting part (%)	95	98.2
d(μm) insulating layer	2	0.7

Detailed analysis of GBBL [11] suggests that the electrical properties of the device depend upon the equilibrium kinetics of oxygen defects in the grain boundary and the diffusion of Ba vacancies into the donor rich grain boundaries during the sintering stages.

3. PTC MATERIALS

Ceramics exhibiting a positive temperature coefficient of resistance in the vicinity of the ferroelectric Curie temperature (Figure 2) are known collectively as PTC materials. The behaviour is most pronounced in doped semiconducting titanates and is exploited, for example, in overload protection applications. Excessive current in a circuit causes self-heating of the device (thermistor) and a reduction of current to a safe value. The device "resets" itself on cooling down after the fault is cleared.

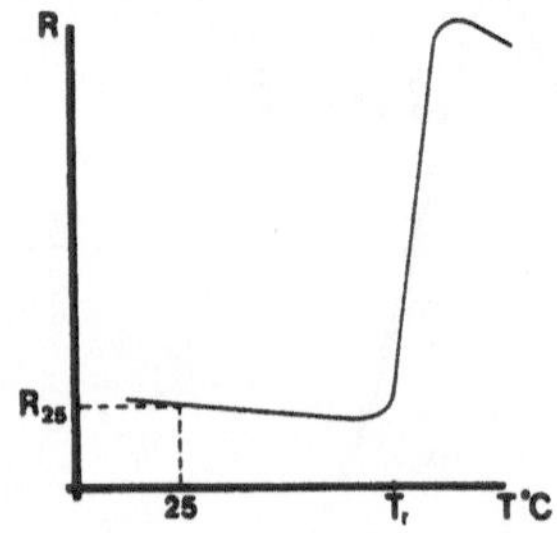

Figure 2. Resistance-temperature characteristics of a material exhibiting PTC behaviour. (Courtesy of J.H. McCartney)

The discovery of the PTC effect emerged from an investigation [12] of controlled valency semiconduction aimed at developing new thermistor compositions. Reports of processing techniques [13] followed, and subsequently there were many studies aimed at understanding the PTC mechanism, e.g. [14,15].

Most interpretations are based on grain boundary behaviour rather than bulk properties, and a barrier layer model was invoked to explain the main features of the resistivity anomaly, initially by Heywang [14] and then in revised forms by Jonker [16] and Heywang [17]. Later work led to further refinements [18,19].

Today PTC devices are used in a wide variety of applications [20,21] including overload protection, honeycomb and static heaters, automotive sensors, TV degaussing, motor start assistors and general purpose temperature sensors. Examples of typical PTC devices are shown in Figure 3.

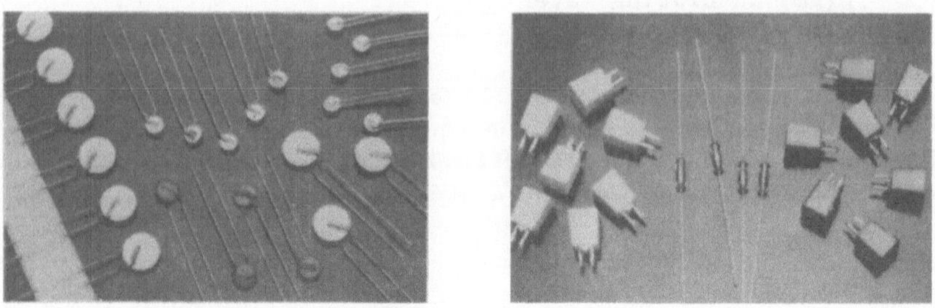

Figure 3. Typical PTC devices: (a) $BaTiO_3$ with wire contacts, (b) other configurations - discs with spring contacts in box packages, chips in diode encapsulation. (Courtesy of J.H. McCartney)

3.1. Materials

Various additives are employed to adjust the properties of barium titanate, which is insulating in its pure form. Donor dopants, such as Y, Nb, La, Er, Ta, W or Sb are used to generate free electrons and render the material semiconductive. Acceptor dopants, e.g. Mn, are usually added to enhance the resistance-temperature characteristics (Figure 4), but this may be at the expense of base level resistance.

The ferroelectric phase transition in La-doped barium titanate, and thus the switching temperature, usually occurs around 125°C. This can be increased by substituting Ba by Pb, or conversely lowered by replacing Ba by Sr. Other additives are employed to modify the microstructure.

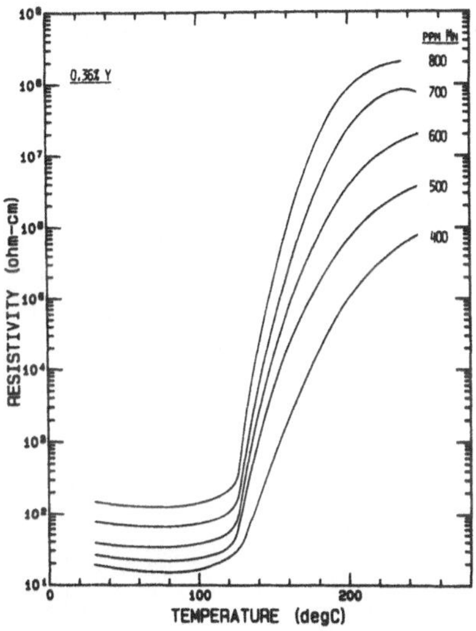

Figure 4. Effect of Mn (acceptors) on PTC resistance behaviour. (Courtesy of J.H. McCartney)

3.2. THE PTC EFFECT

The $BaTiO_3$ is mainly an n-type semiconductor with donors partially compensated by acceptors and Ba vacancies. Grain boundary acceptor states are believed to be generated by the chemisorption of oxygen [16] or halogen [22], and consequently electron depletion layers (i.e. p-type) are formed at the boundaries to yield n-p-n structures. The magnitude of the resulting interface potential depends upon temperature. Support for this basic n-p-n structure comes from thermodynamic modelling [23].

Figure 5 shows schematically the voltage drop across a series of grains and grain boundaries. In the high temperature regime the grain boundaries have a high resistivity, and the grain cores a low resistivity, hence an applied voltage is dropped stepwise across the grain boundaries. The superimposed field at each grain boundary depresses the potential barrier that gives rise to the PTCR effect and reduces the resistance range. The effect is minimised by increasing the number of grain boundaries, i.e. by reducing the average grain size, to reduce the potential across each boundary (Figure 5).

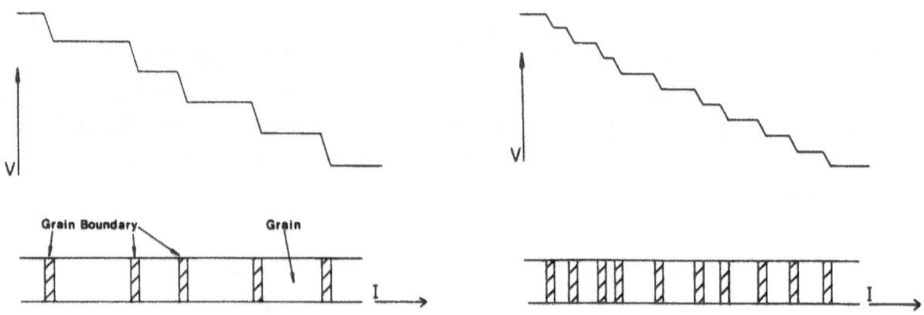

Figure 5. Potential drop across grain boundaries, (a) large grain size, (b) small grain size. (Courtesy of J.H. McCartney)

Figure 6a shows the microstructure of a typical early PTC ceramic. There are many large grains and a spread from 2-15 μm. In contrast Figure 6b shows the microstructure of a more recently manufactured PTC ceramic. The average grain size is much smaller and the size distribution more uniform.

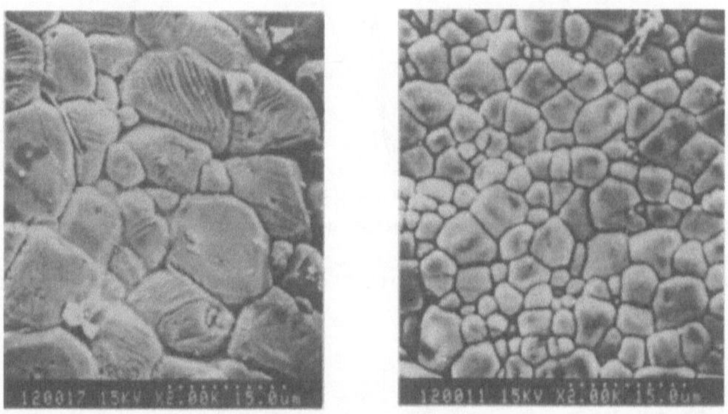

Figure 6(a). Grain structure of early PTC ceramic, (b) Grain structure of more recent PTC ceramic. (Courtesy of J.H. McCartney)

The effect on performance is highlighted by Kulwicki [21] who showed that in 1967, for example, PTC materials typically had a zero-voltage resistivity ratio of perhaps 4 decades. With increasing voltage, the PTC effect was severely restricted, and this limited the range of applications. The modern materials enjoy a zero-voltage resistivity range of ~6 orders of magnitude, which is not severely degraded at moderate (e.g. 100 V/mm) applied fields.

Kulwicki [21] showed that the PTC curve could be explained in the following way:

(i) initial negative temperature coefficient behaviour is due to the temperature-dependence of permittivity.

(ii) the knee is due to a rapid change in polarisation near the phase transformation.

(iii) the initial jump - due to the release of internal stress at the phase transition.

(iv) further PTC - as a result of the temperature-dependence of thermodynamic parameters.

(v) maxima - due to decrease in occupation probability as surface states approach the Fermi level.

In general terms [21]: there is much charge trapped at the grain boundary giving rise to high capacitance; the voltage sensitivity can be understood in terms of Schottky emission across the barrier [17,24]; and strong depletion layers give rise to resistive grain boundaries.

Detailed electron microscopy studies of grain boundaries in doped barium titanate ceramics have found no evidence of continuous second phases ~ few nm width along the grain boundaries, but the second phases tend to be segregated at the interstices of several grains [25]. Kulwicki [21] noted that this ruled out PTC models [26] which require second phase layers to form n-p-n, grain-interface-grain heterojunctions.

4. ZnO VARISTORS

ZnO-based ceramics have grossly non-linear current-voltage characteristics, which makes them useful as varistors for voltage stabilisation or transient surge suppression in electronic circuits and electric power systems [27,28]. The excellent non-ohmic properties and large current withstand capabilities of ZnO varistors are attributed to the behaviour of grain boundaries. It is thought that intergranular regions act as barriers for electrical conduction.

ZnO varistors are formed from the sintering of pressed bodies of zinc oxide containing a number of additives such as Bi_2O_3, BaO, SrO,

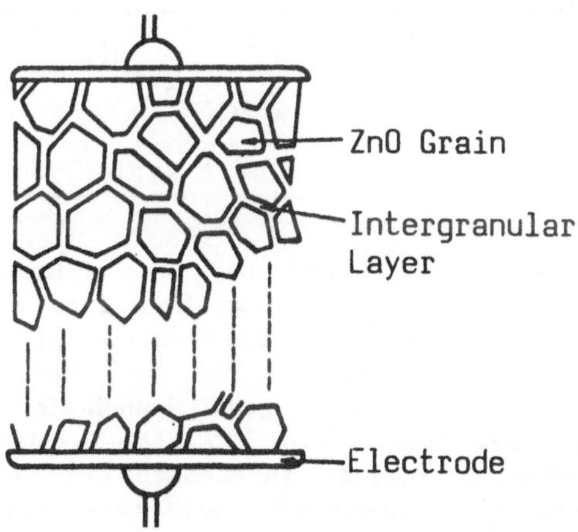

Figure 7. Schematic structure of ZnO varistor. (after Matsuoka [30]).

Pr_2O_3, NiO, CoO, MnO and Cr_2O_3, A typical microstructure is shown schematically in Figure 7. Liquid phase sintering results in the formation of a Bi_2O_3-rich intergranular layer, and within it a spinel phase as localised particles. The spinel is a by-product of the manufacturing process, and only plays a secondary role in the operation of the device [29].

A typical V-I curve for a ZnO varistor is shown in Figure 8. The main features are related to the microstructure of the intergranular layer. The threshold voltage is proportional to the number of intergranular layers in series, whilst the surge withstand capability is proportional to the number of intergranular layers in parallel [30]. Additives such as NiO and Cr_2O_3 are used to improve the reliability of the intergranular layer and stabilise it against changes in electric load and ambient conditions. Transition metal oxide additives, e.g. CoO and MnO, are used to enhance the non-ohmic properties of the device, and these oxides are located within the intergranular layer where they act as traps and surface states [30].

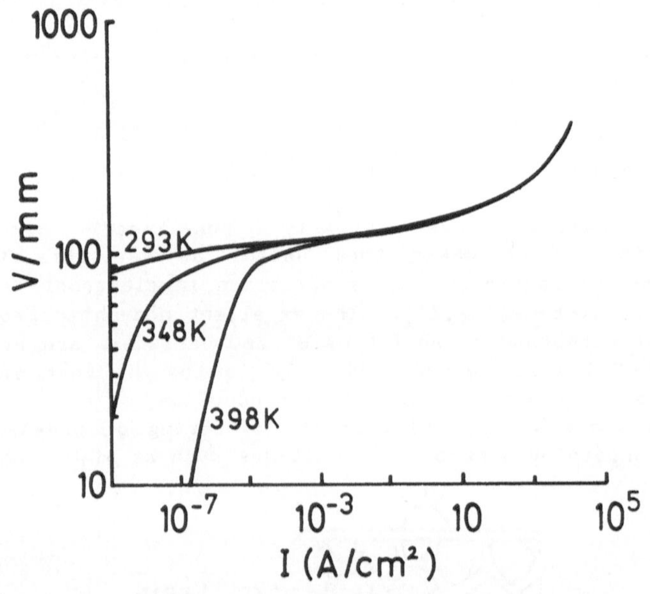

Figure 8. Current-voltage curves for a typical ZnO varistor. (after Matsuoka [30]).

The capacitance and dielectric loss have a characteristic frequency dependence [31] (similar to that shown in Figure 1), and these can be explained in terms of an equivalent circuit [32] which is consistent with the observed microstructure.

Various studies of individual grain boundaries in a number of ZnO varistor materials have demonstrated that the breakdown voltage is ~3.5 V/boundary, and this is largely independent of fabrication conditons [33-35]. In principle, therefore, varistors can be prepared with required breakdown voltages by fabricating devices with the appropriate number of grains between the electrodes [29]. In contrast the leakage current (i.e. current flow through the varistor at voltages below breakdown) are strongly influenced by composition and preparation, and may increase during service, i.e. degradation [33].

The action of the varistor can be understood in terms of the schematic structure (Figure 7). Individual grains of conducting ZnO, typically 10 μm in size, are surrounded by a very thin (few nm) layer [36] which is enriched in additives which have segregated to the grain boundary (e.g. Bi and Pr). An electrically insulating boundary layer (~100 nm thick) is also formed at each ZnO boundary, and this depletion region is wholly within the ZnO grain [29].

A number of models have been proposed to explain grain-grain conduction and other phenomena based upon a double depletion layer concept [30,34,37-39] for closest grain-grain contact. In general the thin intergranular layer (within the ZnO) is not specifically described physically or chemically.

Martzloff and Levinson [29] note that most models are phenomological in their construction, and as such, the physical origin of the surface states and bulk donors that describe the varistor barriers are not considered formally. An alternative approach, based on hole creation [35] as the origin of the highly non-linear conduction in ZnO varistors, has gained some support with the observation of bandgap electroluminescence in some varistor compositions [40].

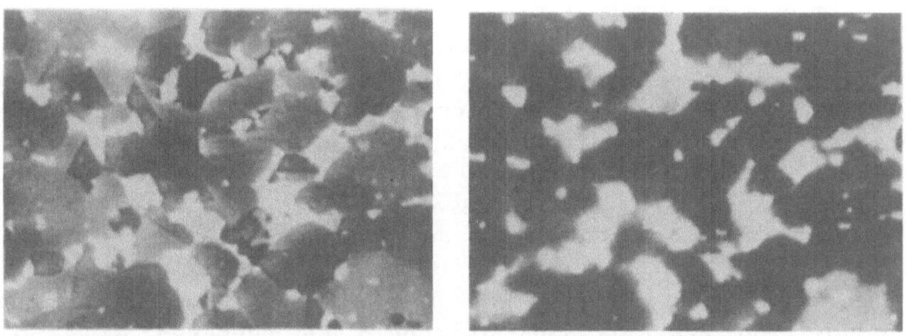

Figure 9. SEM micrographs of ZnO varistor. (a) White regions are Bi_2O_3, light grey is spinel ($Zn_7Sb_2O_{12}$), matrix is ZnO. (b) Backscattered electron image. (Courtesy of M. Gee).

In recent years there has been growing interest in the changes taking place in ZnO varistors as a result of thermal and electrical ageing [33,41]. Figure 9 shows the SEM microstructure of a fresh varistor. After heating for 5 hours at 1200°C most of the main features are still visible, but the intergranular layer is beginning to degrade (Figure 10). After 192 hours at 1200°C the changes become more severe and complete loss of Bi_2O_3 occurs from selected grain boundaries [42]. Binesti et al [41] suggest that ageing at 700°C in air involves the transfer of oxygen between the solid phases, and between zinc oxide and the atmosphere. Some form of diffusion barrier would be required to retard degradation.

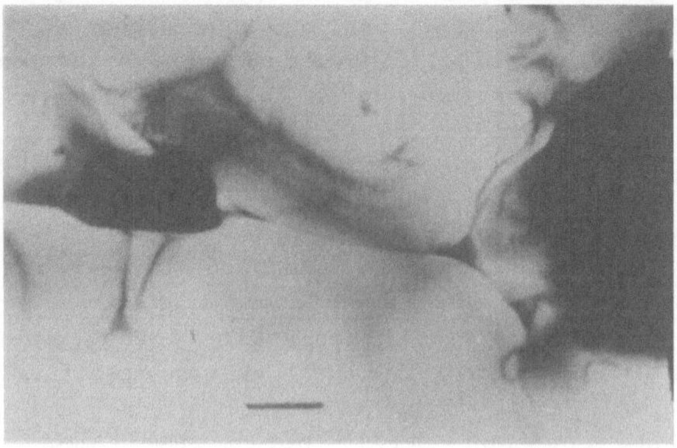

Figure 10. Grain boundary region of ZnO varistor heat treated at 1200°C for 5 hours. Triangular areas running into grain boundary are Bi_2O_3, large dark precipitate is spinel. Scale bar is 1 μm. (Courtesy of M. Gee).

5. ZTS - A LOW LOSS DIELECTRIC CERAMIC

Zirconium titanate-based ceramics have long been used as dielectrics for capacitors. Both zirconium titanate and solid solutions in the system ZrO_2-TiO_2-SnO_2 (ZTS) are temperature stable, and exhibit only a small coefficient of capacitance [43,44]. During the last decade $Zr_xTi_ySn_zO_4$ ceramics have attracted much interest as high Q dielectric resonators for use in filters [45] and frequency-stable oscillators [46] in the GHz range. Single phase ceramics only exist over a limited range of compositions [47], but when y = 1 and z = 0.2 (i.e. $Zr_{0.8}Ti_{1.0}Sn_{0.2}O_4$) the material has a high dielectric constant (>35), a temperature coefficient close to zero, and a high Q value at GHz frequencies [47,48].

Azough and Freer [49] demonstrated that high density ZTS of composition $Zr_{0.8}Ti_{1.0}Sn_{0.2}O_4$ could be prepared by a standard mixed oxide route if suitable additives were employed. However, the sintering aids have a controlling effect on microstructures and electrical properties. For example additions of ZnO yield large rectangular grains, typically 20 μm in size (Figure 11a), whilst Nd_2O_3 gives rise to a more compact microstructure, with small irregular grains <10 μm in size (Figure 11b). TEM studies confirmed that the grain boundaries were generally sharp and clean with no evidence of precipitates (Figure 12a), although a second phase, resulting from liquid phase sintering, did form at triple points and related junctions (Figure 12b). As a result of slow cooling the second phase was generally crystalline, and often exhibited a domain-like structure. The presence of additives could not be detected within individual ZTS grains.

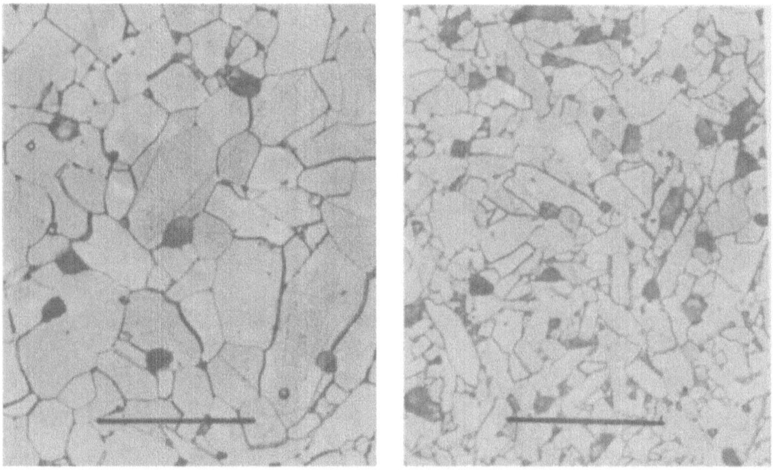

Figure 11. Optical micrographs of ZTS prepared with (a) 2wt% ZnO additions, (b) 2wt% Nd_2O_3 additions. Scale bars are 50μm.

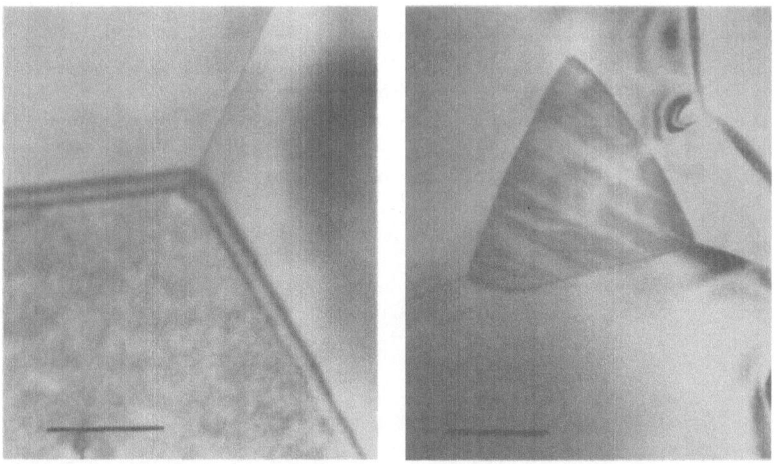

Figure 12. TEM micrographs of ZTS, (a) clean grain boundaries, (b) crystalline region at triple point showing domain-like structure. Scale bars are 1 μm.

The dielectric Q values of three sets of ZTS specimens at 1.5 kHz are summarised in Figure 13. In spite of major differences in microstructure, the Q values of ZTS prepared with ZnO and Nb_2O_3 are very similar for additive levels up to 3 wt%. At frequencies of 1 MHz, Q values for all three specimens are in excess of 10,000, but at frequencies ~5 GHz, the dielectric Q for ZTS prepared with Nb_2O_3 additions is inferior to that for specimens prepared with ZnO additions.

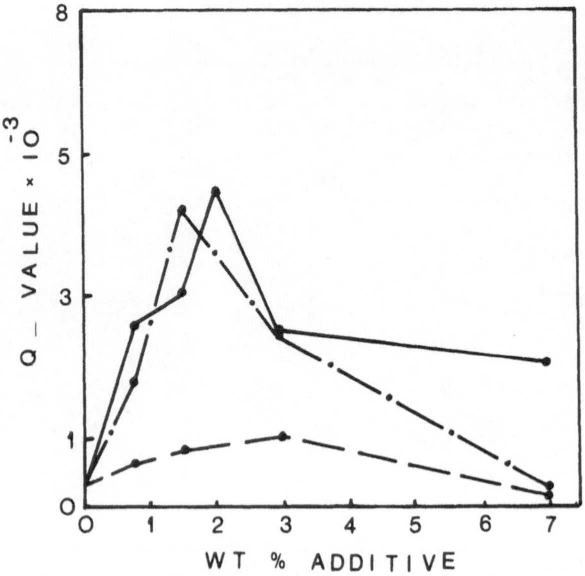

Figure 13. Dielectric Q value at 1.5 kHz for ZTS as a function of additives used for their preparation (solid line Nb_2O_3; even broken line La_2O_3; uneven broken line ZnO).

As Cr_2O_3 additions are known to increase the resistivity of certain titanates, the effect of chromia doping on ZTS was investigated. To the basic mixture $Zr_{0.8}Ti_{1.0}Sn_{0.2}O_4$ plus 2wt% Nd_2O_3 (maximum Q in Figure 13), small amounts ($\leqslant$0.5wt%) of Cr_2O_3 were added. Chromia additions

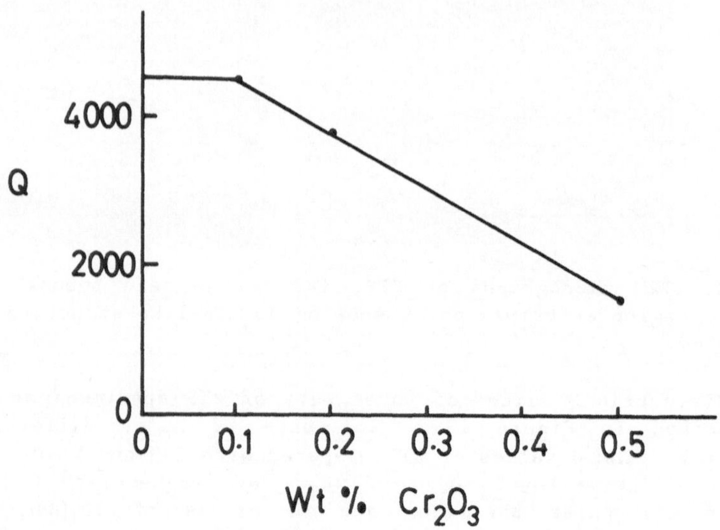

Figure 14. The effect of Cr_2O_3 additions on the dielectric Q value (at 1.5 kHz) of ZTS prepared with 2wt% Nd_2O_3.

improved the density of the product, but did not modify the microstructure. However, the presence of Cr had a deleterious effect on the Q value (Figure 14). Whilst the grain boundaries are again clean, TEM studies confirmed that Cr had diffused into the ZTS grains. The chromium ions therefore appear to be responsible for the deterioration in dielectric loss properties.

6. Mn-Zn FERRITES

Soft ferrites, including Mn-Zn compositions, are used extensively in telecommunications and entertainment electronics [50], and are suitable for wide ranges of frequency. With appropriate formulations and processing routes, ferrites having high permeabilities and low electrical and magnetic losses can be prepared.

Although ferrites are composition-dependent and microstructure-dependent, the magnetic properties depend fundamentally on domain structure [51], and the domain behaviour is strongly influenced by grain boundaries [52]. In particular, boundaries affect equilibrium domain widths, and contribute to pinning of domain walls [52,53].

Three types of boundaries can be identified in ferrites and indeed many other ceramics [54]: grain-grain; grain-pore; grain-aggregate. Livingston [52] suggests that the effects of grain boundaries on properties can be classified into 3 categories: (i) misorientation, due to change in lattice orientation across the boundary; (ii) structural changes due to atomic disorder and strain; (iii) impurity effects, resulting from segregation of impurities to the boundary. Further discussion of grain boundary chemistry and structure are given by Kingery [1], Ghate [54] and Sundahl et al [55].

Chang and Kingery [56] noted that many properties of Mn-Zn ferrites depend upon grain boundary compositions, and rapid grain boundary diffusion of oxygen and subsequent changes in near-boundary stoichiometry have been used to interpret a number of experimental results. For example, the dispersion in the dielectric constant of Mn-Zn ferrites with frequency (similar to Figure 1) was attributed, by Koops [57], to a more oxidized, insulating grain boundary region separating highly conductive grains. Several workers, e.g. [56,58,59], have observed segregation of Ca or Ca and Si in Mn-Zn ferrites, and correlated changes in electrical resistivity with heat treatment atmosphere. Furthermore, initial magnetic permeability [55,60], mechanical strength and fracture properties [61-63] have been correlated with changes in grain boundary and overall stoichiometry.

In this section we briefly consider the role of porosity in highly pure, precipitate-free Mn-Zn ferrite prepared by a gel technique, and then more generally the effect of sintering atmosphere on ferrite properties.

6.1. Mn-Zn Ferrite Prepared by the Citrate Gel and Mixed Oxide Techniques

Traditionally, the industrial preparation of ferrite involves a standard mixed oxide route. Latterly there has been much interest in the use of alternative chemical preparation techniques for the production of highly pure, fine particulate oxides. These include co-precipitation, freeze-drying and various gel techniques [64]. Of these the citrate gel method is attractive because of its comparitive simplicity and the high yields that have been obtained, e.g. with complex perovskite oxides [65].

The citrate gel route offers a means of achieving chemical homogeneity (without the need for repeated firing of oxide powders) by the use of an amorphous precursor. Full details are given elsewhere [64]. Mn-Zn ferrite powder produced by this route was crystalline in

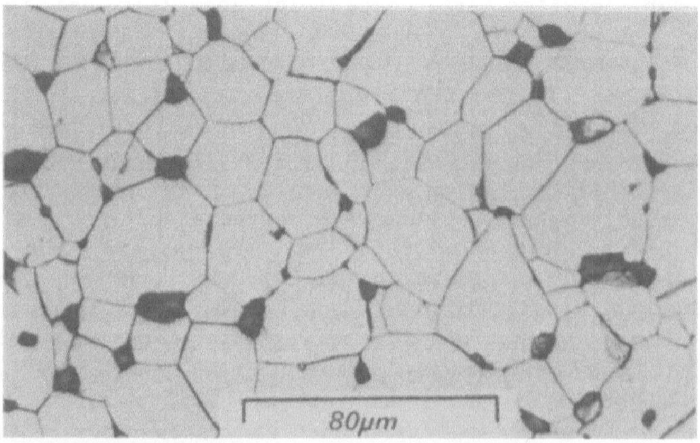

Figure 15. Optical micrograph of high permeability Mn-Zn ferrite prepared by a mixed oxide route.

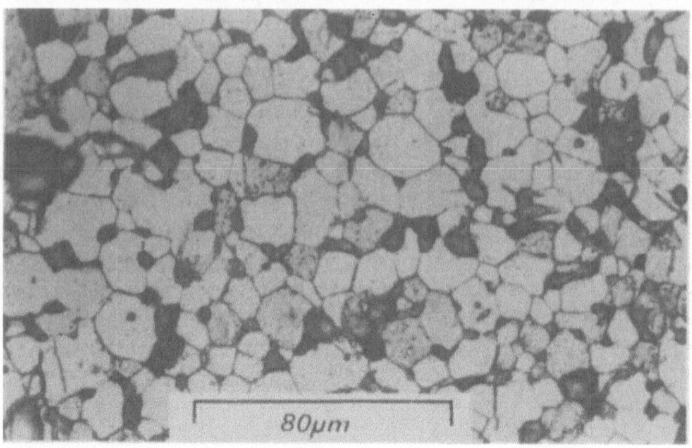

Figure 16. Optical micrograph of Mn-Zn ferrite prepared by the citrate gel route.

nature, but in the form of aggregates, with individual grains 1-2 μm in size.

Microstructures of Mn-Zn ferrites produced by the standard oxide route and the gel route are shown in Figures 15 and 16. It is clear that the gel derived material contains a significant amount of porosity (~75% theoretical density) in comparison with material from oxide processing (85% theoretical density). The differences are reflected in the initial permeabilities (at 100 kHz) of 2400 and 4600 for gel-derived and oxide-derived materials sintered under the same conditions. Improvements in sintering the pure gel-derived powders should lead to major improvements in density, and possibly magnetic properties.

6.2. Effect of Sintering Atmosphere on Mn-Zn Ferrites

It is well known that the sintering atmosphere has a major influence on the properties and microstructures of Mn-Zn ferrites. Blank [66] and Morineau and Paulus [67] discussed the effect of sintering atmosphere on Fe^{2+} content, and Yan [68] examined its effect on microstructure.

The oxygen partial pressure during sintering can be described [69] by

$$\log P_{O_2} = a - b/T$$

where b is a constant, and a is a parameter expressing atmosphere from firing to cooling (so called atmospheric parameter, A.P.). Since the volume diffusion of oxygen is believed to be the rate-determining step in the densification of ferrite, increased densification should be achieved with increased anion-vacancy concentrations. Rikukawa and Sasaki [69] showed that grain growth accelerated as the atmosphere deviates from stoichiometry, and the densification progresses with grain growth. This behaviour had also been observed by Yan [68].

Figure 17 shows the dependence of loss factor, resistivity and Fe^{2+} content on atmospheric parameter. At 100 kHz there is a significant reduction in loss factor, and increase in resistivity as the Fe^{2+} content falls. Since the main conduction mechanism in Mn-Zn ferrite is the transfer of electrons between Fe^{2+} and Fe^{3+} located on B sites, the observed change in resistivity can, in part, be explained by the reduction in available Fe^{2+} in the grain boundary. However, Rikukawa and Sasaki [69] suggest that the resistivity is directly related to the deviation from stoichiometry, and that cation vacancies may cause scattering of conduction electrons or trapping of holes.

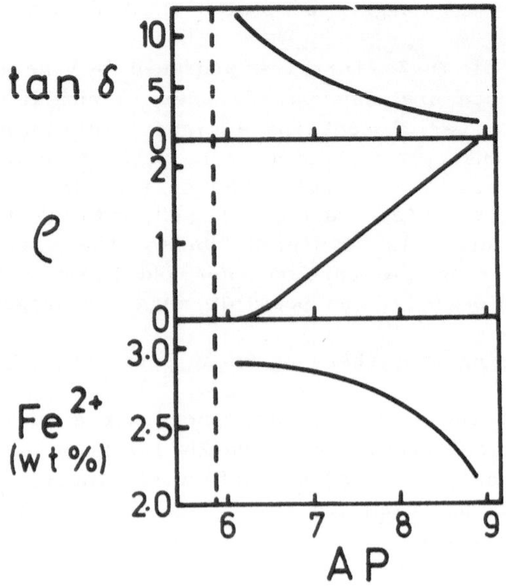

Figure 17. Dependence of tan δ, resistivity and Fe^{2+} content on atmospheric parameter (A.P.) for a Mn-Zn ferrite. Scale of tan δ is $x10^{-3}$, and resistivity $x10^{-3}$ Ωcm. (after reference [69]).

Whilst the control of atmosphere during sintering at elevated temperature is crucial, it is necessary for the oxygen partial pressure to be appropriate for the spinel phase field, e.g. [66], at all stages during processing. Figure 18 shows the microstructure of a Ti-doped, Mn-Zn ferrite which was exposed to an oxygen rich atmosphere during cooling. The outer rim of the ferrite has developed a duplex structure, extending some 400 μm into the body of the specimen. The white 'laths'

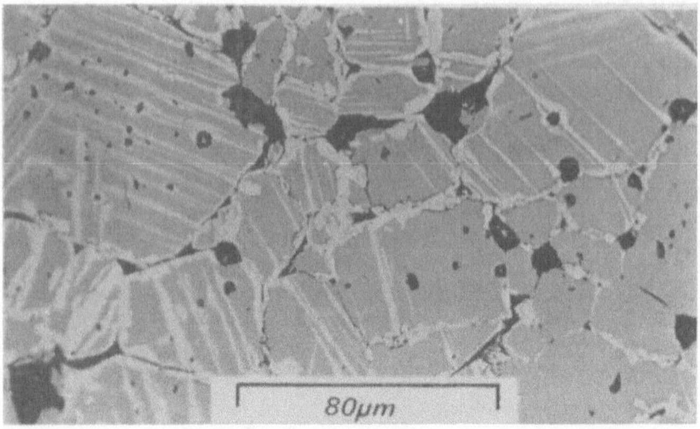

Figure 18. Optical micrograph of a Ti-doped, Mn-Zn ferrite showing a duplex structure which developed near the outer edge of the specimen.

are very rich in Fe oxide and depleted in Mn and Zn oxides with respect to the core of the specimen. Regions between the 'laths' are depleted in iron oxide, but enriched in Mn and Zn oxides. This exsolution-type process is controlled by the diffusion of oxygen along grain boundaries and permeation via pores.

7. CONCLUSIONS

The examples cited here have shown that the properties of many electronic ceramics can be understood in terms of grain boundary processes. It is clear that subtle changes in grain boundary chemistry and structure can give rise to major changes in properties.

Acknowledgements

We gratefully thank Dr. J.H. McCartney and Dr. M. Gee for allowing us to use a number of unpublished figures.

REFERENCES

1. Kingery, W.D., *Advances in Ceramics*, **1**, 1-22, (1981).
2. Kingery, W.D., *J. Amer. Ceram. Soc.*, 57, 1-8, (1974).
3. Henriksen, A.F. and Kingery, W.D., *Ceram. Int.*, **5**, 56-60, (1979).
4. Kingery, W.D., Vander-Sande, J.B. and Mitamura, T., *J. Amer. Ceram. Soc.*, **62**, 221-222, (1979).
5. Santhanam, A.T., Gupta, T.K. and Carlson, W.G., *J. Appl. Phys.*, **50**, 852-859, (1979).
6. Goodman, G., *Advances in Ceramics*, **1**, 215-231, (1981).
7. Schmelz, H. and Schwaen, W., U.S. Patent 4,131,903 (1978).
8. Waku, S., *Electr. Commun. Lab. Tech. J.*, NT & T Corp., **16**, 975-1001 (1967) - cited in *Rev. Electr. Commun. Lab.*, 15, 689-715, (1967).
9. Volger, J., *Progress in Semiconductors*, 4, 207-234, ed. A.F. Gibson, (John Wiley, N.Y., 1960).
10. Waku, S., Ann. Rept. of Study Group on Appl. Ferroelectrics Japan 27-154-984 (1979-4), in Japanese (known through *Advances in Ceramics*, **1**, 23-37, 1981, ref. 28).
11. *Philips Res. Rep.*, **31**, 489-504; 505-515; 516-525; 526-543; 544-559, (1976).
12. Verwey, E.J.W., Haaijman, P.W., Romeijn, F.C. and Van Oosterhout, G.W., *Philips Res. Rept.*, 5, 173-187, (1950).
13. Haayman, P.W., Dam, R.W. and Klasens, H.A., German Patent 929,350, June 23, 1955.
14. Heywang, W., *Solid State Electr.*, **3**, 51-58, (1961).
15. Tennery, V.J. and Cook, R.L., *J. Amer. Ceram. Soc.*, **44**, 187-193, (1961).
16. Jonker, G.H., *Solid State Electr.*, **7**, 895-903 (1964).
17. Heywang, W., *J. Amer. Ceram. Soc.*, 47, 484-490 (1964).
18. Daniels, J., Hartl, K.H. and Wernicke, R., *Philips Tech. Rev.*, **38**, 73-82 (1978/79).
19. Jonker, G.H., *Advances in Ceramics*, **1**, 155-166, (1981).

19. Jonker, G.H., *Advances in Ceramics*, **1**, 155-166, (1981).
20. Hill, D.C. and Tuller, H.L., in *Ceramic Materials for Electronics*, ed. R.C. Buchanan (Marcel Dekker, N.Y., 1986) pp. 326-354.
21. Kulwicki, B.M., *Advances in Ceramics*, **1**, 138-154, (1981).
22. Jonker, G.H., *Mater. Res. Bull.*, **2**, 401-407, (1967).
23. Lewis, B.M., Catlow, C.R.A. and Casselton, R.E.W., *J. Amer. Chem. Soc.*, **68**, 555-558, (1985).
24. Kulwick, B.M. and Purdes, A.J., *Ferroelectrics*, **1**, 253-263, (1970); **2**, 176 (1971).
25. Haanstra, H.B. and Ihrig, H., *J. Amer. Ceram. Soc.*, **63**, 288-291, (1980).
26. Hoffmann, B., *Solid State Electr.*, **16**, 623-628, (1973).
27. Matsuoka, M., *Japan J. Appl. Phys.*, 10, 736-746, (1971).
28. Sakshaug, E.C., Kresge, J.S. and Miske, S.A., *IEEE Trans. Power Appar. Syst.*, **PAS-96**, 647-656, (1977).
29. Martzloff, F.D. and Levinson, L.M., in *Electronic Ceramics*, ed. L.M. Levinson, (Marcel Dekker, N.Y., 1987) pp. 275-305.
30. Matsuoka, M., *Advances in Ceramics*, **1**, 290-308, (1981).
31. Levinson, L.M. and Philipp, H.R., *J. App. Phys.*, 47, 1117-1122, (1976).
32. Matsuura, M. and Yamaoki, H., *Japan J. App. Phys.*, **16**, 1261-1262, (1977).
33. Olsson, E. and Dunlop, G.L., in *High Tech. Ceramics*, ed. P. Vincenzini, (Elsevier, Amsterdam, 1987), pp. 1765-1774.
34. van Kemenade, J.T.C. and Eijnthoven, R.K., *J. App. Phys.*, **50**, 938-941, (1979).
35. Mahan, G.D., Levinson, L.M. and Philipp, H.T., *J. App. Phys.*, 50, 2799-2812, (1979).
36. Clark, D.R., *J. App. Phys.*, **50**, 6829, (1979).
37. Levine, J.D., *Crit. Rev. Sol. State Sci.*, **5**, 597, (1975).
38. Morris, W.G., *J. Vac. Sci. Technol.*, 13, 926-931, (1976).
39. Hower, P.L. and Gupta, T.K., *J. App. Phys.*, **50**, 4847, (1979).
40. Pike, G.E., Kurtz, S.R., Gourley, P.L., Philipp, H.R. and Levinson, L.M., *J. App. Phys.*, **57**, 5512, (1985).
41. Binesti, D., Bonnet, J.P., Onillon, M., Salmon, R. and Tanouti, B., in *High Tech. Ceramics*, ed. P. Vincenzini, (Elsevier, Amsterdam, 1987), pp. 1801-1807.
42. Gee, M., *Brit. Ceram. Proc.*, **42**, (in press).
43. Rath, W., *Keram Radsch*, **49**, 137, (1941).
44. Brit. Pat. Spec., No. 692 468, (1952).
45. Wakino, K., Minai, K. and Tamura, H., *J. Amer. Ceram. Soc.*, **67**, 278-281 (1984).
46. Ishihara, O., Mori, T. Sawano, H. and Nakatani, M., *IEEE Trans. MTT*, **28**, 817, (1988).
47. Wolfram, G. and Gobel, H.E., *Mater. Res. Bull.*, **16**, 1455-1463, (1981).
48. Osbond, P.C., Whatmore, R.W. and Ainger, F.W., *Brit. Ceram. Proc.*, **36**, 167-178, (1985).
49. Azough, F. and Freer, R., *Brit. Ceram. Proc.*, **42**, (in press).
50. Slick, P.I., in *Ferromagnetic Materials*, vol. 2, ed. E.P. Wohlfarth, (North Holland, N.Y., 1980).

51. Hershberg, P.I., *Ferromagnetic Domains*, Electrotech. Sci. and Eng. Ser. No. 37, (C-M Tech. Publications Corp., Jan. 1962).
52. Livingston, J.D., in *Grain Boundaries in Engineering Materials*, ed. J.L. Walter, J.H. Westbrook and D.A. Woodford (Claitors Pub. Division, Baton Roughe, La., 1975).
53. Morrish, A.H., *The Physical Principles of Magnetism*, (John Wiley N.Y. 1965) pp. 340, 407, 529.
54. Ghate, B.B., *Advances in Ceramics*, 1, 477-493, (1981).
55. Sundahahl, R.C., Ghate, B.B., Holmes, R.J., Pass, C.E. and Johnson, W.C., *Advances in Ceramics*, 1, 502-511, (1981).
56. Chang, Y. and Kingery, W.D., *Advances in Ceramics*, **6**, 300-311, (1983).
57. Koops, C.G., *Phys. Rev.*, **83**, 121-124, (1951).
58. Paulus, M., *Mater. Sci. Res.*, **3**, 31-47, (1966).
59. Akashi, T., NEC Res. Div., **19**, 66-82, (1970).
60. Gallagher, P.K., Gyorgy, E.M. and Johnson, D.W., *Amer. Ceram. Soc. Bull.*, **57**, 812-819, (1978).
61. Ishino, K., Makino, H. and Ikehata, E., *Proc. Int. Conf. on Ferrites*, Kyoto, Japan (1970), p. 252-254.
62. Johnson, D.W., in *Processing of Crystalline Ceramics*, ed. H. Palmour, R.F. Davis, and T.M. Hare, (Plenum, N.Y., 1978), p. 381-391.
63. Kosinski, S.G., Vaidya, S., Johnson, D.W. and Tressler, R.C., *Amer. Ceram. Soc. Bull*, 58, 616, (1979).
64. Roberts, V.A., Freer, R. and Sale, F.R., *Brit. Ceram. Proc.*, **41**, (in press).
65. Baythoun, M.S.G. and Sale, F.R., *J. Mater. Sci.*, **17**, 2759, (1982).
66. Blank, J.M., *J. App. Phys.*, 32, 378-379, (1961).
67. Morineau, M. and Paulus, M., *IEEE Trans.*, **Mag. 11**, 1312-1314, (1975).
68. Yan, M.F., *Mater. Sci. Engr.*, **48**, 53-72, (1981).
69. Rikukawa, H. and Sasaki, I., *Advances in Ceramics*, **14**, 215-219, (1985).

SURFACE OXYGEN EXCHANGE IN BISMUTH OXIDE BASED MATERIALS.

B.A. Boukamp, K.J. de Vries and A.J. Burggraaf,
Laboratory of Inorganic Chemistry,
Materials Science and Catalysis,
Faculty of Chemical Technology
University of Twente, P.O. Box 217,
7500 AE Enschede, the Netherlands.

ABSTRACT. The surface oxygen exchange rate has been measured on thin samples of the solid solutions of bismuth oxide with erbium oxide and with terbium oxide. These materials show high surface oxygen exchange values. For the s.s. of Bi_2O_3 with 40 mol% Tb_2O_3 (BT40) a value of 2.5×10^{-8} $mol.cm^{-2}.sec^{-1}$ is found at 973 K. The exchange is thermally activated with E = 110 kJ/mol for BT40, 130 kJ/mol for BE25 and 155 kJ/mol for BE30 (Bi_2O_3 doped with resp. 25 and 30 mol% Er_2O_3). The exchange rate follows a $\sqrt{pO_2}$ dependence. Oxygen exchange measurements on BE25 powder indicates that the dissociative adsorption of oxygen is the rate limiting step in the exchange process.

1. Introduction

The ionic conductivity and the surface oxygen exchange rate are the two most important parameters that govern the process of oxygen transport through solid state oxide devices, such as oxygen pumps, solid oxide fuell cells, mixed conducting semi-permeable oxygen membranes and electrochemical reactors. Of these two parameters the (partial) ionic conductivity or diffusion coefficient has been studied extensively for many solid oxide conductors. The surface oxygen exchange rate has received much less attention as it is less accessible for simple techniques, especially if the exchange rate of small (few cm^2) solid surfaces must be obtained. For purely ionic conductors some information on the exchange rate can be obtained through electrochemical electrode polarisation measurements (I-V measurements). A porous (noble) metal electrode, or a metal point electrode is then used as current collector. Hence the combination electrolyte surface / metal electrode is then probed. The choice of the electrode material however can significantly influence and enhance the exchange current, which is coupled to the surface oxygen exchange rate. This effect has been demonstrated for the zirconia and ceria based electrolyte materials at temperatures below 1070 K [1].

Exchange measurements using the stable ^{18}O isotope have been performed mostly on powdered samples. An extensive overview of the surface oxygen exchange of a number of metal oxides has been given by Boreskov [2] and Winter [3]. Steele, Kilner and co-workers [4-6] were among the first to study the surface oxygen exchange and diffusion kinetics of single and poly-crystalline oxygen ion conducting oxides. With a dynamic Secondary Ion Mass Spectrometer (SIMS) [4], the tracer diffusion profile was measured of solid samples which were exchanged in $^{18}O_2$ enriched oxygen for a fixed period of

J. Nowotny and W. Weppner (eds.),
Non-Stoichiometric Compounds Surfaces, Grain Boundaries and Structural Defects, 299–309.

time. ^{18}O depth profiles could be measured accurately over a depth of more than 10 μm. Assuming a single rate determining step in the surface oxygen exchange and a semi-infinite diffusion condition the exchange rate and the (tracer) diffusion for oxygen could be obtained.

Some remarkable results were obtained from these measurements which have been summarized by Steele, Burggraaf and co-workers [7]. From preliminary bismuth implantation experiments it was found that for scandia stabilised zirconia the exchange rate increased by a factor of 5 after bismuth implantation [6]. The exchange rates for zirconia based materials were found to be a factor 10^3 smaller than deducted from electrochemical measurements using porous platinum electrodes [8], while for bismuth erbium oxides high and identical values were obtained for both methods. This indicates a possible enhancement of the exchange kinetics by the bismuth ion.

The surface oxygen exchange rate is particularly important for thin membranes where the oxygen transport goes from a bulk diffusion rate limiting to a surface exchange rate limiting step when the thickness of the (electrochemical) membrane is decreased. This effect has been demonstrated by Dou et al. [9] for calcia stabilized zirconia. They found a $(pO_2)^{1/4}$ dependence for the oxygen transport through a relatively thick CSZ membrane (bulk controle), which changed to a $(pO_2)^{1/2}$ dependence (surface controle) for thin samples.

In this work ^{18}O exchange rates are reported for the purely ionically conducting solid solutions of Bi_2O_3 with 25 mol% and 30 mol% Er_2O_3 oxide [10], abbreviated BE25 and BE30, and the mixed conducting solid solution $(Bi_2O_3)_{0.6}(Tb_2O_3)_{0.4}$ [11], abbreviated BT40. The isotope exchange is performed on polished disc shaped samples, for BE25 also on powdered material, as function of temperature and partial oxygen pressure. The gas phase ^{18}O ratio is monitored with a quadrupole mass spectrometer. The surface oxygen exchange rate of BE25 is compared with recent electrode polarization (I-V) experiments [12].

2. Instrumental setup and procedure

The exchange measurements were performed on thin disc shaped pellets (0.05 to 0.15 cm thick) which were polished on both sides with a final polish of 0.3 micron Al_2O_3. The relative density of the poly-crystalline samples was better than 98.6% (based on the density obtained from X-ray analysis). The preparation of these dense and machinable bismuth-erbium and bismuth-terbium oxide ceramics has been published elsewhere [13,14]. The total sample surface area was generally between 1 and 2 cm^2, while the oxygen content was about 0.7 to 1.4 mmol per sample.

The ^{18}O exchange is performed in a quartz/stainless steel cell with a fixed volume of 55 cm^3. The exchange cell can be evacuated and flushed with argon. Both natural and ^{18}O enriched oxygen (Amersham International, >97 atom% ^{18}O) can be admitted in the exchange cell to a specified pressure. The gas phase composition is measured continuously with a quadrupole mass spectrometer (Balzers QMS/QMG 112) which is operated under computer controle. The specially designed removable furnace allows rapid heating of the exchange cell (typically to within 10 to 30 degrees from the final temperature within 1 minute. A detailed description of the exchange system has been given in [15].

In order to clean the samples and to ensure identically conditioned surfaces all samples were pre-annealed at 750°C for 0.5 hours in dry pure oxygen at a pressure of about 0.2 atm. After rapid cooling and flushing with argon all the samples were equilibrated for several hours in natural oxygen at the desired exchange temperature and pressure, followed again by rapid cooling and flushing with argon. Next the exchange was performed with the exact same amount of oxygen as used for the previous equilibration step (but now from the $^{18}O_2$ enriched oxygen reservoir) thus ensuring thermodynamical equilibrium after a short initial heating up period. The temperature dependence of the exchange rate was measured with a total oxygen pressure of about 0.2 atm. The pressure dependence was measured between 0.03 to 1.4 atm. Before and during the exchange the mass 32 ($^{16}O_2$), 34 ($^{16}O^{18}O$), 36 ($^{18}O_2$) and 37.5 (background level) quadrupole signals were measured continuously. At the end of the exchange the sample was again rapidly cooled to room temperature.

3. Surface Oxygen Exchange Kinetics

With the previously described experimental precautions the sample may be assumed to be in thermodynamical equilibrium with the gas phase. In first approximation, following Steele et al. [5-7], the isotope exchange is modeled with a single rate determining step, e.g. adsorption and/or dissociation of the oxygen, or charge transfer processes. The overall (exchange) reaction is:

$$O_{2,g} + 2\, V_O^{\cdot\cdot} \rightleftharpoons 2\, O_O^{x} + 4\, h^{\cdot} \tag{1}$$

The exchange rate, i.e. the flux of moles O_2 into and out of the bulk material, k_s, is expressed in mol.cm^{-2}.sec^{-1}. The change of the ^{18}O ratio in the gas phase, R_g^{18}, is then given by:

$$\frac{dR_g^{18}}{dt} = \frac{-S\, k_s}{A} \left(R_g^{18} - R_{bs}^{18}\right) \quad , \tag{2}$$

where S is the surface area of the sample, A is the number of moles of oxygen (^{16}O and ^{18}O) present in the gas phase and R_{bs}^{18} is the ^{18}O ratio at the surface of the bulk material. The ^{18}O ratio in the gas phase is given by:

$$R_g^{18} = \frac{p(^{18}O_2) + 0.5\; p(^{16}O^{18}O)}{p(^{16}O_2) + p(^{16}O^{18}O_2) + p(^{18}O_2)} \quad , \tag{3}$$

where $p(^{m}O_2)$ is the partial pressure of the oxygen species m. The surface ratio R_{bs}^{18} depends on the tracer diffusion coefficient of the bulk and the exchange rate. For short times, that is for $\sqrt{Dt} << d/2$ with d is the sample thickness, the semi-infinite boundary solution [4,5] may be substituted, assuming that the ^{18}O ratio in the gas phase does not change noticably. If the characteristic length $\sqrt{Dt}$ becomes in the order of d the finite length boundary conditions must be used. In general this leads to a very complex and difficult to analyse diffusion equation. When $\sqrt{Dt}$ becomes much larger then d, or if k_s is small compared to D (i.e. exchange controled diffusion) the diffusion profile becomes very flat and R_{bs}^{18} may be replaced by the mean ^{18}O bulk ratio R_b^{18}. The ^{18}O ratios in the gas phase and the bulk are then related to the total amount of ^{18}O present in the closed measuring system:

$$A\, R_g^{18} + B\, R_b^{18} = A\, R_{g0}^{18} + B\, R_{b0}^{18} \tag{4}$$

where R_{g0}^{18} and R_{b0}^{18} are the ^{18}O ratios in the gas phase and the bulk material before the start of the exchange (at t=0). B is the oxygen content of the bulk material expressed in moles. Using eq. (4) the differential equation (2) can be solved, leading to a well known classical diffusion equation:

$$R_g^{18}(t) = R_{g\infty}^{18} + (R_{go}^{18} - R_{g\infty}^{18}) \exp -S\, k_s\, (A^{-1} + B^{-1})\, t \quad (5)$$

where $R_{g\infty}^{18}$ is the equilibrium ^{18}O ratio at infinite time which is the same for both the gas phase and the bulk:

$$R_{g\infty}^{18} = \frac{A\, R_{g0}^{18} + B\, R_{go}^{18}}{A + B} \quad (6)$$

Hence a plot of $\ln[(R_g^{18}(t) - R_{g\infty}^{18})/(R_{g0}^{18} - R_{g\infty}^{18})]$ versus time will yield a straight line with slope $-S\, k_s\, (A^{-1} + B^{-1})$.

Besides the $^{18}O_2$ and $^{16}O_2$ molecules also $^{16}O^{18}O$ molecules will be formed in the gas phase. If dissociative adsorption is predominant the respective equilibrium ratios will be given by:

$$\frac{(R_g^{34})^2}{R_g^{32} \cdot R_g^{36}} = 4 \quad (7)$$

where 32, 34 and 36 stand for the respective masses of the three types of oxygen molecules. If most of the dissociative adsorption of oxygen takes place at the sample surface (as compared to the hot surface of the exchange cell) the changes in these ratios during exchange can give additional information on the exchange mechanism. In the single rate determining step (rds) model a distinction can then be made between (dissociative) adsorption and the charge transfer process as rds. If the adsorption is rate determining, while the charge transfer is fast, the change in the ratios R_g^{32}, R_g^{34} and R_g^{36} can be calculated directly from the experimentally obtained value for k_s. This results then in the following differential equations:

$$\frac{dR_g^{36}}{dt} = \frac{S\, k_s}{A} [(R_b^{18})^2 - R_g^{36}] \quad (8)$$

and:

$$\frac{dR_g^{34}}{dt} = \frac{S\, k_s}{A} [2\, R_b^{18}\, (1 - R_b^{18}) - R_g^{34}] \quad (9)$$

The time dependence of the $^{18}O_2$ ratio is then:

$$R_g^{36}(t) = R_{g0}^{36}\, e^{-p} + (R_{g\infty}^{18})^2\, [1 - e^{-p}] + R_{g\infty}^{18}\, (R_{g\infty}^{18} - R_{g0}^{18})\, [e^{-p} - e^{-q}] + \\ + \frac{B}{B + 2A} (R_{b0}^{18} - R_{g\infty}^{18})^2 [e^{-2p} - e^{-2q}] \quad (10)$$

with $p = S\, k_s/A$, and $q = S\, k_s\, (A^{-1} + B^{-1})$, while the time dependence of the other ratios can be found from the simple relations:

$$R_g^{34}(t) = 2\, [R_g^{18}(t) - R_g^{36}(t)] \quad \text{and} \quad R_g^{32}(t) = 1 - R_g^{34}(t) - R_g^{36}(t) \quad (11)$$

If the charge transfer is rate limiting the mixing of $^{18}O_2$ and $^{16}O_2$ to $^{16}O^{18}O$ will proceed faster then predicted by eqs (10) and (11). The rate of mixing will now be controled by the adsorption rate constant k_a (with the condition that $k_a >> k_s$). For dissociative adsorption the following differential equations can then be derived:

$$\frac{dR_g^{36}}{dt} = \frac{S\, k_a}{A} [(R_g^{18})^2 - R_g^{36}] \tag{12}$$

and:

$$\frac{dR_g^{34}}{dt} = \frac{S\, k_a}{A} [2\, R_g^{18}\, (1 - R_g^{18}) - R_g^{34}] \tag{13}$$

assuming that the ^{18}O ratio of the adsorped oxygen is equal to that of the gas phase because of the rapid adsorption-desorption process. Taking R_g^{18} as almost constant the time dependence of R_g^{36} is given by:

$$R_g^{36}(t) = \{R_{g0}^{36} - [R_g^{18}(t)]^2\} \exp(-S\, k_a\, A^{-1}\, t) + [R_g^{18}(t)]^2 \tag{14}$$

In this case a plot of $\ln[(R_g^{36}(t) - [R_g^{18}(t)]^2)/(R_{g0}^{36} - [R_g^{18}(t)]^2)]$ versus time should yield a straight line with slope $-S\, k_a\, A^{-1}$. If k_a is not much larger than k_s the analysis becomes more complex. A qualitative comparison can still be made, however, by comparing the simulated $R_g^{36}(t)$ using the adsorption rds model (eq.11) and the actually measured value. If the measured ratio changes faster with time than the simulated ratio one may conclude the charge transfer to play a major role in the rds.

One should notice however that these models may only be applied if no parallel (adsorption) processes take place. For the measurement of k_s using eq. (5) this condition is not important as an 'overall' exchange rate is then measured.

4. Experimental Results

4.1. EXCHANGE MEASUREMENTS

Preliminary ^{18}O diffusion profile and surface concentration measurements (NRA and SIMS) indicated a flat profile [15]. Also the analysis of the gas phase exchange measurements, using eq. 5, is sensitive to the validity of the 'flat profile' assumption as has been argumented in [15]. The data analysis, of which a typical example is given in the semi-logarithmic plot of fig. 1, consistently showed this assumption to be fulfilled.

The value of the surface exchange coefficient can be obtained with high precision. The macroscopic surface area is estimated geometrically with an error of 2-5%. Its relation to the true (microscopic) surface area is unknown but it serves as a relative quantity for comparison to identically processed samples. The oxygen content of the exchange cell, A, is calculated from the cell volume, oxygen pressure and temperature at the time of introduction in the cell. The exchange coefficients, k_s, are thermally activated as can be seen from the Arrhenius plot of fig. 2 for k_s values measured at 0.2 atm. The obtained activation enthalpies are 110 ± 10 kJ/mol for BT40, 130 ± 5 kJ/mol for BE25 and 155 ± 10 kJ/mol for BE30. The exchange coefficient of BE25 at 737 K, obtained from SIMS depth profiling by Steele, Burggraaf and co-workers [7] has been included for comparison.

For the pure ionically conducting bismuth erbium oxides the surface exchange coefficient can also be measured electrochemically in a three electrode cell [12]. From the analysis of electrode polarisation measurements, using a Butler Vollmer model [16], an apparent exchange current density, I_0, is obtained. Under the assumption that the

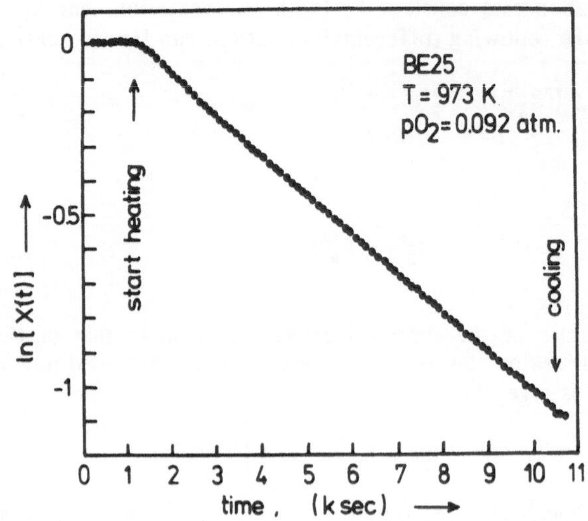

Figure 1. Typical example of ^{18}O gas phase exchange curve plotted as ln[X(t)] versus time, with $X(t) = (R_g^{18}(t) - R_{g\infty}^{18})/(R_{g0}^{18} - R_{g\infty}^{18})$. The slope of the straight section is equal to $-S\ k_s\ (A^{-1} + B^{-1})$.

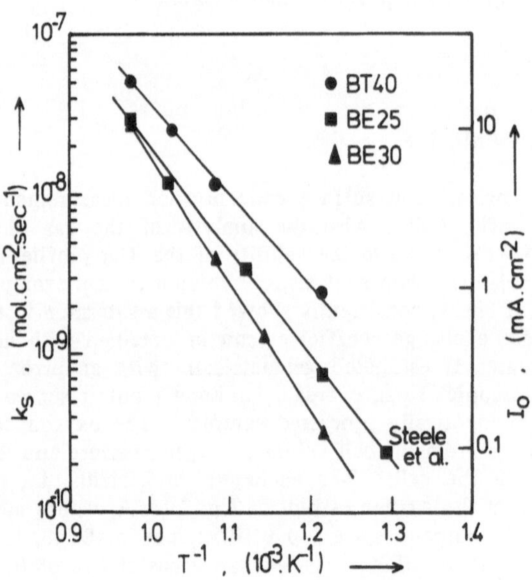

Figure 2. Arrhenius plot of the temperature dependence of k_s for BT40 (●), BE25 (■) and BE30 (▲). The point for BE25 marked 'Steele et al.' has been taken from [7].

electrode material (porous gold) does not activate the exchange process the following relation between the electrochemically measured I_0 and the surface exchange coefficient k_s can be derived:

$$k_s = I_0 / 4F \qquad (15)$$

A comparison of the k_s values, measured electrochemically and from exchange experiments, is given in fig. 3. Taking into account the partial coverage of the surface by the gold electrode (porosity of approx. 50%) the correspondence between the magnitudes is quite remarkable. The electrochemically measured activation enthalpy, however, is somewhat higher (140 ± 5 kJ/mol versus 130 ± 5 kJ/mol).

The pressure dependence of k_s was measured over a range of 0.03 to 1.6 atm., for BE25 at 823 and 973 K and for BT40 at 973 K only. The log-log plot of fig. 4 clearly shows a pO_2^n dependence with n close to 0.5. The same (inverse) pressure dependence was observed by Verkerk et al. [8] for the electrode resistance of BE20 and BE40 (Bi_2O_3 with 20 and 40 mol% Er_2O_3 respectively).

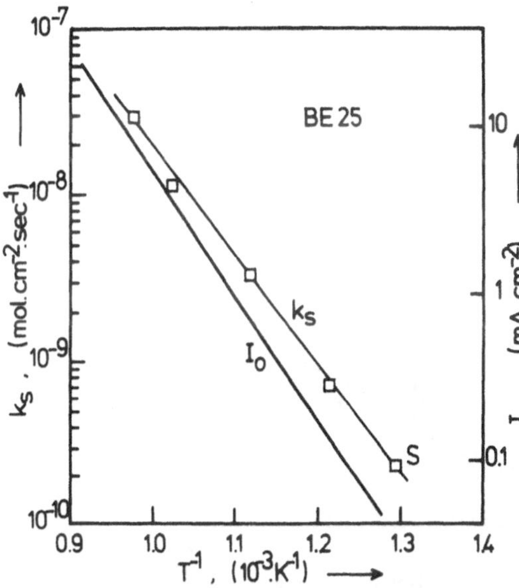

Figure 3. Comparison of the temperature dependence of k_s obtained from the isotope exchange experiments and from the electrochemically measured exchange current density I_0 [12], according to eq. (15).

4.2. EXCHANGE KINETICS ON BE25 POWDER SAMPLES

Sofar only the time dependence of the *total* ^{18}O ratio, $R_g^{18}(t)$ (eq. 3), has been analyzed. The time dependence of the $^{18}O_2$ ratio, $R_g^{36}(t)$, and the related $^{16}O_2$ and $^{16}O^{18}O$ ratios (eq. 11), can give direct information on the rds in the surface oxygen exchange kinetics (eq. 8-14). It is then necessary, however, that the exchange cell itself does not show appreciable dissociative adsorption, as this also leads to the mixing of ^{16}O and ^{18}O:

$$^{16}O_2 + {}^{18}O_2 \rightleftharpoons 2\ {}^{16}O^{18}O \tag{16}$$

At temperatures above about 900 K the quartz cell does have a noticable dissociative adsorption activity towards oxygen (as compared to the sample), *although actual bulk exchange with the quartz wall of the cell does not take place.* This problem can be overcome by measuring the exchange kinetics on a powdered sample at lower temperatures, which has the advantage that the sample surface area is several orders of magnitude larger than for the solid samples used, while at lower temperatures the 'mixing activity' of the quartz cell has decreased considerably. The major assumption one has to make then is that the powder surface may be compared with the solid sample surface. This can in part be justified by giving the powder the same pre-anneal treatment as for the solid samples.

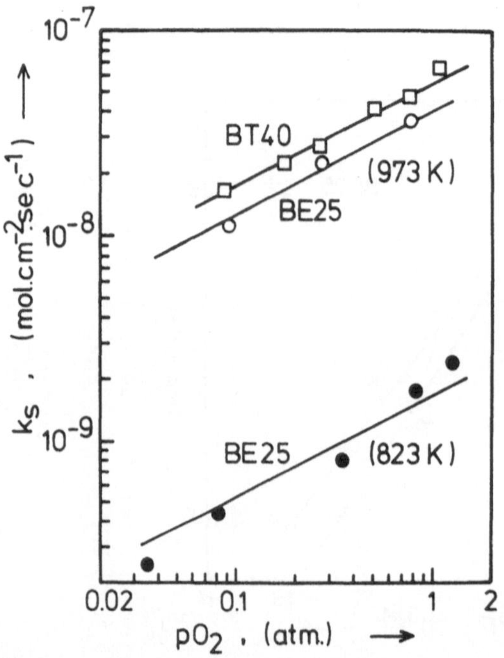

Figure 4. Pressure dependence of k_s for BE25 at 823 (●) and 973 (O) K and for BT40 (□) at 973 K. The straight lines represent the $p(O_2)^{1/2}$ dependence and serve as a guidance to the eye.

An example of an exchange measurement on BE25 powder is given in fig. 5 in which the $R_g^M(t)$ (M = 32, 34 and 36) are shown as function of time. From the analysis of the total ^{18}O exchange the product $S.k_s$ is obtained (eq.5), as well as an extrapolated fictive starting time, t_0^f (which is delayed from the actual starting time, t_0, due to the heating up effects). In case of a dissociative adsorption rds the $R_g^M(t)$ can be simulated by substituting the $S.k_s$ value in eqs (10) and (11) and using t_0^f as the effective starting time. Fig. 5 shows the simulated $R_g^M(t)$ (continuous lines) and the measured values.

Apart from the initial stage the agreement between the measured and simulated values is quite remarkable. An analysis based on the second model with the charge transfer as rds (eq. 14) leads to an inconsistent result.

Comparing the product $S.k_s$ with the value of k_s obtained from solid ceramic BE25 samples yields an estimated accessible surface area of 240 cm^2/gram for the powder material. This relates to a mean particle diameter of about 20 μm (for round particles).

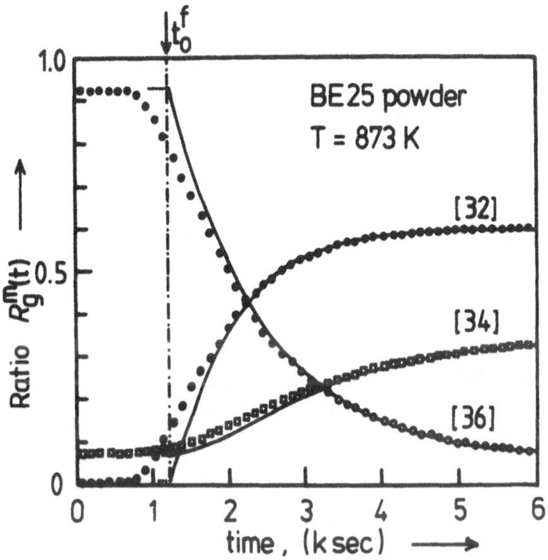

Figure 5. Simulation of the ratios $R_g^M(t)$ as function of time (continuous lines). The actually measured ratios are represented by the symbols. t_0^f is the fictive starting time obtained from extrapolation of eq. (5).

5. Discussion and Conclusion

With an appropriately designed ^{18}O exchange system it is possible to directly measure the surface oxygen exchange coefficient of solid samples of super-ionic oxygen conductors, as has been demonstrated here for the family of bismuth oxide based solid solutions. The important (experimental) condition for these exchange measurements is that the bulk oxygen diffusion must be fast compared to the surface exchange rate and that the sample thickness must be in the order of or smaller than the characteristic diffusion length $\sqrt{Dt}$.

The surface oxygen exchange rate for the bismuth oxide based solid solutions is large (close to a factor 10^3) compared to values for the well known zircoon yttrium oxides [7]. This confirms the notion that bismuth strongly enhances the surface oxygen exchange rate. Further evidence for this notion comes from the lower exchange rate and higher activation energy found for BE30 with respect to BE25 which has a higher bismuth concentration and a higher ionic conductivity than BE30. Apparently the electronic conductivity of BT40 (due to the Tb^{3+}/Tb^{4+} couple) is of importance too, as this compound shows the highest exchange rate with the lowest activation energy of the materials investigated here while it has an even lower bismuth concentration and also a lower ionic conductivity than BE25 and BE30.

While measurements on the solid samples give accurate values for the surface oxygen exchange rates, which may be compared with electrochemically obtained exchange current densities (eq. 15), measurements on powdered material help elucidate the exchange kinetics and can indicate which is the rate determining step. For the bismuth oxide based solid solutions the rds seems to be the dissociative adsorption of oxygen, which fits nicely with the observed $\sqrt{pO_2}$ dependence of the exchange rate.

For the electrochemically measured exchange current density, I_0, of BE25, using porous gold electrodes, also a $(pO_2)^n$ power dependence with $n \approx 0.5$ was observed at 1000 K. With decreasing temperature, however, the value of n decreases to about 0.15 at 650 K, while for k_s the power dependence remains nearly constant. Also the activation energy for the exchange current density is somewhat higher then found from the exchange experiments (140 kJ/mole versus 130 kJ/mole). This can be explained by assuming that the surface diffusion of adsorped oxygen towards the electrode becomes the rds in the electrochemical measurements with decreasing temperature. This is in line with earlier observations by Verkerk et. al [8] for the BE20 and BE40 samples provided with porous platinum electrodes. From the pO_2 dependence of the electrode impedance they concluded the diffusion of adsorbed surface oxygen to the Pt-electrodes to be the rate determining step.

Finally it is interesting to note that, based on an extensive study and analysis of the electrode kinetics, Winnubst and Burggraaf [17] concluded to *associative* adsorption for oxygen on samples of bismuth oxide doped ZY17 (i.e. a solid solution of ZrO_2 with 8.5 mol% Y_2O_3). As in this case the bismuth concentration on the surface is rather low, it is tempting to assume that oxygen preferentially adsorps at a bismuth site because of the Bi lone pair. If the Bi ions are close enough this adsorption will proceed dissociatively, otherwise, at low concentrations, associatively. More research is required however in order to substantiate this model.

Acknowledgement
The author is indebted to Mr. H. Kruidhof and Mr. J. Snoeyenbos who prepared and polished the ceramic samples.

References

[1] D.Y. Wang and A.S. Nowick, *J. Electrochem. Soc.* **126** (1979) 1155, *ibid.* **126** (1979) 1166 and *ibid.* **128** (1981) 55.
[2] G.K. Boreskov, *Discuss. Faraday. Soc.* **41** (1966) 263.
[3] E.R.S. Winter, *J. Chem. Soc. (A)* (1968) 479.
[4] D.S. Tannhauser, J.A. Kilner and B.C.H. Steele, *Nucl. Inst. and Methods in Physical Research* **218** (1983) 504.
[5] J.A. Kilner, B.C.H. Steele and L. Ilkov, *Solid State Ionics* **12** (1984) 12.
[6] B.C.H. Steele, J.A. Kilner, P. Dennis and D. Tannhauser, in *'Ceramic Surfaces and Surface Treatments'*, R. Morrel and M.G. Nicholas Eds, British Ceramic Proc. Vol. **34** (1984) 53.
[7] B.C.H. Steele, J.A. Kilner, P.F. Dennis, A.E. McHale, M. van Hemert and A.J. Burggraaf, *Solid State Ionics* **18&19** (1986) 1038.

[8] M.J. Verkerk, M.W.J. Hammink and A.J. Burggraaf, *J. Electrochem. Soc.* **13** (1983) 70.

[9] S. Dou, C.R. Masson amd P.D. Pacey, *J. Electrochem. Soc.* **132** (1985) 1843.

[10] M.J. Verkerk and A.J. Burggraaf, *J. Electrochem. Soc.* **128** (1981) 75.

[11] B.A. Boukamp, M.P. van Dijk, K.J. de Vries and A.J. Burggraaf, in *'Nonstoichiometric Compounds'* C.R.A Catlow and W. Mackrodt eds, *Advances in Ceramics*, Vol **23** (1987) 447.

[12] I.C. Vinke, J.L. Backiewicz, B.A. Boukamp, K.J. de Vries and A.J. Burggraaf, to be published.

[13] H. Kruidhof, K. Seshan, B.C. Lippens Jr., P.J. Gellings and A.J. Burggraaf, *Mat Res. Bull.* **22** (1987) 1635.

[14] H. Kruidhof, K. Seshan, G.M.H. van de Velde, K.J. de Vries and A.J. Burggraaf, *Mat. Res. Bull.* **23** (1988) 371.

[15] B.A. Boukamp, I.C. Vinke, K.J. de Vries and A.J. Burggraaf, presented at the 11th Int. Symposion 'Reactivity of Solids', Eds S.M. Whittingham, A. Jacobson and S. Bernacek, Proceedings to be published in a special issue of *Solid State Ionics.*

[16] B.A. Boukamp, I.C. Vinke, K. Seshan, K.J. de Vries and A.J. Burggraaf, Proc. 6th Int. Conf. on Solid State Ionics, (Garmisch-Partenkirchen, Fed. Rep. of Germany, Sept. 1987) W. Weppner and H. Schulz Eds., to be published in a special issue of *Solid State Ionics.*

[17] A.J.A. Winnubst, A.H.A Scharenborg and A.J. Burggraaf, *Solid State Ionics* **14** (1984) 319.

DEFECTS AND REACTIVITY AT OXIDE SURFACES : EXPERIMENTAL ASPECTS OF THE INTERACTION OF HYDROGEN, CO AND CO_2 WITH THE NiO{001} SURFACE.

A. BOUDRISS and L.C. DUFOUR *
Réactivité des Solides, UA 23 C.N.R.S.
Université de Bourgogne
B.P. 138
21004 Dijon Cedex
France

ABSTRACT. This paper summarizes some previous or more recent experimental results on chemical interaction, at moderate temperatures, of H, H_2, CO and CO_2 with the surfaces of nickel oxide, mainly the NiO {001} surface. These results are in a general agreement with theoretical predictions. Nearly perfect surfaces of maximal valency oxides cannot react with molecules. Dissociation of molecules is often required for chemisorption. Conversely, chemical reaction is usually observed either when molecules are predissociated or the surface has defects with an energy level high enough to enable both dissociation and chemisorption. On a nearly perfect, "in situ" cleaved, NiO {001} surface, partially atomized hydrogen was found to be easily adsorbed but CO could react only if traces of hydrogen were present. In contrast CO_2 could be adsorbed reversibly.

Morphology, structure and quality of the ultrathin metallic films grown by reduction of the NiO {001} surface were found to be basically dependent on factors involving either the predissociation state of the gas or the presence of defects on the oxide surface. These two factors directly determine the surface coverage of reducing agents and, therefore, the density of nucleation sites.

1. Introduction

According to theoretical predictions nearly perfect surfaces of maximal valency oxides do not react with molecular gases at moderate temperatures. That may be the case when dissociation is required before chemisorption. Nearly perfect surfaces of the rock-salt structured oxides have their crystallographic and electronic structures little different from those of the bulk (ref. 1). Therefore, it may be assumed that concentrations of point defects, such as cationic defects in NiO, are very similar in the surface layers and in the bulk. The maximal density of point defects in these oxides is known to be relatively low. Therefore, the reactivity of the oxide surface requires the presence of other defects than point defects. A surface site having

J. Nowotny and W. Weppner (eds.),
Non-Stoichiometric Compounds Surfaces, Grain Boundaries and Structural Defects, 311–320.

a coordination number lower than in the ideal surface (= 5 for the NiO {001} planes) can be considered as an intrinsic defect. Heterolytic dissociation can also occur on $(O\text{-}Ni)_{LC}$ pair sites corresponding to low coordinated (=LC) surface ions. Also anionic defects can be created by an ion bombardment. Then there is a hierarchic order in these defects according to the energy level of the bond between the defect and species to be adsorbed. If this energy level is greater than that required to dissociate the gas molecules, adsorption is theoretically possible. In other words, atoms behave differently from molecules (ref. 2). Concerning the reactivity between the oxide surfaces and a gas, the following general remarks can be made:

a/ The lattice ionic relaxation around low coordinated sites, such as steps, corners, kinks, ledges, emergences of dislocation..., is theoretically predicted (refs 3). Therefore, the heterogeneous distribution of defect energy becomes strongly modified, leading to a smoother energy spectrum.

b/ The hierarchic order between intrinsic surface defects (excluding impurities or additives) may be attenuated when temperature is increased. Conversely, it is expected that the role of these surface defects will be more important when temperature and chemical potential of gas molecules are lower (ref. 4).

c/ Molecules can be adsorbed without prerequisite dissociation when temperature is low enough. Then the interaction energy and possible charge transfer between the molecules and the surface are mainly depending on both temperature and gas pressure. Both physisorbed and chemisorbed states can occur depending on temperature and the binding energy between solid surfaces and molecules. Experimental methods such as temperature programmed desorption can provide information on the distribution of the interaction energy between molecules and the surface and, therefore, on the different classes of surface sites.

In this paper, we summarize and briefly discuss some experimental results on the interaction between the NiO surface and hydrogen, carbon monoxide and dioxide.

2. Experimental

Most investigations were performed by using a monocrystalline NiO {001} surface in a UHV chamber equipped with Auger electron spectrometry (AES) and reflection high energy electron diffraction (RHEED). Gas pressure and its composition were evaluated by mass spectrometry. The NiO single crystal was grown by the Verneuil technique from high purity (99.995 %) NiO powder. The NiO {001} surface was obtained by "in situ"cleavage in the experimental conditions. The ranges of temperature and gas pressure were :

Temperature : 20 - 300 °C

$P(H_2)$: 10^{-8} Pa ---> 10^{-2} Pa

$P(CO)$: 10^{-9} Pa ---> 10^{-2} Pa

$P(CO_2)$: 0 ---> 2. 10^{-5} Pa

Partial atomization of hydrogen was achieved by using a hot tungsten wire. When hydrogen-carbon monoxide mixtures were used, the $P(H_2)$ / $P(CO)$ ratio ranged from 0.2 to 0.04. When the purpose was to determine the effect of very small amounts of hydrogen (denoted below as traces), the ratio was about 0.001.

Another series of investigations was performed from NiO microcrystals prepared by "in situ" thermal decomposition of $Ni(OH)_2$. The surface area of thus obtained NiO specimens was 172 m^2/g. The mean size of NiO crystallites was 6 nm and they were in the form of very thin platelets oriented perpendicularly to their {111} directions. The amount of adsorbed hydrogen was measured microgravimetrically by using the RG Cahn balance.

3. Interaction of hydrogen atoms with the NiO {001} surface

In partially atomized hydrogen, chemisorption is theoretically possible onto the nearly perfect NiO {001} surface (refs 5). The hydroxylation of any surface site is easy and was observed experimentally (ref. 6). A high density of sites for the nickel nucleation can be calculated by combining the experimental results obtained by RHEED, AES and TEM. We thus obtain 10^{11} - 10^{12} nuclei per cm^2 corresponding to the activation of 10^{-4} - 10^{-3} of surface sites (ref. 2). This favours the growth of ultrathin metallic films. Nickel is found to nucleate in the hexagonal structure below 250 °C and in its usual face centered cubic structure above 250 °C. In all cases very good epitactic relationships can be observed between the metallic film and the oxide matrix. In particular, in the hexagonal structure, very coherent metallic films can be grown by reduction of the clean NiO {100} surface. These films have an excellent chemical stability for a thickness of the order of a few nanometers (refs 7).

4. Interaction of hydrogen molecules with the NiO surfaces

The NiO {001} surface appears to be substantially less reactive when molecular hydrogen is used. This was observed for both single crystals (ref. 3) and annealed microcrystals after a thermal oxygen pretreatment (refs 4, 8). This is in accordance with theoretical predictions (refs 5). As these surfaces have a very limited number of defects able to dissociate hydrogen, its chemisorption leads to a low coverage. Then a limited number of sites is available for nucleation. This number is of the order of 10^6-10^8 nuclei per cm^2 in comparison to the site density of 10^{15} per cm^2. Metallic films grown from these nuclei can form a continuous layer only when their thickness becomes relatively large. This can lead to structural misorientation effects. As the density of nucleation sites is low, their properties can be differentiated, for example according to the magnetic state of NiO (ref. 8).

As it can be expected, the surface reactivity of highly defective NiO microcrystals is totally different. For example, for NiO microcrystals prepared by "in situ" thermal decomposition of $Ni(OH)_2$ at moderate temperature (20-

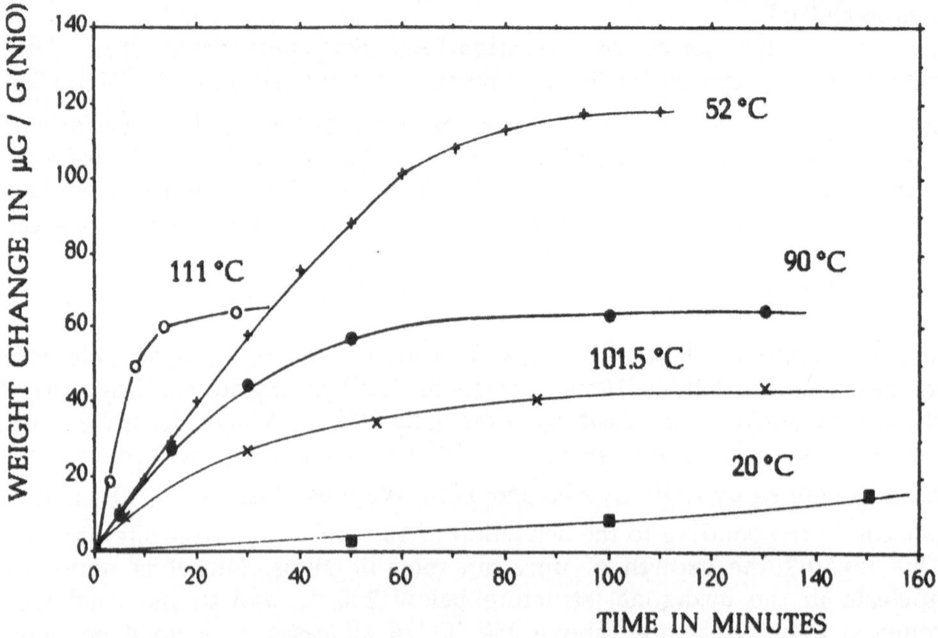

Figure 1. Weight increase during the interaction of purified hydrogen (P = 1KPa) with NiO microcrystals formed "in situ" from nickel hydroxide. The surface area of these microcrystals was 172 m^2/g corresponding to a mean size of about 6 nm. Below 90 °C, adsorption was reversible and no reduction was detected by microgravimetry or magnetic measurements. Above 90 °C, reduction was observed but adsorption and surface reduction were differentiated by their kinetics (after ref. 9).

150 °C), about 1-2 % of all ions are located at the surface. So far the surface structure of these nanometric crystals is not described. It can be assumed that, for not annealed crystals, most of defects are located in the surface layers. Then, hydrogen adsorption can be detected at room temperature (ref. 9). Figure 1 illustrates the adsorption kinetics observed between 20 and 111 °C. The amount of adsorbed hydrogen was higher than that calculated by assuming that all surface sites are active. Therefore, it might be assumed that hydrogen was also incorporated into the bulk. In these experiments, the nickel nucleation was observed at temperatures as low as 90 °C. Then the hydrogen chemisorption and the release of water by surface reduction were

easily differentiated as a long induction period took place before the nickel nucleation. The surface reduction was not observed below 90 °C. Then hydrogen could be desorbed by pumping out.

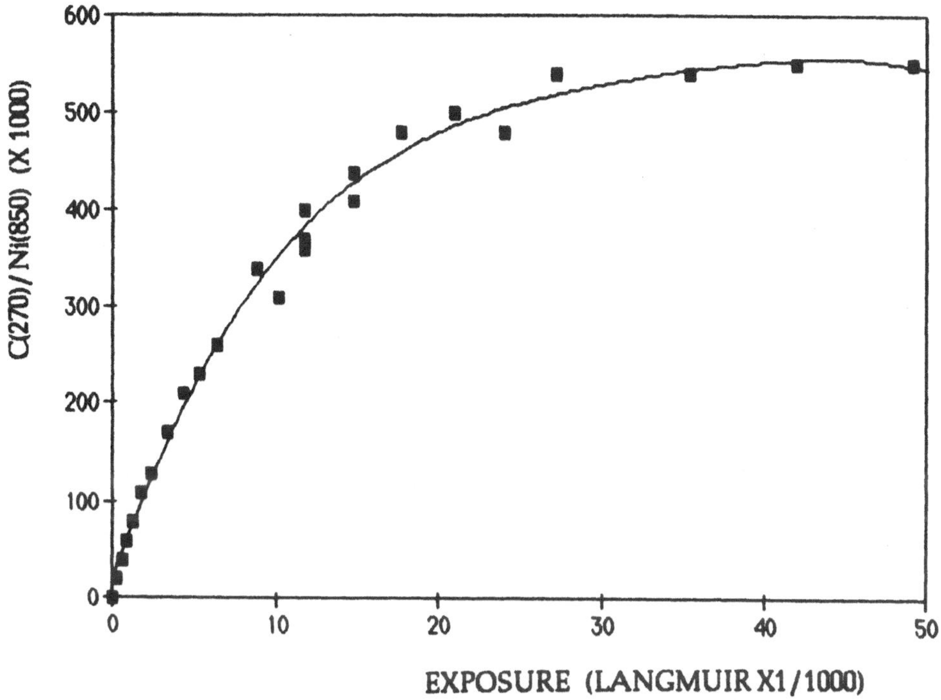

Figure 2. Plot of the Auger peak height ratio for the C270 eV and Ni850 eV emissions as a function of time during the interaction of CO + hydrogen mixtures with the NiO {001} surface at 25 °C. $P(H_2) / P(CO) = 0.04$. $P(CO) = 1.3\ 10^{-4}$ Pa. Total time of treatment : around 50000 s.

5. Interaction of CO and CO_2 with the NiO {001} surface

5.1 INTERACTION OF CO WITH THE NEARLY PERFECT NiO {001} SURFACE

At low temperature (20-100 °C) a non-defective and clean NiO {001} surface does not adsorb pure carbon monoxide. This could be observed unambiguously only when hydrogen partial pressure in the reaction chamber was low enough. For instance, for $P(H_2)/P(CO) = 10^{-6}$ and $P(CO) = 10^{-2}$ Pa, no adsorption of CO could be detected after several tens hours. However, if hydrogen traces were present ($P(H_2)/P(CO) \sim 10^{-3}$), CO could be adsorbed even if the total pressure was below 10^{-4} Pa. Figure 2 gives an example of experimental results obtained when a CO-hydrogen mixture was used at

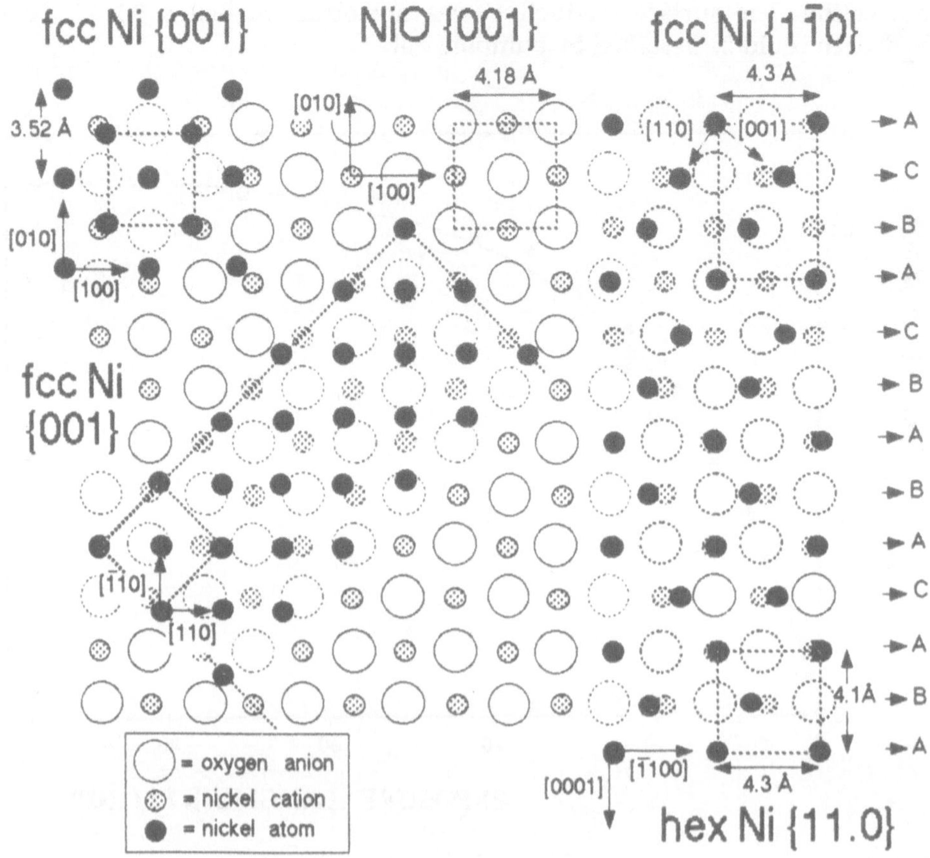

Figure 3. Relative position of the crystallographic planes at the oxide-metal interface as deduced from the RHEED analysis for the first stages of the reduction of a nearly perfect ("in situ" cleaved) NiO {001} surface.

Nickel nucleates in the unusual hexagonal structure in hydrogen or CO at low temperatures (low temperature mode)(bottom right). The ABACAB... sequence of the (0001) hexagonal planes was often observed. Stacking faults were also evidenced. The ABCABC... sequence in the face centered cubic stacking was formed from the hexagonal stacking during the nucleus growth (top right).

At top left, is shown the structural orientation commonly observed in hydrogen and CO in the high temperature mode of the metal nucleation (fcc nickel lattice parallel to fcc oxide lattice). At bottom left, the fcc {001} nickel orientation, at 45° from the previous one, was only detected in CO. An excellent coincidence exists between the two lattices in the NiO <110> directions for this latter orientation.

The position of metallic atoms with respect to the oxide {001} plane is hypothetized. The atomic arrangements are represented with the same parameters as in the bulk. However a slight lattice distortion at the interface can be expected.

room temperature. It can be observed that the CO adsorption is very slow in these experimental conditions. Since CO reacts only in presence of hydrogen, it is expected that adsorption takes place in a form of a compound of H, C and O. It can be assumed that, in the conditions of our experiment, partial atomization of hydrogen occurs on the hot filaments present in the analysis chamber. Due to a very low concentration of the hydrogen atoms and their long diffusion paths to reach the oxide surface, one could expect that a very slow hydroxylation of the oxide surface takes place as mentioned in section 4. Generation of formate groups is possible:

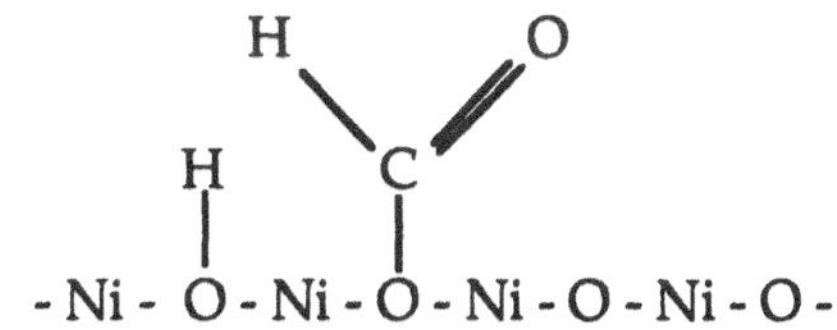

Such surface groups have been found to be formed on non-defective surfaces of oxides in CO - hydrogen mixtures (see e.g. the recent work on ZnO in ref. 10). SIMS (secondary ion mass spectrometry) analysis is very useful in determination of the nature of these surface groups.

It may be expected that the presence of traces of hydrogen results in the formation of nickel atoms scattered on the oxide surface. These atomic species could be considered as adsorption sites for CO molecules which are well-known to be adsorbed in molecular form on low index nickel planes. The process is schematically illustrated in a following way:

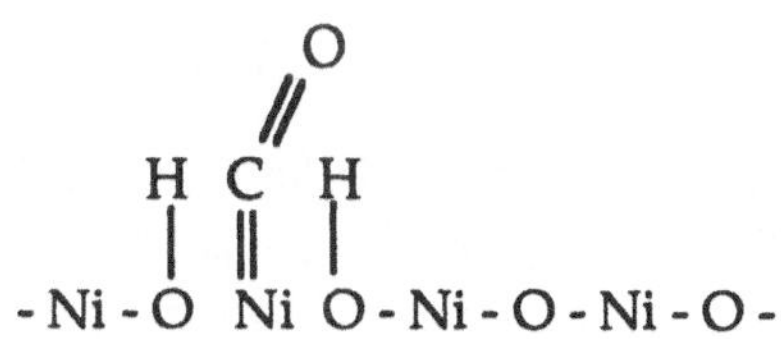

The reducibility of the NiO {001} surface in CO at moderate temperatures is directly related to the previous features. No reduction could be detected below 270 °C in pure CO. Nevertheless, the presence of hydrogen in small amount enables the nickel nucleation from 220 °C.

The processes of nucleation and growth have been analyzed by RHEED. The structural results obtained in CO with hydrogen (typically $P(H_2) / P(CO) = 0.04$) were close to those observed in pure hydrogen (refs 2, 7, 11 and 12). They are schematically illustrated in Fig. 3. Between 220 and 250 °C, nickel nucleates in the "hexagonal" stacking. The (0001) planes are perpendicular to the initial oxide surface and the ABAC... sequence was often observed. Both this unusual stacking and the structural relationship at the interface enable to minimize the interfacial energy (ref. 2). When metallic nuclei grow, the "hexagonal" ABAC stacking is transformed into the usual nickel fcc stacking (sequence ABCABC..). Above 250 °C, the nickel nuclei are formed and grow directly in the fcc structure as previously observed in hydrogen. In this high temperature growth mode, the crystallographic axes of the two fcc lattices remain parallel during the nucleus growth. Nevertheless,

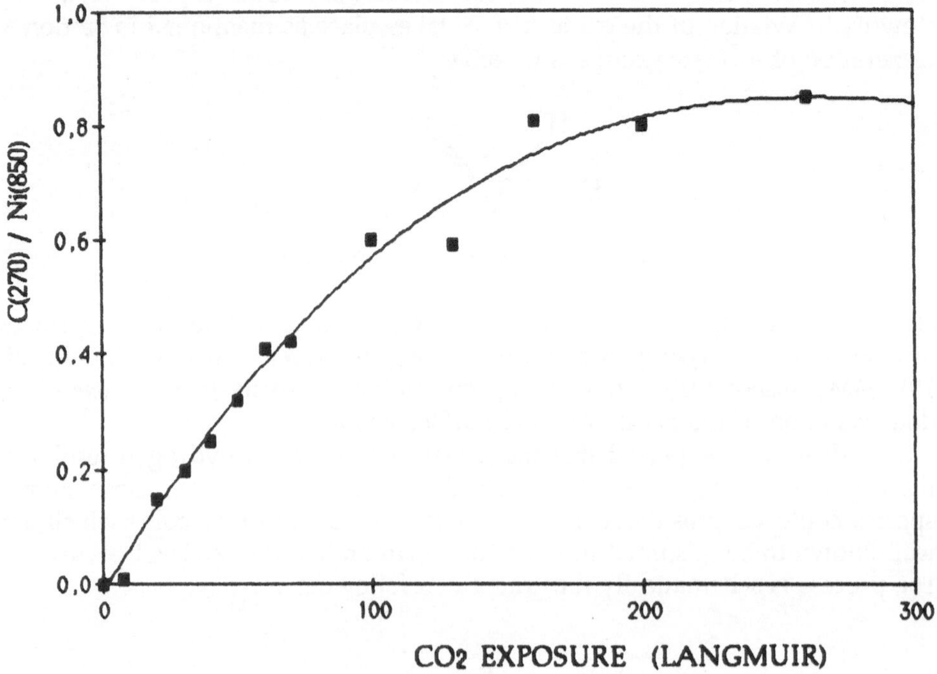

Figure 4. Plot of the Auger peak height ratio for the C270 eV and Ni850 eV emissions as a function of exposure during the adsorption of CO_2 onto the NiO {001} surface at 25 °C. $P(CO_2) = 1.3\ 10^{-5}$ Pa.

a second structural orientation was found in CO. The {001} planes of the two lattices are still parallel but the <110> rows of the first lattice remain collinear to the <100> rows of the second one. This arrangement enables a better coincidence of the two lattices in their {001} planes.

5.2 INTERACTION OF CO_2 WITH THE NEARLY PERFECT NiO(001)SURFACE

A reversible adsorption of CO_2 onto the non-defective and clean NiO {001} surface takes place between 20 and 200 °C . The evolution of the carbon Auger peak height is plotted in Fig. 4 as a function of CO_2 exposure at room temperature. The maximum concentration of carbon is higher than that corresponding to the adsorption of CO (Fig. 2). CO_2 desorption is instantaneous above 180-200 °C. It may be assumed that, in this temperature range, adsorption of CO_2 corresponds most likely to surface carbonation. Nevertheless, it is surprising that CO_2 can adsorb on a nearly perfect surface. This effect could explain why carbonaceous impurities are always detected on air-cleaved surfaces of the rock-salt type oxides.

6. Concluding remarks

Carbon monoxide and molecular hydrogen as well as oxygen are not chemisorbed on the clean and nearly perfect NiO {001} surfaces. Instead, chemisorption takes place easily on surfaces exhibiting defects. The studies performed on highly defective microcrystals and polycrystalline surfaces have clearly shown that CO adsorbs on extended and point defects (refs 13, 14 and 15 and ref. therein). Although more complex, oxygen chemisorption obeys the same rule (ref. 8 and ref. therein). It is also clear that atomic hydrogen can be chemisorbed on any oxide surface. All these experimental results are in agreement with theoretical predictions (refs 5, 16). On the other hand, the behaviour of carbon dioxide is not well understood. Additional work is needed to investigate why carbonatation of oxide surfaces appears to be so easy. The question whether or not chemisorption is a sensitive tool to characterize the nature and concentration of defects at surfaces of NaCl-type oxides still remains open.

References

(1) V.E. HENRICH, 'Electronic and geometric structure of defects on oxides and their role in chemisorption', in: J. Nowotny and L-C. Dufour (Eds), Surface and Near-Surface Chemistry of Oxide Materials, Elsevier, Amsterdam, 1988, pp. 23-60

(2) L-C. DUFOUR, N. FLOQUET and B. DE ROSA, 'Progress in the understanding of the mechanisms of nucleation and growth of metals from oxides. A study from the NiO-Ni model system', in: P. Barret and L-C. Dufour (Eds), Reactivity of Solids, Materials Science Monographs 28A, Elsevier, Amsterdam, 1985, pp. 47-52

(3) (a) E.A. COLBOURN and W.C. MACKRODT, 'Irregularities at the {001} surface of MgO: topography and other aspects', Solid State Ionics, 8 (1983) 221-231
(b) P.W. TASKER and D.M. DUFFY, 'The structure and properties of the stepped surfaces of MgO and NiO', Surface Sci., 137 (1984) 91-102

(4) L-C. DUFOUR and B. DE ROSA, 'Chemical nucleation and growth of metals from microcrystals and monocrystalline surfaces of oxides', Proc. 3rd Round Table on Wustite, Jadwisin-Varsovie, 1986, Metalurgia i Odlewnictwo, 13 (1987) 177-191

(5) (a) G.T. SURRATT and A.B. KUNZ, 'Theoretical study of H chemisorption on NiO. Perfect surfaces and cation vacancies', Phys. Rev., B19 (1979) 2352-2358
(b) G.G. WEPFER, G.T. SURRATT, R.S. WEIDMAN and A.B.KUNZ, 'Theoretical study of H chemisorption on NiO. II. Surface and second-layer defects, Phys. Rev., B21 (1979) 2596-2601
(6) S. BOURGEOIS and M. PERDEREAU, 'SIMS study of the interaction of Ni-O surfaces with H_2 and H_2O', Surface Sci., 117 (1982) 165-168
(7) (a) J.M. RICKARD, M. PERDEREAU and L-C. DUFOUR, 'Croissance épitaxique de films de nickel par réduction d'une face NiO(100) dans l'hydrogène à basse pression', J. Microsc. Spectrosc. Electron., 4 (1979) 95-110
(b) L-C. DUFOUR and N. FLOQUET, unpublished results
(8) B. DE ROSA, L-C. DUFOUR and J. NOWOTNY, 'Nickel monoxide-oxygen interaction and its impact on the NiO reduction mechanism', Reactivity of Solids, 4 (1987) 53-72
(9) L-C. DUFOUR and C. GAY-LANCERMIN, unpublished results
(10) C.T. AU, W. HIRSCH and W. HIRSCHWALD, 'Adsorption of carbon monoxide and carbon dioxide on annealed and defect zinc oxide (000-1) surfaces studied by photoelectron spectroscopy (XPS and UPS)', Surface Sci., 197 (1988) 391-401
(11) N. FLOQUET, P. DUFOUR and L-C. DUFOUR, 'Caractérisation par résolution de plans de cristallites de nickel cfc obtenus par réduction de NiO, J. Microsc. Spectrosc. Electron., 6 (1981) 473-481
(12) N. FLOQUET and L-C. DUFOUR, 'Stability and reactivity of (001) and (111) NiO : a RHEED-AES investigation of Si segregation and Ni formation by gas reduction, Surface Sci., 126 (1983) 543-549
(13) P.C. GRAVELLE and S.J. TEICHNER, 'Carbon monoxide oxidation and related reactions on a highly divided nickel oxide', Adv. Catal., 20 (1969) 167-266
(14) M.W. ROBERTS and R. St.C. SMART, 'XPS studies of donator and acceptor chemisorption of NO and CO on nickel oxide surfaces', Surface Sci., 100 (1980) 590-604
(15) E. ESCALONA PLATERO, S. COLLUCCIA and A. ZECCHINA, 'CO and NO adsorption on NiO: a spectroscopic investigation', Langmuir, 1 (1985) 407-414
(16) J.M. BLAISDELL and A.B. KUNZ,'Theoretical study of O chemisorption on NiO. Perfect surface and cation vacancies', Phys. Rev. B, 29 (1984) 988-995

NUCLEAR TECHNIQUES IN SURFACE STUDIES OF CERAMICS

Hj. MATZKE

Commission of the European Communities, Joint Research Centre, Karlsruhe Establishment, European Institute for Transuranium Elements,
Postfach 2340, D-7500 Karlsruhe, Federal Republic of Germany

ABSTRACT

Nuclear techniques applying energetic ion beams to study structural defects, surfaces and near surface changes in non-stoichiometric ceramics are described. Most examples are for uranium oxide and nitride and for titanates. New results on leaching of UO_2, formation of U_3O_7, leaching of SYNROC and radiation damage effects due to ion implantation of UO_2 and UN are treated. The techniques described in detail are Rutherford backscattering, RBS (with and without channeling), nuclear reactions such as $^{23}Na(p,\alpha)^{20}Ne$, elastic recoil deflection analysis, ERDA and high resolution α-spectroscopy. Elements between H and actinides and gradients in their concentration near surfaces can easily and accurately be measured.

1. INTRODUCTION

Nuclear techniques have frequently and increasingly been used in the recent past as a fast and non-destructive means to investigate surfaces and changes that occur at and near surfaces of solids. Most applications are for semiconductors and metals where techniques employing energetic ion beams and related nuclear methods have been used for about 2 decades, and international conferences have been held on this topic (1,2). Some work has

J. Nowotny and W. Weppner (eds.),
Non-Stoichiometric Compounds Surfaces, Grain Boundaries and Structural Defects, 321–336.

also been done with non-metals and the methods have been used to study surface changes in ceramics due to physical or chemical treatments. Rutherford-backscattering (RBS) of 2-3 MeV He-ions of a van-de-Graaff accelerator, applied to single crystals and if combined with channeling techniques, can be used to define surface states and purities as well as surface-near structural defects or defects induced by mechanical treatment (e.g. polishing or grinding), by ion implantation or by chemical reactions or attack (e.g. by leaching in aequous solutions). Changes in chemical composition near surfaces can be measured with RBS also on polycrystalline ceramics or amorphous glasses. Information on many elements with masses between those of C and O up to the actinides is obtained fast and non-destructively. Gradients in the concentration of specific elements can also be detected with high depth resolution by using nuclear reactions, e.g. that of Na with the reaction $^{23}Na(p,\alpha)^{20}Ne$, that of Al with the reaction $^{27}Al(p,\gamma)^{28}Si$ etc. by using high energy p-beams. Even very light elements such as H can be measured with high depth resolution with nuclear reactions, e.g. $H(^{15}N,\alpha)^{12}C$ using nitrogen beams, or with elastic recoil deflection analysis (ERDA) using different types of beams including the 2 MeV He-beams of RBS analysis. Examples of recent applications of all these techniques are given. Following the scope of the workshop, emphasis is put on nonstoichiometric compounds (uranium oxide and nitride). An example of determining structural defects is given, and new results on leaching of complex ceramics (titanates and silicates) and glasses for long term storage of radioactive waste are reported as well as exact determinations of near-surface non-stoichiometry of UO_2 following exposure to water at increased temperatures. The non-destructive nature of the techniques enables kinetics (time- and temperature dependences) to be followed easily even with a very limited number of specimens. The specimen size can also be limited and very small diffusion rates can easily be measured. Examples with non-stoichiometric ceramics are given.

Only nuclear techniques employing energetic ion beams are described. Techniques to study surfaces with low energy ions and with electrons are treated in other contributions to this workshop, as is secondary ion mass spectrometry, SIMS, which is therefore not dealt with here either.

The application of a specific nuclear technique, i.e. using high resolution α-spectrometry to measure diffusion and near-surface gradients of (α-particle emitting) actinides, was extensively described at previous conferences and NATO ARW's on transport in non-stoichiometric compounds, the topics being Pu-diffusion in $(U, Pu)O_{2\pm x}$ (3), thermally activated diffusion in carbides (4), radiation-enhanced diffusion in nuclear carbides (5), and diffusion in nitrides (6).

Specific examples of new results in applying such nuclear techniques to non-stoichiometric compounds which are reported here are leaching of UO_2 and of SYNROC, diffusion processes involving actinides and radiation damage in near-surface layers of ion-bombarded UO_2 and UN.

2. Experimental and Physical Principles of Ion Beam Techniques

Basic and experimental principles of ion beam techniques have been described in monographs (e.g. 7) and in topical papers (e.g. 8). The proceedings of a series of conferences entitled "Ion Beam Analysis" contain up-to-date summaries of techniques and applications. Ref. (2) gives the proceedings of the last conference "IBA 7".

To apply ion beam techniques, access to an accelerator (e.g. van-de-Graaff accelerator) with energies in the MeV range is necessary. The well collimated beam of energetic ions is shot into a high vacuum scattering chamber and on the specimen. For channeling measurements, the specimen must be mounted on a goniometer for controlled tilting and rotation. A suitable detector, e.g. a solid state surface barrier-type detector, is used to measure backscattered particles or particles (or γ- or X-rays) originating from nuclear or excitation reactions produced within the target by the energetic ion beam.

2.1 Rutherford backscattering and channeling

Fig. 1 shows a typical experimental arrangement to measure Rutherford backscattering, RBS, spectra, or also channeling. The backscattered particles are measured with a solid state detector. A pre-amplifyer and a multichannel pulse-height analyser are used for energy analysis and data storage (see Fig. 1b for a RBS spectrum). The method is non-destructive, the beam area is typically ~1mm Ø, an average measurement takes some minutes. If rotating target supports are used, the influence of the time needed to introduce the samples into the accelerator and to evacuate the target chamber is negligible and a number of samples can be analysed within 1 h. The channeling phenomenon arises because rows or planes of atoms can "steer" energetic ions by means of a correlated series of gentle, small-angle collisions. Channeling is thus not a "transparency effect", but rather particles arriving within a certain critical angle Ψ_c to a low index direction of a single crystal will be steered into this "channel" (Fig. 1c). If combined with backscattering (or nuclear reactions), strikingly large changes in the yield of backscattered particles (or of reaction products) are produced as the orientation of the single crystalline target is changed with respect to the incident beam. The yields can easily decrease by factors of ≥ 50 (see Figs. 1 d and 2).

As indicated in Fig. 1c, displaced lattice atoms and randomly located impurity atoms show no directional effect, whereas substitutional impurity atoms are "shielded" and follow the yield curve of the lattice atoms. The aligned spectrum therefore contains peaks for the unshielded surface atoms of the specimen (surface peak, s, in Fig. 2a), for displaced lattice atoms (damage peaks, d, in Fig. 2 b, c) and for impurity atoms (impurity peaks, i). The location of the damage and of the impurity peaks depends on the distance of the atoms from the surface and on their atomic number. The energy E' of a

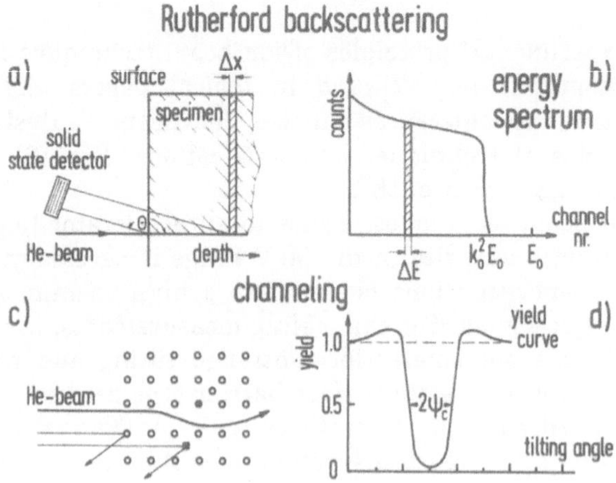

Fig. 1: Schematic presentation of a) RBS geometry and b) RBS energy spectrum for a monatomic specimen. For single crystals, channeling occurs: c) shows a 2-dimensional picture with a shielded (full dot) and an unshielded atom (full square) and one displaced lattice atom d) shows the yield curve for tilting a crystal through a low index direction.

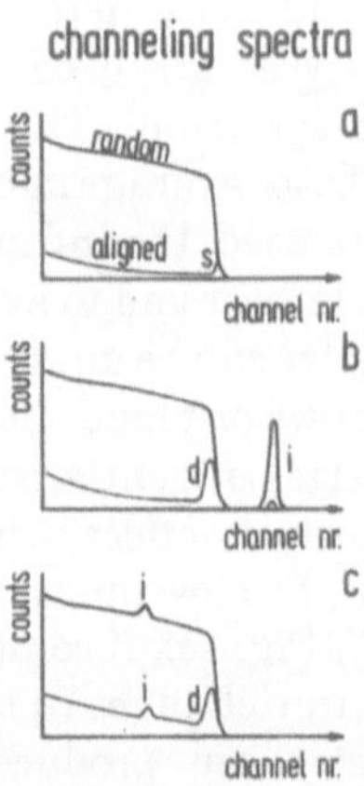

Fig. 2: Principle of channeling spectra for a) unbombarded crystal showing a surface peak (s), and crystals implanted with b) a heavy largely substitutional impurity and c) a light impurity with random location.
i = impurity peak, d = damage peak

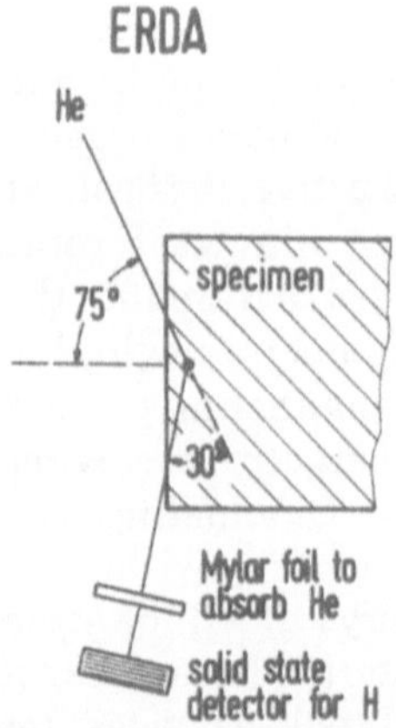

Fig. 3: Schematic presentation of the geometry of elastic recoil deflection analysis, ERDA to measure concentrations and depth profiles of H at and near surfaces.

backscattered particle (original energy E_0, mass M_1) depends on the scattering angle Θ and on the mass of the scattering atoms, M_2. The heavier the backscattering atom, the smaller the energy loss during backscattering. E' is related to E_0 via $E' = k_0^2 E_0$. For not too light targets, $k_0 \simeq (M_1\cos\Theta + M_2)/(M_1 + M_2)$. For $\Theta = 165°$ and 2.0 MeV He-ions, the energy E' of the He-ions following scattering off U, Al, Mg and O atoms located at the surface are 1.87, 1.11, 1.04 and 0.74 MeV, respectively. Therefore, heavy elements give rise to counts at high energies, hence to counts on the right side of the spectrum, whereas light elements fall in the left part of the spectrum (low energies) (see Fig. 2b, c).

The scattering probabilities for the above four elements (relative to oxygen) are 1.0, 3.3, 11.8 and 149, respectively. Heavy elements give thus a much more pronounced contribution to the spectrum than light elements. The ratios are, for fixed Θ and E, essentially given by Z^2, though a minor effect of M_2 is also present. The mass resolution becomes poor for high M_2 but the sensitivity becomes high. Fractions of a surface monolayer of a very heavy impurity on a light specimen can be detected. Mass resolution, however, is best at low masses. As an example, the energy difference of 2.8 MeV He-ions backscattered off U-238 and off Pb-208 is only 24 keV though the mass difference is 30. In contrast, the energy difference for the heavy and the light Si-isotopes (i.e. Si-28 and Si-30; mass difference of only 2) is more than twice as large with 63 keV.

For thick targets, some of the particles penetrate into the sample and are scattered at a given depth x. They reach the detector with a lower energy due to energy loss by ionization and excitation of target atoms before and after the scattering process on their way into and out of the crystal. The energy loss proceeds at a rate determined by the stopping power, dE/dx, i.e. the energy loss per unit path length. These values have frequently been measured for many materials and/or can reliably be calculated. As shown in Fig. 1b, a layer Δx at depth x corresponds to an energy interval ΔE in the energy spectrum. A thick monatomic target gives thus rise to a continuum in the RBS-spectrum extending to zero energy. For more details, the reader is referred to refs. (7, 8).

2.2 Elastic recoil deflection analysis, ERDA

Light elements are difficult to detect with RBS. A forward scattering eleastic recoil detection (see Fig. 3) can be used to measure light atoms down to H (e.g. 9-11). Separation of H, D, and T is easily possible. If energetic beams of rather heavy ions (e.g. Cl or Si at 30 MeV) are used, detection limits for H are as low as 10^{15} H/cm^2 with a depth resolution of better than 30nm. Such beams are available at a very limited number of laboratories. Fortunately, the He-beams used for RBS can be used for ERDA as well, although there is some loss in sensivity (~0.1 at%) and depth resolution (typically 50 nm). The advantages of He-beams of 1-2 MeV energy are their easy availability and the smaller rate of damage formation.

2.3 Nuclear reactions

A large number of nuclear reactions induced by ion beams can be used to detect small concentrations of impurity atoms or gradients in matrix atom concentrations near surfaces. Depth resolutions of the order of 100 Å can be obtained. Fig. 4 shows the method for H analysis with high energy N-atoms achieved by measuring the 4.4 MeV γ-rays of the $^{1}H(^{15}N, \alpha\gamma)^{12}C$ reactions and shifting the "resonance window" into different depths by varying the N-energy. Similary, H can also be analysed with the nuclear reaction $^{1}H(^{19}F, \alpha\gamma)^{16}O$ using a 6.4 MeV ^{19}F-beam. The resonance yields a depth resolution of ~14 nm in Si_3N_4 and a sensitivity of ~200 ppm (10). Other ion beam induced nuclear reactions that have been successfully used are

- Na^{23} (p, αγ)Ne^{20} with a p-beam to measure depth profiles of Na (12, 13)
- Al^{27} (p, γ)^{28} Si with a p-beam to measure depth profiles of Al
- O^{16}(d, p)O^{17} and C^{12} (d,p)C^{13} with a d-beam to measure depth profiles of O and C(14, 15).

Fig. 5 shows an example, a leached glass: the data prove convincingly the opposite direction of the diffusion of Na and H during leaching.

Nuclear microprobe: in normal work with RBS or with nuclear reactions, with and without channeling, beam sizes of typically 0.5 to 1mm Ø are used. If the high-energy beam of the accelerator is focussed to a small beam spot of < 10 μm, a nuclear microprobe is obtained. Since the first set-up at Harwell (16), more than a dozen such nuclear microprobes have been set up. Nearly all methods of ion beam analysis (such as backscattering, ion-induced X- and γ-ray analysis, etc.) have been used (e.g. 17) though rarely with ceramics. With nuclear reactions and PIXE (particle induced X-ray emission (e.g. 18)), the whole periodic table can be measured with detection limits down to the ppm (wt.) level.

For space reasons, these limited descriptions must suffice as explanations of the most important aspects of ion beam techniques. Readers interested in more details are referred to the quoted references. Rather, applications of ion beam techniques to the solid state physics and technology of non-stoichiometric and other ceramics are explained in the following with a number of typical examples.

3. APPLICATIONS TO NON-STOICHIOMETRIC COMPOUNDS AND CERAMICS

In this section, five typical applications of ion-beam methods are given.

3.1 Structural defects in U_4O_9

An early example is that of channeling experiments on U_4O_{9-y} (14). Because of the large mass difference between uranium and oxygen,

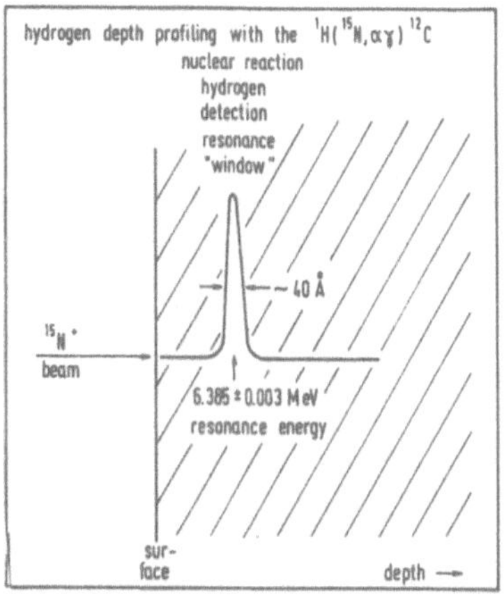

Fig. 4: Schematic presentation of the N^{15} profiling method. The resonance energy is well defined. Thus, by varying the N^{15}-energy, the H concentration can be measured as function of depth.

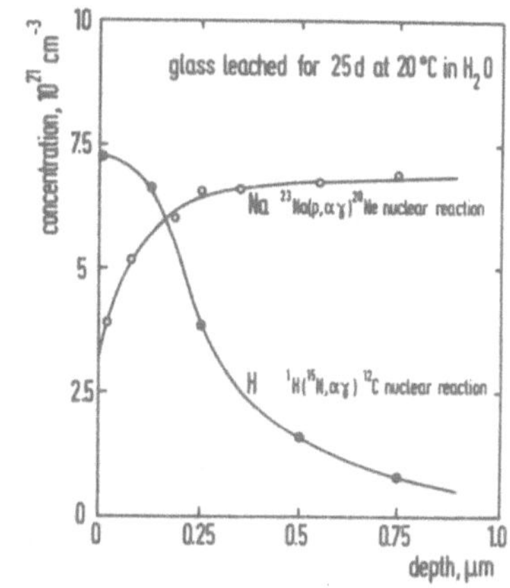

Fig. 5: An example for nuclear reaction analysis (12): Comparison of the Na and H concentration profiles in the first µm of a leached glass (25 d, 20 °C, deionized water)

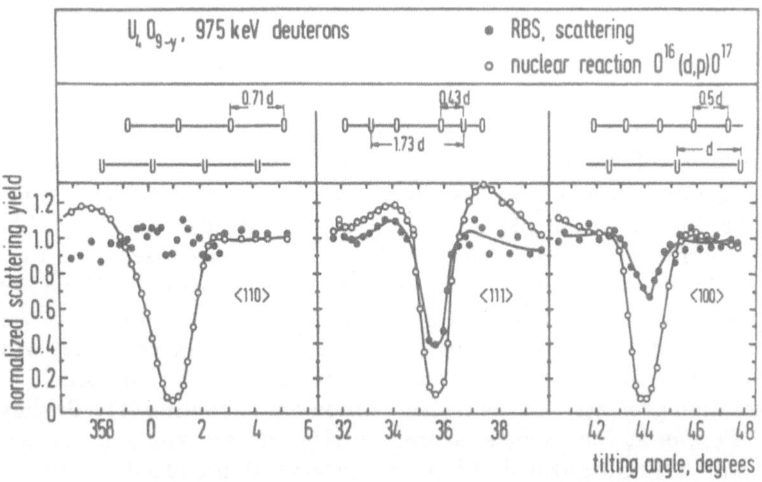

Fig. 6: An early example for analysis of structural defects: Channeling-RBS results (d for U-sublattice) and a nuclear reaction (O^{16}(d, p)O^{17}) for the O-sublattice showing disordering of O-atoms due to cluster formation. The channel-configurations are shown in the upper part of the fig. The <100> channel is practically completely blocked by O-atoms displaced from their normal sites.

Rutherford backscattering shows almost exclusively the metal lattice, wheras nuclear reactions, e.g. the reaction O^{16} (d, p)O^{17}, can be used to study selectively the non-metal sublattice. In this way, information was obtained on the differences in perfection of the two sublattices. In particular, for the complex structure of U_4O_9, direct experimental evidence was obtained for the fact that the U-sublattice is nearly undisturbed whereas significant cluster formation occurs in the O-sublattice, as originally deduced by Willis (19) from neutron-diffraction measurements. These clusters contain the excess oxygen atoms (compared to UO_2) which are displaced by ~0.1 nm from the regular interstitial sites and displace, in addition, some normal oxygen atoms. The channeling study (14) (see Figs. 6 and 7) indicated also that the U-atoms may be slightly (~0.025 nm) displaced by the presence of oxygen atoms, as indicated by the slightly smaller critical channeling angle for U_4O_9 as compared with UO_2 (Fig. 7). Otherwise, the U-sublattice remains perfect. The disorder in the O-sublattice is clearly indicated in Fig. 6. The Willis-model predicts a 25 % shielding of displaced oxygen and rather open channels along <111>, and very filled channels in <100>, with <110> in between, as is confirmed by the channeling data.

3.2 Catalytic properties of Pt in and on single crystalline MgO and Al_2O_3 supports

Ceramic oxides are a good support for active metals in heterogeneous catalysis. To obtain basic insights in the dependence of catalytic behavior on the shape of Pt, on surface state and orientation of the support, different low index surfaces of MgO and Al_2O_3 single crystals were implanted with 30 keV Pt-ions (range ~16 nm). Radiation damage and its recovery on annealing as well as Pt-location were measured with RBS and channeling (20-22). Fig. 8 shows an RBS spectrum for <100> MgO. The "damage peaks" at the O-and Mg-shoulders (see arrows for surface position of O and Mg) can be used to deduce the number of displaced atoms per Pt-ion: 20 oxygen and 20 Mg atoms were permanently displaced per Pt-ion for the dose used. The Pt-impurity peak is well separated due to the high atomic number of Pt. Upon annealing, the damage is recovered and Pt migrates to the surface. Pt was not substitutional following implantation, but it formed precipitates which were coherent with the single-crystalline matrix. A careful study in "high resolution geometry" of the Pt-peak following annealing (Fig. 9) showed that the Pt was at the surface, a prerequisite for catalytic activity which was measured simultaneously. Tilting the crystal produced hardly any peak change for Al_2O_3 indicating a near-spherical shape of the Pt-precipitates whereas broadening and shift of the peak in MgO indicates the Pt-precipitates to be located at the bottom of pores or cracks in the recovered surface layers.

Following these two older examples, three recent sets of application of nuclear techniques and ion beams to study surfaces and near-surface layers of non-stoichiometric ceramics are given, describing work that was recently started and is still ongoing.

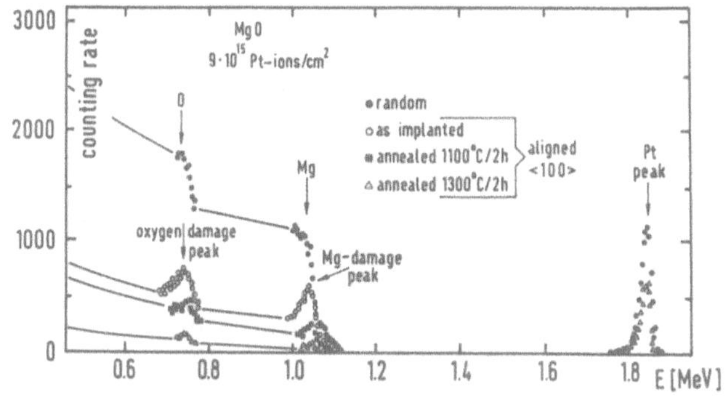

Fig. 8: RBS spectra for 2 MeV He-ions on random and aligned < 1 0 0 > MgO crystals implanted with Pt at 30 keV energy (20).

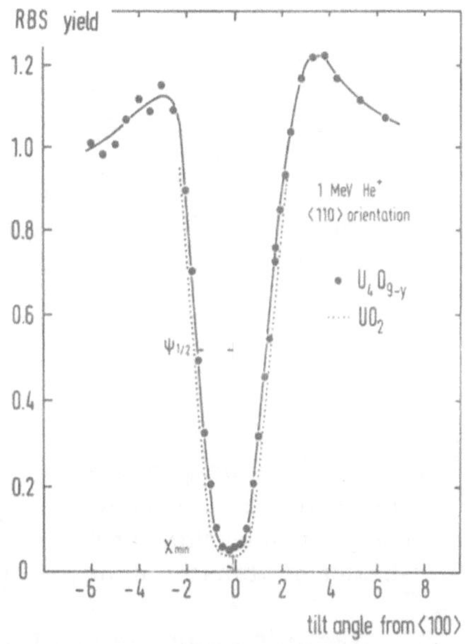

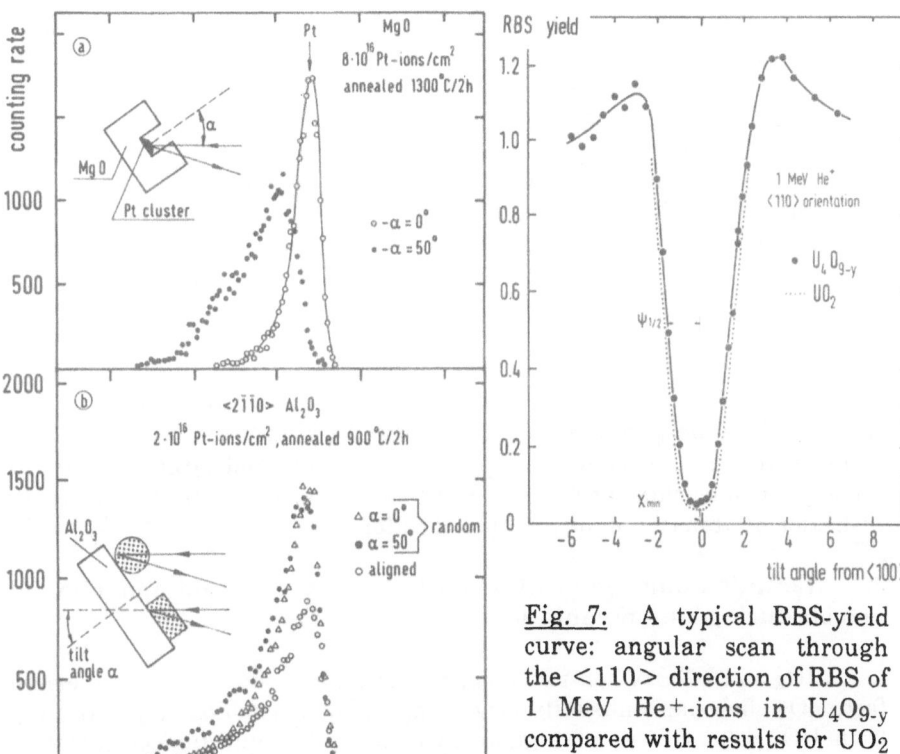

Fig. 7: A typical RBS-yield curve: angular scan through the <110> direction of RBS of 1 MeV He+-ions in U_4O_{9-y} compared with results for UO_2 (14).

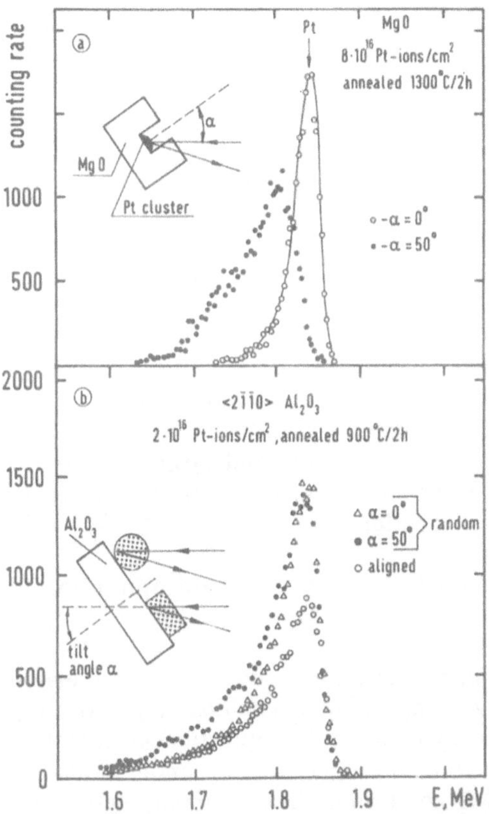

Fig. 9: RBS spectra for 2 MeV He-ions on Pt-implanted MgO and Al_2O_3 single crystals for different tilt angles. Only the Pt-peaks are shown. The inserts give possible geometrical arrangements of the Pt diffused to (or near to) the surfaces of the crystals (20).

3.3 Leaching of titanates

Titanates are very insoluble in water. A titanate-based tailor-made ceramic called SYNROC (synthetic rock) was therefore developed as possible matrix to solidify high-level liquid radioactive waste for safe long-time storage in deep dry geological formations (23). SYNROC consists (predominantly) of four titanate minerals, perovskite $CaTiO_3$, Ba-hollandite $BaAl_2Ti_5O_{14}$, zirconolite $CaZrTi_2O_7$, and substoichiometric rutile, TiO_{2-x}. Fig. 10 shows the complex RBS spectrum of SYNROC (15). The measured spectrum can easily be reproduced by calculating the spectrum based on the known components, using the theory of RBS. A Pb peak located at about 50 nm below the surface is added in some specimens as depth marker for congruent leaching. Other unmarked and leached specimens showed depletion of Ca and Ba in RBS, and enrichment of fission products (24). If present, actinides (Np, Pu, Cm) were also enriched at the surface (leaching conditions 150 to 200 °C, H_2O, times of 1 to 4 weeks), as demonstrated by *high resolution α-spectroscopy* on the leached specimens (25). Similar to the shoulders in RBS spectra of specimens or specimen components of extended thickness, shoulders are found in the α-energy spectra. Knowing the stopping power, the energy scale can be converted to a depth scale. The resolution of this technique is about 10 nm. The measured depth values for actinide enrichment in SYNROC were in the range of 100 to 1000 nm.

The implanted depth marker was observed to move its position relative to the surface, providing a direct non-destructive means of measuring dissolution rates very accurately. For instance, for 150 °C and H_2O leachant, the SYNROC dissolution rate was 1.5 Å/day.

ERDA with He-ions was used to investigate whether hydrogen uptake occurred during leaching (25). Fig. 11 shows the measured H-profile. It is very shallow, despite the long leaching time and the high leaching temperature. Glasses showed deeper H-penetration and larger H-contents already for 2h leaching at 150 °C (26). These results are the first evidence for hydration of titanates in the presence of water.

3.4 Radiation damage in UO_2 and UN. Formation and annealing of defects during ion-implantation.

UO_2 and UN are non-stoichiometric ceramics used as nuclear fuels. In fact, UO_2 shows an unusually large range of non-stoichiometry extending from $UO_{1.65}$ to $UO_{2.25}$ at high temperatures. The question of radiation damage and fission product behavior is important to understand the materials' behavior under operating conditions, hence during continuous damage formation by fission. Ion bombardment of single crystals with fission products and RBS-channeling examination of damage formation and recovery upon annealing are an ideal means to selectively investigate the behaviour of U-defects (see section 2.1). Metal atom mobility is rate-determining for most high-temperature kinetic processes in both UO_2 and UN because of the high mobility of the non-metal atoms. The diffusion

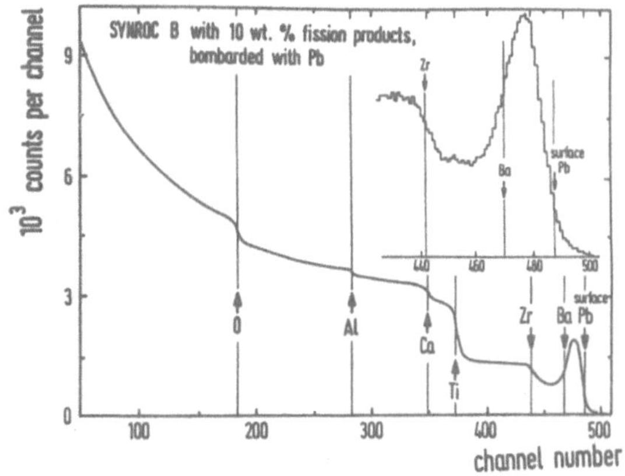

Fig. 10: RBS-spectrum of Pb-implanted SYNROC containing 10 wt.% (simulated, inactive) fission products before leaching. The arrows show the surface location of the different elements.

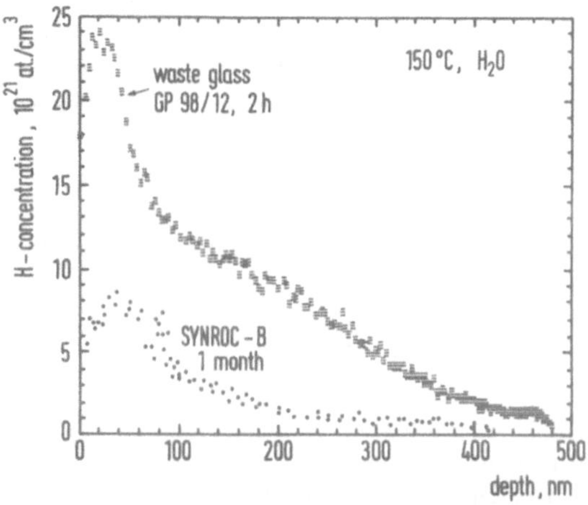

Fig. 11: Hydrogen profile measured with ERDA and He-ions in SYNROC-titanate-based waste ceramic following leaching in H_2O at 150 °C for one month (26).

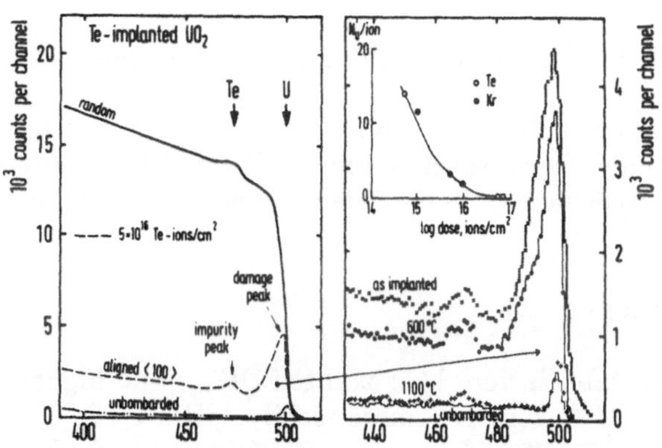

Fig. 12: RBS-channeling spectra of Te-implanted UO_2 showing surface (s), impurity and damage peaks. The right part shows annealing of the damage, and the insert shows the number of displaced U-atoms per Te-ion of 40 keV energy (28).

coefficients for N in UN and for O in $UO_{2\pm x}$ are higher by a factor $>10^5$ (depending on temperature) than those for U. Previous work on radiation damage used only unspecific properties (lattice parameter changes, changes in electrical resistivity etc.), to follow damage formation and recovery (27).

It was thus, for instance, not unequivocally possible to attribute recovery stages to the non-metal or the metal sublattice. As shown in Section 2.1, the RBS-channeling technique enables to follow the formation of U-defects, to determine the number of U-defects formed per incoming ion, and to investigate specifically recovery stages for U-defects. The range of the 40 keV Te-ions used to produce the damage is 13 nm. As a by-product, the important information is obtained that most of the implanted Te-ions occupy substitutional sites in the as-implanted state. This fact is deduced from a comparison of the impurity peaks for random and aligned crystals. The number of displaced U-atoms N_U increases with the implantation dose, though less than linearly: N_U *per ion* decreases with dose, as shown in the insert in the right half of Fig. 11. The total number of U-defects produced can be calculated according to the Kinchin-Pease model (29) to be 250. Obviously, important defect recombination takes place during implantation. Another interesting and new feature of the radiation damage in UO_2 is its strong orientation dependence: N_U for <111> oriented crystals is substantially higher than N_U for the <100> direction. This indicates that the athermal rearrangement of the U-defects in the <111> direction is much less pronounced than in the <100> direction.

Fig. 13 shows RBS-channeling spectra of UN single crystals implanted with Xe-ions. In contrast to (largely ionic) UO_2, which shows only a pronounced damage peak within the range of 13 nm of the implanted ions, a characteristic "dechanneling knee" is found in (metallic) UN. A large dechanneling rate is obvious behind the damage peak, extending to a dose-dependent depth of up to about 10x the range of implanted ions. In still larger depths, the crystals are of perfect quality. The conclusion is that ion-implantation effects in metallic UN are significantly different from those in largely ionic UO_2. In UN, near-surface regions form containing extended defects, most probably dislocation loops, which extend well beyond the range of the implanted ions. This shows a high mobility of the created U-defects at ambient temperature in UN.

These few remarks must suffice to highlight these interesting investigations which will be published elsewhere in more detail. The results given here are mainly meant to demonstrate the potentials of the techniques used.

3.5 Leaching of UO_2 in water

The solubility of UO_2 in water is very low; oxidized UO_{2+x} and higher oxides containing U^{5+} and U^{6+} dissolve more easily. XPS-UPS etc. techniques had previously shown surface formation of U_3O_7 on UO_2 during leaching (31). Depth and time dependence of U_3O_7 formation cannot be measured with these techniques and were unknown. Again, RBS-channeling

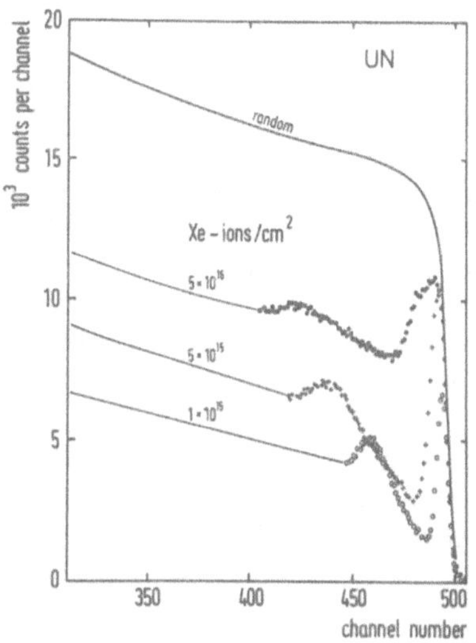

Fig. 13: RBS-channeling spectra of Xe-implanted UN showing "dechanneling knees" characteristic for metallic bonding (30).

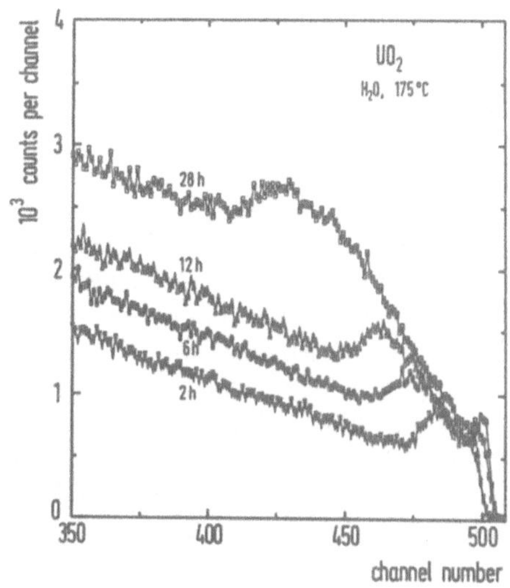

Fig. 14: RBS-channeling spectra of UO_2 single crystals leached in water at 175 °C for 2, 6, 12 or 28 h. Only the spectra for aligned crystals are shown (30).

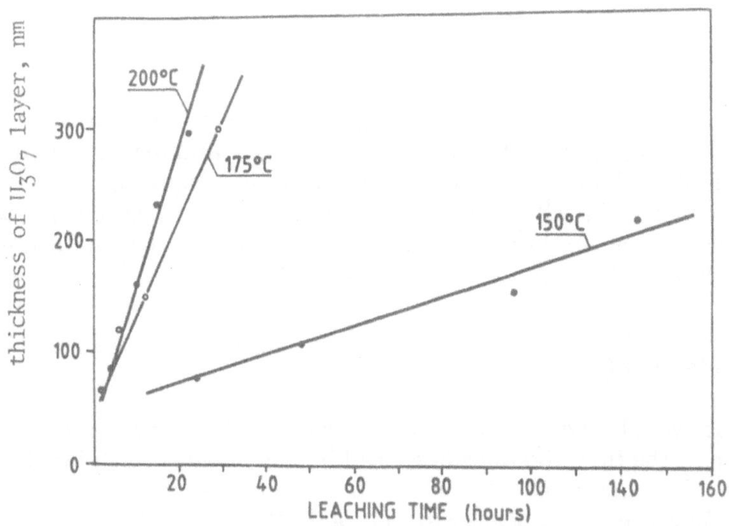

Fig. 15: Increase of the thickness of the U_3O_7-layer on leached UO_2 with time for 3 leaching temperatures (30).

techniques are ideal means to study these phenomena occurring at and near surfaces. Exact knowledge on mechanisms and rates of UO_2 dissolution in water or aqueous solutions are important for the concept of direct storage of spent nuclear UO_2 fuel in deep geological repositories.

Fig. 14 shows typical RBS-channeling spectra of leached UO_2. Peaks reflecting the displacement of U-atoms from their regular lattice position are seen. Prolongued leaching produces broadening of these peaks. The totality of the data (random spectra, aligned spectra, critical angles for channeling, various temperatures, various times) leads to the conclusion that U_3O_7 layers form. The time dependence of their thickness for three temperatures is shown in Fig. 15. ERDA-measurements (see section 2.2) confirmed the absence of hydrogen (water). The linear kinetics in Fig. 15 indicate that the oxidation reaction takes place at the interface between the distorted and the undistorted parts of the crystals. Again, space limitation prevents a more detailed description of this further application of nuclear techniques to surface studies of nonstoichiometric ceramics.

4. SUMMARY

Nuclear techniques involving ion beams are described in their application to surface studies and characterization of structural defects of non-stoichiometric ceramics. Following the tradition of previous conferences on nonstoichiometric compounds, both basic aspects and aspects of technological importance are treated. Rutherford backscattering, channeling and elastic recoil deflection analysis with MeV-He beams and nuclear reactions suitable for analysing shallow surface-near gradients of low atomic number elements are described. The specific examples given are structural defects and formation of surface layers of nonstoichiometric uranium oxide, leaching of titanate-based SYNROC and investigation of ion-implantation-induced radiation damage in UO_2 and UN. Detailed knowledge on physical mechanisms and on the chemistry of surface reactions can be obtained non-destructively with the methods described.

Acknowledgements

The author would like to thank a number of colleagues for help and cooperation in the investigations described in this paper. Particular thanks are due to G. Linker, O. Meyer and A. Turos, Institut für Nukleare Festkörperforschung, KfK, Karlsruhe, G. Della Mea and V. Rigato, Univ. of Padova, A. Solomah, KFA Jülich, and S. Fritz and V. Meyritz, TU Karlsruhe.

5. REFERENCES

1) Modern Nuclear Methods in Materials Science, Eds. M.J. Fluss and Y.C. Yean, Mater. Sci. Forum **2** (1984)

2) Ion Beam Analysis, Eds. J.P. Biersack and K. Wittmaack, Nucl. Instr. Methods in Phys. Research, spec. vol. **B 15** (1986)

3) Hj. Matzke, in Proc. Int. Conf. Transport in Nonstoichiometric Compounds, Ed. J. Nowotny, North Holland, Elsevier, Mater. Sci. Monographs **15** (1982) 203

4) Hj. Matzke, in Transport in Nonstoichiometric Compounds, Eds. G. Petot-Ervas, Hj. Matzke and C. Monty, North Holland, Spec. Vol. Sol. State Ionics **12** (1984) 25

5) Hj. Matzke, in Transport in Nonstoichiometric Compounds, Eds. G. Simkovitch and V.S. Stubican, Plenum Publ. Corp. (1985) p. 331

6) Hj. Matzke, in Nonstoichiometric Compounds, Eds. R.A. Catlow and J. Harding, Advances in Ceramics **23** (1987) 617

7) W.K. Chu, J. W. Mayer and M.A. Nicolet, Backscattering Spectrometry, Academic Press, N.Y. (1978)

8) Hj. Matzke, Fresenius Z. Anal. Chem., **319** (1984) 801

9) J.F. Ziegler et al., Nucl. Instrum. Methods **149** (1978) 19

10) B.L. Doyle and P.S. Peercy, Appl. Phys. Letters **34** (1979) 811

11) A. Turos and O. Meyer, Nucl. Instrum. Methods **B4** (1984) 92

12) P. Trocellier, B. Nens and Ch. Engelmann, Nucl. Instrum. Methods **197** (1982) 15

13) G. Battaglin, G. Della Mea, G. De Marchi, P. Mazzoldi and D. Puglisi, Radiation Effects **64** (1982) 99

14) Hj. Matzke, J.A. Davies and N.G.E. Johansson, Can. J. Phys. **49** (1971) 2215

15) Hj. Matzke and G. Linker, unpublished results

16) J.A. Cookson, A.T.G. Ferguson and F.D. Pilling, J. Radioanal. Chem. 12 (1972) 39

17) R. Nobiling, Nucl. Instrum. Methods **218** (1983) 197

18) D. Heck, Atomkernenergie - Kerntechnik **46** (1985) 187

19) B.T.M. Willis, Nature **167** (1963) 153; Proc. Roy. Soc. **A 274** (1963) 122 and 134; J. Phys. Fr. **25** (1964) 431; Proc. Brit. Ceram. Soc. 1 (1964) 9

20) Hj. Matzke, A. Turos and P. Rabette, Radiation Effects **65** (1982) 1

21) A. Turos, Hj. Matzke and P. Rabette, Phys. Stat. Sol. (a) **64** (1981) 585

22) A. Turos, O. Meyer and Hj. Matzke, Appl. Phys. Letters **38** (1981) 910

23) A.E. Ringwood, Safe disposal of high level nuclear reactor waste: a new strategy, Austral. Nat. Univ. Press, Canberra, Austr., and Norwalk, Conn., USA (1978)

24) Hj. Matzke, J.C. Dran and A. Solomah, unpublished results

25) Hj. Matzke, E. Toscano, C.T. Walker and A.G. Solomah, Advanced Ceram. Mater. **3** (1988) 285

26) G. Della Mea, V. Rigato and Hj. Matzke, unpublished results

27) Hj. Matzke, Radiation Effects **53** (1980) 219

28) Hj. Matzke, A. Turos and S. Fritz, to be reported at Int. Conf. Lattice Defects in Insulating Crystals, Parma, Italy (1988)

29) G.H. Kinchin and R.F. Pease, Rep. Progr. Phys. **18** (1955) 1658

30) Hj. Matzke, S. Fritz and A. Turos, unpublished results, and S. Fritz, PhD Thesis to be submitted at Université de Strasbourg (1988)

31) D.W. Shoesmith, S. Sunder, M.G. Bailey, G.I. Wallace and L.H. Johnson, 36. Meeting Int. Soc. Electrochem., Salamanca, Spain Abstr. 06240 (1985)

CALCULATED GRAIN BOUNDARY STRUCTURES IN NiO; COMPARISON WITH EXPERIMENT

J.H. HARDING[1], S.C. PARKER[2] AND P.W. TASKER[1]
[1]Theoretical Physics Division, Harwell Laboratory, Oxon, U.K.
[2]Dept of Chemistry, University of Bath, Bath, U.K.

ABSTRACT. Recent high resolution electron microscopy studies have revealed structures for the Σ5, (310), symmetrical tilt grain boundary in NiO that differ from the earlier predictions based on static lattice calculations. Possible reasons for the discrepancy are discussed and two possible differences are examined in more detail. First, the static calculations can miss alternative stable structures. Further searching has revealed at least one new structure that is stable. It accords with the experimental observation but is of higher energy than the earlier configuration. Second, the calculations are performed at 0 K while the experiment is quenched from high temperature. The effect of temperature on the calculated structures is calculated. It is found that the original structure is stable to high temperature and that the entropy of both structures is low.

1. Introduction

Grain boundaries in oxide materials have a profound effect on many important properties in the real polycrystalline material. However, there have been fewer studies of their structures by theoretical or experimental methods than for the corresponding boundaries in metals. Most computational studies have concentrated on nickel oxide and magnesium oxide[1,2]. In order to simplify the computer calculations, almost all the published work has considered symmetrical, coincidence boundaries with small periodic unit cells in the interface plane. For example, Duffy and Tasker[1] have reported structures for the series of [001] coincidence tilt boundaries in NiO. This work was stimulated by a desire to understand the apparent enhanced grain boundary diffusion observed in nickel oxide scales growing on corroding nickel[3]. At that time there was little information on grain boundary structure and prediction of the structure was a necessary precursor to studying the boundary properties. However, recently high-resolution electron microscopy (HREM) has been applied to these coincidence tilt boundaries[4] and we are now able to compare directly experiment and calculation.

The latest generation of HREM's have enabled atomic structures to be deduced directly from experimental observation. This has been applied by Merkle and Smith[4] with fascinating results for nickel oxide. Their

J. Nowotny and W. Weppner (eds.),
Non-Stoichiometric Compounds Surfaces, Grain Boundaries and Structural Defects, 337–349.

observation have revealed a number of features including,

a) several different structures exist for a particular grain boundary misorientation,

b) grain boundary structures are often asymmetric, and asymmetric units may occur even within a symmetrical boundary,

c) boundaries can deviate on an atomic scale from a planar configuration.

In particular, they have made a detailed analysis of the Σ5 and Σ13 symmetric tilt boundaries which were also studied by Duffy and Tasker. Although some features are similar, there are some distinct differences between the calculation and observation. There is asymmetry in the Σ13 boundary and evidence of a high vacancy concentration. There are two distinct structures observed for the Σ5 boundary. One structure has no mirror symmetry and can be derived by a translation in the boundary plane. The second structure is more symmetric and closer to the calculated configuration.Both Σ5 boundaries show a substantially smaller volume expansion than calculated (0.3A or 0.4A as compared with 1.1A calculated). There are several possible reasons for the differences.

a) The interpretation of HREM images is not straightforward so it is difficult to invert the data to obtain a reliable atomistic structure. However, the qualitative differences between the images and the calculation indicate that the discrepancy is real.

b) The static lattice calculation can miss alternative structures and settle for the most accessible energy minimum. The calculations of Duffy and Tasker included an extensive search for alternative minima without success but it is impossible to show that an alternative does not exist.

c) The static calculations are carried out at absolute zero temperature while the experiment is on a structure frozen in by quenching from 1400 K. Entropy effects may not only perturb structures but also may lead to other structures being more stable at elevated temperature.

d) The model for interatomic forces used in the calculation may be wrong. The results of these calculations are not very dependent on the short-range potential used but the structures are determined by the long-range Coulomb potential. Thus if the model is at fault, one would have to question the ionic description of nickel oxide.

e) The presence of additional defects generated by non-stoichiometry in the oxide or impurities segregated to the boundary may influence the boundary structure. Merkle and Smith report that care was taken to ensure that the samples were both pure and stoichiometric so changes would have to be induced by very strongly segregating species present only in low amounts in the bulk.

In this paper, we consider b) and c) above in more detail in an attempt to understand the differences between the observation and calculation.

2. Alternative Structures

We start by reconsidering the static simulations of boundary structure. These were carried out with the Harwell MIDAS code which represents the crystal as a stack of planes periodic in two dimensions. Thus although the crystal can dilate and shear, it is constrained by its two dimensional periodicity. In addition, local symmetry at an ion site may further reduce the space sampled by the relaxation unless the symmetry is specifically reduced. In the calculations reported here, we used the same Sangster and Stoneham[5] empirical potentials used in the original work. We started by confirming the structure of Duffy and Tasker for the structure of the Σ5 (310) boundary. This is stable with an energy of 1.95 Jm^{-2} and shows a dilatation across the interface of 1.1 A. (The energy is slightly larger than that reported previously due to the use of somewhat smaller relaxation regions in this work). This structure is shown in Figure 1. Merkle and Smith's observations clearly do not agree with the calculation, particularly with respect to the dilatation. They observed both a symmetric and an asymmetric configuration. These are considered in more detail below.

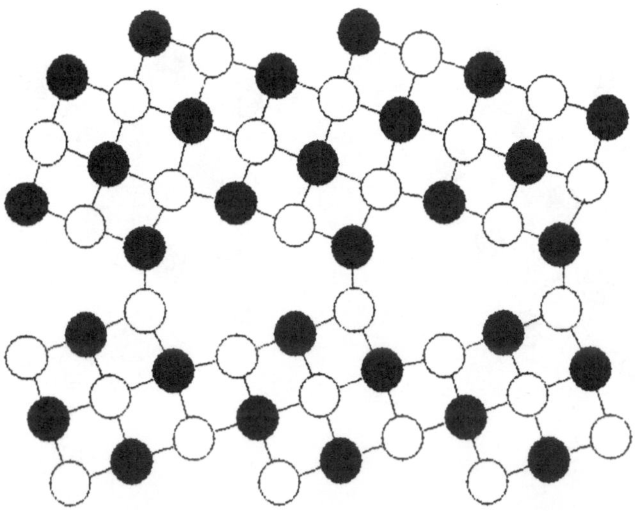

Figure 1: Structure for the Σ5, (310) tilt grain boundary in NiO predicted by Duffy and Tasker[1].

2.1 SYMMETRIC STRUCTURE B

We start by considering the symmetric structure observed by Merkle and Smith. They suggest an atomic configuration and this provides a suitable starting point for a calculation. Since they have no information on the distribution of ionic charges, there are two distinct possible arrangements for their proposed structure. These are shown in Figure 2. The figure uses an idealised structure to represent Merkle and Smith's suggestion. All ions are on unrelaxed lattice sites and the dilatation has not been applied. this tends to make the structures look less realistic than was intended by Merkle and Smith. However, these structures enable us to examine the symmetry of the configuration and are a suitable starting point for a lattice relaxation calculation. If the Merkle and Smith structure is stable, one of the two configurations will come to equilibrium with only minor lattice distortion. We can easily see that one of the structures looks more plausible than the other, since the first brings like charged ions into close proximity.

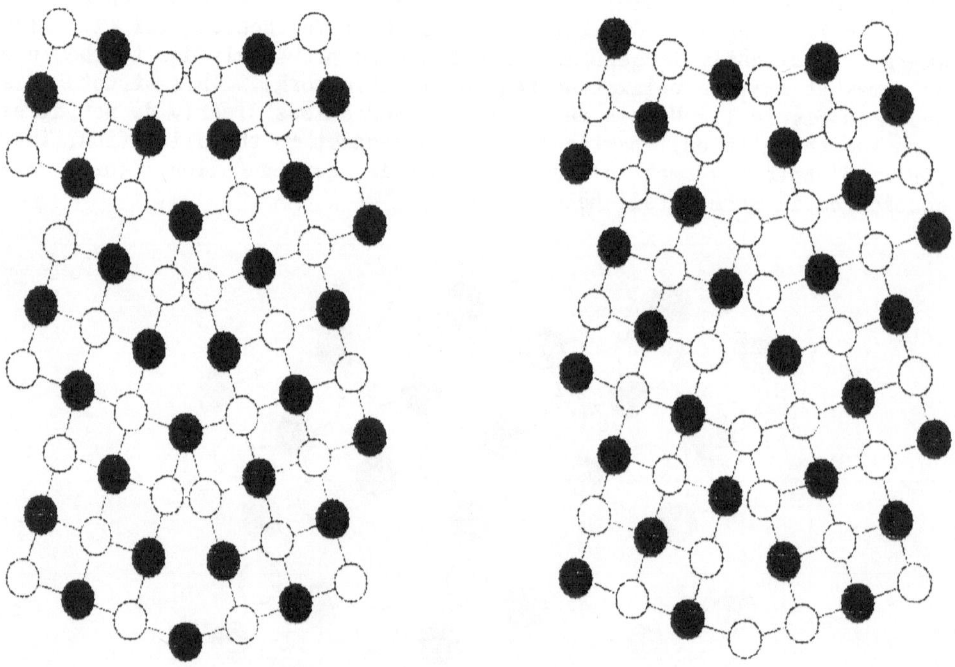

(a) (b)

Figure 2: Possible symmetric structures for the tilt boundary in NiO as suggested by high resolution electron diffraction[4].

When these structures are relaxed to equilibrium, they behave differently. The first, and less stable, starting configuration relaxes to the Duffy and Tasker structure (Figure 1). The second starting configuration also undergoes major relaxation and comes to equilibrium at a new structure. This new structure is shown in Figure 3. It is clearly different from the suggested structure of Merkle and Smith but retains some features. In particular, it achieves stability and a reduced boundary density with minimal dilatation. The calculated dilatation is 0.46 A. Even though this calculation is effectively at 0K while the experiment is quenched from 1400 K, the dilatation is more consistent with observation. The calculated core structure is markedly different with a reconstruction to distorted hexagonal units giving the stability. The calculated energy of this boundary is 2.45 Jm^{-2}, substantially higher than that of the Duffy and Tasker structure.

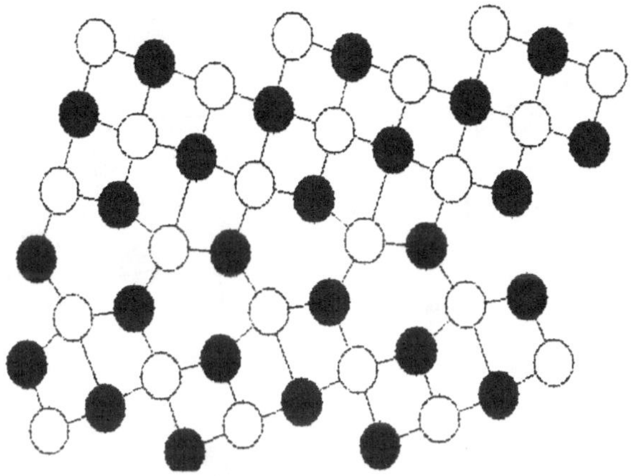

Figure 3: Stable grain boundary structure obtained by relaxing to equilibrium the idealised structure in Figure 2 (b).

In conclusion, we have demonstrated that the Merkle and Smith symmetric structure is not stable without reconstruction. If that structure does exist, it would imply that our model of the oxide is incorrect. The lattice calculations of boundary structure are largely independent of the short-range potentials used but depend primarily on the Coulomb interactions. Thus, if they deny the stability of a structure that is observed, we would have to question the use of the ionic model in this case. However, the calculations have revealed a restructured, Merkle and Smith configuration that agrees with the most straightforward observation. It now needs to be compared, in detail, with the electron microscopy image to decide if it is a plausible interpretation of the observation. We are still left with the problem that at 0 K, at least, the new structure is metastable with respect to the earlier Duffy and Tasker structure.

2.2 ASYMMETRIC STRUCTURE A

Merkle and Smith also observed an asymmetric region of (310) boundary. This they interpreted with a structure (they designated A) consisting of half the bicrystal displaced parallel to the boundary. Again, there are two possible configurations depending on the distribution of ionic charges. These are shown in Figure 4. Once more, one of the starting configurations looks less stable than the other. When this (figure 4a)

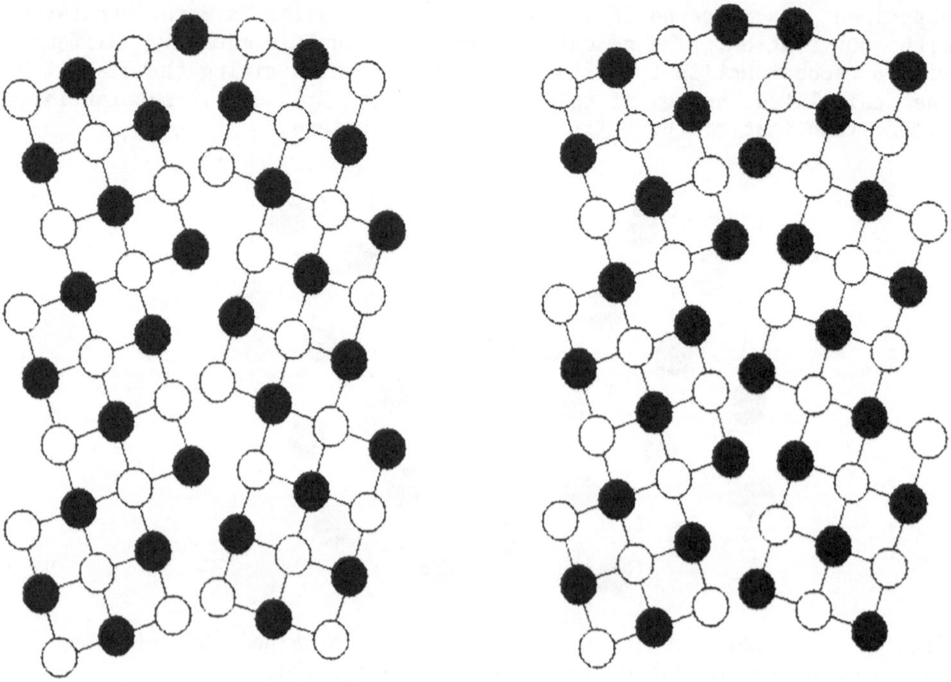

(a) (b)

Figure 4: Possible asymmetric structures for the tilt grain boundary in NiO as suggested by high resolution electron microscopy[4].

is relaxed to equilibrium in a lattice statics program, it reconstructs to the Duffy and Tasker structure. The other structure also substantially relaxes to a new equilibrium configuration. This new structure is shown in Figure 5. It is remarkably similar to the other new structure discussed above and shown in Figure 3. However, it is not identical since it is more symmetrical. Nevertheless, its energy is very similar to the alternative configuration. It is not clear whether this should be treated as an additional grain boundary structure. It may be that the relaxation can be trapped in very shallow local minima around a true metastable configuration. More work is needed to see if there are genuinely several closely related structures or only one, particularly at elevated temperatures. In this work, we have considered only one new structure since in any case they are clearly similar. An important, and intriguing point, is that the relaxed structure shown in Figure 5 has recovered the mirror symmetry lacking in the starting configuration. This is despite the fact that the bicrystal did not shear during the relaxation. Indeed, the structure derived from the asymmetrical starting configuration is more symmetrical than that derived from the symmetrical start. We have not, as yet, found any stable structure showing the asymmetry observed in the electron microscope.

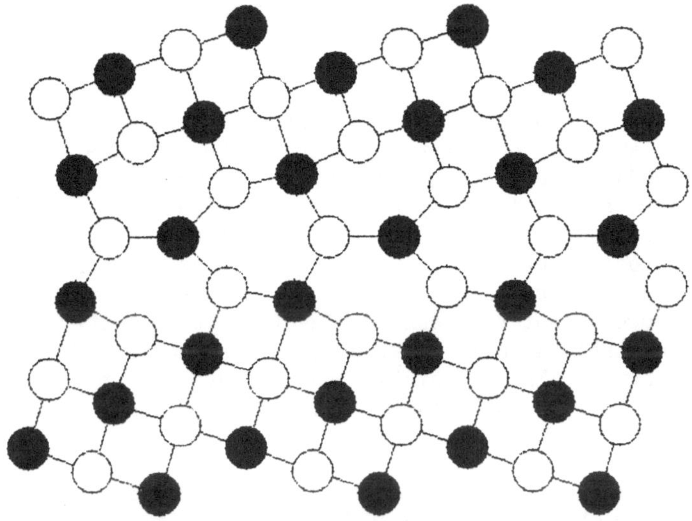

Figure 5: Stable grain boundary structure obtained by relaxing the ideal structure shown in Figure 4(b).

2.3 OTHER STRUCTURES

All the structures so far considered, including (and perhaps especially) that of Duffy and Tasker, can be considered as two (310) surfaces pushed together to construct the boundary (examine Figures 1, 2 and 4). This is an unnecessary constraint but often imposed in calculations unwittingly. The boundary must be overall symmetrical in its tilt angle to be designated (310) but does not have to be able to dissociate into perfect (310) surfaces. The most striking example of this concerns the Σ5, twist boundary in NiO. Here the most stable structure was only found after the constraint that the boundary should be able to dissociate into perfect (001) surfaces was removed[6]. This was achieved by removing ions from the boundary plane before relaxation, thus lowering the density of ions in the boundary core. This should not be considered as a vacancy structure but merely a device for finding alternative boundary configurations with lower ion density. Since Merkle and Smith have evidence that the boundary, at least in some cases, consists of a lower ion density but does not have a large dilatation, this would seem a likely way to find alternative structures. The only calculation of this type that we have tried so far consisted of removing ions from the Duffy and Tasker structure. We removed a row of ions from the step corner in every unit cell in the structure. This is, of course, exactly equivalent to displacing half the bicrystal relative to the other half. The resulting structure relaxed to the identical structure obtained from the Merkle and Smith structure B (Figure 3). This is not too surprising since the removal of ions makes the starting configuration similar to a dilated Merkle and Smith structure. Of more interest will be structures resulting from the removal of separated ion pairs in the boundary plane. This may reveal more stable boundaries with small dilatations.

3. Temperature Effects

As discussed above, one of the main differences between the experiment and the calculation is one of temperature. The relative stabilities of different boundary structures is dependent on their free energies which vary with temperature. The temperature coefficient of the free energy is called the boundary entropy and there is little information of its magnitude. Studies of micro-crystallites have suggested that this entropy could be very large. Thus it is possible that the Duffy and Tasker structure is not the most stable at elevated temperature. Indeed, it may not be stable at all, restructuring as the lattice expands to a new configuration. These points can be addressed by extending the existing calculations to include temperature effects. We are pursuing this at three levels of sophistication.

1) The Harmonic Approximation involves calculating the lattice and interface phonons from the 0 K structure. It neglects, therefore, the consequences of lattice expansion and the anharmonicity at the lattice sites. It cannot give us any information about interface stability at elevated temperature but can reveal instabilities in the 0 K structure through the emergence of soft phonon modes. These have been used to refine surface structures and hence improve the static search for the equilibrium configuration. The method is simple to apply and can be used to derive thermodynamic functions such as entropy, albeit approximately.

2) Free Energy Minimisation. A direct calculation and minimisation on the free energy of the system will clearly enable us to calculate both the thermodynamic functions and the structural stability of the interface. This is achieved within the quasi-harmonic approximation described in more detail below.

3) Molecular dynamics simulation provides, in principle, the complete classical description of the system. Within this calculation, temperature is directly incorporated and its consequences observed. However, computational expense limits it applicability, particularly in ionic systems with a long-range Coulomb potential. Thus it is generally restricted to rigid ion potentials

(i.e. no ionic polarizability) and to modest periodicities in three dimensions. In addition, extracting the thermodynamic functions is not entirely straightforward.

3.1 FREE ENERGY MINIMISATION

We have used all three methods in this study but we have studied the boundary stability primarily with the free energy minimisation technique. This has been described in detail elsewhere[7]. There are two major limitations of the approach when applied to interface simulations. First, we must assume a three dimensional periodicity. Thus, we had to develop a supercell containing a periodic array of boundaries which, in this case, were separated by approximately 14 A. This separation was chosen by comparing the results at 0 K with calculations on the isolated boundary using the MIDAS code and ensuring that there were negligible differences in the predicted structures and energy. The periodic structure naturally contains two opposite boundaries to ensure no net rotation of the structure. Once the structure at 0K had been determined, the effect of temperature on the grain boundary energy can be calculated. This involves calculating the phonon dispersion relations and the resulting density of states of the pure and defective structures. This is the harmonic approximation (method 1 above). The computational cost for evaluating the phonon dispersion relations for such a large unit cell is high so there has been considerable effort spent on efficient ways to obtain an accurate description of the density of states for a material from a limited sampling of its phonon dispersion spectrum[8-10]. We found that sampling points described by Chadi and Cohen[8] were most appropriate for the cubic non-defective NiO while the points specified by Filippini et al[10] were more suitable for the orthorhombic supercell containing the grain boundaries. Our criterion for choosing the number of points to be sampled was that there should be no further significant change in free energy on increasing the number.

We first used the harmonic approximation to calculate the grain boundary free energy as a function of temperature. As stated above, this approach does not include the influence of temperature on the grain boundary structure. Our second approximation in the free energy minimisation technique is to use the quasi-harmonic method. Now lattice expansion is explicitly allowed for since the phonon frequencies vary with lattice volume. Thus the method for calculating the structure at a given temperature is simply to modify the cell dimensions until the free energy is minimised. In the case of a non-cubic material, the cell dimensions must be adjusted anisotropically. On each adjustment of the cell volume, the atom positions are relaxed to remove mechanical strain. This ensures that each atom is at a potential energy minimum thereby reducing the possibility that an atom will be moved to a position where the quasi-harmonic approximation will break down.

3.2 LATTICE EXPANSION

Previous experience has shown that the free energy minimisation method and the quasi-harmonic approximation yields very similar results to molecular dynamics in ionic systems. This is due to the deep potential wells created by the Coulomb potential. However, although the lattice expansion is treated in this approximation, the anharmonic contribution to the lattice phonons and hence the entropy is neglected. A check on the likely errors can be achieved through a direct comparison of the lattice expansion for the non-defective crystal calculated both by lattice statics (free energy minimisation) and molecular dynamics. We have made this comparison for NiO and the results are given in Table 1.

Temperature (K)	Lattice Statics primitive cell (A^3)	Molecular Dynamics primitive cell (A^3)
0	18.05	18.10
500	18.41	18.38
1000	18.66	18.82
1500	18.95	18.91
2000	19.28	19.32

Table 1: A comparison of lattice expansion with temperature for non-defective NiO, calculated with the quasi-harmonic approximation and by molecular dynamics.

The results show that the two methods are in very close agreement for this case and we can be confident that the lattice statics, free energy minimisation will not be seriously in error.

3.3 BOUNDARY STRUCTURE

We have applied the free energy minimisation technique to examine the stability of the Duffy and Tasker structure. In raising the temperature from 0 K to 2000 K, there is no discernable alteration in the structure apart from the inevitable lattice expansion. This result is similar to that obtained by Kalonji[11] who applied molecular dynamics to the Tasker and Duffy structure for the Σ5 twist boundary in NiO. She also found that the structure was stable up to the melting temperature. This, again, is due to the depth of the potential wells created by the Coulomb potential in an ionic system. It produces exceptionally stable lattice configurations. The dilatation across the boundary was observed to increase from 1.1 A at OK to 1.3 A at 2000 K.

We have not yet studied the new boundary configuration at elevated temperature but we have no reason to doubt that it will be similarly stable. Furthermore, the likely increase in the 0 K dilatation of 0.15 A will increase the correspondence with Merkle and Smith's observation.

3.4 BOUNDARY ENTROPY

The remaining important point is whether the Duffy and Tasker structure remains more stable than the alternative structure at high temperature. The free energies of both boundaries should be calculated as a function of temperature within the quasi-harmonic approximation. We present here the results of an harmonic calculation of the free energy for both structures. This gives a good estimate of the relative entropies but the anharmonic corrections are likely to increase the entropy values somewhat. The results are shown in Figure 6. The first point to note is that the calculated entropies are very small. Indeed, if a rigid ion model is used rather than the shell model applied here, the entropy of the boundary is negligible. As it is, the entropy for the Duffy and Tasker structure is approximately an order of magnitude less than for the free surface. Despite speculation that the boundary entropy might be large, this is perhaps not unreasonable. The entropy is a difference between that for the defective and non-defective lattices. The boundary can be considered as two surfaces pushed together and so the entropy should be between that of the surface and that of the bulk (which is zero in the difference defined above). Studies of defect energies at these symmetrical boundaries found that the diversity of sites meant that some were raised in energy while others were lowered. Thus, we may expect some phonon frequencies to be increased while others will be lowered. The entropy, which is a sum over such frequencies, may therefore be little perturbed from the bulk value as is seen in these calculations.

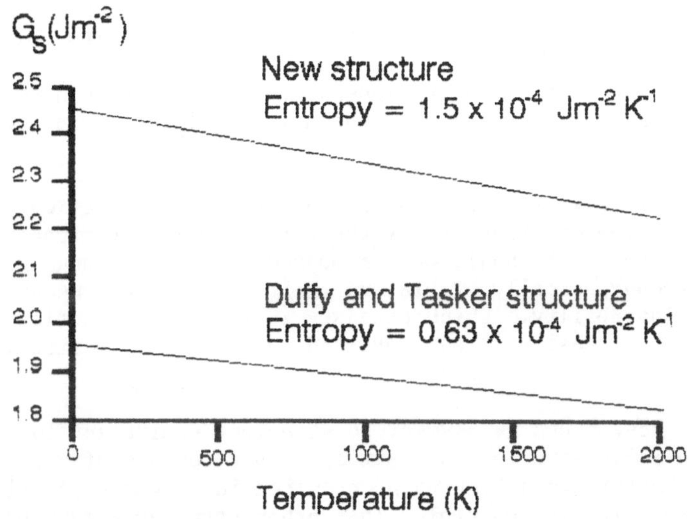

Figure 6: Calculated free energies, G_s, for two possible structures of the $\Sigma 5$ grain boundary in NiO.

The second point is that the entropy of the new structure is indeed higher than for the Duffy and Tasker structure. However, the difference is not sufficient to make it more stable at any reasonable temperature. We are left with the conclusion that within our calculation the Duffy and Tasker structure is the most stable at all temperatures. The features of the two structures are summarized in Table 2.

Boundary	Energy (Jm^{-2})	Entropy ($Jm^{-2}K^{-1}$)	Dilatation (A)
Duffy and Tasker Structure	1.95	0.63×10^{-4}	1.1 - 1.3 (0 - 2000 K)
New Structure	2.45	1.5×10^{-4}	0.46 (0 K)

Table2: Summary of boundary parameters for two stable structures of the Σ5 (310) tilt boundary in NiO.

4. Conclusion

1. We have now found two stable grain boundary structures for the Σ5, (310) tilt boundary in NiO. One is the original Duffy and Tasker structure while the other is a new configuration based on the observation of Merkle and Smith.

2. The new structure consists of distorted hexagons in the boundary core and has a lattice dilatation consistent with observation.

3. It has a significantly higher internal energy than the Duffy and Tasker structure and does not have the lowest free energy even at 1400 K.

4. If this new structure corresponds to the observation, it leaves open the question of why the Duffy and Tasker structure is not observed. It should be remembered that grain boundaries are not thermodynamically stable and their energies are measuring degrees of instability. Which boundary is formed in practice may depend on how easily it can move and anneal out of a structure or on other kinetic factors.

5. We have found no asymmetric structure equivalent to the Merkle and Smith structure A. Attempts to relax a configuration from an asymmetric starting point resulted in a symmetric final structure. It is possible that other structures may be found by reducing the number of ions in the boundary region.

6. The calculated grain boundary entropy for these structures is very low (about an order of magnitude less than for a free surface). This suggests that low temperature calculations are an adequate description of the boundaries even at higher temperature.

Acknowledgment

This work has been carried out as part of the Underlying Research Programme of the U.K. Atomic Energy Authority.

References

1 D.M. Duffy and P.W. Tasker, *Phil Mag.* **A47** 817 (1983)
2 D. Wolf, *J. Phys (Paris) Colloq.* **46** C4-197 (1985)
3 A. Atkinson, R.I. Taylor and A.E. Hughes, *Phil Mag.* **A45** 823 (1982)
4 K.L. Merkle and D.J. Smith, *Phys Rev Lett,* **59** 2887 (1987)
5 M.J.L. Sangster and A.M. Stoneham, *Phil Mag.* **B43** 597 (1981)
6 P.W. Tasker and D.M. Duffy, *Phil Mag.* **A47** L45 (1983)
7 S.C. Parker and G.D. Price, in *Advances in Solid State Chemistry* ed. C.R.A. Catlow, (J.A.I.Press; 1988)
8 D.J. Chadi and M.L. Cohen *Phys Rev* **B8** 5747 (1973)
9 A. Baldereschi, *Advances in Physics* **20** 609 (1973)
10 G. Filippini, C.M. Gramacciolli, M.Simonetta and G.B. Suffritti, *Acta Cryst.* **A32** 259 (1976)
11 G. Kalonji, *Proc of Workshop on Interfaces '87* ,Lake Placid, (1987)

III. Transport Properties

OXYGEN SELF DIFFUSION IN SYNTHETIC RUTILE UNDER HYDROTHERMAL CONDITIONS

P.F. Dennis* and R. Freer**
*Dept. Geological Sciences, University College London, Gower Street, London, WC1 6BT, U.K.
**Materials Science Centre, University of Manchester/UMIST, Grosvenor Street, Manchester, M1 7HS, U.K.

ABSTRACT. Oxygen self diffusion coefficients in a synthetic rutile crystal have been measured under hydrothermal conditions at 100MPa total water pressure in the temperature range 700-1100°C. The diffusion coefficients are lower than the results from dry gas studies would predict. The results can be represented by a linear Arrhenius relationship having $D_o(m^2s^{-1})$ of 2.4×10^{-12} and ΔH of 172.5kJ mol^{-1}. The experiments are interpreted in terms of a defect model involving dissolution of water in rutile as substitutional hydroxyl defects on oxygen lattice sites.

1. INTRODUCTION

Rutile is one of the most extensively studied binary metal oxides as regards point defects, defect structure and chemical and physical properties, including oxygen self-diffusion (see reviews by Kofstad [1] and Matzke [2]).

The most complete diffusion data are due to Haul and Dümbgen [3,4] and Haul et al [5]. The results, determined using bulk exchange techniques, for TiO_2 powders annealed in O_2 are well described by a single Arrhenius relationship covering the temperature range 710-1300°C:

$$D_{ox} = 2.0 \times 10^{-7} \exp(-251 \pm 6.3\text{kJ mol}^{-1}/RT)\,m^2s^{-1} \qquad (1)$$

Subsequent single crystal studies using direct ^{18}O isotope tracer profiling methods (proton activation [6,7] and secondary ion mass spectrometry, SIMS, [8]) have yielded results substantially in agreement with this relationship. These studies have also reported a small diffusion anisotropy with $D_{\perp c}/D_{//c} \simeq 1.5$ at 1620K [4] and 1353K [8]. Haul and Dümbgen [4] further observed no change in D_{ox} with oxygen activity between 10^5 and 1.3×10^{-1} Pa at 897°C. This result is compatible with extrinsic diffusion behaviour due to the relatively high Al_2O_3 content (100-200 ppm) of their diffusion samples and suggests an impurity controlled vacancy diffusion mechanism. Oxygen vacancies are introduced by the reaction (using Kröger-Vink

J. Nowotny and W. Weppner (eds.),
Non-Stoichiometric Compounds Surfaces, Grain Boundaries and Structural Defects, 353–362.

notation),

$$Al_2O_3 = 2Al_{Ti}' + V_O^{\cdot\cdot} + 3O_O \quad (2)$$

This model is supported by the study of Arita et al [8], who recorded an increase in D_{ox} on doping rutile with Cr_2O_3.

Preliminary experiments carried out in water saturated air [4] show that the oxygen self-diffusion coefficient is slightly reduced (by a factor <2) at both 815° and 897°C. The cause of this reduction is not understood, though it was suggested by Haul and Dümbgen [4] that water dissolves in the rutile lattice according to the reaction:

$$O_O + V_O^{\cdot\cdot} + H_2O = 2OH_O^{\cdot} \quad (3)$$

The net effect of incorporating hydroxyl defects then is to reduce the oxygen vacancy concentration and thus the oxygen self-diffusion coefficient. Infrared spectroscopic studies do indicate that the dominant hydrogen speciation in rutile is OH^- [9]. At higher water activities (≃ 100MPa) and 1050°C, Freer and Dennis [10] reported an oxygen diffusion coefficient of $3.2 \times 10^{-19} m^2 s^{-1}$ for a natural crystal of rutile. This is about 100x slower than the dry gas results at the same temperature [3-8]. Whilst this observation is qualitatively consistent with equation (3), our understanding of the role of water in the defect structure of rutile, and other oxides, is too limited to uniquely define this as a correct model.

In order to contribute to a better understanding of the diffusion mechanism for oxygen in rutile, and the role of water-related defects, we have determined oxygen self-diffusion coefficients as a function of temperature (700-1100°C) under hydrothermal conditions at 100MPa total pressure. From a single synthetic crystal prepared fragments were isotope exchange annealed with water enriched in ^{18}O, and the tracer diffusion profiles subsequently determined using SIMS.

2. EXPERIMENTAL PROCEDURE

2.1 Samples and Sample Preparation

The samples used in this study were prepared from a large synthetic crystal. This was in the form of a colourless, transparent plate (approximately 10 x 4 x 1mm), cut from a boule grown by Dr. D. Jones of the Centre for Materials Science, University of Birmingham. The principal impurity contents are: Si($1320/10^6$ Ti); Mg($400/10^6$ Ti); Al($160/10^6$ Ti), and; Fe($100/10^6$ Ti).

Small pieces (2 x 2 x 1mm) were cut from the parent crystals, with the large faces oriented perpendicular to the c-axis. These were then polished using SiC and diamond paste down to 1μm.

2.2 Hydrothermal Isotope Exchange Experiments

Charges for the diffusion experiments consisted of a crystal fragment sealed by carbon arc welding with approximately 8mg of water enriched in ^{18}O ($^{18}O/(^{18}O + ^{16}O)$ ≃ 40%) in 3mm O.D. Pt tubes. The capsules were run in internally heated, gas medium pressure vessels at temperatures to 1100°C and a pressure of 100MPa. Anneal durations ranged from approximately 23 hours at 1100°C to 70 hours at 700°C.

The oxygen fugacity of all the runs is thought to be close to the natural Ni-NiO buffer of the pressure vessels. A summary of the run conditions is given in Table 1. Heating and cooling periods from 100°C below the desired run temperatures are less than 1% of the total anneal durations. Reported temperatures are believed to be accurate to ±5°C, and pressures to ±50MPa. At the termination of the run the charges were checked for leaks and the presence of excess water.

2.3 SIMS Analysis

Oxygen isotope compositions as a function of depth beneath the sample surfaces were determined by SIMS ion-microprobe analysis. The instrument used in this study was an Atomika A-DIDA II quadrupole equipped ion-microprobe [11]. Full details of the technique are given in Freer and Dennis [10] and Dennis [12].

SIMS operating parameters during the analyses were as follows: the primary ion beam was mass filtered $^{40}Ar^+$, accelerated to 10keV, with a maximum total curent of 150nA focussed into a 50μm spot at the sample surface; to ensure flat bottomed craters and improve depth resolution, by eliminating contributions from the crater edge to the determined profiles, the beam was rastered over an area up to 350 x 350μm, with the data acquisition electronics gated to accept secondary ions from the central 18% of the crater floor. Depending on the size of the rastered area, sputter rates ranged from 8Å sec^{-1} to 1.5Å sec^{-1}.

Charge compensation of the primary ion beam was achieved using an electron flood gun operated at 250-500eV energy and 10-20mA total current.

Negative secondary ions at m/e equal to 16, 18 and 47, corresponding to the species $^{16}O^-$, $^{18}O^-$ and $^{47}Ti^-$, were sequentially monitored by rapid peak switching. Count rates for the total oxygen signal exceeded 1 x 10^5cps. No corrections were made to the raw data, with maximum background count rates less than 10 cps. ^{18}O isotope abundances recorded in the tails of the diffusion profiles and on unexchanged samples are 0.2%, in agreement with natural abundances, and indicate that there was no $H_2{}^{16}O$ mass spectral interference with ^{18}O.

Sputtered craters were examined optically using interference contrast reflected light, and their depths measured using a Talystep. The determined crater depths are accurate to ±0.05μm or better (Table 1).

2.4 Computation of Diffusion Coefficients

In the experiments self-diffusion of the oxygen isotope may be modelled by transport into a semi-infinite medium from a fluid phase held at a constant isotropic composition. The surface concentration of ^{18}O on the crystals always closely approached that of the initial starting fluid ($^{18}O/(^{18}O + ^{16}O) \simeq 40\%$), indicating a rapid phase

Table 1. Summary of experimental conditions and results.

Sample	T(°C)	$t(x10^{-5}s)$	d(μm)	$D(m^2s^{-1})$
OX104B	700	2.5290	0.34±0.02	$1.49x10^{-21}$
OX109B	800	2.5368	0.35±0.05	$9.40x10^{-21}$
OX107B	900	2.3754	0.72±0.05	$4.50x10^{-20}$
OX105B	1000	1.5384	0.85±0.03	$1.64x10^{-19}$
OX108B	1100	0.8160	1.45±0.005	$8.52x10^{-19}$

boundary exchange reaction for the transfer of oxygen.

The general solution to Fick's first law under these boundary conditions is given by [13]:

$$\left[\frac{C_x - C_1}{C_o - C_1}\right] = \text{erfc}\left[\frac{x}{2\sqrt{Dt}}\right] \qquad (4)$$

where C_x, C_o and C_1 are respectively the ^{18}O concentration at a distance x from the crystal surface, in the fluid phase, and at $x = \infty$ in the crystal ($\simeq$ 0.002, the natural composition); t is the hydrothermal anneal time; D is the diffusion coefficient; and erfc is the error function complement.

Plots of erfc^{-1} $[(C_x - C_1)/(C_o - C_1)]$ vs. x give a straight line with slope $1/2\sqrt{Dt}$. An example of a determined diffusion profile is given in Figure 1(a) and (b) for run OX108A (annealed at 1100°C for almost 23 hours). In general diffusion coefficients are believed to be reliable to ±30%.

3. RESULTS

Results for the ^{18}O tracer diffusion coefficients are listed, with a summary of the experimental conditions, in Table 1. The values of D are plotted against reciprocal temeprature in Figure 2. In the temperature range 700-1100°C the results for oxygen diffusion parallel to c-axis of synthetic rutile are well described by a standard Arrhenius relationship:

$$D = 2.41 \times 10^{-12} \exp(-172.5 \pm 23.6 \text{kJ mol}^{-1}/RT) m^2s^{-1} \qquad (5)$$

where R is the gas constant ($8.314 J\ mol^{-1}K^{-1}$) and T is the absolute temperature.

4. DISCUSSION

4.1 Comparison With Other Data.

A clear difference exists between the present results, determined under hydrothermal conditions, and those of previous studies which

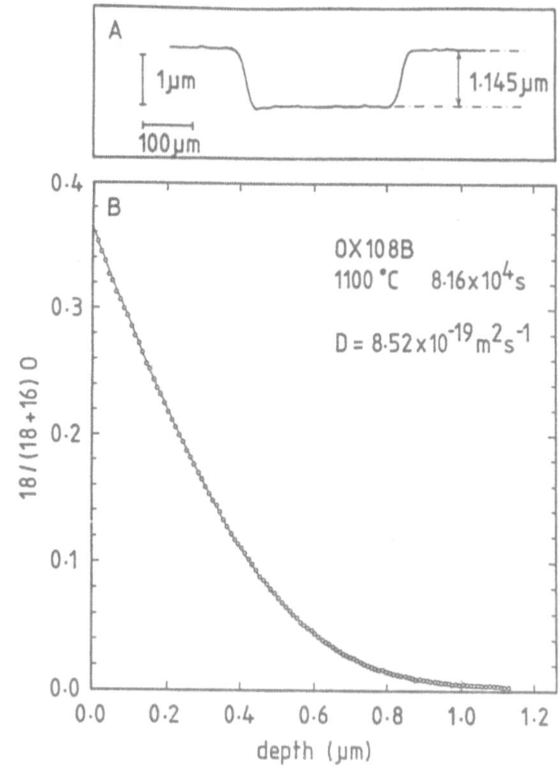

Figure 1. Sample OX108B: (a) Talystep profile of the SIMS sputter eroded crater; (b) The concentration profile of ^{18}O as a function of depth.

were all carried out in pure O_2, at activities in the range 3×10^{-1}-10^{5}Pa [3-8]. These results are compared in the Arrhenius diagram (Figure 3), where a marked reduction in the oxygen diffusivity is observed when rutile is annealed in water at moderate to high pressures (≃ 100MPa), and at temperatures greater than 600-700°C. For example, at 1100°C the oxygen self-diffusivity reported here is a factor of ~100 lower than the dry gas results at comparable temperatures. This result confirms the preliminary work of Haul and Dümbgen [4] who noted a small decrease in D_{ox} (by a factor <2) on annealing TiO_2 in water saturated air at 815° and 897°C. In the only other study that has been carried out under hydrous conditions, Freer and Dennis [10] reported an oxygen diffusivity of $3.2 \times 10^{-19} m^2 s^{-1}$ at 1050°C and 100MPa $P(H_2O)$ for a natural crystal containing in excess of

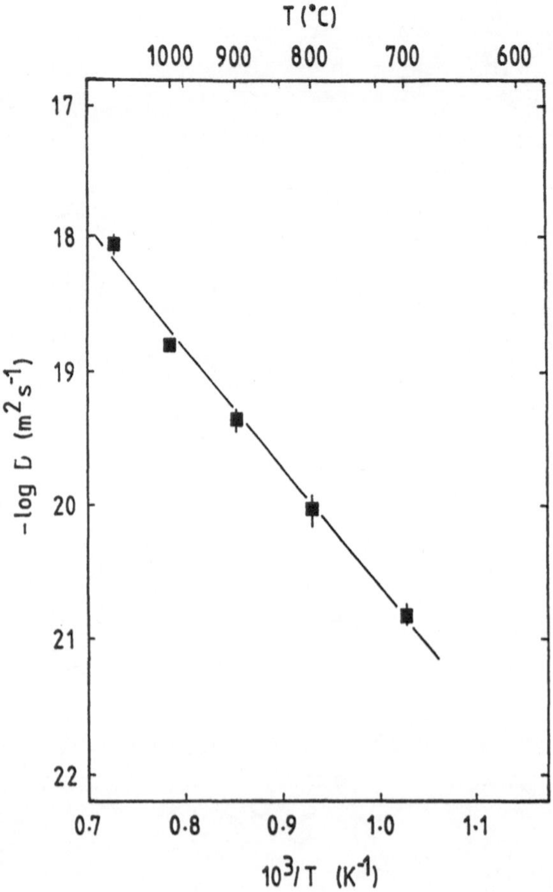

Figure 2. Log D_{ox} as a function of reciprocal temperature. The solid line represents a least squares fit to the data in the temperature interval 700-1100°C.

1.2wt% oxide impurities. This value is comparable with the present data.

Within the temperature range 710-1400°C the activation enthalpies, ΔH, determined for oxygen self-diffusion in rutile under dry O_2 conditions are all in good agreement and lie between 251 and 276kJ mol^{-1} [4,6-8]. The pre-exponential constants, D_o, and hence the absolute magnitudes of the reported diffusion coefficients are also in good agreement (Figure 3). Slight variations in the results, if significant, may be due to different impurity contents of the samples used. For example, Arita et al [8] reported an increase in D_{ox} after doping a sample with 0.08 mol% Cr_2O_3. There have been no systematic studies of the effect of oxygen activity on the activation parameters. Haul and Dümbgen, however, observed no change in D_{ox} with

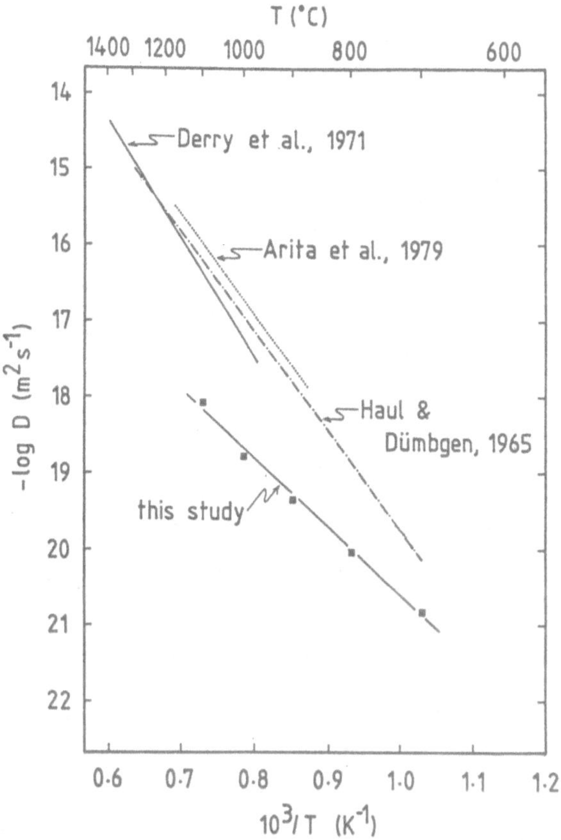

Figure 3. Compilation Arrhenius diagram for the available oxygen self diffusion data in rutile.

oxygen activity between 10^5 and 1.3×10^{-1}Pa at 897°C.

In contrast to the dry gas studies, ΔH for oxygen self-diffusion under hydrothermal conditions is significantly reduced to approximately 170kJ mol^{-1}, equation (5), and Figures 2 and 3. Moreover, the pre-exponential constant for the synthetic rutile is also greatly reduced, c.f. equation (1) and equation (5).

Because of the lower ΔH and D_o for oxygen diffusion under hydrothermal conditions the present results approach the dry gas data towards lower temperatures, intersecting in the range 600-700°C (Figure 3). However, it is not clear whether the two sets of data do cross over to give faster oxygen self-diffusivity under hydrothermal conditions than dry O_2 conditions at the lower temperatures. It is interesting to note that in a compilation of oxygen diffusion data, Freer and Dennis [10] showed that studies performed under dry conditions are generally associated with high activation energies and comparatively low diffusion coefficients, whilst hydrothermal studies are associated with lower activation energies and faster diffusion rates.

4.2 Defect Chemistry

The dominant disorder in rutile is still under active discussion. On the basis of the avilable thermogravimetric data Kofstad [15] proposed that the defect structure simultaneously comprises doubly charged oxygen vacancies, $V_O^{\cdot\cdot}$, and interstitial titanium ions with three and four effective charges, Ti_I^{3+} and Ti_I^{4+} respectively. The model predicts that $V_O^{\cdot\cdot}$ predominate at high oxygen fugacities and low temperatures, whereas Ti_I^{3+} and Ti_I^{4+} are the dominant defects towards the lower limit of the homogeneity range, at higher temperatures and lower oxygen fugacities.

On the basis of the empirical defect equilibria [15], at temperatures below 1700K and near ambient oxygen fugacities (10^5Pa) the intrinsic defect concentrations in rutile are $<10^{-3}$ mol%. Since most commercially available rutile crystals contain $>10^{-3}$ mol% of aliovalent impurities extrinsic defect concentrations are likely to dominate. For the present hydrothermal experiments, run at the Ni-NiO buffer, the maximum intrinsic departure from stoichiometry occurs at the highest temperature (1373K), and is of the order of 3×10^{-2} mol%. This is below the concentration (6.6×10^{-2} mol%) of aliovalent impurities in the synthetic crystal used in this study.

The dry gas data at near ambient oxygen activities have readily been interpreted in terms of a vacancy diffusion mechanism in an extrinsic defect regime [2,4,8]. Oxygen vacancies are introduced by reactions of the form,

$$Al_2O_3 = 2Al_{Ti}' + 3O_O + V_O^{\cdot\cdot} \quad (6)$$

As an initial model for diffusion under hydrothermal conditions we assume that a vacancy mechanism still dominates, and that water dissolves in rutile according to the reaction [4],

$$O_O + V_O^{\cdot\cdot} + H_2O = 2OH_O^{\cdot} \quad (7)$$

The substitutional hydroxyl groups are consistent with the available IR spectroscopic results [9], and the interpretation of enhanced electronic conductivity in rutile on annealing in hydrogen atomspheres [16]. A similar dissolution mechanism has also been proposed for other nonstoichiometric oxides [17,18,19].

5. CONCLUSIONS

Experimental studies of oxygen diffusion in rutile (parallel to the c-axis) under hydrothermal conditions at 700-1100K, and Ni-NiO buffer, yield an activation energy which is significantly less than that obtained under dry gas conditions, and diffusion coefficients which are a factor 10-100 less than those obtained under dry gas conditions. At temperatures less than 1400K, and oxygen fugacities greater than the Ni-NiO buffer, most commercially available rutile is dominated by extrinsic defects due to aliovalent substituional cations. These cations charge compensate the substitutional hydroxyl defects.

The net effect of this reaction is to reduce the oxygen vacancy concentration and thereby reduce the oxygen self-diffusivity.

We note that a solution type mechanism has been proposed to account for the increased electronic conductivity of rutile when annealed in hydrogen [16]. Further, substitutional hydroxyl ions have been observed to influence the defect chemistry of Y_2O_3 [16], and ThO_2 [19].

REFERENCES

1. Kofstad, P., Nonstoichiometry, Diffusion and Electrical Conductivity in Binary Metal Oxides, (Wiley Interscience, New York), 382 pp, (1972).
2. Matzke, Hj., 'Diffusion in nonstoichiometric oxides', in Nonstoichiometric Oxides, ed. O. Toft Sorensen, (Academic Press, New York), 155-232, (1981).
3. Haul, R., and Dümbgen, G., Z. Elektrochem., **66**, 636-641, (1962).
4. Haul, R., and Dümbgen, G., 'Self diffusion of oxygen in rutile crystals', J. Phys. Chem. Solids, **26**, 1-10, (1965).
5. Haul, R., Just, D., and Dümbgen, G., in Reactivity of Solids, Proc. of the 4th International Symposium on the reactivity of solids, (Elsevier, Amsterdam), (1960).
6. Derry, D.J., Lees, D.G., and Calvert, J.M., 'A study of oxygen diffusion in titanium dioxide', Proc. Brit. Ceram. Soc., **19**, 77-83, (1971).
7. Gruenwald, T.B., and Gordon, G., 'Oxygen diffusion in single crystals of titanium dioxide', J. Inorg. Nucl. Chem., **33**, 1151-55, (1971).
8. Arita, M., Hosoya, M., Kobayashi, M., and Someno, M., 'Depth profile measurement by secondary ion mass spectrometry for determining the tracer diffusivity of oxygen in rutile', J. Amer. Ceram. Soc., **62**, 443-446, (1979).
9. Cathcart, J.V., Perkins, R.A., Bates, J.B., and Manley, L.C., 'Tritium diffusion in rutile', J. Appl. Phys., **50**, 4110-18, (1979).
10. Freer, R., and Dennis, P.F., 'Oxygen diffusion studies, I: A preliminary ion microprobe investigation of oxgyen diffusion in some rock forming minerals', Mineral. Mag., **45**, 197-192, (1982).
11. Wittmaack, K., 'DIDA - A multipurpose scanning ion microprobe', in Proceedings 8th International Conference on X-ray Optics and Microanalysis, ed. by D. Beaman, R. Ogilvy, and D. Wittry, (Science, Princeton N.J.), 32-38, (1978).
12. Dennis, P.F. 'Oxygen self-diffusion in quartz under hydrothermal conditions', J. Geophys. Res. (red), **89**, 4047-57, (1984).

13. Crank, J., The Mathematics of Diffusion, 2nd ed., (Oxford University Press, New York) 414 pp. (1975).
14. Colby, J.W., 'Ion microprobe mass analysis', in Practical Scanning Electron Microscopy, ed. by J.I. Goldstein, and H. Yakowitz, (Plenum, New York) 529-572, (1975).
15. Kofstad, P., 'Note on the defect structure of rutile (TiO_2)', J. Less Common Mat., **13**, 635-638, (1967).
16. Hill, G.J., 'The effect of hydrogen on the electrical properties of rutile', Br. J. Appl. Phys., Ser. 2, **1**, 1151-62, (1968).
17. Stotz, S., and Wagner, C., Ber. Bunsenges. Physik. Chem., **70**, 781-788, (1966).
18. Norby, T., and Kofstad, P., 'Electrical conductivity and defect structure of Y_2O_3 as a function of water vapour pressure', J. Amer. Ceram. Soc., **67**, 786-792, (1984).
19. Shores, D.A. and Rapp, R.A, 'Hydrogen ion (proton) conduction in thoria base solid electrolytes', J. Electrochem. Soc., **119**, 300-305, (1972).

SHORT-CIRCUIT DIFFUSION IN α-Al_2O_3.

E. MOYA and F. MOYA
Université de Droit, d'Economie et des Sciences
d'Aix-Marseille III
Faculté des Sciences et Techniques - Centre St Jérôme
Avenue Escadrille Normandie Niemen
13397 - MARSEILLE CEDEX 13 - FRANCE

ABSTRACT. Results concerning diffusion in αAl_2O_3 are reviewed, compared and discussed in terms of lattice defects, purity, doping, grain-boundary structure and chemistry. Some of the difficulties encountered in determining diffusion parameters are pointed out. Attention is paid to short-circuit diffusion, as penetration in the volume of αAl_2O_3 is negligible in most cases. Recent data for the silver-αAl_2O_3 system providing all parameters relative to bulk and dislocation and grain boundary diffusion, are analysed to gain a better understanding of the diffusion mechanisms involved in this system and, from analogy, in cationic diffusion in α-Al_2O_3.

1. INTRODUCTION

This paper concerns diffusion in αAl_2O_3 . Owing to its high melting point (2045°C) and low electrical conductivity, this oxide is widely used as refractory and insulating material up to very high temperatures. Consequently, knowledge of diffusion parameters is important for the understanding of the various processes which can occur at high temperatures : sintering, creep, grain growth, or diffusion of foreign elements present in the atmosphere or bonded to the ceramic. Bulk diffusivity in this oxide is rather slow compared with diffusivity in metals so dislocations and grain boundaries may be the main contributors to transport of matter.

The purpose of this paper is to look at the present state of knowledge concerning these short-circuit diffusion mechanisms, considering the effects of various parameters such as desorientation, impurity content, doping and segregation. Recent results obtained by the authors in the Ag-Al_2O_3 system will be more detailed as they give a whole set of data about bulk, dislocation and grain boundary diffusion.

J. Nowotny and W. Weppner (eds.),
Non-Stoichiometric Compounds Surfaces, Grain Boundaries and Structural Defects, 363–385.

2. BULK DIFFUSION

2.1. Theoretical values for defect formation and migration energies.

Determination as well as interpretation of short-circuit diffusion data involves knowledge of bulk diffusion : αAl_2O_3 is characterized by very small concentrations of intrinsic point defects. This is linked to high Schottky and Frenkel disorder formation energies, computed values of which are given in Table I, along with experimental values estimated by Kröger and co-workers from conductivity measurements in α-Al_2O_3.

TABLE I

Defect energies per defect in αAl_2O_3

ΔH_f = energy of formation (kJ/mol)

ΔH_m = energy of migration (kJ/mol)

	DIENES et al (1)	CATLOW et al (2)	MACKRODT (3)	El-AIAT and KROGER (4)
1/5 ΔH_f Schottky	549	402/495	482	630 ± 135
1/2 ΔH_f anion Frenkel	674	365/796	800	530 ± 116
1/2 ΔH_f cation Frenkel	963	502/682	674	790 ± 170
ΔH_m (V_{Al}''')	Jump A→1* 366 Jump A→2* 640			
ΔH_m $(V_O^{\bullet\bullet})$	279			
ΔH_m $(Al_I^{\bullet\bullet\bullet})$	Jump I→I'* 462			

* See Fig. 1 for signification of jumps A→1, A→2 , I→I'

Kröger et al's experimental values are not conclusive concerning which disorder mechanism is dominant in α-Al_2O_3 (5), whereas computed values suggest that the Schottky disorder is dominant.

Both results lead to the conclusion that the concentrations of intrinsic defects would be extremely small, even close to melting point (about 10^{-13}). Impurities are always present at higher concentrations, so that it is assumed that an intrinsic behaviour can never be observed. Diffusion data are then interpreted by taking into account extrinsic defects, the concentration of which is controlled by the aliovalent impurity amount. For the cation, diffusion activation energy is then equal to migration energy of the defect involved ; computed values of which are given in Table I. To understand the nature of the jump, indicated as A→1, A→2, I→I' a schematic view of the structure of α-Al_2O_3, neglecting the distorsion of the lattice, is given on Fig. 1. It should

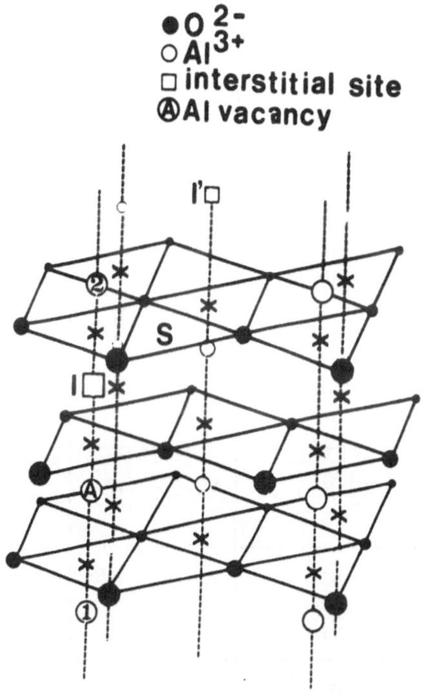

Figure 1 . Schematic representation of α-Al_2O_3 structure from (1).

be noted that the interstitial sites I and I' are rather similar in size to substitutional Al sites, since the structure of alumina may be described as a hexagonal close packing of O^{2-} ions, where 2/3 of octahedral sites are occupied by Al^{3+} ions, and 1/3 of them are left vacant to preserve the electroneutrality of the crystal.

Before attempting to link experimental results to these theoretical calculations, it is important to point out the difficulty of diffusion measurements in Al_2O_3.

Owing to the low bulk diffusivity, penetration depths obtained with reasonable diffusion times are excessively small. With diffusion coefficients as low as 10^{-18} to 10^{-20} m^2/s at 1500°C,diffusion anneals performed during $t = 10^6 s$ (about 12 days) would lead to mean penetration depths of 10^{-6} to 10^{-7}m, and even smaller at lower temperatures. These measurements imply the use of refined sectioning and analysing techniques.

2.2. Experimental diffusion data.

Oxygen diffusion parameters are presented in Table II. Small diffusion coefficients obtained by indirect measurements at low temperatures (11) were in good agreement with the tracer data except that obtained by Reed and Wuensch (9).

TABLE II Oxygen bulk diffusion in α-Al_2O_3

Authors (references)	Year	Method	Temp. range (°C)	D_o (m^2/s)	Q (kJ/mol)
OISHI and KINGERY (6)	1960	Gas-solide isotope exchange	1450-1780	0.19*	636 ± 105
OISHI et al (7)	1980	"	1504-1776	0.112	648
OISHI et al (8)	1983	"	1500-1700	$5.62\ 10^{-2}$	665
REED and WUENSCH (9)	1980	SIMS	1585-1840	64	787 ± 30
REDDY and COOPER (10)	1982	Nuclear reaction analysis	1477-1677	$2.77\ 10^{-2}$	615 ± 41
LAGERLOF et al (11)	1983	Dislocation loop anneals	1200-1500	$6.8\ 10^{-4}$	587
D. PROT and C. MONTY (work in progress-private communication)	1988	SIMS	1400-1700	$1.3\ 10^{2}$	694

* Recalculation by OISHI et al (8) gave $D^o = 5.6\ 10^{-3}\ m^2/s$

Cation diffusion data are presented in Table III.

TABLE III

Cation bulk diffusion in α-Al_2O_3

Element	Authors (references)	Year	Method	Temp. range (°C)	D_o (m^2/s)	Q (kJ/mol)
Al	PALADINO and KINGERY (12)	1962	Tracer/ polycrystal	1670-1905	$2.8\ 10^{-4}$	477 ± 63
Al	CANNON et al (13)	1980	Diffusional creep/poly-crystal	1200-1700	13.6	577
Cr	LESAGE et al (14)	1983	Tracer/single crystal	1200-1700	$6.9\ 10^{-11}$	265
Fe	LESAGE et al (14)	1983	Tracer/single crystal	1200-1700	$3.6\ 10^{-9}$	300
Ni	LESAGE et al (14)	1983	Tracer/single crvstal	1200-1700	$2.5\ 10^{-10}$	280
Ag	BADROUR et al (15)	1986	Tracer/single and polycrys-tal	800-1500	$2.4\ 10^{-4}$	331

* Direct aluminium self diffusion measurements were only performed by Paladino and Kingery (12) in the high temperature range.

* Cr, Fe and Ni (14) presented lower activation energies than Al but comparable kinetics at 1200°C whereas Ag (15) exhibited lower activation energy but higher diffusion coefficients.

* Ag diffusivity was measured at low temperatures (down to 825°C) due to the high diffusion rate of this element (15).

Bulk diffusion data are plotted as Arrhenius curves in Fig. 2.

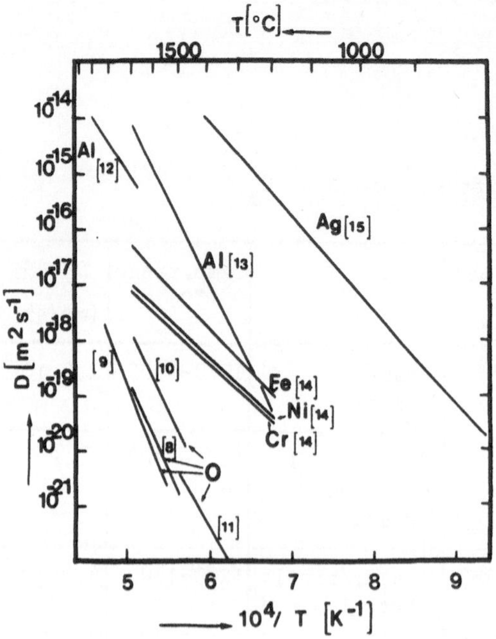

Figure 2 . Arrhenius curves for bulk diffusion in α-Al_2O_3.

2.3. Diffusion mechanisms

Some experimenters have investigated the effect of purity and doping to elucidate diffusion mechanisms. To understand the following discussions, fundamental defect reactions must be written

SCHOTTKY disorder :

(I) $\text{null} \rightleftharpoons 2\,V_{Al}^{'''} + 3\,V_{O}^{\cdot\cdot}$ $\qquad K_S = \left[V_{Al}^{'''}\right]^2 \cdot \left[V_{O}^{\cdot\cdot}\right]^3$

Anion FRENKEL disorder :

(II) $O_o^x \rightleftharpoons O_i^{''} + V_O^{\cdot\cdot}$ $\qquad K_{AF} = \dfrac{\left[O_i^{''}\right]\left[V_O^{\cdot\cdot}\right]}{\left[O_o^x\right]}$

Cation FRENKEL disorder :

(III) $Al_{Al}^x \rightleftharpoons Al_i^{\cdot\cdot\cdot} + V_{Al}^{'''}$ $\qquad K_{CF} = \dfrac{\left[Al_i^{\cdot\cdot\cdot}\right]\left[V_{Al}^{'''}\right]}{\left[Al_{Al}^x\right]}$

Acceptor dopant (for example MgO)

(IV) $2\ MgO \xrightarrow{Al_2O_3} 2\ Mg'_{Al} + 2\ O^x_o + V^{\bullet\bullet}_O$

(V) $3\ MgO \xrightarrow{Al_2O_3} 2\ Mg'_{Al} + 3\ O^x_o + Mg^{\bullet\bullet}_i$

(VI) $3\ MgO + Al^x_{Al} \xrightarrow{Al_2O_3} 3\ Mg'_{Al} + 3\ O^x_o + Al^{\bullet\bullet\bullet}_i$

Donor dopant (for example TiO_2)

(VII) $3\ TiO_2 \xrightarrow{Al_2O_3} 3\ Ti^{\bullet}_{Al} + 6\ O^x_o + V'''_{Al}$

(VIII) $2\ TiO_2 \xrightarrow{Al_2O_3} 2\ Ti^{\bullet}_{Al} + 3\ O^x_o + O''_i$

2.3.1. Oxygen diffusion. It is generally assumed that oxygen diffusion takes place by oxygen vacancies $V^{\bullet\bullet}_O$. The activation energy for an intrinsic mechanism would be :

$$Q = \Delta H_{fS}/5 + \Delta H_{m\ V^{\bullet\bullet}_o} \cong 761 \text{ to } 838 \text{ kJ/mol}$$

where ΔH_{fS} is the enthalpy of formation of a Schottky quintet.

Oishi, Ando and Matsuhiro (16) observed no influence of purity on oxygen diffusion coefficients. Reddy and Cooper (10) using nuclear reaction analysis found a decrease in oxygen diffusivity of a factor 5 on 800 ppm Ti-doped alumina samples. Lagerlof et al (17) observed a similar decrease in 600 ppm TiO_2 doped sapphire by dislocation loop anneals. Recent work by Haneda and Monty (18) in 1000 ppm Ti-doped alumina samples has given a similar oxygen diffusion coefficient. The agreement between these three results (Figure 3) is satisfactory given the differences between methods, sample purities and range of temperatures.

This could be accounted for by an extrinsic mechanism resulting from reactions (VII) or (VIII). Both of these reactions result in a decrease in the oxygen vacancy concentration. Considering only the Schottky disorder, the following may be written :

$$\left[Ti^{\bullet}_{Al}\right] = 3\left[V'''_{Al}\right]$$

and according to |I|

$$\left[V'''_{Al}\right]^2 \left[V^{\bullet\bullet}_O\right]^3 = \exp\ (-\Delta G_{fS}/RT)$$

where ΔG_{fs} is the free energy of formation of a Schottky quintet.

Then $$\left[V^{\bullet\bullet}_O\right] = \left(\frac{3}{\left[Ti^{\bullet}_{Al}\right]}\right)^{2/3} \exp\ (-\ \Delta G_{fS}/3RT)$$

The theoretical activation energy deduced from this extrinsic vacancy mechanism would be equal to $\Delta H_{fS}/3 + \Delta H_{mV^{\bullet\bullet}_O} \geqslant 1000$ kJ/mol according to Table I. This value is higher than the experimental acti-

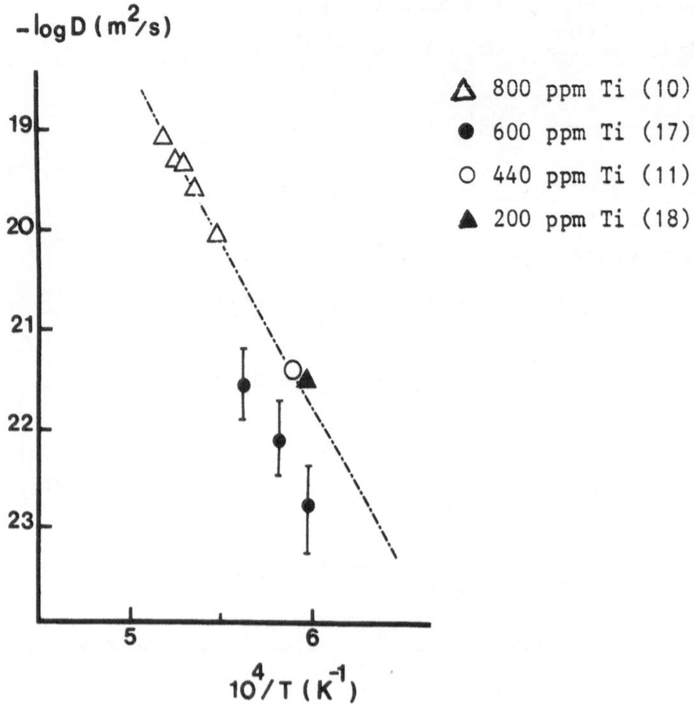

Figure 3. Oxygen self-diffusion coefficients in Ti-doped Al_2O_3.

vation energy ≅ 630 kJ/mol (10). However, Schottky disorder predominance might be ruled out, since in Ti^{4+} doped sapphire Phillips, Mitchell and Heuer (19) found that O_i'' would be the dominant defect, probably associated with $Ti_{Al}^{\bullet}$ in a cluster $(Ti_{Al}^{\bullet}\ O_i''\ Ti_{Al}^{\bullet})^x$. Unfortunately no comparison can be made since migration energies of O_i'' and associated defects do not appear in the literature. In 200 ppm Mg^{2+} doped sapphire Lagerlöf et al (17) found an increase in D of about 80 times at 1300°C. But, when Al_2O_3 was doped with 100 ppm MgO no oxygen diffusion coefficient enhancement was observed by Reddy and Cooper (10) in contrast to was expected. In 30 ppm and 130 ppm Mg doped alumina, Haneda et Monty (18) obtained similar diffusion coefficients. These results seem to indicate that the solubility limit is lower than 30 ppm for Mg at 1800°C. In the same way, it should be noted that dopant insensitive oxygen diffusivity was observed in MgO : in this material high valence dopants would theoretically decrease oxygen diffusion coefficients through the decrease of oxygen vacancy concentration ; however no significant effect was observed by doping MgO with 310 ppm Fe or 2300 ppm Fe (20). It was suggested that diffusion could occur by vacancy pairs $(V_O^{\bullet\bullet}\ V_{Mg}'')^x$, the concentration of which would be higher than that of free oxygen vacancies.

From the foregoing discussion the question of oxygen diffusion mechanisms in αAl_2O_3 still remains to be clarified.

2.3.2. Cation diffusion. In the same way, interpretation of cationic diffusivity is still ambiguous, but certain information about diffusion mechanisms may be given by considering the effect of impurities.

The influence of doping on electrical conductivity and on chemical diffusion has been extensively studied by Huntz and Coworkers (14, 21 to 26). It was found that titanium and yttrium act as donors and increase the cation vacancy concentration V_{Al}'''. For chemical diffusion coefficients, breaks in the Arrhenius curves in Ti-doped and Y-doped alumina were attributed to precipitations of compounds : indeed microstructural studies of yttria-doped alumina showed the presence of $Y_3Al_5O_{12}$ precipitates. These heterogeneities dissolved after heat treatment at T>1600°C. Lesage et al (14) observed a slight increase for Fe diffusivity in αAl_2O_3 doped with 800 ppm and 8000 ppm Cr_2O_3 or 1000 ppm Y_2O_3. This effect suggested an extrinsic vacancy mechanism controlled by a donor character of these impurities according to (VII). However Lesage did not exclude the possibility of diffusion occuring through octahedral innocupied sites (27).

In a recent paper, Koripella and Kröger (28) reanalyzed Lloyd and Bowen's data about diffusion of iron in αAlumina (29). Taking into account their own conductivity results, they attributed diffusion of Fe in Al_2O_3 to $Fe_i^{\bullet\bullet\bullet}$. The same interstitial mechanism, first proposed by Paladino and Kingery (12) for Al self diffusion, would be a particulary suitable interpretation for silver diffusion (15) for the following reasons :

- the bulk diffusion coefficients measured in single crystals and polycrystals were identical, although the impurity contents of the two samples were very different ;
- oxygen pressure did not influence bulk diffusion coefficients ;
- the Al vacancies and Ag atoms (or Ag^+ ions) being negatively charged, would exhibit repulsive interactions ; so a very low diffusivity would be expected by a vacancy mechanism, contrary to our observation :

$$D_{Ag}/D_{Al} \cong 10^4 \text{ to } 10^5$$

In this interpretation the difference between 331 kJ/mol (silver migration energy) and 462 kJ/mol (Dienes et al (1) interstitial migration energy) or 477 ± 63 kJ/mol (Paladino and Kingery (12) self diffusion measurements) could be attributed to the large distorsion induced in the Al_2O_3 lattice by the size of the silver atom or ion. The difference of 130 to 140 kJ/mol could account for high values of silver diffusion coefficients compared to those of Al.

When the interstitial migrating silver atom arrives close to an aluminium vacancy, a jump to this defect cannot be excluded ; the migration energy would be about half the migration energy given by Dienes for the jump A→2 (see Table I and Fig. 1), i.e. 1/2×(640) = 320 kJ/mol. A contribution from this mechanism could perhaps explain the low silver activation energy.

2.4. Comments.

Considering the difficulties encountered in the above interpretations, it must be recalled that discrepancies between activation energies are often compensated by preexponential factors, in experiments which are obviously controlled by the same diffusion mechanism. This compensating law, clearly shown in MgO diffusion experiments (30), suggests to be cautious with interpretations based only on experimental activation energy values calculated from Arrhenius curves.

A second point should be recalled when a defect is considered as the most likely involved in diffusion, that does not mean that this defect is dominant, as diffusion only allows migration of the more mobile species, or more precisely, that of the species presenting the higher (concentration × diffusion coefficient) product to be observed. For example even if $V_O^{\bullet\bullet}$ is predominant in MgO doped alumina, migration could occur by O_i'' if this species migrates faster. In the same way we can imagine that a majority of silver atoms are in substitution in the Al sublattice but that these atoms jump from time to time in interstitial positions ; this would account for the observed diffusivity.

A third important point arises from this overview : not only simple defects may be involved in diffusion mechanisms : complex defects such as cation-anion vacancy pairs, impurity-vacancy complexes, defect clustering (2,3) might take part in diffusion processes, involving lower formation but higher migration energies. From the number of equilibrium reactions, a kind of "buffer effect" would explain that the effect of impurities is much less than expected.

3. SHORT-CIRCUIT DIFFUSION

3.1. Determination of short circuit parameters

Considering a dislocation as a pipe and a grain boundary as a slab, the main parameters for short-circuit diffusion are indicated in Table IV. Most of quantitative data for short-circuit diffusion are deduced from penetration profiles obtained in the so called "type B" kinetic regime

$$\alpha_d\, a \ll \sqrt{Dt} \ll 1/\sqrt{d} \quad \text{or}\ \alpha'\delta \ll \sqrt{Dt} \ll g$$

When these conditions are fulfilled, two part penetration profiles are observed : the first part represents bulk diffusion and the second part contribution from dislocations or grain-boundaries. The second part or "tail" is practically independent of the effective boundary condition at the specimen surface (constant concentration source or instantaneous source)

- For dislocations, Le Claire et Rabinovitch analysis (31) gives the parameter $D_d\, \alpha_d\, a^2$, through the slope of log c = f(y) curves.

- For grain boundaries, using to Le Claire analysis (32), the parameter $P = D'\alpha'\delta$ is usually extracted from the slopes of $\log c = f(y^{6/5})$ curves. However it should be pointed out that the gradient $\partial(\log c)/\ \partial(\eta\beta^{-1/2})$ varies slowly with the parameter $\eta\beta^{-1/2}$ as can be

TABLE IV
Parameters involved in short-circuit diffusion measurements

	Dislocation	Grain boundary
Geometry	a = radius d = dislocation density	δ = width g = grain size
Diffusion coefficient	D_d	D'
Enhancement of diffusivity	$\Delta = D_d/D$	$\Delta = D'/D$
Reduced penetration depth along the y axis	$\eta = y/\sqrt{Dt}$	$\eta = y/\sqrt{Dt}$
Segregation ratio	$\alpha_d = C_d/C$	$\alpha' = C'/C$
Characteristic parameters		$\beta = \dfrac{(\Delta-1)\delta\alpha}{2\sqrt{Dt}}$
Parameter experimentally determined	$D_d\alpha_d a^2$	$P = D'\alpha'\delta$

seen on Fig. 4 plotted from Suzuoka's computed values (33). Sectioning is generally limited to $\eta\beta^{-1/2} < 4$ or 5 corresponding to a decrease of about 2 orders of magnitude for the average concentration.

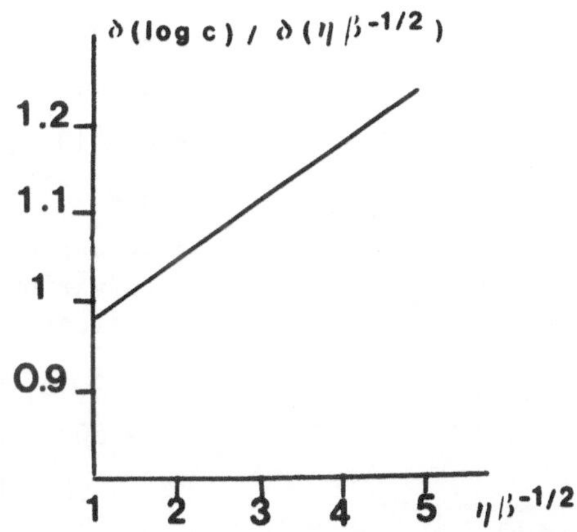

Figure 4. Dependence of $\partial(\log c)/\partial(\eta\beta^{-1/2})$ on $\eta\beta^{-1/2}$, from (33).

So the curves log c = f(y) are almost linear and the parameter β may be extracted by using the appropriate $\partial(\log C)/\partial(\eta\beta^{-1/2})$ value within the range of sectioning (34).

Comparisons have been extensively carried out on analysis of Ag-alumina penetration profiles (34) and it has been checked that the two methods lead to results in complete agreement.

Plotting Log c = f(y) curves enables dislocation and grain boundary penetrations to be compared directly. It can be shown that the ratio of the gradients of these curves is given by :

$$\frac{p_d}{p_l} = \frac{(\partial \log c/\partial\eta)\ \text{dislocations}}{(\partial \log c/\partial\eta)\ \text{grain boundaries}} \cong k\left[\frac{\sqrt{Dt}}{a}\right]^{1/2}$$

with k of the order of 0.3 to 0.6, assuming $D' \cong D_d$ and $\delta \cong a$.

As $\sqrt{Dt}$ is necessarily at least one order of magnitude larger than a, the dislocation curve is steeper than the grain boundary one, every other parameter being equal. In "type B" kinetic regime, a distinction between bulk and short-circuit contributions in the concentration profiles cannot be made if β is not large enough. In Fig. 5 we have plotted theoretical curves calculated for the constant source case with the following values : $\Delta = 10^4$; $\alpha = 1$; $a = \delta = 10^{-9}$ m ; $g = 10^{-5}$ m ; $d = 5.10^{10}$ m^{-2}. It appears that bulk and short-circuit curves become clearly distinct when β > 50 for dislocations and β > 5 for grain boundaries. The parameter β may be increased by decreasing $\sqrt{Dt}$, however the

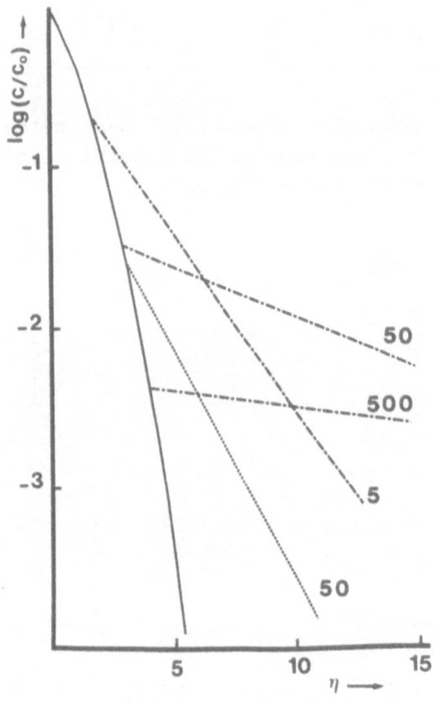

Figure 5. Theoretical concentration profiles.

_____ bulk
..... dislocation
._._._.grain boundary

* constant concentration Co at the surface
* $a = \delta = 10^{-1}$m
* $\alpha = 1$
* $\Delta = 10^4$
* $g = 10^{-5}$m, $d = 5.10^{10}$m^{-2}
* values of β corresponding to various $\sqrt{Dt}$ values are indicated on the curves

β = 5 $\sqrt{Dt} = 10^{-6}$ m

β = 50 $\sqrt{Dt} = 10^{-7}$ m

β = 500 $\sqrt{Dt} = 10^{-8}$ m

lower the $\sqrt{Dt}$ values, the lower the concentrations and the more difficult the measurements. The same problem of detection arises when experiments are carried out on samples with large grains or on bicrystals.

3.2. Experimental data

Experimental data on short-circuit diffusion in alumina are even scarcer than bulk diffusion data.

Influence of dislocations was often proposed to explain some high diffusivities. For example Oishi et al (7) observed a pronounced enhancement of oxygen diffusion when single crystal samples were pre-annealed at 1547°C, but no enhancement for preannealing at 1900°C. In the same way crushed particles of alumina exhibited markedly high diffusion coefficients. This enhancement was presumably due to short-circuit diffusion via dislocation networks created by slicing and polishing in the case of disk samples or crushing in the case of particles. In spite of these observations, there is a lack of quantitative studies on dislocation diffusion. Lesage et al (14) observed that penetration profiles of Fe, Cr and Ni in alumina presented two distinct parts : the high-penetration part of the curves were attributed to a fast diffusion along dislocations However, as the analysis given by Le Claire and Rabinovitch (31) was not applied to these curves, the information they give remains qualitative. In the case of silver diffusion in single crystals of alumina (35), analysis of the second part of the curves resulted in the determination of dislocation diffusion parameters at different temperatures and led to the Arrhenius equation :

$$D_d a^2 \alpha_d \text{ (Ag)} = 4.3\ 10^{-16} \exp(-\ 321\ \text{(kJ/mol)}/RT)\ m^4/s$$

Using Le Claire and Rabinovitch's method (31) again , it was checked that the absolute magnitude of concentrations in the "tail" part of the curves corresponded to a reasonable dislocation density ($\cong 5\ 10^7\ cm^{-2}$). Moreover the slopes of log c = f(y) plots, at a given temperature, were found to be almost independent of anneal time (Fig. 6). This fact characterizes a diffusion along isolated, or randomly dispersed, dislocations (31

For grain boundaries, some data are available ; the most general of them are presented in Table V and plotted as Arrhenius curves on Fig. 7.

For self diffusion direct measurements are still lacking. Oxygen grain boundary diffusivity was deduced from Oishi and Kingery's results on polycrystalline alumina (6). Mistler and Coble (36) related experimental diffusion coefficients to grain boundary diffusion coefficients by the following relation deduced from Hart analysis (38) :

$$D\ \text{(experimental)} = \pi\delta D'/g$$

We must note that creep data were interpreted on the basis of a fast oxygen grain boundary diffusion. This fast diffusion is now supposed to occur by neutral interstitial oxygen O_i^x, as shown by conductivity and creep experiments (39, 40) as well as chemical diffusion measurements (41).

Aluminium diffusivity, assumed to be the rate controlling step, was determined from creep measurements (13).

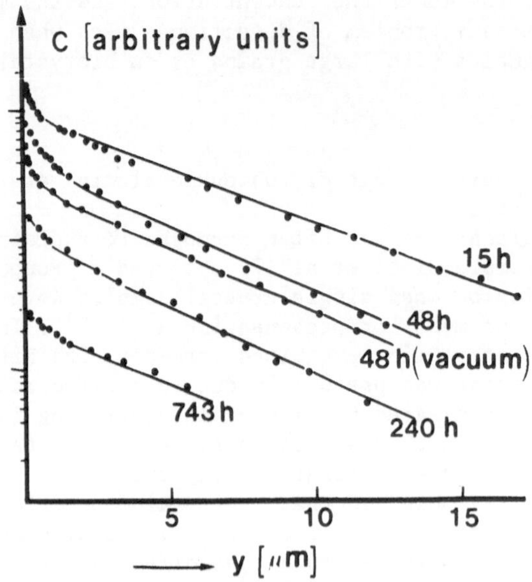

Figure 6. Penetration profiles log c = f(y) for diffusion of Ag along dislocations of single crystal Al_2O_3, at 916°C, for different diffusion times.

TABLE V
Grain boundary diffusion parameters in α-Al_2O_3

Element	Authors (reference)	Year	Method	Temp. range (°C)	$D'_o\ \delta\alpha'$ (m^3/s)	Q (kJ/mol)
O	Oishi and Kingery (6)	1960	Gas-solid Isotope exchange Technique	1450/1780	$1.6\ 10^{-9}$ *	460 ± 63
Al	Cannon et al (13)	1980	Creep	1200/1700	$8.6\ 10^{-10}$	419
Cr	Lagrange et al (37)	1982	Radiotracer	1200/1500	$5\ 10^{-12}$	341
Fe	Lagrange et al (37)	1982	Radiotracer	1200/1500	$4\ 10^{-15}$	212
Ag	Badrour et al (35)	1986	Radiotracer	1000/1460	$9.2\ 10^{-6}$	321

* Value calculated by Mistler and Coble (36) from apparent lattice diffusion coefficients determined by Oishi and Kingery in polycrystalline alumina with 25 μm grain size.

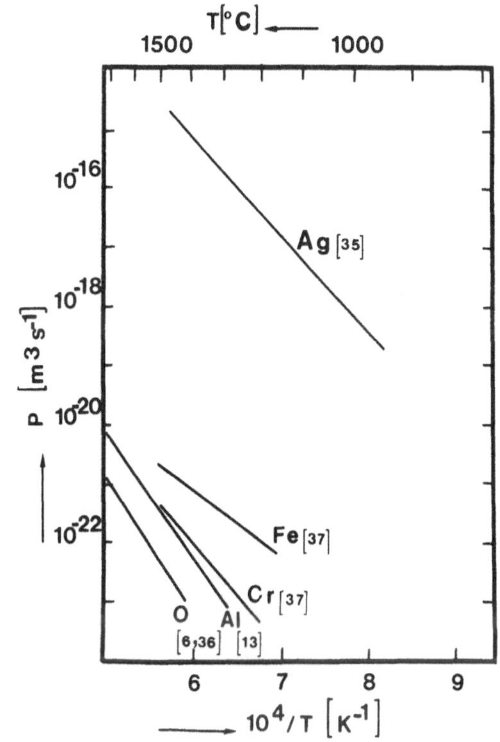

Figure 7. Arrhenius curves for grain boundary diffusion in α-Al_2O_3.

The corresponding Arrhenius curve lies close to that of chromium (37). This fact can be related to the complete solubility of Cr_2O_3 and Al_2O_3, giving to Cr^{3+} and Al^{3+} a similar behaviour. It can be noticed that the grain boundary activation energy for Cr was relatively large compared to the bulk activation energy.

The same remark applies to silver diffusion, where similar values were found for bulk, dislocation and grain boundary diffusion. In this system, the validity of the short-circuit model was checked in a way similar to that used previously for dislocation diffusion (35). The grain size was calculated from the absolute level of concentration measured in the grain boundary field, and found in agreement with the effective grain size.

Given that the diffusional processes in the Ag-Al_2O_3 system were investigated by the same authors, reliable comparisons could be made for all the data.

* $Q_d = Q'$

* The ratio of slopes of penetration profiles log c = f(y) in single crystals and polycrystals were in agreement with the previous evaluation. At 827°C for t = 8.682 10^5 s the theoretical evaluation gave pd/p' = 4 to 8, and the experimental value was pd/p' = 5 (Fig. 8).

* $\alpha_d D_d$ and $\alpha' D'$ calculated by the assumption a $= \delta = 10^{-9}$m, differ

only by an order of magnitude.

In view of these assumptions, it was suggested that these values were in agreement and representative of the same diffusion mechanism. A possible explanation for the enhancement of diffusivity along these defects could be the contribution of jumps type I→2 (Figure 1) due to an increase of vacancy concentration (remember that for an Alatom in the bulk $\Delta H_m \simeq 320$ kJ/mol).

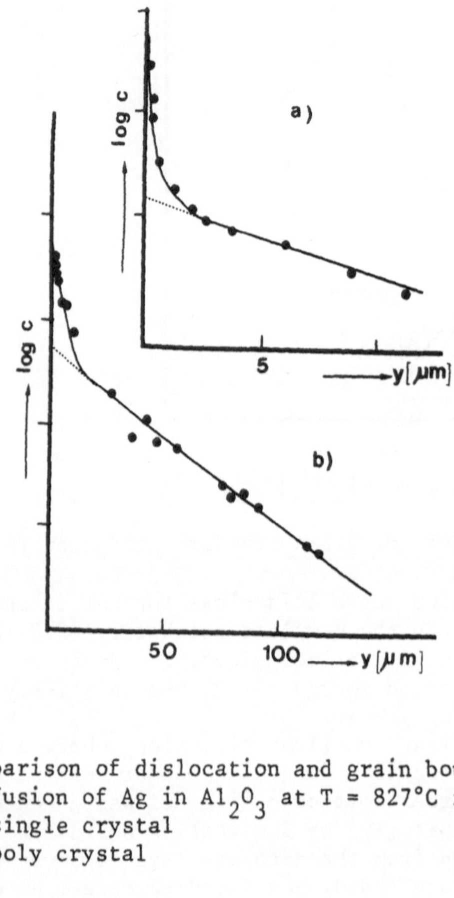

Figure 8. Comparison of dislocation and grain boundary penetration profiles. Diffusion of Ag in Al_2O_3 at T = 827°C for t = 8.682 x 10^5 s
a) single crystal
b) poly crystal

3.3. Influence of misorientation.

Diffusion of chromium along low angle tilt boundaries was studied by Stubican and Osenbach (42) as a function of misorientation and anisotropy in the temperature range 1200-1600°C. The absolute parameters $D'\delta\alpha$ were higher than those determined by Lagrange (37) but the relative enhancement of diffusion indicated by the authors $D'/D \cong 10^6$ (if $\delta \cong 10^{-9}$m) was similar to that calculated from data given in Tables I and II ($D'/D = 2\ 10^5$ at T = 1200°C).

A strong anisotropy in boundary diffusion was observed for low-angle grain boundaries but not for angles above 10°. This result was consistent with the dislocation pipe model. Activation energies calculated for diffusion parallel and perpendicualr to the growth direction of the bicrystal were the same ≅ 170 ± 15 kJ.mol again in good agreement with the pipe diffusion model. Contrary to polycrystalline sample results (37) the ratio Q'/Q was found to be lower than 1 (Q'/Q   0.56).

3.4. Influence of purity and doping.

3.4.1. Oxygen diffusion. Chemical diffusion in polycrystalline α-Al_2O_3 doped with Ti or Fe was studied by Wang and Kröger (41) through the movement of a colour front on oxidation of reduced samples : oxygen grain boundary diffusion coefficients were higher, by several orders of magnitude, than those found in undoped alumina. The reason for these large values was not clear ; the authors suggested that grain-boundaries with precipitates of $FeAl_2O_4$ or $TiAl_2O_5$ would transport oxygen faster than clean boundaries.

3.4.2. Cation diffusion. The effect of Y_2O_3 dopant on Cr and Fe grain boundary diffusion was investigated by Lagrange (37,43). It was found that the parameter P was higher (by a factor 10 at 1300°C) in 0.1 at% doped alumina than in undoped alumina ; this increase might be related to the following facts :

- bulk diffusivity was found to increase by about 4 times in 0.1% doped alumina at 1600°C (14). So a similar increase can be expected in grain boundaries if the same mechanism is involved in both cases.
- precipitates of yttro-garnets (3 Y_2O_3 - 5 Al_2O_3) were observed along grain boundaries. These precipitates induced an increase in cationic diffusion coefficients when compared with those obtained in undoped alumina.

Therefore the results of Wang and Kröger (41) and Lagrange (43) underline the importance of chemistry and microstructure on grain boundary diffusion processes ; impurity segregation and precipitation along grain boundaries are supplementary parameters to be considered in relation with transport of matter along short-circuits.

In alumina polycrystals, used for silver diffusion studies, such precipitation did not occur. It was found that dislocation diffusion coefficients measured in single crystals with an impurity content of about 0.01% were very close to grain boundary coefficients measured on samples with a larger impurity content (3%) (35). This comparison showed that the nature and the amount of impurities contained in this material had a negligible influence on short-circuit diffusion, as previously observed for bulk diffusion (15).

3.5. Influence of segregation.

Equations for grain boundary diffusion must necessarily be written by taking into account the enrichment factor, or segregation ratio, α, of the diffusing element in the grain boundaries. This ratio is close to 1 when the impurity does not introduce important perturbations in the host lattice ; this fact is generally related to a high solubility. On the contrary, there is a link between low solubilities, highly disturbing impurities and high segregation ratios. In alumina, it has been found that experimental segregation results could be interpreted by a size misfit effect. For example, Ca^{2+}, with a radius of 0.099 nm (against 0.05 nm for Al^{3+}) would present a strong tendency to segregate, characterized by segregation ratios ≅ 1300 to 3500 according to the evaluations (44 to 46) and a free energy of segregation of ≅ 120 kJ/mol (45).

Therefore, experimental parameters P, including this segregation ratio α, must reflect the tendency to segregate.

Unfortunately, there does not exist any data giving information both on segregation and on diffusion of the same element in α-Al_2O_3. To examine the influence of segregation, we have calculated P/D values at a given temperature (1250°C) for self and heterodiffusion data. Assuming $\delta = 10^{-9}$m we have deduced $D'\alpha'/D$ values, presented in Table VI.

TABLE VI

$D'\alpha'/D$ for different elements T = 1250°C ($\delta = 10^{-9}$m)

O	Al	Cr	Fe	Ag
10^8	$1.6\ 10^4$	$1.6\ 10^5$	10^6	10^8

It was then tentatively assumed that differences in $D'\alpha'/D$ ratios were due to differences in segregation. Considering that the values for non segregating species, Al and Cr, could be averaged at ≅ $5\ 10^4$, comparison of Ag results lead to $\alpha'(Ag)$ ≅ 2000.

Though no direct experiment has yet been performed to confirm this value, it appears reasonable when examining published segregation data : in agreement with the size misfit effect, a segregation ratio as high or even higher than that of Ca^{2+} may be expected for the species Ag^+ or Ag.

Moreover, the solubility of Ag in α-Al_2O_3 was measured by a radiotracer technique (35) and was found equal to 0.44 at% at 900°C. By extending the relation determined by Hondros and Seah on metals (47) : α ≅ 1 to 10 x 1/(solubility), to other materials, a segregation ratio of 200 to 2000 must be expected.

It may be asked whether such a large segregation, which implies a large energy decrease when an atom from the bulk moves into the grain boundary, does not result in a low grain boundary activation energy, according to a simplified relation :

$$Q'_{(experimental)} = Q'_{(real)} - \Delta H$$

Segregation isotherms, plotted from Mc Lean's equation (48) :

$$C' = \frac{C \exp - \Delta G/RT}{1 + C \exp - \Delta G/RT}$$

exhibit a "saturation effect" much more pronounced as ΔG (and ΔH) is higher (Fig. 9).With ΔG 150 kJ/mol, C'/C is nearly independant of temperature for $C > 10^{-3}$; this saturation effect results then in :

$$Q'_{(experimental)} = Q'_{(real)}$$

Another effect of segregation, which is worth considering, is its intervention in the parameter β. When segregation occurs, high β values are obtained without the need to decrease $\sqrt{Dt}$. Thus, experimental curves display the short-circuit diffusion part more obviously, in a range of concentrations within the measurable field , whereas similar enhancement of diffusivity D'/D unaccompanied by a segregation ratio may be more difficult to observe.

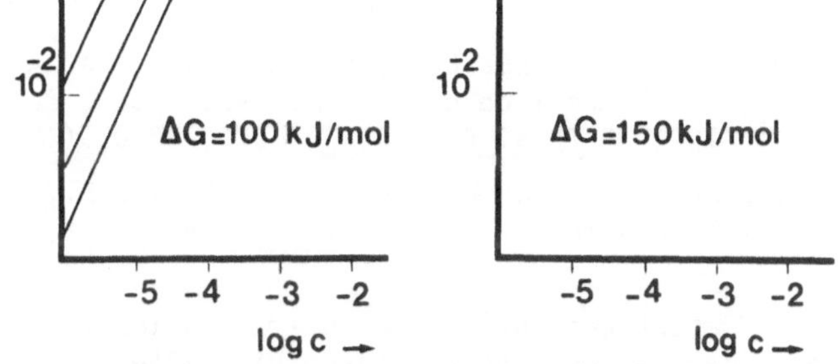

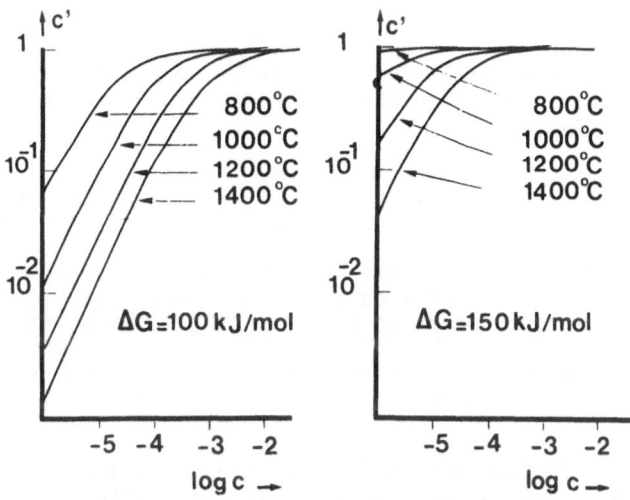

Figure 9. Relation between grain-boundary concentration c' and bulk concentration c, from (48).

4. CONCLUSION

Although more experimental data in diffusion and theoretical values of formation and migration energies of defects along dislocations and grain boundaries would be required for a complete understanding of diffusional processes in α-Al_2O_3, some ideas are forthcoming from this overview :

a) like metals, α-Al_2O_3 presents short-circuit paths, the effect of which can be obviously displayed as well for metallic diffusing elements as for oxygen. Assuming the width of the defect to be $\cong 10^{-9}$ m, the ratio of short-circuit to bulk diffusion coefficients (for non-segregating cationic species) is 5.10^4 at 1250°C.

In alumina the ratios Q'/Q are most generally in the range $0.7 < Q'/Q < 1$, whereas in metal $Q'/Q \cong 0.5$.

Dislocations and grain boundaries induce the same increase in diffusivity, though for geometric reasons the penetration along the latter defects extends further.

b) contrary to metals, the presence of impurities and the increase of dopant concentration seems to have a limited influence on diffusion kinetics. Several explanations may be proposed for this doping insensitive behaviour :

- the amount of impurity, even in the purer crystals, is sufficient to control the whole defect chemistry ;
- "active" elements present a low solubility, so that they precipitate at very small concentrations, limiting their possible influence ;
- defects created by impurities are more complex than generally assumed, and through the different equilibrium involved a kind of a "buffer effect" limits variations of defect concentration.

c) octahedral unoccupied sites in the anion sublattice, which are interstitial sites for oxygen as well as for aluminium atoms, are to be considered in a number of bulk diffusion processes ; they could explain, in some cases, the relatively insensitive impurity behaviour of alumina with respect to diffusion.

REFERENCES

1 DIENES G.J., WELCH D.O., FISHER C.R., HATCHER R.D., LAZARETH D. and SAMBERG M., "Shell-model calculation of some point defect properties in α-Al_2O_3" - Phys. Rev. B, 11, 3060-3070, (1965)

2 CATLOW C.R.R., JAMES R., MACKRODT W.C. and STEWART R.F., "Defect energies in α-Al_2O_3 and rutile TiO_2" - Phys. Rev. B, 25, 1006-1026, (1982)

3 MACKRODT W.C., "Defect energetics and their relation to non stoichiometry in oxides" - Solid State Ionics, 12, 175-188, (1984)

4 EL AIAT M.M. and KROGER F.A., "Determination of the parameters of native disorder in α-Al_2O_3" - J. of the Am. Ceram. Soc., 65, 162-166 (1982)

5 KROGER F.A., "Experimental and calculated values of defect parameters and the defect structure of -Al_2O_3" - Advances in Ceramics, 10, 100-118, (1984)

6 OISHI Y. and KINGERY W.D., "Self diffusion of oxygen in single crystal and polycrystalline aluminium oxide" - J. Chem. Phys., 33, 480-486, (1960)

7 OISHI Y., ANDO K. and KUBOTA Y., "Self diffusion of oxygen in single crystal alumina" - J. Chem. Phys., 73, 1410-1412, (1980)

8 OISHI Y., ANDO K., SUGA N. and KINGERY W.D., "Effect on surface condition on oxygen self-diffusion coefficients for single crystal Al_2O_3" - J. Amer. Ceram. Soc., 66, C 130-131, (1983)

9 REED D.J. and WUENSCH B.J., "Ion probe measurement of oxygen self-diffusion in single crystal Al_2O_3" - J. Am. Ceram. Soc., 63, 88-92, (1980)

10 REDDY K.P.R. and COOPER A.R., "Oxygen diffusion in Sapphire" - J. Amer. Ceram. Soc., 65, 634-638, (1982)

11 LAGERLOF K.D.D., PLETKA B.J., MITCHELL T.E. and HEUER A.H., "Deformation and diffusion in sapphire" - Radiation effects, 74, 87-107, (1983)

12 PALADINO A.E. and KINGERY W.D., "Aluminium ion diffusion in aluminium oxide" - J. Chem. Phys., 37, 957-962, (1962)

13 CANNON R.M., RHODES W.H. and HEUER A.H., "Plastic deformation of fine grained alumina : I, Interface controlled diffusional creep" - J. Amer. Ceram. Soc., 63, 46-53, (1980)

14 LESAGE B., HUNTZ A.M. and PETOT-ERVAS G., "Transport phenomena in undoped and chromium-or yttrium-doped alumina" - Radiation effects, 75, 283-299, (1983)

15 BADROUR L., MOYA E.G., BERNARDINI J. and MOYA F., "Bulk diffusion of ^{110}Ag tracer in Al_2O_3" - Scripta Met., 20, 1217-1222, (1986)

16 OISHI Y., ANDO K. and MATSUHIRO K., "Self diffusion coefficient of oxygen in vapor-grown single crystal alumina" - Yogyo-Kyokai-Shi, 85, 54-56, (1977)

17 LAGERLOF K.P.D., MITCHELL T.E. and HEUER A.H., "Defect-dislocation interactions in sapphire (α-Al_2O_3)" - Solute defect interactions, Pergamon Press, Toronto Ont., 152-161, (1986)

18 HANEDA H. and MONTY C., "Oxygen self diffusion in Mg or Ti doped Al_2O_3 single crystals" - to be published

19 PHILLIPS D.S., MITCHELL T.E. and HEUER A.H. - Phil. Mag., 42, 417, (1980)

20 ANDO K., KUROKAWA Y. and OISHI Y., "Oxygen self diffusion in Fe-doped MgO single crystals" - J. Chem. Phys., 78, 6890-6892, (1983)

21 LESAGE B., HUNTZ A.M., OCHIN P., SAADI B. and PETOT-ERVAS G., "Influence of chromium and yttrium doping on transport phenomena in monocrystalline alpha-alumina" - Solid State Ionics, 12,243-251,(1984)

22 SAADI B., PETOT-ERVAS, OCHIN D., LESAGE B., HUNTZ A.M., "Chemical diffusion in α alumina, titanium and yttrium influence" - Physical chemistry of the solid state : applications to metals and their compounds - Elsevier Science Publishers, Amsterdam, 389-395, (1984)

23 BEN ABDERRAZIK G., MILLOT F., MOULIN G. and HUNTZ A.M., "Determination of transport properties of alumina oxide scale" - J. of Am. Ceram. Soc., 68, 307-314, (1985)

24 LOUDJANI M., HUNTZ A.M. and PETOT-ERVAS G., "Microstructure and transport properties of Y_2O_3 doped or undoped polycrystalline alumina in relation with its elaboration" - J. de Physique C1, 47, 323-328, (1986)

25 HUNTZ A.M., MOULIN G., BEN ABDERRAZIK G., "Influence des impuretés sur l'oxydation à haute température des alliages Fe-Cr-Al : propriétés de transport de la couche d'alumine" - Ann. Chim. Fr., 11, 291-307, (1986)

26 LOUDJANI M.K., ROY J. and HUNTZ A.M., "Study by extended X-ray absorption fine-structure technique and microscopy of the chemical state of yttrium in α-polycrystalline alumina" - J. Am. Ceram. Soc. 68, 559-562, (1985)

27 LESAGE B., "Contribution à l'étude des mécanismes de transport dans l'oxyde de nickel NiO et l'alumine Al_2O_3 alpha. Influence de dopants" - Thesis, Université Paris-Sud, centre d'Orsay (1985)

28 KORIPELLA C.R. and KROGER F.A., "Electrical conductivity, diffusion of iron and the defect structure of α-Al_2O_3 : Fe" - J. Phys. Chem. Solids, 47, 565-576, (1986)

29 LLOYD I. and BOWEN H.K., "Iron tracer diffusion in aluminium oxide" J. Amer. Ceram. Soc., 64, 744-747, (1981)

30 DOSDALE T. and BROOK R.J., "Comparison of diffusion data and of activation energies" - J. of the Am. Ceram. Soc., 66, 392-395,(1983)

31 LE CLAIRE A.D. and RABINOVITCH A., "A mathematical analysis of diffusion in dislocations" - J. Phys. C : Solid State Phys., 14, 3863-3879 (1981) ; 15, 3455-3471, (1982) ; 16, 2087-2104, (1983) ; 17, 991-1000, (1984)

32 LE CLAIRE A.D., "The analysis of grain boundary diffusion measurements" - Brit. J. Appl. Phys., 14, 351, (1963)

33 SUZUOKA T., "Exact solutions of two ideal cases in grain boundary diffusion problem and the application to sectioning method" - J. of the Phys. Soc. of Japan, 19, 839-850, (1964)

34 MOYA E.G., BADROUR L., BERNARDINI J. and MOYA F., "Study of silver penetration into polycrystalline alumina" - Grain boundary structure at related phenomena, Proc. of JIMIS-4, Suppl. to Trans. of the Japan Institute of Metals, 27, 517-524, (1986)

35 BADROUR L., MOYA E.G., BERNARDINI J. and MOYA F., "Fast diffusion of silver in single and polycrystals of α-alumina" - to be published

36 MISTLER R.E. and COBLE R.L., "Grain boundary diffusion and boundary widths in metals and ceramics" - J. Appl. Phys., 45, 1507, (1974)

37 LAGRANGE M.H., HUNTZ A.M., DAVIDSON J.H., "The influence of Y, Zr or Ti additions on the high temperature, oxidation resistance of Fe-Ni-Cr-Al alloys of variable purity" - Corros. Sci., 24,613,(1984)

38 HART E.W. - Acta Metal., 5, 597, (1957)

39 HOU L.D., TIKU S.K., WANG H.A. and KROGER F.A., "Conductivity and creep in acceptor-dominated polycrystalline Al_2O_3" - J. of Mat. Science, 14, 1877-1889, (1979)

40 EL AIAT M.M., HOU L.D., TIKU S.K., WANG H.A. and KROGER F.A., "High temperature conductivity and creep of polycrystalline Al_2O_3 doped with Fe and/or Ti" - J. Am. Ceram. Soc., 64, 174-182, (1981)

41 WANG H.A. and KROGER F.A., "Chemical diffusion in polycrystalline Al_2O_3" - J. Amer. Ceram. Soc., 63, 613-619, (1980)

42 STUBICAN V.C. and OSENBACH J.W., "Influence of anisotropy and doping on grain boundary diffusion in oxide systems" - Solid State Ionics, 12, 375-381, (1984)

43 LAGRANGE M.H., "Rôle des additions Y, Zr, Ti sur l'oxydation à haute température d'alliages Fe-Ni-Cr-Al. Relation entre la microstructure de l'alumine dopée ou non en Y et la diffusion cationique" Thèse 3ème cycle, Université Paris XI, Orsay (1982)

44 MARCUS H.L. and FINE M.E., "Grain boundary segregation in MgO-doped Al_2O_3" - J. Amer. Ceram. Soc., 55, 568-570, (1972)

45 JOHNSON W.C. and STEIN D.F., "Additive and impurity distributions at grain boundaries in sintered alumina" - J. Amer. Ceram. Soc. 58, 485-488, (1975)

46 JUPP R.S., STEIN D.F. and SMITH D.W., "Observations on the effect of calcium segregation on the fracture behaviour of polycrystalline alumina" - J. Mater. Sci., 15, 96-102, (1980)

47 HONDROS E.D. and SEAH M.P., "The theory of grain boundary segregation in terms of surface adsorption analogues" - Met. Trans. A, 8A, 1363, (1977)

48 Mc LEAN D., Grain boundaries in Metals , Oxford University Press, London, 118-119, (1957)

OXYGEN SELF-DIFFUSION IN VOLUME AND IN GRAIN-BOUNDARIES OF $Cu_{2-x}O$

F. PERINET, J. LE DUIGOU, C. MONTY
CNRS, Laboratoire de Physique des Matériaux
1, place A. Briand, Bellevue
92195 Meudon Cedex
France

ABSTRACT. Oxygen self-diffusion has been measured both in the bulk and in grain-boundaries of $Cu_{2-x}O$ as a function of temperature T (712°C-1100°C) and oxygen partial pressure, p_{O2}(10^{-7}-0.21 atm). The measured diffusion coefficients can be described by the expressions

$$D_O(cm^2s^{-1}) = 2.2\ 10^{-3} p_{O2}(atm)^{0.48\mp0.2} \exp[-\frac{1.46\mp0.18(eV/at)}{kT}]$$

in the bulk and

$$D_O'(cm^2s^{-1}) = 8.13\ 10^{-2} p_{O2}(atm)^{0.59\mp0.18} \exp[-\frac{1.0\mp0.45(eV/at)}{kT}]$$

in grain-boundaries (if grain-boundary thickness δ = 1 nm).
These results indicate that the same main defect, a neutral oxygen interstitial O_i^x , is responsible for the diffusion of oxygen both in the bulk and in grain-boundaries and that the sum of the formation and migration enthalpies for O_i^x is 1.5 eV in the bulk and around 1 eV in grain-boundaries.

1. Introduction

Diffusion measurements have been extensively used to characterize point defects in non stoichiometric oxides /1/. The oxygen sublattice has been less studied because of the lack of a good radiotracer. But now the appearance of new techniques such as nuclear reaction analysis or secondary ion mass spectrometry (SIMS) has opened this field and the defects in the oxygen sublattice are now characterized in classical systems such as $CO_{1-x}O$ /2/.
Diffusion can also yield information about extended defects such as grain-boundaries or dislocations and several studies have been recently performed in oxides /3/, /4/, /5/ which are beginning to increase knowledge in this field. Nevertheless problems remain which seem to indicate that grain boundary behaviour in oxides may not be as straightforward as for metals /6/.

J. Nowotny and W. Weppner (eds.),
Non-Stoichiometric Compounds Surfaces, Grain Boundaries and Structural Defects, 387–397.

The present work is a study of the oxygen sublattice point defects in a covalent oxide $Cu_{2-x}O$, both in bulk and in grain boundaries. Using improved techniques /8/, the ^{18}O selfdiffusion has been studied as a function of the two parameters defining thermodynamic equilibrium in non stoichiometric binary oxides : the oxygen partial pressure, p_{O2}, and the temperature T. The studies have been carried out on single crystals and on polycrystals in the temperature range 712°C-1100C and in the p_{O2} range 10^{-7}-0.21 atm. The aim was to identify the type and the charge of the main defects responsible for the oxygen diffusion mechanism in the bulk and in grain boundaries using the p_{O2} dependence at constant T, and to deduce the enthalpies and entropies of formation and migration of these defects from the temperature dependence of D at constant p_{O2}.

2. Experimental

2.1. SAMPLE PREPARATION AND DIFFUSION EXPERIMENTS

To study volume diffusion, Cu_2O single crystals were grown from the melt by the floating zone technique in an arc image furnace /7/. Samples were prepared as slices mechanically polished (diamond paste down to 1 µm grain size). To study grain-boundary diffusion, polycristalline Cu_2O has been obtained by oxidizing Cu foils (thickness ≅ 50 µm) surrounding Cu_2O monocristalline slices. The grain size of the polycristalline Cu_2O ranged between 50 µm and 500 µm (oxidation temperature : 1035°C in air for 3 hours), and did not change on further annealings. The samples were annealed before the diffusion experiments in p_{O2} and T conditions to establish a thermodynamic equilibrium at the values selected for the diffusion annealing.
The tracer ^{18}O was introduced either by the solid/gas isotopic exchange technique or as a thin film of radiofrequency sputtered $Cu_2{}^{18}O$, depending on the p_{O2} selected for the diffusion annealing (gas/solid exchange is the better technique only for high p_{O2} values).

2.2. ANALYSIS OF DIFFUSION PROFILES /8/

Analysis of the diffusion profiles was performed by secondary ion mass spectrometry (SIMS) using a CAMECA SMI 300 or, more recently, a CAMECA IMS 3F. The masses 16 and 18 were simultaneously analysed using the "crater method" (a flat bottomed hole is produced by the etching of the primary ion beam). When the depth of the diffusion zone was larger than 30 µm, the ^{18}O concentration was analysed on the surface of bevel cut samples (angles ranging between 1 and 4 degrees). At room temperature Cu_2O is an insulator, so a thin film of gold was deposited on the surface before the analysis. Primary ions used were Ar^+ or $C_s{}^+$ and secondary ions, O^-. Possible interference of water vapor $H_2{}^{16}O$ on the ^{18}O counting was controlled using high resolution spectroscopy ($M/\Delta M \gtrsim 10000$). Cratering depths or bevel angles were measured using a Talystep roughness meter or sometimes by optical techniques. Distances on the bevel cut samples were converted to distances measured normally to the initial surface with an uncertainty of about 5%. In the crater method we had to convert a sputtering time into a distance, knowing the total depth of the crater (accuracy 5 to 10%). We assumed the sputtering rate to be proportional to Q_{16}, the counting on mass 16 during a given time. Concentrations were deduced from the ratio $Q_{18}/(Q_{16}+Q_{18})$ of the counts on masses 16 and 18, taking into account a background on each mass. The natural abundance of ^{18}O (2‰) gives an internal calibration of the measurements. In the case of polycristalline materials, difficulties come from the difference in sputtering rates from one grain to another one. This anisotropy can be large for special cristallographic orientations. We have taken into account only the experiments in which the sputtering process could be considered as homogeneous which means the crater depth is known with an accuracy of the order of 10%.

2.3. DETERMINATION OF DIFFUSION COEFFICIENTS

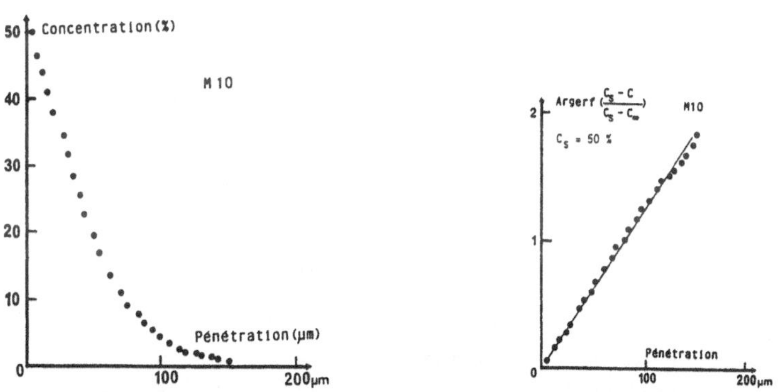

Figure 1 : Example of diffusion profile obtained by the gas/solid exchange method.

The diffusion coefficients are deduced from the diffusion profiles using solutions of Fick's law corresponding to the boundary conditions of the experiments. Three solutions have been fitted to the concentration/penetration experimental curves depending on the experimental conditions /9/.

2.3.1. *Gas/solid isotopic exchange*

$$\frac{C_s - C}{C_s - C_\infty} = \text{erf}\left(\frac{x}{2\sqrt{Dt}}\right) \qquad (1)$$

where C is the measured concentration of the tracer ^{18}O, t the time of

annealing, x the penetration distance normal to the initial surface (assumed to be an infinite plane), C_s is the concentration at the initial surface (assumed to be constant), C_∞ is the natural abundance of the tracer in the bulk (2‰ for ^{18}O).
An example of such a profile obtained on a single crystal is given in figure 1.

2.3.2. *Thin film*

$$C - C_\infty = C_0 \exp\left(-\frac{x^2}{4Dt}\right) \qquad (2)$$

where C_0 is the surface concentration, now depending on time. Figure 2 shows examples of such behaviour in single crystals.

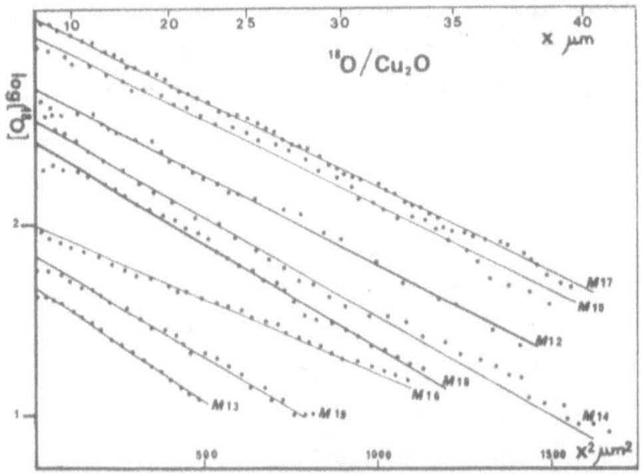

Figure 2 : Diffusion profiles obtained by the thin film method in log C/x^2 coordinates.

2.3.3. *Diffusion in grain-boundaries*. The solutions have the general form :

$$C = C_V + C_{GB} \qquad (3)$$

where C_V is the solution in volume ; (it is one of the preceedings ones, depending on the boundary conditions) and C_{GB} is the solution in the grain boundary. It is a contribution leading to a "tail" to the bulk solution. To deduce the grain boundary diffusion coefficient, we used the slope of the Leclaire's asymptotic forms of the Suzuoka solution corresponding to the thin film conditions :

$$D'\delta = \left(\frac{4D}{t}\right)^{1/2} \left(-\frac{\partial \ln \bar{C}}{\partial x^{6/5}}\right)^{-5/3} (0.72\ \beta^{0.008})^{5/3} \qquad (4)$$

where δ is the grain- boundary thickness, $\bar{C}$ is the average concentration in a slice parallel to the initial surface, β is a relative normalized penetration ($\beta = \frac{D'\delta}{2D\sqrt{Dt}}$).

This solution assumes that $x \gg \sqrt{Dt}$.
D can be deduced from the first part of the profiles or from the study

on single crystals. Fig. 3 shows a profile plotted to illustrate both the bulk part and the grain boundary diffusion tail.

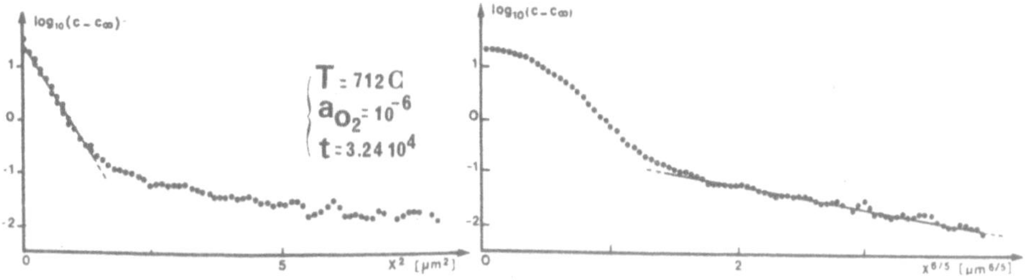

Figure 3 : Example of diffusion profile obtained in polycristalline sample

3. Results

3.1. OXYGEN VOLUME SELF-DIFFUSION

The volume diffusion coefficients deduced from the experiments carried out on single crystals are collected in the table I. The results are reported in an Arrhenius diagram (fig.4) where each straight line refers to a different constant p_{O2} and in a diagram log D/log P_{O2} at constant T (figure 5).

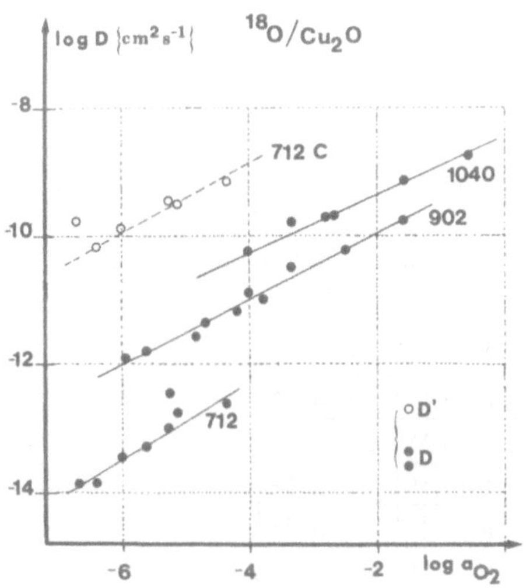

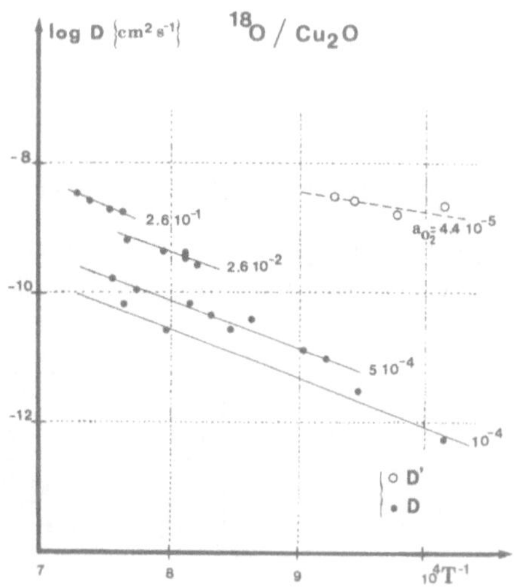

Figure 4 : Arrhenius diagram for volume diffusion (full dots) and grain-boundary diffusion (open dots).

Figure 5 : Oxygen activity, a_{O2}, dependence ($a_{O2}=p_{O2}$ in atm.) of the volume diffusion coefficients D and of the grain boundary diffusion coefficients D'.

TABLE I

Results obtained in Cu_2O single crystals (M series)

Ref.	T(C)	a_{02}	D($cm^2 s^{-1}$)
M1	1036	2.6 10^{-1}	1.9 10^{-9}
M2	1055	-	2.0 10^{-9}
M3	1080	-	2.7 10^{-9}
M4	1098	-	3.6 10^{-9}
M5	1031	2.6 10^{-2}	7.3 10^{-10}
M6	984	-	4.8 10^{-10}
M7	958	-	3.4 10^{-10}
M8	958	-	4.2 10^{-10}
M9	958	-	3.6 10^{-10}
M10	958	-	4.1 10^{-10}
M11	945	-	2.7 10^{-10}
M12	1043	5 10^{-4}	1.7 10^{-10}
M13	1020	-	1.15 10^{-10}
M14	951	-	6.7 10^{-11}
M15	930	-	4.75 10^{-11}
M16	907	-	2.9 10^{-11}
M17	887	-	4.0 10^{-11}
M18	832	-	1.4 10^{-11}
M19	812	-	1.0 10^{-11}
M20	787	10^{-4}	3.1 10^{-12}
M21	984	-	2.6 10^{-11}
M22	1036	-	7.0 10^{-11}
M23	1051	3.2 10^{-5}	3.6 10^{-11}
M24	1035	6.0 10^{-5}	3.5 10^{-11}
M25	1073	1.7 10^{-3}	2.1 10^{-10}
M26	1038	1.3 10^{-3}	2.3 10^{-10}
M27	1053	10^{-3}	3.5 10^{-10}
M28	990	10^{-5}	1.13 10^{-11}
M29	904	1.1 10^{-6}	1.26 10^{-12}
M30	907	2.5 10^{-6}	1.76 10^{-12}
M31	907	1.4 10^{-5}	3.18 10^{-12}
M32	902	2 10^{-5}	4.9 10^{-12}
M33	898	6.5 10^{-5}	7.0 10^{-12}
M34	902	1.7 10^{-4}	1.04 10^{-11}
M35	900	3.4 10^{-3}	6.05 10^{-11}
M36	712	2 10^{-7}	1.3 10^{-14}

TABLE II

Results obtained in polycristalline Cu_2O (P series) :
D has been deduced from the first part of the diffusion profiles, D' has been deduced from D values computed using the relation (12) and taking $\delta = 1$ mm

Ref.	T(C)	a_{02}	D($cm^2 s^{-1}$)	D'($cm^2 s^{-1}$)
P 1	712	2 10^{-7}	1.4 10^{-4}	1.8 10^{-10}
P 2	-	4 10^{-7}	1.4 10^{-14}	6.6 10^{-11}
P 3	-	1 10^{-6}	2.1 10^{-14}	1.4 10^{-10}
P 4	-	1 10^{-6}	3.8 10^{-14}	4.6 10^{-11}
P 5	-	5.5 10^{-6}	(3.9/1.0) 10^{-13}	3.8 10^{-10}
P 6	-	6.6 10^{-6}	1.8 10^{-13}	3.3 10^{-10}
P 7	-	4.4 10^{-5}	2.6 10^{-13}	2.4 10^{-9}
P 8	750	-	9.3 10^{-13}	1.6 10^{-9}
P 9	785	-	2 10^{-12}	2.7 10^{-9}
P 10	804	-	5.3 10^{-12}	3.3 10^{9-}

The results can be represented by the relations :

at T = 902°C

$$\log . D(cm^2 s^{-1}) = (0.52 \mp 0.02) \log . p_{O2}(atm) - (8.9 \mp 0.1) \quad (5)$$

at T = 990°C

$$\log . D(cm^2 s^{-1}) = (0.48 \mp 0.02) \log . p_{O2}(atm) - (8.5 \mp 0.09) \quad (6)$$

at T = 1040°C

$$\log D(cm^2 s^{-1}) = 0.45 \mp 0.02 \log . p_{O2}(atm) - (8.45 \mp 0.07) \quad (7)$$

at p_{O2} = 2.6 10^{-2}atm

$$D(cm^2 s^{-1}) = (2.7^{37}_{0.9}) 10^{-3} \exp\left[- \frac{1.61 \mp 0.3 (eV/at)}{kT} \right] \quad (8)$$

p_{O2}= 2.6 10^{-1}atm

$$D(cm^2 s^{-1}) = (1.75^{19.7}_{0.22}) 10^{-4} \exp\left[- \frac{1.39 \mp 0.2 (eV/at)}{kT} \right] \quad (9)$$

p_{O2} = 5.10^{-4}atm

$$D(cm^2 s^{-1}) = (5.47^{15}_{0.20}) 10^{-5} \exp\left[- \frac{1.45 \mp 0.1 (eV/at)}{kT} \right] \quad (10)$$

p_{O2} = 10^{-4}atm

$$D(cm^2 s^{-1}) = (1.69^{12.6}_{0.22}) 10^{-5} \exp\left[- \frac{1.42 \mp 0.2 (eV/at)}{kT} \right] \quad (11)$$

All these results can be gathered in one relation :

$$D(cm^2 s^{-1}) = 2.2 \; 10^{-3} \; p_{O2} \; (atm)^{0.48 \mp 0.2} \exp\left[- \frac{1.46 \mp 0.18 (eV/at)}{kT} \right] \quad (12)$$

3.2. OXYGEN GRAIN-BOUNDARY SELF-DIFFUSION

Table II shows the results of the measurements of diffusion coefficients on polycristalline Cu_2O. D is the bulk diffusion deduced from the profiles obtained on large grains far from the boundaries or deduced from the first part of the profile. The measured quantity is $D'\delta$, the product of the GB diffusion coefficient and the effective grain-boundary thickness. D' values have been deduced taking δ = 1 nm. On the figure 5 the variation of D' with p_{O2} has been reported at T=712°C.

The values of D at the same temperature have been also reported on the figure. The variation of D can be represented by

$$D(cm^2 s^{-1}) = 1.12 \; 10^{-2} \; p_{O2}(atm)^{0.67 \mp 0.04} \exp\left[- \frac{1.46 \mp 0.18 (eV/at)}{kT} \right] \quad (13)$$

The p_{O2} dependence of D' gives a p_{O2} exponent of 0.59∓0.18 and the study of the dependence on T at p_{O2} = 4.4 10^{-5}atm (see figure 4) gives an activation energy of about 1 eV. The results can be gathered in one relation :

$$D'(cm^2 s^{-1}) = 8.13 \; 10^{-2} \; p_{O2}(atm)^{0.59 \mp 0.18} \exp\left[- \frac{1.02 \mp 0.45 (eV/at)}{kT} \right] \quad (14)$$

4. Diffusion mechanisms

4.1. OXYGEN VOLUME SELF-DIFFUSION

If the point defect population in $Cu_{2-x}O$ is mainly dominated by only two types of charged species (for exemple α times charged copper vacancies, $V_{Cu}^{\alpha'}$ and holes $h^{\cdot}$) the concentration of a given defect d being simply related to the two parameters p_{O2} and T which define a thermodynamic equilibrium :

$$[d] = A\ p_{O2}(atm)^{l} \exp(-\frac{\tilde{\Delta H}_d^f}{kT}) \tag{15}$$

where A contains entropic terms and a numerical constant, l is a number characterizing the type and charge state of the defect, $\tilde{\Delta H}_d^f$ is an apparent formation enthalpy of the defect d, k is the Boltzmann constant.

A self diffusion coefficient D is related to [d] and to D_d , the diffusion coefficient of the defect itself, by the relation $D = [d]D_d$. D_d is thermally activated (free energy of migration ΔG_d^m) and therefore finally D can be written /10/ :

$$D = D_o\ p_{O2}(atm)^{l} \exp(-\frac{\tilde{\Delta H}_d^f + \Delta H_d^m}{kT}) \tag{16}$$

in which $D_o = g\ \bar{\nu} a^2 \exp(\frac{\tilde{\Delta S}_d^f + \Delta S_d^m}{k})$

where g is a numerical constant, $\bar{\nu}$ is an average frequency (Debye frequency), a the lattice parameter.

These relations hold in the case of oxygen self-diffusion. We know the l exponent and are able to identify the defect responsible for the oxygen diffusion in the lattice. Considering only simple defects, l values positive and close to 0.5 can only be attributed to a neutral oxygen interstitial, O_i^x. The formation of O_i^x is indeed described by the reaction :

$$\frac{1}{2}\ O_2(g) \rightleftarrows O_i^x \tag{17}$$

and applying the mass action law, we can write :

$$[O_i^x] = K^f_{O_i^x}\ p_{O2}{}^{1/2} = p_{O2}{}^{1/2} \exp(\frac{\Delta S^f_{O_i^x}}{k}) \exp(-\frac{\Delta H^f_{O_i^x}}{kT}) \tag{18}$$

The concentration of oxygen interstitiels γ times charged can be written :

$$[O_i^{\gamma'}] = K^f_{O_i^{\gamma'}}\ p_{O2}{}^{1/2} [h^{\cdot}]^{-\gamma} \tag{19}$$

where $[h^{\cdot}]$ is the concentration of holes which depends on p_{O2} and T. Cu_2O is a p-type semiconductor, its electrical conductivity σ is proportional to $[h^{\cdot}]$ and to the hole mobility μ_p which depends on T but is generally considered as independent on p_{O2}. The dependence on P_{O2} of $[O_i^{\gamma'}]$ is hence given by : $[O_i^{\gamma'}] \propto p_{O2}^{1/2-\gamma r}$ where r is the characteristic exponent of p_{O2} dependence of σ. r values are always positive and range between 0.2 and 0.1 /12/. We can explain values of 1 such as 0.45 by a contribution of O_i' to the diffusion mechanism but not values of 1 higher than 0.5 (i.e. 0.52 at 902°C or 0.64 at 712°C) cannot be explained on this basis.

A possibility to explain 1 values higher than 0.5 is to consider the formation of di-interstitials, which are a sort of molecular oxygen, following the reaction :

$$O_i^x + O_i^x \rightarrow (O_2)_i^x \tag{20}$$

Applying the mass action law to this equilibrium leads to :

$$[(O_2)_i^x] = K^b_{(O_2)_i^x} [O_i^x]^2 = K^b_{(O_2)_i^x} (K^f_{O_i^x})^2 p_{O2} \tag{21}$$

where $K^b_{(O_2)_i^x}$ is the constant of the reaction (20).

A contribution of such a defect to the diffusion of oxygen would mean that the activation energy for diffusion is not constant. The diffusion coefficient D is a sum D1 + D2 of coefficients of diffusion related to, respectively O_i^x and $(O_2)_i^x$. If there is no curvature in the Arrhenius plot, the activation energies Q1 and Q2 must be close, which in turn means that the binding enthalpy of $(O_2)_i^x$ must be of the order of the formation enthalpy for O_i^x and negative. It is interesting to note that the activation energy for diffusion has been found to increase slightly when p_{O2} increases; that is to say when the relative concentration of $(O_2)_i^x$ would increase. Again, the p_{O2} exponent indicates a higher contribution of $(O_2)_i^x$ at low temperature (712°C) than at high ones.

When O_i^x dominates, the activation energy has been found to be close of 1.5 eV/at. Unambiguously in such a case we can write :

$$\tilde{\Delta H}^f_{O_i^x} + \Delta H^m_{O_i^x} = 1.5 \text{ eV/at} \tag{22}$$

where $\tilde{\Delta H}^f_{O_i^x}$ is identical to $\Delta H^f_{O_i^x}$ characterizing the mechanism of formation of O_i^x described by the reaction (17) and $\Delta H^m_{O_i^x}$ characterizes a jump frequency of an oxygen interstitial perhaps by an interstitialcy mechanism.

4.2. OXYGEN GRAIN-BOUNDARY SELF-DIFFUSION

The dependence on p_{O2} observed for the oxygen diffusion coefficient in grain-boundaries is very close of that observed for volume diffusion. The activation energy appears smaller ($\simeq$ 1 eV compared to 1.5 eV) but uncertainties are larger.

As far as the point defect chemistry developed for the bulk may be valid in the case of a quasi-bidimensional system, we can conclude that the same oxygen diffusion mechanism holds in the bulk and in grain-boundaries of Cu_2O . That means a mechanism involving mainly O_i^x and, as a minor contribution, perhaps $(O_2)_i^x$. A simple behaviour such as that observed for the p_{O2} dependence of D' shows also that there is no spread on the results due to diffusion anisotropy. The main reason for this is the small probability of having "special" boundaries in the polycrystals. Most of the boundaries are "general" and behave similarly.
There are no systems in which the use of the p_{O2} parameter has provided the possibility to identify directly the diffusion mechanism in GB diffusion, apart from the diffusion on Ni in NiO /11/. Moreover in this system the authors worked close to an intrinsic/extrinsic transition which does not make an accurate characterization easy. It is interesting to notice that in NiO also, the same defect (V_{Ni}) seems to be responsible for bulk or GB diffusion.

5. Conclusion

The dependence of the concentrations of point defects on p_{O2} in binary non stoichiometric oxides provides a powerful tool to characterize the point defects. The diffusion experiments are able to explore the oxygen sublattice as well as the cationic sublattice despite the fact that in oxides such as $Cu_{2-x}O$ the concentrations of point defects in the oxygen sublattice are very low compared with the concentration of defects responsible for the nonstoichiometry and electrical properties.
We have shown that in $Cu_{2-x}O$ the main defects responsible for the oxygen diffusion both in volume and in grain boundaries, are neutral oxygen interstitials O_i^x which could perhaps be partly associated as molecular interstitials $(O_2)_i^x$ at low temperature. Oxygen interstials have been seen in other systems such as $Co_{1-x}O$ /2/. They are favoured in $Cu_{2-x}O$ by its less dense structure (the oxygen sublattice in Cu_2O is not compact, but is body centered). This would explain the relatively low values obtained for the sum $\Delta H^f_{O_i^x} + \Delta H^m_{O_i^x}$ (1.5 eV in bulk, $\simeq$ 1 eV in grain boundaries).
Computations of point defect energies in this system and in CuO (so important for superconductors) would be usefull.

6. References

/ 1/ P. KOFSTAD
Non stoichiometry, Diffusion and Electrical Conductivity in Binary Metal Oxides, Ed. Wiley Interscience, New York (1972).

/ 2/ R.J. TARENTO, C. MONTY
"Influence of non stoichiometry on the oxygen self-diffusion in $Co_{1-x}O$ *single crystals"* Sol.Stat.Ionics (1988) to be published.

/ 3/ A. ATKINSON
"Grain-boundary diffusion-Structural effects and mechanisms"
Journal de Physique C4, 46 (1985).

/ 4/ A. ATKINSON, F.C.W. PUMMERY, C. MONTY
"Diffusion of ^{18}O tracer in NiO grain boundaries"
in Transport in non stoichiometric compounds, Ed. G. Simkovitch, V.S. Stubican, Plenum (1985).

/ 5/ V.S. STUBICAN, J.W. OSENBACH
"Influence of anisotropy and doping on grain boundary diffusion in oxide systems", Solid State Ionics 12, 375-381 (1984).

/ 6/ F. BARBIER, C. MONTY, M. DECHAMPS
"On the grain boundary diffusion of Co in NiO bicrystals"
Phil. Mag. A58, 3, 475-490 (1988).

/ 7/ R.D. SCHMIDT-WHITLEY, M. MARTINEZ-CLEMENT, A. REVCOLEVSCHI
"Growth and microstructural control of single crystal cuprous oxide Cu_2O", J. of Crystal Growth 23, 113-120 (1974).

/ 8/ C. DUBOIS, C. MONTY, J. PHILIBERT
"Oxygen self-diffusion in NiO single crystals"
Phil. Mag. A, 46, 3, 419-433 (1982).

/ 9/ J. PHILIBERT
Diffusion et transport de matière dans les solides
Ed. de Physique, France (1985).

/10/ C. MONTY
"Diffusion in stoichiometric and non-stoichiometric cubic oxides", Radiations effects, 74, 29-55 (1983).

/11/ A. ATKINSON, R.I. TAYLOR
"The diffusion of ^{63}Ni along grain boundaries in nickel oxide"
Phil. Mag. A 43, 979 (1981).

/12/ P. OCHIN, C. PETOT, G. PETOT-ERVAS
"Thermodynamic study of point defects in $Cu_{2-\delta}O$. Electrical conductivity measurements at low oxygen partial pressures".
Solid State Ionics, 12, 135-143 (1984).

7. Acknowledgements

We are indebted to A. Atkinson who read the manuscript and suggested several language corrections.

DEFECT STRUCTURE AND SELF DIFFUSION IN Mg_2SiO_4 (FORSTERITE) AT HIGH TEMPERATURE

K. Andersson, G. Borchardt
FB Metallurgie und Werkstoffwissenschaften,
Technische Universität Clausthal
D-3392 Clausthal-Zellerfeld (F.R.G.)

ABSTRACT. For the explanation of creep behaviour and solid state reactions in crystalline silicates exact values of the atomic mobilities of the constituent elements are of major importance. The use of Neutral Primary Beam Secondary Ion Mass Spectrometry (NPB-SIMS) makes it possible to analyse the isotopic concentration profiles of all constituent elements in Mg_2SiO_4 simultaneously. The results indicate that $D_{Si} \ll D_O \ll D_{Mg}$. This can be used as a basis for model calculations by both earth scientists and materials scientists.

1. INTRODUCTION

Orthosilicates of the olivine type are favourite model substances for earth scientists who are concerned with deformations in the upper earth mantle and for high temperature materials scientists. To explain deformation kinetics and solid state compound formation kinetics, point defects, diffusion mechanisms, and precise values of the self diffusion coefficients of the constituent elements must be known. Because of the absence of suitable radioactive tracers for silicon, magnesium, and oxygen, and because of the very low mobilities of these elements even at high temperatures reliable data are difficult to obtain. Available data on the diffusivities of the three constituent elements were very contradictory [1-7], s. Fig. 1. The contradictions may be due to the fact that the quoted authors used samples in different thermodynamic states. For this work, single crystals were prepared with $a_{MgO} = 1$ or with $a_{MgSiO_3} = 1$ and the tracer diffusion profiles were obtained by a secondary ion mass spectrometer working with a neutral primary beam (NPB-SIMS) which had been developed for the analysis of non-conducting surfaces [8]. Especially interesting is the possibility of measuring the simultaneous diffusion of silicon, magnesium, and oxygen. The results establish the basis for a general understanding of the defects and the transport mechanisms in Mg_2SiO_4 as a basis for model calculations for both earth scientists and materials scientists.

J. Nowotny and W. Weppner (eds.),
Non-Stoichiometric Compounds Surfaces, Grain Boundaries and Structural Defects, 399–409.

2. DEFECT STRUCTURE

Electrochemical measurements [9] and solubility experiments (Fig. 2) show that at high temperatures Mg_2SiO_4 should dissolve a certain amount of excess silica or of excess magnesia, depending on the respective activities of these oxides. In an earlier paper [10] we presented a point defect model consisting of interstitial silicon $Si_i^{\cdot\cdot\cdot\cdot}$ and magnesium vacancies V''_{Mg} for SiO_2 rich Mg_2SiO_4, and interstitial magnesium interstitial $Mg_i^{\cdot\cdot}$ and silicon vacancies V''''_{Si} for MgO rich Mg_2SiO_4. Solution energies for MgO and SiO_2 were calculated using the Born-Haber process for the formation of the different point defects [11]. For example, for the solution reaction written according to the Kröger-Vink notation

$$Mg^x_{Mg} + SiO_2 \rightarrow 0.5\ (Si_i^{\cdot\cdot\cdot\cdot},\ 2V''_{Mg})^x\ +\ 0.5\ Mg_2SiO_4 \qquad (1)$$

the energy of solution is

$$U_s = 0.5\ U_D(Si_i^{\cdot\cdot\cdot\cdot},\ 2V''_{Mg})^x\ +\ 0.5\ U_D(Mg_2SiO_4) - U_L(SiO_2) \qquad (2)$$

The different lattice energies were calculated using the CASCADE computer programme [12] which is based on the Born-Mayer ionic model of ionic lattices with integral ionic charges and paired short range interactions represented by a potential of the form

$$U_L = \frac{1}{2} \cdot \sum_j \sum_i \frac{k \cdot q_i \cdot q_j}{r_{ij}} + A \exp(-r_{ij}/\rho) - C/r_{ij}^6$$

where r_{ij} is the distance between a given pair of ions and the parameters A, k, and C were derived empirically for each ion-ion potential. Ion polarisation is included using the shell model which represents each oxygen ion in terms of a core and a shell charge which are harmonically coupled. For silicates which are highly covalent, angular dependent forces have to be included as an explicit 'bond bending' term of the type

$$F_\Theta = \varkappa_{ijk}(\Theta-\Theta_0)^2 \qquad (3)$$

where Θ is the O-Si-O bond angle and $\Theta_0 = 109.5^\circ$ is the ideal tetrahedral angle. The defect energy U_D was calculated by subtracting the lattice energy of the ideal system and the lattice energy of the system with the defect after relaxation in a region surrounding the defect. The results are listed in table 1 and support the above mentioned complex defect models in a qualitative way. Values of the activation energy and the length of a single diffusion jump were calculated by simulating an atomistic difffusion jump as a series of fixed point defects of the moving species (Fig. 3). The difference between the highest defect energy and the defect energy of the starting and final positions of the moving atom is assumed to be the activation energy. The results of the diffusion simulations are listed in table 2.

3. DIFFUSION OF ^{18}O, ^{26}Mg, AND ^{30}Si: EXPERIMENTAL PROCEDURE

Czochralski-grown monocrystalline forsterite (Mg_2SiO_4) samples [13] with a dislocation density of $2\text{-}8 \cdot 10^4$ cm^{-2} [14] were mechanically polished and orientated according to the crystallographic main axes. The samples were etched for 30 s in dilute hydrofluoric acid and cleaned in distilled water before they were heat treated in a MgO or $MgSiO_3$ powder at the same temperature which was chosen for the diffusion experiment, but ten times longer to obtain a constant SiO_2 activity in the near surface region. A thin film of about 500 Å thickness containing the rare stable isotopes in > 50% enrichment was prepared by evaporating a compressed target consisting of $^{30}SiO_2$ and ^{26}MgO in a vacuum chamber using a Nd-YAG Laser [15]. Because of the fast oxygen isotope exchange between an isotope enriched thin film and the atmosphere, this method is not suitable for the determination of the oxygen tracer diffusion coefficient. The effect could, however, be used to perform heterogeneous diffusion experiments in a platinum capsule in which the monocrystalline diffusion samples were added to ^{18}O enriched magnesium silicate which supplied enough ^{18}O to guarantee constant boundary conditions during the high temperature diffusion annealing [16]. Platinum capsules were carefully plasma welded without being heated too much. The diffusion heat treatment was performed under the same SiO_2 activities as the buffer heat treatments. After the diffusion heat treatment, the tracer diffusion profiles were analysed using our Neutral Primary Beam-SIMS technique [8]. Fig. 4 shows a typical isotope concentration profile for the determination of the silicon tracer diffusivity in Mg_2SiO_4.

4. RESULTS AND DISCUSSION

Silicon is clearly identified as the slowest species in forsterite. The experimentally determined values of the tracer diffusion coefficients are listed in table 3. There exist no strong dependencies of the silicon diffusion on the silica activity (Fig. 5). The small influence is in good agreement with the computer simulations where for both SiO_2 rich Mg_2SiO_4 and MgO rich Mg_2SiO_4 complex defects consisting of interstitials and vacancies were calculated to be the most probable ones. The oxygen diffusion coefficients of earlier publications agree well with our own data (Fig 6). Magnesium diffuses 10 times faster in the crystallographic [001] direction than in the [100] or [010] direction (Fig. 7). This can be understood taking into consideration magnesium Frenkel pairs and the fact that the closest distances between octahedrally coordinated holes occur in the [001] direction.

Table 1: Calculated solution energies of various solution reactions for MgO and SiO_2 in Mg_2SiO_4

Reaction	U_s/eV 2B P	3B P
$MgO \rightarrow Mg_{i,octa.}^{\bullet\bullet} + O_i''$	12,30	11,45
$MgO \rightarrow Mg_{i,tetra}^{\bullet\bullet} + O_i''$	8,13	7,11
$MgO \rightarrow \frac{1}{2}Mg_{i,octa}^{\bullet\bullet} + \frac{1}{4}V_{Si}'''' + \frac{1}{4}Mg_2SiO_4$	7,13	6,69
$MgO \rightarrow \frac{1}{2}Mg_{i,tetra.}^{\bullet\bullet} + \frac{1}{4}V_{Si}'''' + \frac{1}{4}Mg_2SiO_4$	5,04	4,58
$MgO \rightarrow \frac{1}{3}Mg_{i,tetra.}^{\bullet\bullet} + \frac{1}{3}(V_{Si}'''', V_O^{\bullet\bullet})'' + \frac{1}{3}Mg_2SiO_4$	5,82	5,88
$MgO \rightarrow \frac{1}{2}(V_{Si}'''', 2V_O^{\bullet\bullet})^{\times} + \frac{1}{2}Mg_2SiO_4$	3,66	4,14
$MgO \rightarrow \frac{1}{2}(V_{Si}'''', 3V_O^{\bullet\bullet})^{\bullet\bullet} + \frac{1}{2}O_i'' + \frac{1}{2}Mg_2SiO_4$	5,18	7,00
$MgO \rightarrow \frac{1}{2}(V_{Si}'''', 4V_O^{\bullet\bullet})^{\bullet\bullet\bullet\bullet} + O_i'' + \frac{1}{2}Mg_2SiO_4$	8,76	10,35
$MgO \rightarrow \frac{1}{4}(V_{Si}'''', 2Mg_{i,Octa.}^{\bullet\bullet})^{\times} + \frac{1}{4}Mg_2SiO_4$	2,74	2,43
$MgO \rightarrow \frac{1}{4}(V_{Si}'''', 2Mg_{i,Octa.}^{\bullet\bullet})^{\times} + \frac{1}{4}Mg_2SiO_4$	2,00*	1,61*
$SiO_2 \rightarrow Si_i^{\bullet\bullet\bullet\bullet} + 2O_i''$	24,78	24,14
$SiO_2 \rightarrow \frac{1}{2}Si_i^{\bullet\bullet\bullet\bullet} + V_{Mg(I)}'' + \frac{1}{2}Mg_2SiO_4$	12,37	10,82
$SiO_2 \rightarrow 2V_{Mg(I)}'' + 2V_O^{\bullet\bullet} + Mg_2SiO_4$	15,32	13,31
$SiO_2 \rightarrow \frac{1}{2}(Si_i^{\bullet\bullet\bullet\bullet}, V_{Mg(I)}'', V_{Mg(II)}'')^{\times} + \frac{1}{2}Mg_2SiO_4$	7,34	2,59
$SiO_2 \rightarrow \frac{1}{2}(Si_i^{\bullet\bullet\bullet\bullet}, 2V_{Mg(I)}'')^{\times} + \frac{1}{2}Mg_2SiO_4$	6,92	2,21
$SiO_2 \rightarrow \frac{1}{2}(Si_i^{\bullet\bullet\bullet\bullet}, 2V_{Mg(I)}'')^{\times} + \frac{1}{2}Mg_2SiO_4$	6,67*	1,88*
$SiO_2 \rightarrow 2(V_{Mg}'', V_O^{\bullet\bullet})^{\times} + Mg_2SiO_4$	10,62	8,68

* These defect complexes were allowed to relax in order to find the minimum energy configuration. All other defects were fixed at their positions and only the surrounding spheres were allowed to relax.

2BP: using two body potential parameters
3BP: using three body potential parameters

Table 2: Activation energy of simulated atomistic diffusion jumps of Si in Mg_2SiO_4

mechanism	majority defect	atomistic diffusion jump	jump distance [Å]	energy of activation [eV]	approximate diffusion direction
1. ring exchange mechanism of two Si-atoms along octohedrally coordinated vacancies	$V''_{Mg(I)}$, $Mg_i^{\cdot\cdot}$	$2\,Si_{Si} \rightarrow 2\,Si_{Si}$	3,6	14,6	$[1\bar{1}2]$
2. vacancy mechanism along tetrahedrally coordinated vacancies	V''''_{Si}	$Si_{Si} \rightarrow V''''_{Si}$	5,4	50,1	$[\bar{1}20]$
3 vacancy mechanism	V''''_{Si}	$Si_{Si} \rightarrow V''''_{Si}$	3,0	11,0	[100]
4. vacancy mechanism along octahedrally coordinated vacancies	V''''_{Si}	$Si_{Si} \rightarrow V''''_{Si}$	3,6	2,4	$[1\bar{1}2]$
5. vacancy mechanism along tetrahedrally and octahedrally coordinated vacancies	V''''_{Si}	$Si_{Si} \rightarrow V''''_{Si}$	5,4	> 25,0	$[\bar{1}20]$
6. interstitial mechanism: $(Si_i^{\cdot\cdot\cdot\cdot}, 2\,V''_{Mg})^x$ clusters	$(Si_i^{\cdot\cdot\cdot\cdot}, 2\,V''_{Mg})^x$	$Si_i \rightarrow Si_i^{\cdot\cdot\cdot\cdot}$,	2,8	7,68	[221]
7. vacancy mechanism: $(2\,Mg_i^{\cdot\cdot}, V''''_{Si})^x$ clusters	$(2\,Mg_i^{\cdot\cdot}, V''''_{Si})^x$	$2\,Mg_i^{\cdot\cdot} \rightarrow 2\,Mg_i^{\cdot\cdot}, Si_{Si} \rightarrow V''''_{Si}$	3,6	8,06	$[1\bar{1}2]$
8. vacancy mechanism of a SiO_2-molecule	$(V''''_{Si}, 2\,V_O^{\cdot\cdot})^x$	$SiO_{2SiO_2} \rightarrow (V''''_{Si}, 2\,V_O^{\cdot\cdot})^x$	3,6	5,87	$[1\bar{1}2]$

Table 3: Summary of experimentally determined tracer diffusion coefficients in Mg_2SiO_4

tracer	cristallographic direction	D_0 [$cm^2 \cdot s^{-1}$]	ΔH [eV]	buffer	T [°C]	reference
Si-30	*	$1{,}5 \cdot 10^{-6}$	3,9 ± 0,4	MgO	1320-1680	[2]
Si-30	*	$2{,}19 \cdot 10^{-5}$	3,1 ± 0,4	?	1030-1504	[1]
Si-30	[100]	$6{,}72 \cdot 10^{-5}$	4,3 ± 0,67	MgO	1250-1520	this work
Si-30	[010]	$2{,}2 \cdot 10^{-3}$	5,0 ± 1,1	MgO	1250-1520	this work
Si-30	[001]	$7{,}8 \cdot 10^{-14}$	1,5 ± 1,1	MgO	1250-1520	this work
Si-30	[100]	29,35	6,2 ± 0,84	$MgSiO_3$	1250-1520	this work
Si-30	[010]	$5{,}0 \cdot 10^{-3}$	4,9 ± 0,81	$MgSiO_3$	1250-1520	this work
Si-30	[001]	$3{,}2 \cdot 10^{-13}$	1,5 ± 1,2	$MgSiO_3$	1250-1520	this work
Si-30	*	$2{,}67 \cdot 10^{-8}$	3,32 ± 0,41	MgO	1250-1520	this work
Si-30	*	$7{,}58 \cdot 10^{-5}$	4,33 ± 0,39	$MgSiO_3$	1250-1520	this work
O-18	*	3,5·10	3,84 ± 0,13	?	1275-1625	[4]
O-18	*	8,3	4,80	?	870-1280	[1]
O-18	*	$2{,}85 \cdot 10^{-2}$	4,29	?	1472-1734	[5]
O-18	*	$2{,}3 \cdot 10^{-6}$	3,02 ± 0,13	?	1300-1600	[3]
O-18	*	$6{,}86 \cdot 10^{-6}$	3,13 ± 0,21	*	1250-1450	this work
Mg-26	*	1,35	3,72 ± 0,21	?	944-1403	[1]
Mg-26	[100], [010]	$2{,}81 \cdot 10^{-4}$	2,25	?	1000-1200	[7] +
Mg-26	[001]	1,99·10	1,47	?	1000-1200	[7] +
Mg-26	[001]	$1{,}54 \cdot 10^{3}$	4,58	?	1300-1400	[15]
Mg-26	[100]	$1{,}94 \cdot 10^{-4}$	3,13 ± 0,69	MgO	1100-1385	this work
Mg-26	[010]	$1{,}47 \cdot 10^{5}$	4,69 ± 1,72	MgO	1100-1450	this work
Mg-26	[001]	$4{,}58 \cdot 10^{3}$	4,98 ± 1,2	MgO	1100-1350	this work
Mg-26	[100]	22,73	4,52 ± 0,5	$MgSiO_3$	1100-1350	this work
Mg-26	[010]	0,353	3,88 ± 0,16	$MgSiO_3$	1100-1385	this work
Mg-26	[001]	0,18	3,64 ± 0,32	$MgSiO_3$	1100-1450	this work
Mg-26	[100], [010]	0,506	4,01 ± 0,30	*	1100-1450	this work
Mg-26	[001]	1,388	3,91 ± 0,37	*	1100-1450	this work

* experimentally determined but no influence on D_i^*

? not determined

+ from interdiffusion experiments in $(Fe_x, Mg_{1-x})_2SiO_4$, 0,3490 < x < 0,8665, $p_{O_2} = 10^{-12}$ atm

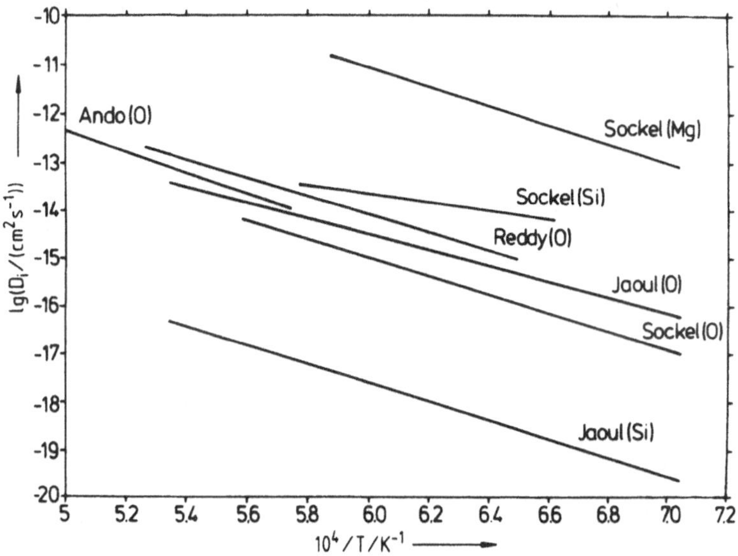

Fig. 1 Experimental values of the tracer diffusivities of the constituent elements in Mg_2SiO_4 according to earlier publications: magnesium [1]; silicon [1,2]; oxygen [1,3-5]

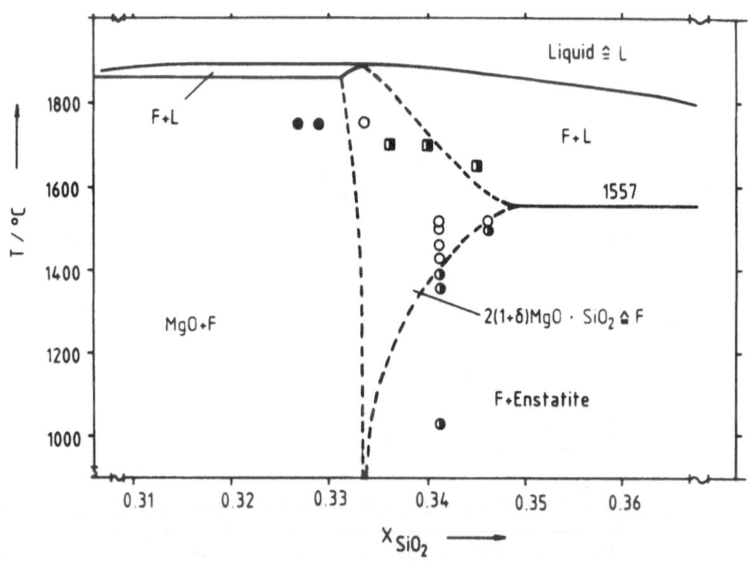

Fig. 2 Experimental values of the homogeneity range in Mg_2SiO_4 as determined by x-ray diffraction spectroscopy

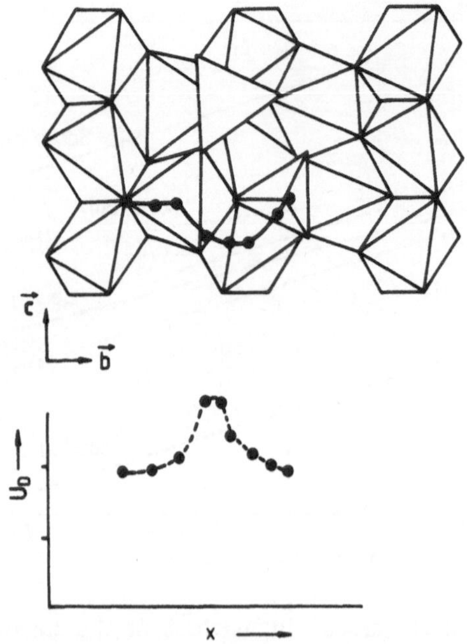

Fig. 3 Simulation of elementary diffusion jumps (schematic) (s. text)

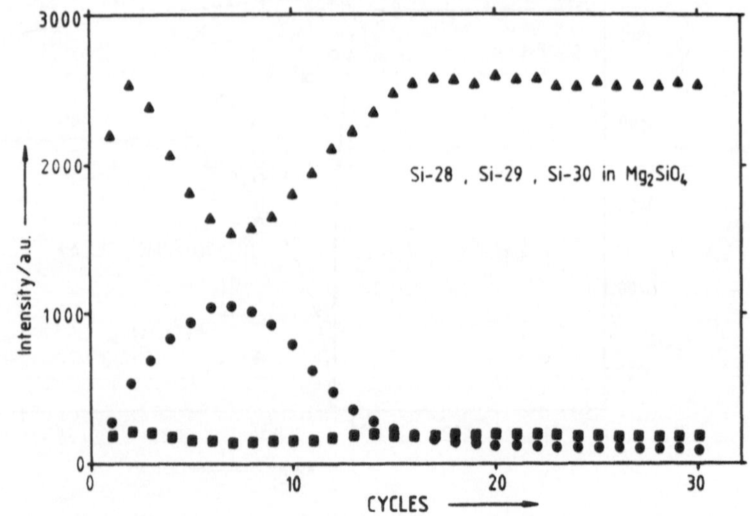

Fig. 4a Silicon isotope intensity profiles in a Mg_2SO_4 single crystal: [010], T = 1350 °C, t = 46 h, ● = Si-30, ■ = Si-29, ▲ = Si-28

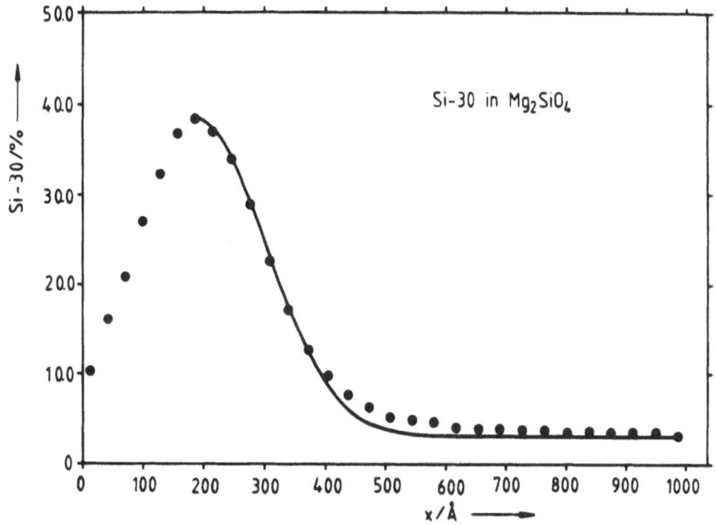

Fig. 4b Si-30 isotope concentration profile in a Mg_2SO_4 single crystal: solution of Fick's second law fitted to experimental points (D^*_{Si} = 3.30 · 10^{-18} cm^2s^{-1}, parameters as in Fig. 4a)

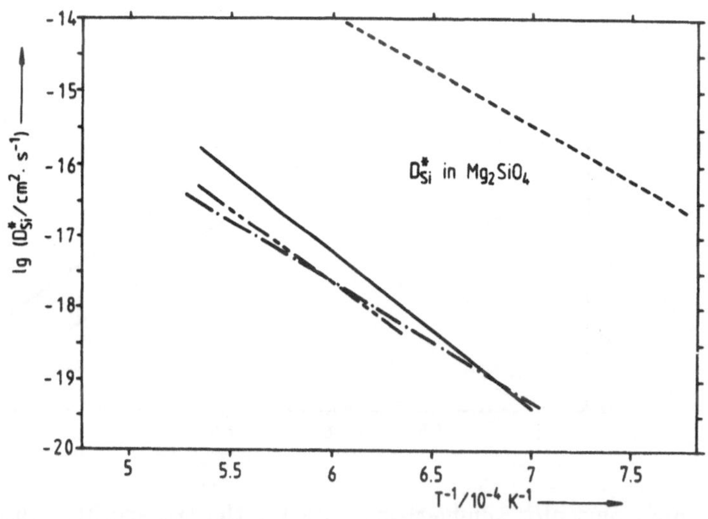

Fig. 5 Arrhenius plot summarizing the data for the tracer diffusivity of silicon in Mg_2SiO_4:
- - -Sockel et al. [1];— - - -Jaoul et al. [2];—·— this work, a_{MgSiO_3}=1;
——this work, a_{MgO} = 1

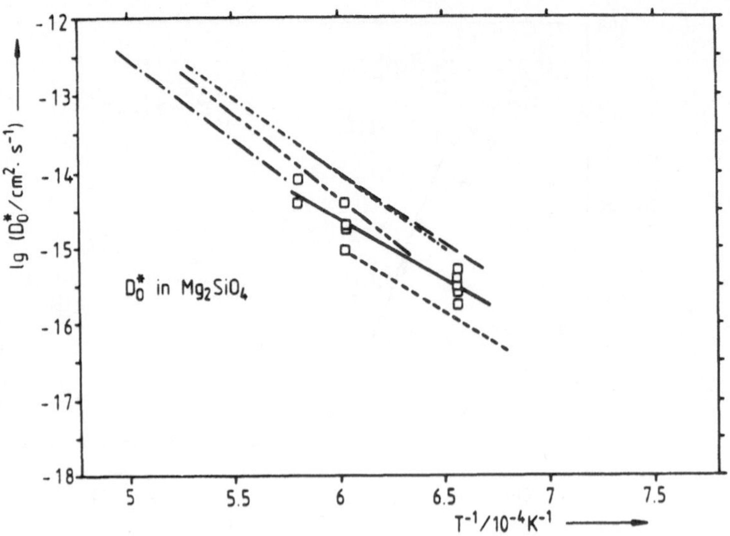

Fig. 6 Arrhenius plot summarizing data for the tracer diffusivity of oxygen in Mg_2SiO_4:
—-—-— Jaoul et al. [3]; — — Jaoul et al. [2]; - - - - Sockel et al. [1]; —·— Ando et al. [5]; -·-·- Reddy et al. [4]; —□— this work

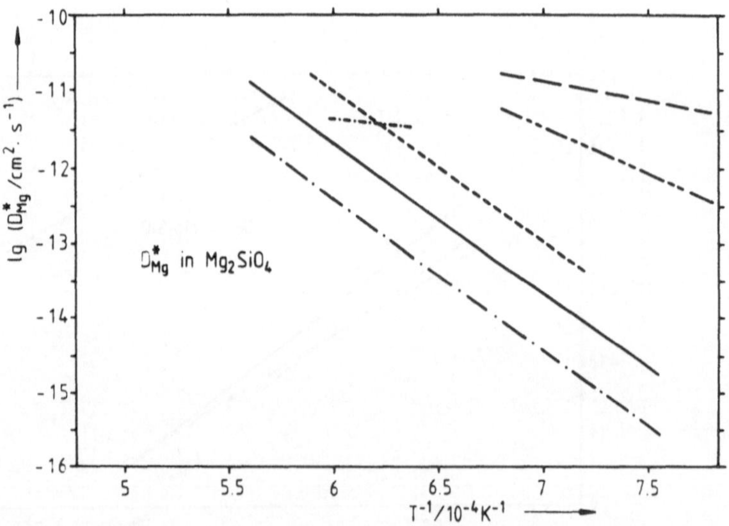

Fig. 7 Arrhenius plot summarizing data for the tracer diffusivities of magnesium in Mg_2SiO_4 for some crystallographic orientations.
The values from Buening and Buseck (B+B) were extrapolated from interdiffusion experiments in $(Fe_x, Mg_{1-x})_2SiO_4$.
— — — B+B [7] [001]; —··— B+B [7] [010] , [100];
- - - - Sockel et al. [1]; -·-·- Morioka [6]; —— this work, [001];
—·— this work , [010] , [100]

Acknowledgements. We gratefully appreciate the help of S. Scherrer and S. Weber (Ecole des Mines, Nancy, France) with the SIMS analyses, of H. Takei who kindly supplied the Mg_2SiO_4 single crystals, of C. R. A. Catlow and co-workers (University of Keele, U.K.) with the computer simulations, and of C. Esling (Universitè de Metz, France) with the orientation of the crystals. P. Shipley worked on the typescript and helped with language problems. Financial support of the Deutsche Forschungsgemeinschaft made this work possible.

References

1. Sockel HG, Hallwig D, Schachtner R (1980) Mater Sci Eng 42:59-64
2 Jaoul O, Poumellec M, Froidevaux C, Havette A (1981) In: Stacey FD et al. (eds.) Anelasticity of the earth, Geodyn Ser Vol 4, AGU, Washington DC, p 95-100
3. Jaoul O, Houlier B, Abel F (1983) J Geophys Res 88:613-624
4. Reddy KPR, Oh SM, Major Jr LD, Cooper Jr AR (1980) J Geophys Res 85:322-326
5. Ando K, Kurokawa H, Oishi Y, Takei H (1981) J Am Ceram Soc 64:C 30
6. Morioka M (1981) Geochim Cosmochim Acta 45:1573-1580
7. Buening DK, Buseck PR (1973) J Geophys Res 78:6852-6862
8. Borchardt G, Scherrer S, Weber S (1987) Fresenius' Z Anal Chem 329: 129-132
9. Rög G, Borchardt G (1984) J Electrochem Soc 131:380-384
10. Andersson K, Borchardt G, Müller O, Rög G (1987) In: Catlow CRA, Mackrodt WC (eds.) Nonstoichiometric Compounds, Advances in Ceramics, Vol 23:399-407
11. Andersson K, Borchardt G, Catlow CRA, Jackson RA, Doherty M (in preparation)
12. Catlow CRA (1982) In: Catlow CRA, Mackrodt WC (eds.) Computer Modelling of Solids, Lecture Notes in Physics 166:3-20
13. Takei H, Kobayashi T (1974) J Crystal Growth 23:121-124
14. Inoue T, Komatsu H, Takei T, Ozima M, Shimizu M (1980) Kristall Tech 15:1295-1302
15. Dearnley PA, Andersson K (1987) J Mater Sci 22:679-682
16. Andersson K, (1987) Dissertation, Technische Universität Clausthal

INFLUENCE OF CHROMIUM SEGREGATION ON THE TRANSPORT PROPERTIES OF IRON MONOXIDE

G. Petot-Ervas[°], H. Klimczyk[°°], C. Monty[x], C. Petot[xx], J. Janowski[xxx]

[°]ISMA, Laboratoire de Métallurgie, Bât. 413, Université Paris-Sud, 91405 - ORSAY (France).

[xx]CNRS, LIMHP, Université Paris-Nord, Av. J.B. Clément, 93430 - VILLETANEUSE (France)

[x]CNRS, Laboratoire de Physique des Matériaux, 1 place A. Briand, 92190 - MEUDON (France).

[xxx]Institute of Metallurgy, Academy of Mining and Metallurgy, 30059 - KRAKOW (Poland)

Summary

The diffusion coefficients of chromium and iron in wüstite have been determined in the temperature range 800 - 1200°C at an O/Fe ratio of 1.05. The diffusion of chromium has been found to be two orders of magnitude lower than the diffusion of iron. Assuming matter transport via "free mobile vacancies", a model has been developped that allows to analyse the effect of chromium on the transport properties of wüstite under chemical potential gradients. It follows that a decrease of the oxidation or reduction kinetics must be observed when wüstite is doped with chromium. This decrease is due to the difference in mobilities of chromium and iron in wüstite and to the demixing tendency of chromium when an oxygen potential gradient exists through the sample. Experimental results from the literature confirm these previsions.

[°]permanent address : CNRS-LIMHP
[°°]present address : Institute of Metallurgy, KRAKOW (Poland)

J. Nowotny and W. Weppner (eds.),
Non-Stoichiometric Compounds Surfaces, Grain Boundaries and Structural Defects, 411–421.

1. INTRODUCTION

The purpose of this work is to analyse the effect of chromium on the oxidation or reduction kinetics of iron monoxide, $Fe_{1-x}O$. In a previous paper [1] an analysis was made of the effect of impurities on the kinetics of oxidation or reduction of p-type semiconducting oxides within their range of stability. This analysis shows the effect of impurities on these kinetics in connection with segregation phenomena which appear in a material exposed to a chemical potential gradient if the diffusion coefficients of the solvent and solute cations are different.

Reduction kinetic studies of pure and chromium doped wüstite within its range of stability have been performed by Wagner et al.[2] These authors have observed that chromium decreases the rate of reduction. On the basis of our treatment it is of prime importance to know the relative mobilities of the cations in the oxide in order to understand the observations of Wagner et al.[2] and to analyse the effect of chromium on the oxidation or reduction rate of wüstite. In the present study we have then determined the diffusion coefficients of iron and chromium in wüstite single crystals. It follows from the literature data that there is a tendency for chromium to segregate under a chemical potential gradient [3-6]. According to these results it has been possible to discuss the effect of chromium on the transport properties of wüstite.

2. IRON AND CHROMIUM DIFFUSION IN WUSTITE

2.1. Defects and diffusion processes in wüstite

$Fe_{1-x}O$ is the transition metal oxide which shows the greatest deviation from stoichiometry in the series NiO, CoO, MnO, FeO[7]. While the highest value of x is about 0.001 for $Ni_{1-x}O$, x is between 0.05 and 0.15 for $Fe_{1-x}O$. Wüstite is stable only above ~ 560°C and the deviation from stoichiometry decreases with increasing temperature at constant oxygen partial pressure in contrast with many other oxides. The analysis of self diffusion experiments [8], of electrical conductivity [9] and thermogravimetry measurements [10] and of Seebeck coefficient values [9] have shown that wüstite is a p-type semiconducting oxide and that the prevailing point defects are iron vacancies α times ionized $V_{Fe}^{\alpha'}$. Nevertheless, due to the high concentration of defects, a simple defect model that does not consider complex defects would appear to be unappropriate for this oxide[11-12]. Diffraction techniques [13], in agreement with computer simulation studies, have confirmed this complex defect structure in which the basic cluster usually considered consists of four vacant neighbouring octahedral cation sites grouped around a tetrahedral site occupied by an interstitial iron ion. These defect aggregates seem to persist to at least 1150°C although the long range periodicity of the clusters is probably lost at high temperature. In view of this defect clustering and also of their own experimental results on iron self diffusion and isotopic effect, Peterson et al. [8] proposed that diffusion of iron occurs via "free mobile vacancies" coexisting with the defect clusters. One can point out that Kofstad et al. [14] have introduced the concept of excluded sites around the clusters. This idea has been developped more generally by Madouri et al. [15]. As a consequence, the macroscopic diffusion rate of "free mobile vacancies" decreases with an increase of x because the number of available neighbouring sites for a vacancy jump decreases as a result of an "excluded site" effect.

2.2. Sample preparation.

The wüstite single crystals used in this work were prepared by the floating zone method in an arc image furnace. The starting material was a polycrystalline rod of wüstite prepared by total oxidation of high purity rods of iron (Johnson and Matthey) in CO/CO_2 atmospheres at 1200°C. In order to stabilize any remaining stresses and damages introduced during the elaboration, the crystals were annealed at 1200°C in CO/CO_2 mixtures $\left(P_{CO_2} / P_{CO} \simeq 1\right)$ for several days. Then rectangular parallelepipeds of about 6 x 5 x 4 mm^3 were cut with a diamond saw and one side carefully polished to obtain a flat smooth surface. These samples were pre-annealed in a wüstite holder for several hours at the same temperature and oxygen partial pressure as those used later for the diffusion annealings. Then they were quenched as fast as possible in the CO/CO_2 atmosphere. The partial pressure of oxygen was established with mixtures of CO/CO_2 flowing through the laboratory tube at a rate of approximately 1 cm s^{-1}. Composition of the gas mixture was controlled near the sample with a zirconia electrochemical gauge.

2.3. Experimental techniques

a. Iron diffusion experiments

The radioactive tracer ^{59}Fe in the form of $FeCl_3$ in 0.5 N HCl was deposited on the polished surface of the sample by drying a drop of solution under a small heater. The temperature of the diffusion annealing was monitored with a Pt - Pt Rh 10 % thermocouple placed under the sample. At the end of the annealing treatment the edges of the samples were ground to eliminate the possibility of additional activity diffusing from the sides of the sample. The tracer concentration profile was determined by grinding off thin layers of the crystal and measuring the rest activity of the sample using a well type NaI crystal scintillation counter. The concentration - penetration curves follow the Fick's law solution corresponding to a thin film. The relationship between tracer concentration, C, and depth, X, is given by [16] :

$$C = \frac{K}{(\pi\ Dt)^{1/2}} \exp - \frac{X^2}{4\ Dt} \qquad (1)$$

where K is the concentration of tracer per unit area initially deposited onto the surface and t the diffusion annealing time.

b. Chromium diffusion experiments

During the diffusion annealing the polished surface of the pre-annealed wüstite single crystal was put in contact with a sintered tablet of FeO containing approximately 20 % by weight of Cr_2O_3. As was the case for the wüstite single crystals, this polycrystalline sample had been previously annealed at the same temperature and oxygen partial pressure that were used for the diffusion annealing.

After the diffusion annealing the samples were cut perpendicularly to the diffusion front. One of these surfaces was polished and the chromium diffusion profile was determined on this surface by electron probe microanalysis (EMA). Some samples were also analysed by secondary Ion Mass Spectrometry (SIMS). A perfect agreement of the two techniques was found.

The diffusion profiles were analysed using the solution of Fick's law assuming a constant chromium concentration C_s at the sample surface

$$C_s - C = C_s \text{ erf } X/2\ (Dt)^{1/2} \qquad (2)$$

where C is the concentration at the penetration depth X, D the diffusion coefficient and t the diffusion annealing time. According to our experimental conditions the concentration of chromium at the surface is relatively small. At 1000°C, C_s was found to be about 600 ppm in weight.

2.4. Results

The total penetration depths ($\approx (Dt)^{1/2}$) obtained in these experiments are found to be between 100 and 300 µm. The diffusion coefficients for iron and chromium have been determined for a ratio O/Fe = 1.05 [10] as a function of temperature. The Arrhenius plots are shown in Fig. 1.

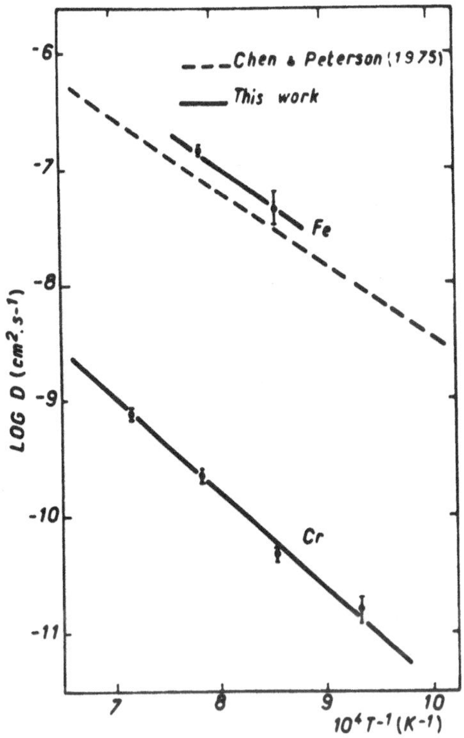

Fig. 1 : Diffusion of chromium and iron in iron monoxide

The iron tracer diffusion data obtained by Peterson et al. [8] are also reported in Fig. 1. Our results are in good agreement with these data. To our knowledge there are no values in the literature on chromium diffusion in FeO. Our results show that chromium ions diffuse much more slowly that iron ions in FeO. A similar behavior of chromium has been observed in the literature in pure CoO and NiO[17].

3. INFLUENCE OF CHROMIUM ON THE OXIDATION OR REDUCTION KINETICS OF WUSTITE

In the following section we will analyse the effect of chromium on the transport properties of wüstite when an oxygen potential gradient exists across the sample. Such a situation is encountered during the oxidation or reduction of the material.

One can recall that wüstite is an electronic conductor, and matter transport occurs via "cationic vacancies"[7-15]. In a previous paper an analysis was made of the effect of impurities on the transport processes under oxygen potential gradients in p-type semiconducting solid solution $AO-BO_{\Upsilon}$ [1]. The formalism will be briefly outlined here. Indeed, when an oxygen potential gradient appears in a material such as wüstite it gives rise to a flux of vacancies toward the direction of the lower oxygen partial pressure accompanied by a flux of cations in the opposite direction [18]. Away from vacancy sources and sinks lattice sites are conserved. The different fluxes are then coupled by the condition :

$$J_V + \Sigma J_C = 0 \tag{3}$$

Furthermore, the shift velocity of the oxidation or reduction front (with respect to the laboratory reference frame) is given by the following expression [16] :

$$v = J_V V_M \tag{4}$$

where V_M is the molar volume of the oxide and J_V the flux of cationic vacancies.

The flux of vacancies in a pure oxide under an oxygen potential gradient is given by the following expression [19-20]:

$$J_V = - C_M D_A \delta \ln a_0 / \delta x \tag{5}$$

In order to understand the effect of impurities on the cationic vacancy flux responsible for the oxidation or reduction kinetics it is then necessary to express the flux of cations A and B in the solid solution $AO - BO_{\Upsilon}$ as a function of the oxygen potential gradient.

a. Derivation of the vacancy cationic flux in an oxide solid solution $AO - BO_{\Upsilon}$ - General treatment.

In electronic conductors it has been shown, assuming local chemical equilibrium $\left(d\mu_A = d\mu_{A^{2+}} - 2d\mu_{\bar{e}} \text{ and } d\mu_B = d\mu_{B^{2\Upsilon+}} - 2\Upsilon d\mu_{\bar{e}}\right)$ that the flux of cations is proportional to the chemical potential gradient of the metal [1,21]:

$$J_{A^{2+}} = - C_M D_A (1-m) \delta \ln a_A/\delta x \tag{6}$$

$$J_{B^{2\Upsilon+}} = - C_M D_B m \delta \ln a_B/\delta x \tag{7}$$

where C_M is the overall concentration (in moles cm^{-3}) of cation sites in the host lattice AO, m the molar concentration of the impurity B, D_A and D_B the diffusion coefficients of the cations A and B in the solid solution.

In order to compare these two fluxes we have expressed them as a function of the oxygen potential gradient. It follows from the local chemical equilibrium between the different species ($1/2\ O_2 + A \rightleftarrows AO$, $\gamma/2\ O_2 + B \rightleftarrows B\,O_\gamma$) that :

$$d \ln a_A = d \ln a_{AO} - d \ln a_O \qquad (8)$$
$$d \ln a_B = d \ln a_{BO_\gamma} - \gamma\, d \ln a_O \qquad (9)$$

Furthermore, in the case of low enough impurity concentration that Raoult law may be applied to the quasibinary system $AO - BO_\gamma$ one can show, according to Gibbs-Duhem equation $\left(x_{AO}\ d \ln a_{AO} + x_{BO_\gamma}\ d \ln a_{BO_\gamma} = 0\right)$, that :

$$d \ln a_{AO}\ /\ d\,m = -\ 1/(1-m) \qquad (10)$$
$$d \ln a_{BO_\gamma}\ /\ dm = 1/m \qquad (11)$$

Inserting Eqs 8, 9, 10, 11 in Eqs 6 and 7 we get for the cationic vacancy flux (Eq. 3), assuming that no correlation effect exists between different fluxes :

$$J_V = -\ C_M\ (\gamma D_B - D_A)\ [m\ \delta \ln a_O\ /\ \delta x - \delta m/\delta x] + C_M\ D_B\ (1 - \gamma)\ \delta m/\delta x - C_M\ D_A\ \delta \ln a_O/\delta x \qquad (12)$$

where a_O is the oxygen activity in the material, related to the oxygen partial pressure by the relation $a_O = P_{O_2}^{1/2}$.

As we have indicated previously the shift velocity of the oxidation or reduction front when the sample is exposed to a change in P_{O_2} is proportional to the cationic vacancy flux (Eq. 4). The vacancy flux through the surface then controls the oxidation or reduction kinetics of the material. Consequently, the effect of impurities on the oxidation or reduction kinetics of the material may be given by the difference of cationic vacancy fluxes in the doped and pure oxide :

$$\Delta J = J_V\ (\text{doped}) - J_V\ (\text{pure}) \qquad (13)$$

From Eqs 5 and 12 it follows :

$$\Delta J = -\ C_M\ (\gamma D_B - D_A)\ [m\ \delta \ln a_O\ /\ \delta x - \delta m/\delta x] + C_M\ D_B\ (1 - \gamma)\ \delta m/\delta x \qquad (14)$$

b. Demixing tendency under oxygen potential gradients

Simultaneously to the flux of vacancies, a demixing tendency is observed in the solid solution $AO\text{-}BO_\gamma$ if the mobilities of the cations

A and B are different [18]. Indeed, the exchange frequency between vacancies and neighbouring cations is lower for the less mobile cations. During oxidation or reduction reactions, vacancies move from the higher to the lower oxygen partial pressure side. Consequently, an enrichment of the less mobile cations must be observed at the side of the lower oxygen activities. In view of the chromium and iron diffusion results an enrichment of chromium must be observed in wüstite at the lower oxygen partial pressure side (that is, on the surface of the crystal during a reduction process).

c. Experimental results and discussion

From our experimental results it follows that :

$$\gamma D_{Cr} < D_{Fe}$$

with $\gamma = 1.5$ in the solid solution considered.
In addition, as we have previously indicated the chromium concentration must decrease near the lower oxygen partial pressure side. The gradients $\delta m/\delta x$ and $\delta \ln a_0/\delta x$ have then opposite signs. Consequently, it follows from Eq. 14 that a decrease of the oxidation or reduction kinetics must be observed when wüstite is doped with chromium.
To verify this analysis we have plotted (fig. 2) re-equilibration kinetics curves deduced from experimental results obtained by Wagner et al. [2] for pure and chromium-doped wüstite samples using a gravimetric technique. The authors [2] have previously checked that wüstite doped with 0.67 wt pct Cr forms a complete solid solution. During these experiments, the crystal initially in thermodynamical equilibrium is submitted to a sudden change of the equilibrium oxygen partial pressure. The results shown on Fig. 2 have been obtained at 950 and 1050°C, respectively. They correspond to the reduction of a wüstite sample from an O/Fe ratio of 1.125 to a ratio of 1.050. In agreement with our analysis the results of Fig. 2 show that the reduction rates are lower when wüstite is doped with 0.67 wt pct Cr.
Furthermore, according to our general treatment (Eq. 14) the decrease of the re-equilibration kinetics when the material is doped suggests a demixing tendency for chromium at the crystal surface during the reduction process. Experiments to check this trend are in progress. Nevertheless, examination of wüstite scales formed during the oxidation of Fe-Cr and Fe-Cr-Al alloys shows an increase of the chromium concentration near the metal/oxide interface [3-5,23]. Similar observations have also been made with Ni-Cr alloys[22,23]. Let us remark that the oxygen potential gradient is maintained continuously throughout the oxide scale during corrosion. The segregation phenomena then increase with time until formation of a second phase. During corrosion of Fe-Cr or Ni-Cr alloys authors generally observed the formation of a Cr_2O_3 layer at the alloy/oxide interface after few hours of oxidation.

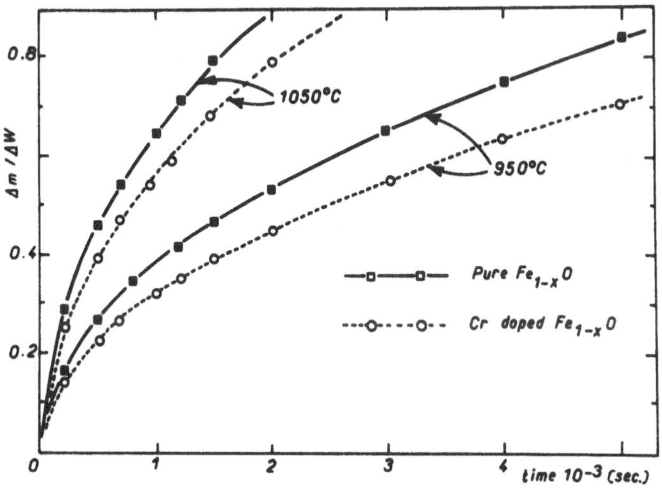

Fig. 2 : Re-equilibration kinetics of pure and chromium doped iron monoxide [2].

In our formal analysis we have assumed that no interaction effect exists between the different fluxes wich appear in the oxide solid solution under an oxygen potential gradient. The good agreement observed between our analysis and the experimental results obtained for pure and chromium doped wüstite samples seems to confirm our hypothesis. This then suggest that the chromium-vacancy binding energy is weak in this material.

4. CONCLUSION

The diffusion coefficients of iron and chromium in wüstite single crystals have been measured in the temperature range 800 - 1200°C and for an O/Fe ratio of 1.05. Assuming matter transport via "free mobile vacancies" as suggested by other experiments it has been possible to analyse the effect of chromium on the transport properties of wüstite under an oxygen potential gradient. From our experimental work we have shown that chromium diffuses more slowly than iron in wüstite. A chromium enrichment must then be observed at the lower oxygen partial pressure side when an oxygen potential gradient exists across the sample. The difference of mobilities of chromium and iron in wüstite and the demixing tendency of chromium allow to predict a decrease of the oxidation or reduction kinetics of wüstite when the material is doped with chromium. These previsions are confirmed by re-equilibration kinetic results and by chromium enrichment at the alloy/oxide interface of the FeO layers formed during the corrosion of Fe-Cr alloys.

Acknowledgements

The authors are grateful to Prof. J. Philibert for his supporting interest in our work. Thanks are also due to Dr. B. Lesage for his help with the tracer diffusion experiments and to B. Picinin for his technical assistance during the experimental studies.

REFERENCES :

1. G. PETOT-ERVAS, C. PETOT to be published

2. R.L. LEVIN, J.B. WAGNER, Trans. Met. Soc. AIME, 233, 159, 1965.

3. J. BENARD, Oxid. des métaux, Gauthier-Villars 1962.

4. A.V. MALIK, D.P. WHITE, Oxid. Met. 16, 339, 1981.

5. Y. IKEDA, K. NII, Oxid. Met., 12, 487, 1978.

6. M. HAJDUGA, J. KUCERA, Oxid. Met. 25, 121, 1988.

7. P. KOFSTAD, Nonstoichiometry, Diffusion and Electrical Conductivity in binary metal oxides, Wiley Interscience, N.Y., 1972.

8. W.K. CHEN, N.L. PETERSON, J. Phys. Chem. Solids, 36, 1097, 1975.

9. E. GARSTEIN, T.O. MASON, J.B. COHEN, J. Phys. Chem. Solids, 47, 759, 1986.

10. R.A. GIDDINGS, R.S. GORDON, J. Am. Cer. Soc., 56, 111, 1973.

11. S.M. TOMLINSON, C.R.A. CATLOW, J.H. HARDING, Transport in Nonstoichiometric Compounds Ed. G. Simkowitch, V.S. Stubican, Plenum Press, 539, 1985.

12. R.W. GRIMES, A.B. ANDERSON, A.H. HEUER, J. Am. Cer. Soc. 69, 619, 1986.

13. J.R. GAVARRI, C. CAREL, D. WEIGEL, Solid. St. Phys. 12, 337, 1979.

14. P. KOFSTAD, A.Z. HED, J. Electrochem. Soc., 115, 102, 1968.

15. A. MADOURI, C. MONTY, Adv. in ceramics, 23, 55, 1987.

16. J. PHILIBERT, Diffusion et Transport de Matière dans les Solides, Ed. Phys. 1985.

17. N.L. PETERSON, Transport in Nonstoichiometric Compounds, Ed. G. Petot-Ervas, C. Monty, Hj. Matzke, North Holland, Sol. St Ionics, 201, 1984.

18. W. LAQUA, H. SCHMALZRIED, High Temp. Corrosion, Ed. R.A. Rapp 115, 1983.

19. P. OCHIN, G. PETOT-ERVAS, C. PETOT, J. Phys. Chem. Solids 46, 695, 1985.

20. G. PETOT-ERVAS, C. PETOT, F. GESMUNDO, J. Phys. Chem. Solids 48, 767, 1987.

21. C. WAGNER, Corr. Sci. 9, 91, 1969.

22. F.H. STOTT, J.S. PUNNI, G.C. WOOD, G. DEARNALEY, Transport in Nonstoichiometric Compounds, Ed. G. Simkowitch, V.S. Stubican, Plenum Press, 463, 1985.

23. G. BEN ABDERRAZIK, Thesis University Paris XI, 1986.

INFLUENCE OF POINT DEFECTS ON THE NEAR-SURFACE DIFFUSION IN SOME OXIDE SYSTEMS

V. S. Stubican and C. M. Lin
The Pennsylvania State University
University Park, PA 16802
U.S.A.

ABSTRACT. Diffusion of the ^{57}Co isotope on Fe_3O_4 (110) and NiO (100) surfaces was investigated by the edge-source method. The surface diffusion parameter, $\alpha D_s\delta$, where α is the segregation factor, D_s, the surface diffusion coefficient, and δ the thickness of the high-diffusivity layer, was determined at 750° for different partial pressures of oxygen. In both cases point defects strongly influenced surface diffusion. For Fe_3O_4 the vacancy mechanism is dominant at high oxygen activities and interstitial (or interstitialcy) mechanism dominates at low oxygen activities. For NiO the surface diffusion at low oxygen activities is influenced by aliovalent impurities and at high oxygen activities the strong influence of the intrinsic defects, nickel vacancies, on surface diffusion was observed. It was concluded that the similar mechanisms operate during surface diffusion in the near surface layer and during diffusion in the lattice.

1. Introduction

In a recent article Bonzel (1) has summarized extensive work on surface diffusion in metals. To understand the mechanisms of surface diffusion it is helpful to think in terms of an atomistic model e.g. the terrace-ledge-kink model (2). Two important defects exist on the surface: adatoms (adsorbed atoms), and terrace vacancies. The adatom diffusion mechanism was treated theoretically by Blakely (3), Gjostein (4) and Choi and Shewmon (5). Birchendall (6), however, has proposed a surface vacancy mechanism for surface selfdiffusion in metals. He argues that the concentration of adatoms should be far less than the concentration of vacancies. In metals vacancy concentrations in the bulk approach 10^{-3} - 10^{-4} per atom. Vacancies are favored near the surface since relaxation can occur by outward displacement allowing reduced interference by the nearest neighbors.

The purpose of our research was to determine the influence of point defects on the near surface diffusion in Fe_3O_4 (magnetite) and NiO where the concentration of defects can be varied by changing the oxygen activity.

2. Experimental

Diffusion of ^{57}Co in the surface layers of Fe_3O_4 and NiO was studied by using the edge-source technique (7). The single crystals of Fe_3O_4 were grown by the skull method. By using this method clusters of crystals were obtained. Monocrystals of Fe_3O_4 were extracted from the clusters. The crystals obtained were analyzed by semi-quantitative emission spectroscopy and the following impurities were detected, Si 100, Al 20, Cr100, Mn 200 ppm by weight. The method of preparing crystals for the surface diffusion studies was described previously (7).

J. Nowotny and W. Weppner (eds.),
Non-Stoichiometric Compounds Surfaces, Grain Boundaries and Structural Defects, 423–431.

To equilibrate the magnetite crystal prior to a diffusion run it was annealed in a furnace at a prefixed oxygen partial pressure at 750°C for ~ 100 hrs in CO-CO_2 gas mixtures. The oxygen partial pressure, and the temperature were measured close to the sample simultaneously with a help of a solid state galvanic cell using a ZrO_2-cell and a Pt-Pt 10% Rh thermocouple respectively. After an equilibration anneal the high specific activity isotope ^{57}Co (specific activity of 0.1 m Ci/g) was applied to one of the surfaces of the crystal by drying a drop of ^{57}Co-sulfate solution with an IR lamp. Surface diffusion runs were performed at 750°C for 3-5 hrs, at the partial pressures of oxygen, under which the crystals had been previously equilibrated. After diffusion annealing, all surfaces, except the tracer-deposited and diffusion surfaces, were polished with 1 μm diamond paste to a depth of $>10\sqrt{Dt}$. A serial sectioning technique was used to determine concentration profiles.

The volume diffusivity of Co in Fe_3O_4 was determined by sputtering metallic Co onto a surface of the Fe_3O_4 crystal which was previously equilibrated for 72 hours under the oxygen partial pressure under which diffusion was to be performed. After Co was oxidized at ~250°C for a short time volume diffusion runs were carried out under different oxygen partial pressures. Volume diffusion profiles were determined by using SIMS.

The single crystals of NiO were cut from a solidified NiO-rod obtained by melting in an arc image furnace. The crystals were analyzed by semi-quantitative emission spectroscopy and the following impurities were detected; Si50, Al<10, Cr20, Fe20, Mn 10 ppm. by weight. The concentrations of Na, Li and Ca were not determined.

The preparation of crystals for the surface diffusion and volume diffusion runs was the same as for Fe_3O_4. Diffusion runs at 750°C were performed under the same oxygen partial pressures as the equilibration runs, using CO-CO_2, Ar+Ar 1% O_2 and O_2 gas mixtures.

3. Results and Discussion

The surface diffusion data were evaluated by using Whipple's formula (8,9) which contains a factor of 1/2 and can be expressed as:

$$\alpha D_s\,\delta = 1/2\,(\partial \ln \bar{c}/\partial y^{6/5})^{-5/3}\,(4D_v/t)^{1/2} 0.78^{5/3}\ (1) \qquad (1)$$

where α is the ^{57}Co segregation factor, D_s the surface diffusion coefficient, δ the thickness of the high diffusivity layer, $\bar{c}$ the average specific activity in a section of thickness Δy from $(0 \leq x \leq \infty)$ at a distance y from the original surface, D_v the volume diffusion coefficient of the tracer, and t is the time of anneal.

3.1 MAGNETITE, $Fe_{3-y}O_4$.

To calculate the surface diffusion product, $\alpha D_s \delta$, the values of Dv, the volume diffusion coefficient of ^{57}Co in Fe_3O_4 should be know. Dieckmann and Schmalzried (10) have determined cobalt tracer diffusion coefficients in magnetite for the temperature range 906-1210°C, and the activities of oxygen between 10^{-3} and 10^{-15} atm. From their results the following expression can be derived:

$$D^*_{Co} = 1.35 \times 10^{-11} a_{O2}^{2/3} \exp\left(\frac{149.2\ KJ/mol}{RT}\right) +$$

$$1.12\,.10^{8}\ a_{O2}^{-2/3} \exp\left(\frac{-615.5\ KJ/mol}{RT}\right) cm^2/s \qquad (2)$$

To verify if this expression is valid for the temperatures below 906°C e.g. for 750°C two volume diffusion runs for Co-ion in Fe_3O_4 were made. The calculated and experimental results are shown on Fig. 1. The agreement obtained is quite good. Eq. 2 was then used to calculate the volume diffusion coefficients of the Co-ion which were necessary to evaluate surface diffusion products using Eq. 1.

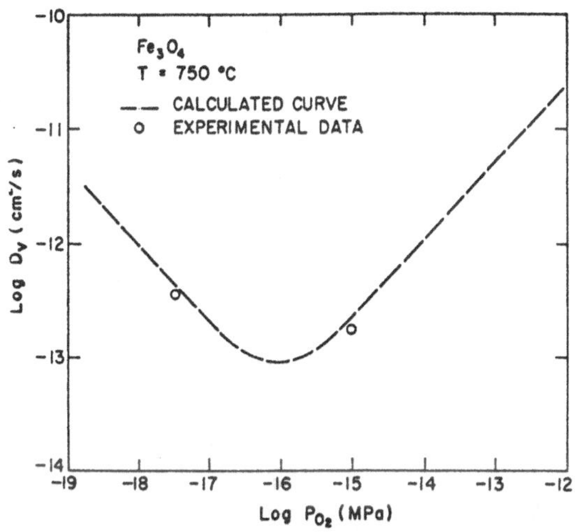

Fig. 1. The volume diffusivity of ^{57}Co in Fe_3O_4 as a function of the partial pressure of oxygen. Dashed line shows calculated values (Ref. 10), the open circles are the experimental data.

The values for the slope $\partial \ln \bar{c}/\partial y^{6/5}$ in Eq. 1 were obtained from the surface diffusion plot shown in Fig. 2. The first part of this plot corresponds to the volume diffusion and the linear part to the surface diffusion.

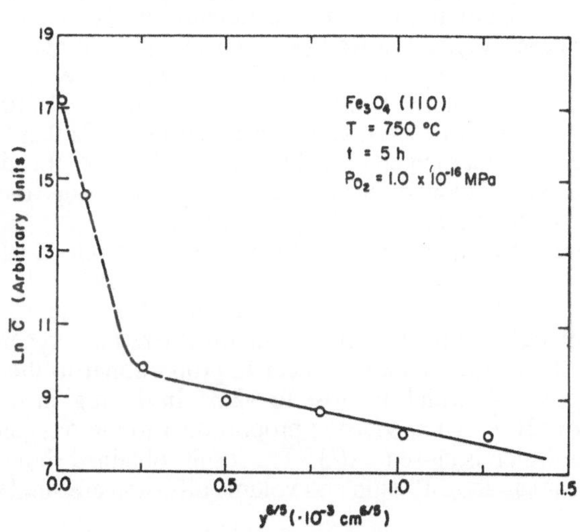

Fig. 2. Concentration profile for diffusion of ^{57}Co in Fe_3O_4 surface layer.

Several surface diffusion runs at 750°C under different oxygen partial pressures were made and the results obtained are shown in Fig. 3. It is evident that the plot log $\alpha D_s\delta$ vs logP_{O_2}, for the surface diffusion shows a minimum at $P_{O_2} = 10^{-17}$-10^{-16} MPa. It is reasonable to assume that the segregation factor of the tracer, α, and the thickness of the high diffusivity layer, δ, do not change with P_{O_2}. Changes in $\alpha D_s\delta$ shown on Fig. 3 reflect the changes in the surface diffusion coefficient, D_s.

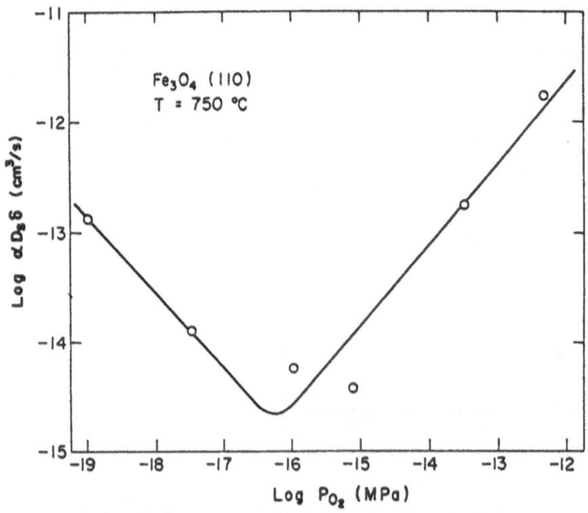

Fig. 3. Surface diffusion product, $\alpha D_s\delta$, of ^{57}Co in Fe_3O_4 surface layer as a function of partial pressure of oxygen at 750°C.

Dieckmann and Schmalzried (11) have shown that at high oxygen potentials magnetite exists with a cation deficit (cation vacancies); at low oxygen potential it has a cation excess (interstitial cations). Furthermore, if it is assumed the point defects form an ideal solution with the crystal, the point defect thermodynamics then describes volume diffusion and nonstoichiometry quantitatively. At a sufficiently low concentration of defects, at high oxygen activities $(V_O)\alpha(V_T)\alpha a_{O_2}^{2/3}$, where index T and O marks tetrahedral and octahedral sites, respectively. At low oxygen activities, $(Fe_i^{2+})\ \alpha\ (Fe_i^{3+})\ \alpha a_{O_2}^{-2/3}$, where, i, marks interstitial sites. Dieckmann and Schmalzried (10) found experimentally for the volume diffusion of Cr, Co and Fe isotopes in Fe_3O_4 that the slope d log D*/d log a_{O_2} is close to 2/3 at high oxygen partial pressures where volume diffusion proceeds by the vacancy mechanism. At low oxygen partial pressures the slope d log D*/d log a_{O_2} is -2/3, indicating an interstitial (or intersticialcy) mechanism.

The curve which shows the change of the surface diffusion product, $\alpha D_s\delta$, for ^{57}Co in the near surface layer of the Fe_3O_4 (110) plane as a function of P_{O_2} at 750°C (Fig 3) exhibits a minimum between 10^{-17} and 10^{-16} MPa. The results of best fitting indicate that there are two sets of correlated data points. In the region where the oxygen partial pressure is above 10^{-16} MPa, the $\alpha D_s\delta$ of the Co-tracer is proportional to the oxygen partial pressure to the power of 0.736, which is close to +2/3. In the region where the oxygen partial pressure is below 10^{-17} MPa, $\alpha D_s\delta$ is proportional to the oxygen partial pressure to the power of -0.689 which is close to -2/3. The results obtained show that the diffusion mechanisms for the near surface diffusion and volume diffusion are similar.

3.2 NICKEL OXIDE, $Ni_{1-y}O$.

To calculate the surface diffusion product, $\alpha D_s\delta$, the values of, D_V, the volume diffusion coefficient of Co ion in NiO had to be determined. Since the oxygen partial pressure dependence of Co-ion diffusion in NiO at 750°C was not measured previously, it was necessary to obtain experimentally the volume diffusion coefficients at 750°C over the same P_{O_2} range as for the surface diffusion experiments. The results obtained are shown in Fig. 4. It can be observed that the volume diffusivity of the Co-ion in NiO remains constant for $P_{O_2} <10^{-7}$ MPa indicating that in this region the volume diffusion was impurity controlled (extrinsic region). A strong increase in the volume diffusivity is observed above $\sim 10^{-7}$ MPa showing that at high oxygen partial pressures the volume diffusivity is increasingly dependent on the intrinsic cation vacancy concentration.

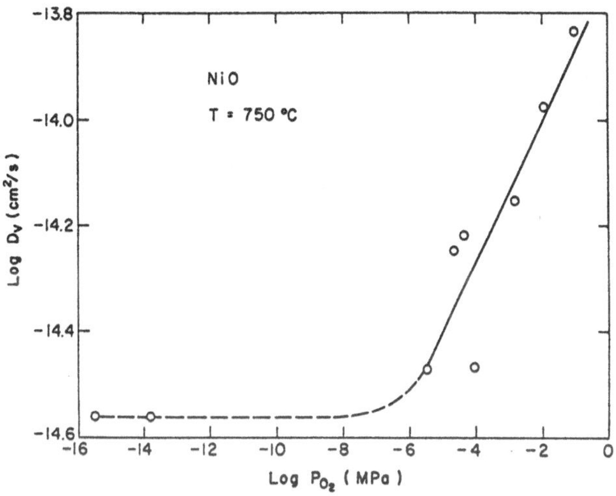

Fig. 4. Volume diffusivity of ^{57}Co in NiO as a function of partial pressure of oxygen at 750°C.

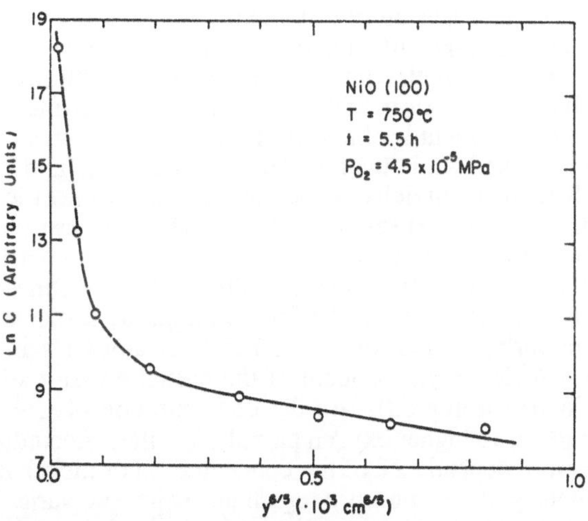

Fig. 5. Concentration profile for diffusion of ^{57}Co in NiO surface layer.

The surface diffusion concentration profiles of Co-tracer were then obtained as described previously and the values for the slope $\partial \ln\bar{c}/\partial y^{6/5}$ were calculated from the similar plots shown on Fig. 5. The oxygen partial pressure dependence of the surface diffusion product, $\alpha D_s\delta$, is illustrated in Fig. 6. Below P_{O_2} ~10^{-6} MPa, the surface diffusion product is independent of the oxygen partial pressure, and above P_{O_2} ~10^{-6} MPa, there is a strong dependence of $\alpha D_s\delta$ on P_{O_2}. Again, it is reasonable to assume that α and δ do not change with P_{O_2}. Changes in $\alpha D_s\delta$ shown in Fig. 6 reflect the changes in D_s.

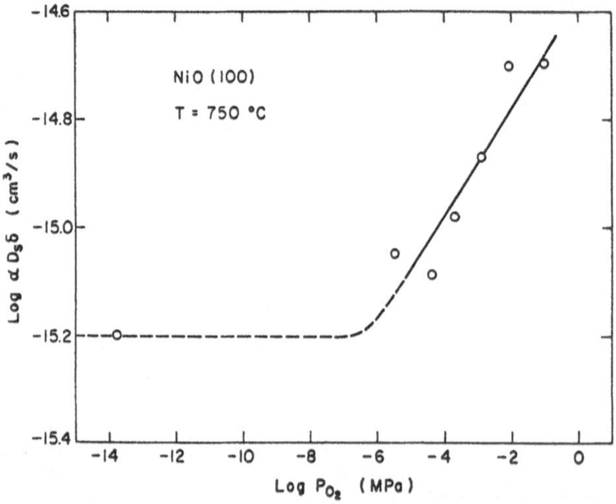

Fig. 6. Surface diffusion product, $\alpha D_s\delta$, in NiO surface layer as a function of the partial pressure of oxygen.

Volpe and Ready (12) measured the volume diffusion of ^{63}Ni in NiO single crystals at oxygen partial pressures 10^{-8} - 10^{-1} MPa at 1380° and 1245°C. They concluded that both singly and doubly charged vacancies contribute to the selfdiffusion. In NiO the oxygen partial pressure dependencies of the electrical conduction (13) and the cation selfdiffusion (14) require at least two types of charged atomic defects. Following these studies, Peterson and Wiley (15) further analyzed the diffusion model for NiO. They also proposed that $[V_{Ni}']$ and $[V_{Ni}'']$ are point defects which contribute to the volume selfdiffusion and these defects have different mobilities. If their model is applied to the experimental results of Volpe and Ready, a reasonable agreement is obtained. However, Peterson and Wiley's model is based on measurements carried out in the high temperature region and at oxygen partial pressures $>10^{-7}$ MPa, where the point defect structure is primarily intrinsic. At low temperatures and low oxygen partial pressures, this may not be the case due to the influence of aliovalent impurities. From the measurements of the Seebeck coefficient for pure NiO and Cr-doped NiO at 872°C, Nowotny and Rekas (16) calculated concentrations of defects at different partial pressures of oxygen and aliovalent impurity contents. At low temperatures, e.g. 872°C, for impurity content of only 1.7×10^{-2} at% of Cr^{3+} in NiO the total concentration of cation vacancies is independent of the partial pressure of oxygen below P_{O_2} ~3.2×10^{-6} MPa (extrinsic region). As the concentration of Cr^{3+} in NiO increases, the extrinsic region extends to higher oxygen partial pressures. Accordingly, the exponent of oxygen partial pressure dependence of the concentration of nickel vacancies can vary from $1/\infty$ to 1/4 continuously. In the intrinsic region this exponent varies between 1/6 and 1/4. In the transient region (between extrinsic and intrinsic regions) it changes from 1/∞ to 1/6. To evaluate semi-quantitatively the concentrations of point defects in our samples, we used the value of 0.9 eV for the Fermi energy which corresponds to a concentration of 8.4 x 10^{-3} atom % Cr^{3+}-ions in Nowotny and Rekas' work. According to estimations this concentration of Cr^{3+} ions is approximately equivalent to the concentration of aliovalent impurities in our samples. The other assumption used in these calculations was that the Seebeck coefficient at 750°C is approximately equal to that at 872°C. The calculated results are presented in Fig. 7 and indicate that the oxygen partial pressure dependence of the volume diffusion observed in our study (Fig. 4) at P_{O_2} greater than 10^{-7} MPa is for the transient region.

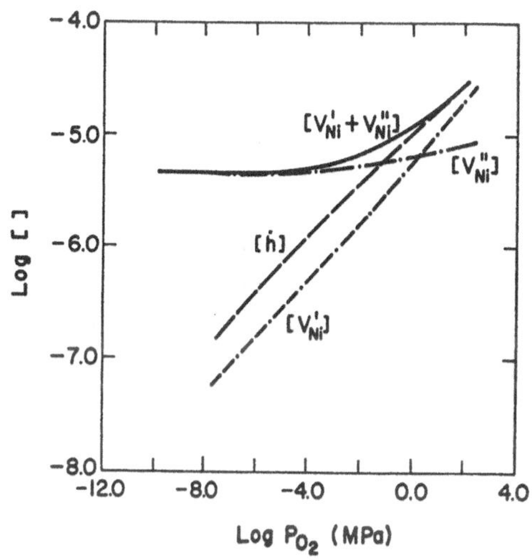

Fig. 7. Calculated concentrations of defects at 750°C as a function of the partial pressure of oxygen for NiO doped with ~8.4 x 10^{-3} at % of trivalent impurities.

Similar reasoning seems to be applicable to surface diffusion (Fig. 6). However, aliovalent impurity segregation to the surface layer influences the concentration of nickel vacancies and consequently the surface diffusivity. The qualitative results of impurity segregation in our samples at 750°C are shown in Fig. 8.

The consequence of the segregation of the higher valent impurities to the near surface layer is that in the transient region surface diffusion coefficient should increase more slowly with the increase in the partial pressure of oxygen than the volume diffusion coefficient. Figs. 4 and 6 show that such a behavior was observed in our experiments.

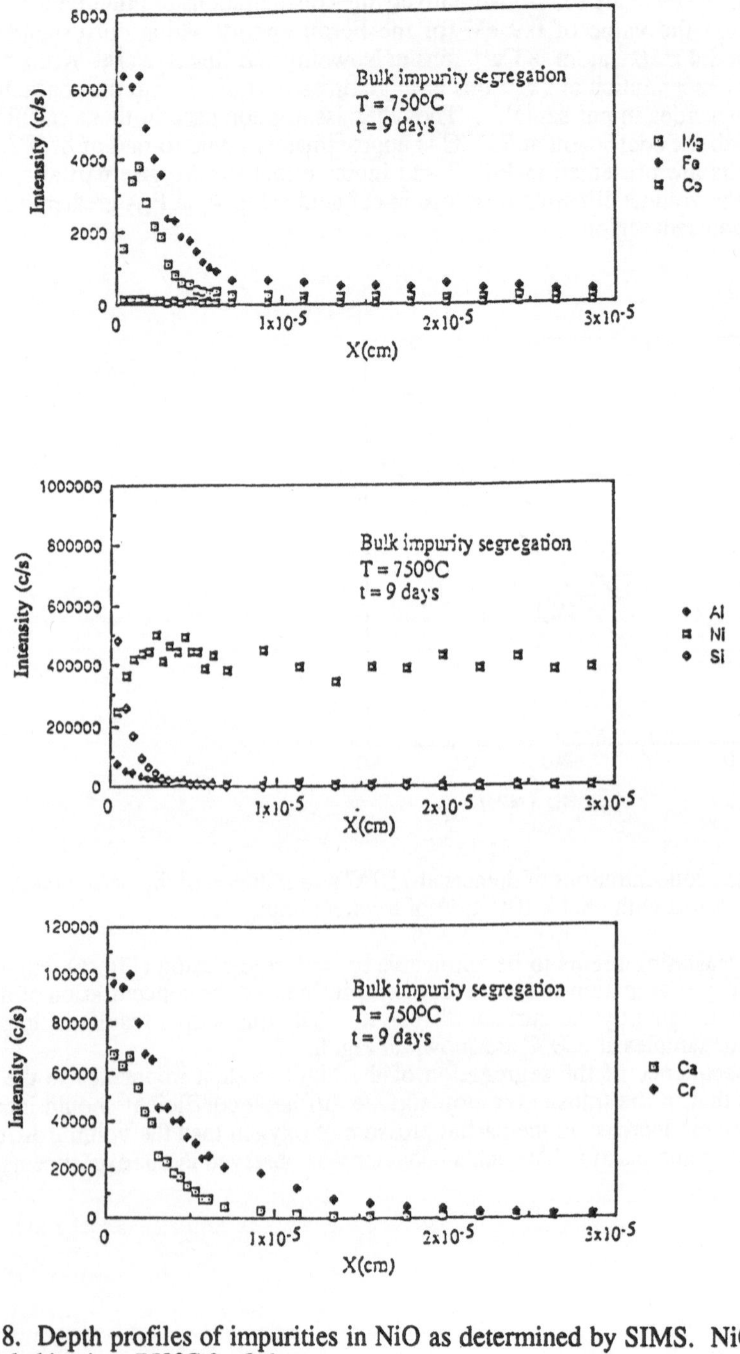

Fig. 8. Depth profiles of impurities in NiO as determined by SIMS. NiO single crystal annealed in air at 750°C for 9 days.

ACKNOWLEDGEMENTS

Supported by the U.S. Department of Energy under grant No. DE-FG02-85ER45180. The authors thank Prof. G. H. Honig for Fe_3O_4 single crystals and Prof. A. Revcolevschi for NiO single crystals.

REFERENCES

1. H. P. Bonzel in *Surface Mobilities on Solid Materials*, Ed. Vu Thien Binh, Plenum Press, New York, NY, 1083, p. 195.
2. I. N. Stranski, Z. Phys. Chem., **136**, 259 (1928).
3. J. M. Blakely, in *Progress in Materials Science*, Vol. 10, Ed. B. Chalmers, Pergamon Press, New York, NY, 1961, p. 395.
4. V. A. Gjostein, in *Surface and Interfaces I*, Ed. T. T. Burke, Syracuse University Press, Syracuse NY, 1967, p. 271.
5. J. Y. Choi and P. G. Shewmon, Trans. AIME, **224**, 589 (1962).
6. C. E. Birchendall, Trans. AIME, **227**, 781 (1963).
7. V. S. Stubican, G. Huzinec and D. Damjanovic, J. Am. Ceram. Soc., **68**, 181 (1985); C. M. Lin and V. S. Stubican, J. Am Ceram. Soc., **70**, C-73 (1987).
8. R. T. Whipple, Phil. Mag., **45**, 1225 (1954).
9. A. D. LeClaire, Br. J. Appl. Phys., **14**, 351 (1963).
10. R. Dieckmann and H. Schmalzried, Ber. Bunsenges. Physik. Chem., **81**, 344 (1977).
11. R. Dieckmann and H. Schmalzried, Ber. Bunsenges. Physik. Chem., **81**, 414 (1977).
12. U. L. Volpe and J. Reddy, J. Chem. Phys., **53**, 1117 (1970).
13. R. Fahri and G. Petot-Ervas, J. Phys. Chem. Solids, **39**, 1169 (1978); **39**, 1175 (1978).
14. A. Atkinson, A. E. Hughes and A. Hammon, Phil. Mag. **A43,** 1071 (1981).
15. N. L. Peterson and C. L. Wiley, J. Phys. Chem. Solids, **46**, 43 (1985).
16. J. Nowotny and M. Rekas, Solid State Ionics, **12,** 253 (1984).

IV. High T_c Oxide Superconductors

STATIC SIMULATION STUDIES OF La_2CuO_4

C.R.A. CATLOW and S.M. TOMLINSON

Department of Chemistry,
University of Keele,
Staffordshire, ST5 5BG, U.K.

Abstract

In this paper we describe the application of classical static lattice simulation techniques to the study of the lattice dynamics, dopant incorporation and valence states of La_2CuO_4. We also discuss the mechanisms by which 'chemical' bipolaron states may form, and present results which suggest the stability of inter-layer Cu^{3+} bipolarons on the copper sub-lattice, and peroxy like bipolarons on the oxygen sub-lattice.

INTRODUCTION

It is now recognised that questions relating to defect structure and valence states are of vital importance both for our fundamental understanding and for the technological applications of the recently discovered superconducting oxides. Such problems have been examined with success in a wide range of oxides by computer modelling techniques (see for example the articles by Colbourn et al[1], Tomlinson et al[2] and Cormack et al[3] in previous proceedings in this conference series). In this paper, we summarise the results of recent applications of the technique to La_2CuO_4[4]. We will show that structural and lattice dynamical properties of this oxide can be successfully modelled; formation energies for vacancies and hole states are calculated from which we may estimate the energies of redox, solution and disproportionation reactions. In the final section of the paper we will discuss hole pairing mechanisms where our calculations may be used to estimate the energetics of a variety of possible models.

J. Nowotny and W. Weppner (eds.),
Non-Stoichiometric Compounds Surfaces, Grain Boundaries and Structural Defects, 435–450.

TECHNIQUES

We use standard static lattice simulation techniques based on energy minimisation procedures employing ionic Born model potentials. Both perfect and defect lattice calculations were undertaken, with the latter employing Mott-Littleton procedures; the computer programs THBREL and CASCADE were used. Detailed discussion of the techniques are available in references (5) and (6). In applying these methodologies to new materials, the most important consideration is the development of requisite interatomic potentials. In modelling La_2CuO_4, potential parameters were derived for La_2O_3 and CuO, which were transferred to the ternary oxide. For La_2O_3 we used the potential model derived by Lewis [7]. No previous model was available for CuO and parameters for this material were therefore derived by empirical fitting procedures. We found that to achieve an adequate reproduction of the structural properties of this oxide, it was necessary to include two types of angle-dependent forces in the CuO_6 octahedra. These forces are described by simple potential functions of the type

$$E(\theta) = \tfrac{1}{2}K(\theta-\theta_0)^2$$

where θ_0 is the equilibrium bond angle and K the appropriate force constant. The two bond angles around which these forms were included are shown in figure (1).

Since our calculations also examined the Sr^{2+} and Ba^{2+} doped material, potential parameters were needed for the interactions between dopant and lattice ions. These were again taken from the Lewis[7] compilation of binary oxide potentials. Our potentials employ a shell model treatment of ion polarisability; for details of these and of all the potential parameters used in this study we refer to Islam et al[8] and Catlow et al[9].

RESULTS AND DISCUSSION

Perfect Lattice Properties

We first examined the ability of our potential models to reproduce the observed crystallographic properties of La_2CuO_4 using the program THBREL. We performed full energy minimisation calculations in which unit cell co-ordinates were varied. The calculated and observed bond lengths are reported in table (1). The orthorhombic structure of the crystal (illustrated in figure (2)) is obtained with lattice parameters very close to those observed experimentally (within 1%). The atomic co-ordinates are also close to the measured values, although there are small

Figure 1. Axially distorted CuO_6 unit in La_2CuO_4. Angles about which bond-bending potentials operate are shown.

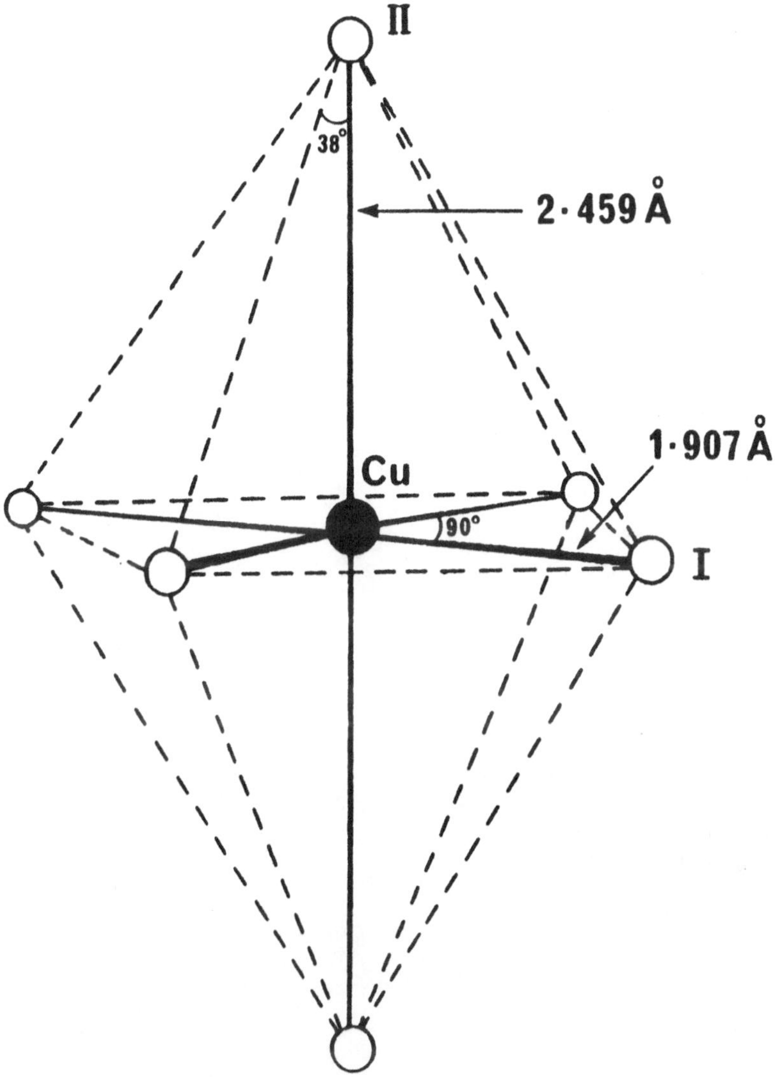

Figure 2. The structure of La_2CuO_4 (after Yu et al[14])

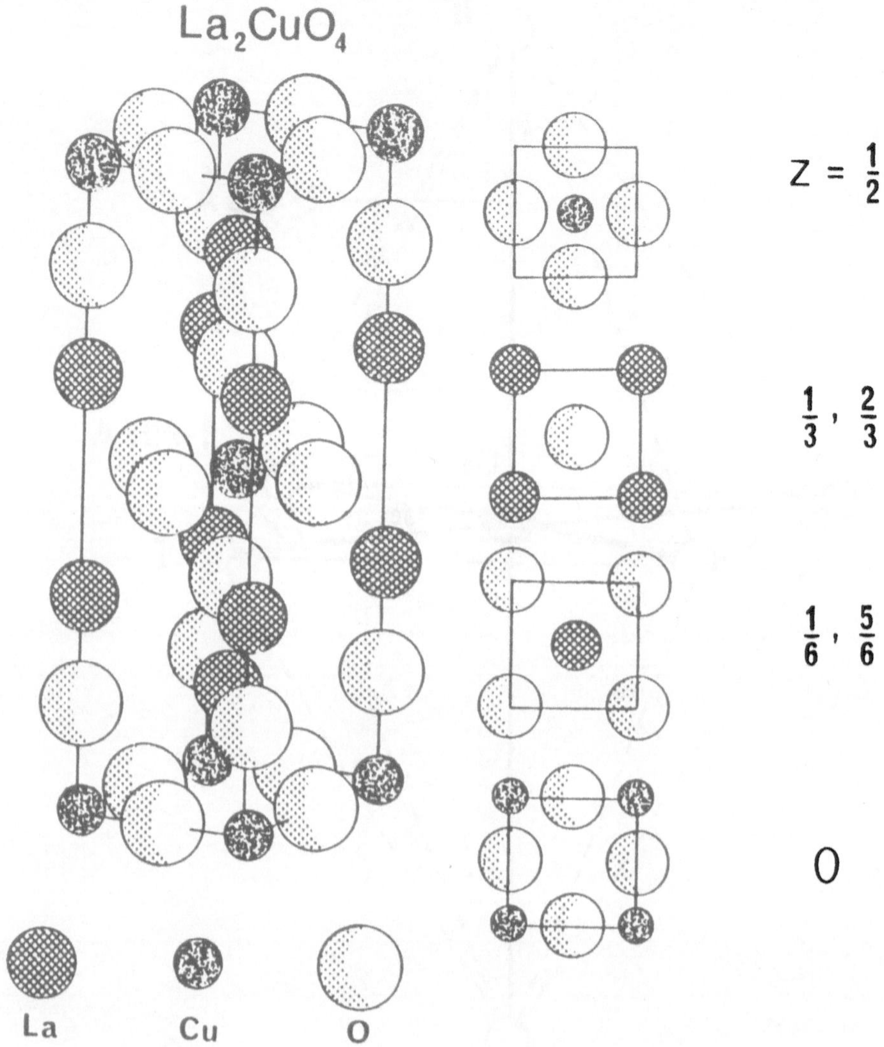

discrepancies; minor refinement of the potential parameters would remove these. In table (2), we give the calculated crystal properties for the phase. To date, elastic and dielectric data have not been reported. Such measurements would be a useful check of the accuracy of our models.

Calculations of the phonon dispersion curves have also been undertaken. Positive phonon frequencies are obtained for all branches (unlike the results of an earlier theoretical study[10]). Of greatest interest however, is the observation of a soft phonon mode for phonon propagation along the O–Cu–O axis, the calculated dispersion curve for which is shown in figure (3). The results of these calculations compare well with experiment[11] which has detected phonon mode softening as illustrated in figure (4). Indeed, there is good quantitative agreement between the calculated and experimental phonon properties. (The experimental wavevector is defined with respect to the tetragonal cell, whereas ours refers to the orthorhombic). We should note that soft phonon modes almost certainly play an important role in the mechanism of superconductivity in these oxides.

Thus it appears that where comparison between theory and experiment is possible the calculations perform well in reproducing the observed perfect lattice properties of La_2CuO_4. This success gives us greater confidence in using the potentials in the study of defect and valence properties, which will be our next concern.

Table (1)

Composition of calculated and observed bond lengths in La_2CuO_4

Bond	r_{expt}/Å (a)	r_{calc}/Å	Δ/Å
Cu-O(1)	1.907	1.906	0.001
Cu-O(2)	2.459	2.366	0.093
La-O(1)	2.681	2.658	0.023
	2.592	2.599	0.007
La-O(2)	2.304	2.409	0.105
	2.652	2.584	0.068
	2.766	2.743	0.023
	2.900	2.934	0.034

(a) after Grande et al (1977)

Figure 3. Calculated phonon dispersion curves for La_2CuO_4 along (0ξ0) direction of orthorhombic cell, showing mode softening

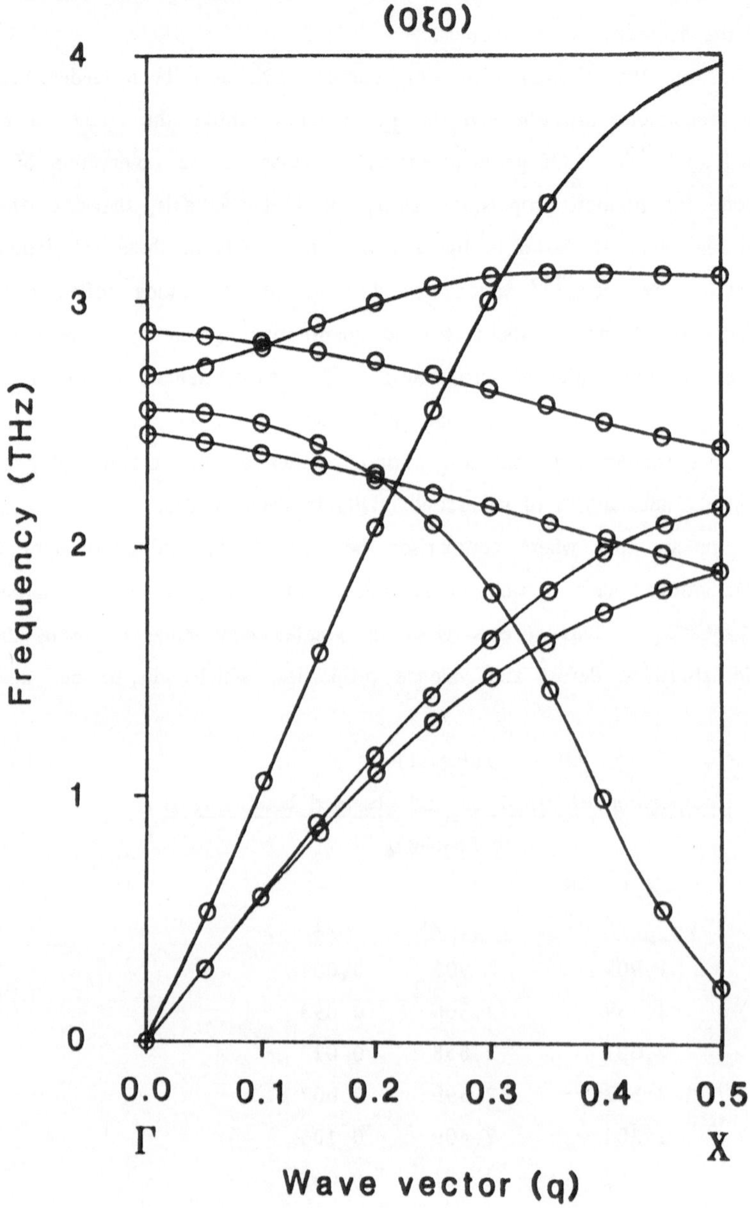

Figure 4. Experimental phonon dispersion curves for La_2CuO_4 (after Birgeneau et al[10])

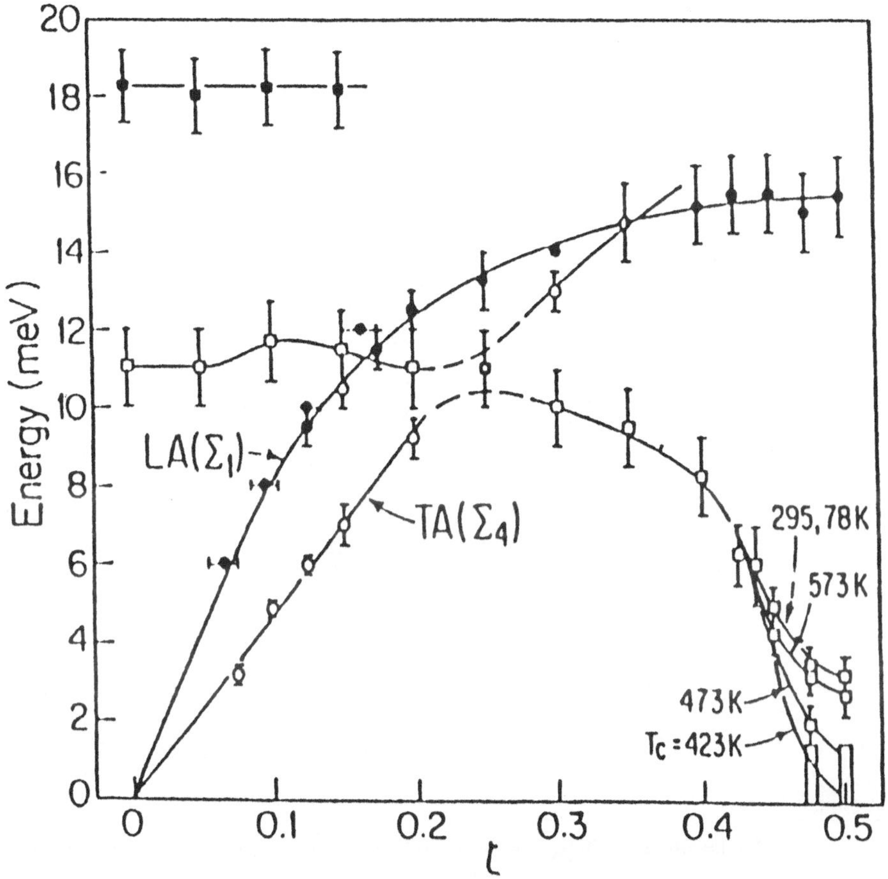

Table 2

Calculated crystal properties of La_2CuO_4 orthorhombic phase

Lattice Energy = −326.59 eV

Elastic constants/Dyne.$10^{11}cm^{-2}$

C_{11}	28.24
C_{12}	11.88
C_{13}	2.97
C_{22}	34.61
C_{23}	12.53
C_{33}	13.52
C_{44}	11.87
C_{55}	9.03
C_{66}	11.31

Static dielectric constant

ϵ_{11}	42.10
ϵ_{22}	16.69
ϵ_{33}	25.07

Hole States

La_2CuO_4 becomes a high Tc superconductor on doping with divalent ions. The divalent ion substitutes for the La^{3+} and except at very low values of $p(O_2)$ the negative charge of these substitutionals is compensated by hole formation. The nature of these hole states is controversial, with there being no consensus at present as to whether they are best represented as Cu^{3+} or O^- ions. To investigate this problem we have performed simple calculations in which the Cu^{3+} and O^- species are treated as substitutional defects; the stability of the Cu^+ species was also investigated in this way. We performed the calculations in two ways: first, a full relaxation of the lattice, including both cores and shells of the shell model potentials; secondly a restricted minimisation in which only shells are allowed to relax. Since shells represent the polarisable valence shell electrons, the latter describe 'optical' processes in which electron but not nuclear relaxation may occur; the former simulate 'thermal' processes in which nuclear relaxation occurs, and the hole species correspond to small polarons.

The results of the calculations are presented in table (3a). We may then combine the calculated defect energies with the appropriate ionisation potentials in order to obtain estimates of the formation energies of the Cu^{3+}, Cu^{+} and O^{-} species with respect to electrons in the vacuum; the resulting energies are given in table (3b).

From the results of these calculations, we may readily make two deductions. First, we find that the formation energies for Cu^{3+}(3d) holes are considerably less than for the O^{-}(2p) holes. We should be, of course, aware of the limitations of a fully ionic approach. There will be extensive mixing of Cu(3d) and O(2p) orbitals, and the hole state must therefore involve both 3d and 2p character. The present calculations, however, would not favour models in which the hole is predominantly located on the oxygen sub-lattice. Secondly, we find that the energy of the disproportionation reaction:

$$2Cu^{2+} \longrightarrow Cu^{3+} + Cu^{+}$$

is unfavourable by 4.20 eV for a thermal process and 9.59 eV for an optical process. We consider that such disproportionation reactions do not play a significant role in this material. Other disproportionation reactions will be considered later in this paper.

Table (3a)

Formation energies of isolated electronic defects

Species	Energy/eV		Electronic polarisation energy/eV	Displacement polarisation energy/eV
	Thermal	Optical		
Cu^{+}	21.68	24.37	3.38	2.69
Cu^{3+}	-33.92	-31.22	3.07	2.70
$O^{-}(1)$	14.19	17.19	2.03	3.00
$O^{-}(2)$	14.75	18.22	1.82	3.47

Table (3b)

Energies of hole and electron state formation and charge transfer process

Process	Energy/eV Thermal	Energy/eV Optical
Total formation energy for Cu^{3+} ion	2.91	5.61
Total formation energy for O^- ion	5.44	8.44
Total formation energy for Cu^+ ion	1.29	3.98
$2Cu^{2+} \rightarrow Cu^+ + Cu^{3+}$	4.20	9.59

Oxygen vacancies; oxidation and dopant solution reactions

The energies of the two types of oxygen vacancy at the O(1) and O(2) sites were calculated. The former correspond to equatorial and the latter to axial positions in the tetragonally distorted CuO_6 octahedra observed in these structures. The calculated energies are reported in table (4). We note that the equatorial species have lower energy. We may now calculate the energy of the oxidation reaction:

$$V_o^{\cdot\cdot} + \tfrac{1}{2}O_2 \longrightarrow O_o + 2Cu_{cu}^{\cdot}$$

for which we obtain a value of −0.25 eV, a result that is consistent with the ease with which the doped material is oxidised with the creation of holes.

By calculating the energies of the substituted dopants (also shown in table (4)), Sr_{La} and Ba_{La}, we may also obtain estimates of the solution energies of SrO and BaO in La_2CuO_4. The solution reaction is written:

$$MO \longrightarrow M_{La}' + \tfrac{1}{2}V_o^{\cdot\cdot} + \tfrac{1}{2}La_2O_3$$

(where M = Sr or Ba). The calculated energies of −0.29 eV and 0.25 eV for the solution of SrO and BaO respectively are consistent with the observed high solubility of these oxides in La_2CuO_4.

The results of our calculations of oxidation and solution reactions thus accord at least qualitatively with experiment, and encourage the further application of these techniques to study defect reactions in superconducting oxides.

Table (4a)

Formation energies of isolated oxygen vacancies and dopant substitutionals

Species	Energy/eV
$V_o^{\cdot\cdot}(1)$	15.93
$V_o^{\cdot\cdot}(2)$	17.73
Ba_{La}'	24.09
Sr_{La}'	21.38

Table (4b)

Energies of solution and oxidation processes

Process	Energy/eV
$BaO \rightarrow Ba_{La}' + \frac{1}{2}V_o^{\cdot\cdot} + \frac{1}{2}La_2O_3$	0.25
$SrO \rightarrow Sr_{La}' + \frac{1}{2}V_o^{\cdot\cdot} + \frac{1}{2}La_2O_3$	-0.29
$V_o^{\cdot\cdot} + \frac{1}{2}O_2 \rightarrow O_o + 2Cu_{Cu}^{\cdot}$	-0.25

Coupling Mechanisms

One of the central theoretical problems posed by the phenomenon of high Tc superconductivity is the nature of the electron coupling mechanisms. Encouragement for the interpretation of the superconducting pairs in 'chemical' terms is given by the evidence that they have only small extents (approximately 5–10 Å) in real space(12). There are two obvious types of process giving rise to closely coupled pairs: the first are charge disproportionation processes in which the energy (or Hubbard 'U') associated with accommodating pairs of electrons on the same site becomes negative; alternatively holes on different sites may couple to form a bipolaron.

(i) Charge Disproportionation

The following two types of disproportionation reaction are possible on the copper sub-lattice:

(1) $Cu^{2+} + Cu^{2+} \longrightarrow Cu^{3+} + Cu^{+}$

(2) $Cu^{3+} + Cu^{3+} \longrightarrow Cu^{4+} + Cu^{2+}$

The first of these was, we note, considered earlier. On the oxygen sub-lattice the following reaction is possible:

(3) $O^- + O^- \longrightarrow O^o + O^{2-}$

(ii) Bipolaron Formation

Again, we may envisage bipolarons on both the copper and the oxygen sub-lattices. In the former case, we consider the two models shown in figure (5): one in which the two component Cu^{3+} ions are on neighbouring sites in the same layer; the second in which the two Cu^{3+} ions are in adjacent layers with the shortest possible separation. A bipolaronic pair of oxygen holes can be equated to a distinct O_2^{2-} species; calculations were therefore performed for such a species along the equatorial edge of the CuO_6 octahedron (compared with which other of these peroxy species were less stable).

Calculated energies for both the charge disproportionation and bipolaron formation energies are given in table (5). It is immediately clear that all disproportionation processes are energetically unfavourable. Of the bipolaron models, the peroxy species is clearly bound, although we note that our earlier calculations favoured the localisation of holes of the Cu sub-lattice. The intra-layer Cu^{3+} bipolaron is strongly unbound and the magnitude of the repulsive interaction energy rules out this species. In contrast the inter-layer bipolaron is only slightly repulsive. It is likely that there will be additional small elastic terms favouring the binding of this species; in addition de Jongh(13) has argued that bipolarons may be stabilised by magnetic terms. It is plausible therefore to suggest the inter-layer bipolaron as a stable coupled hole species.

We consider that bipolaron rather than charge disproportionation is the more likely mechanism for hole coupling in high Tc superconductors, and there are energetically feasible models for such species on both the copper and oxygen sub-lattices.

Figure 5a. Intra-layer Cu^{3+} bipolaron

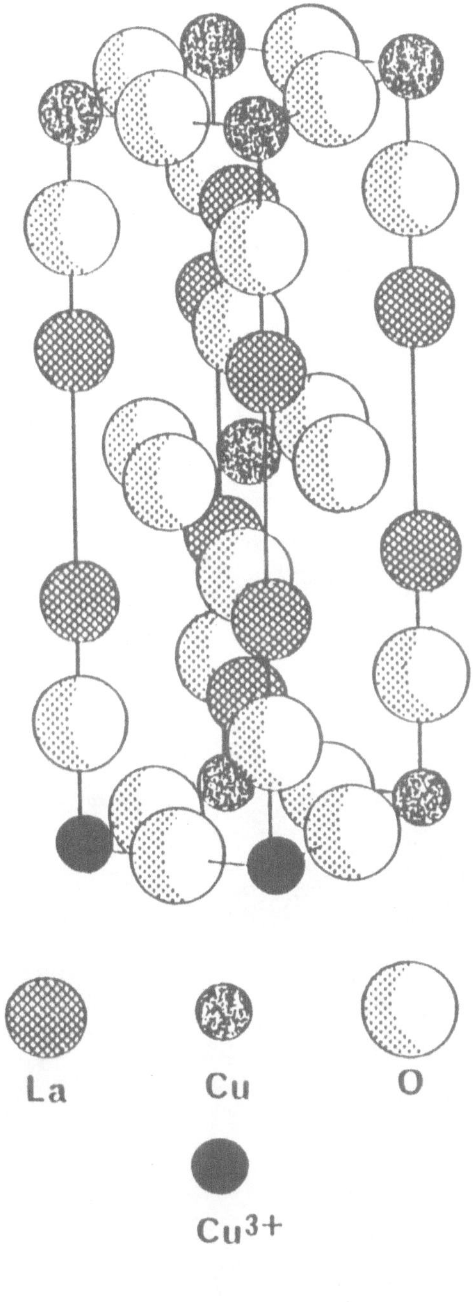

Figure 5b. Inter-layer Cu^{3+} bipolaron

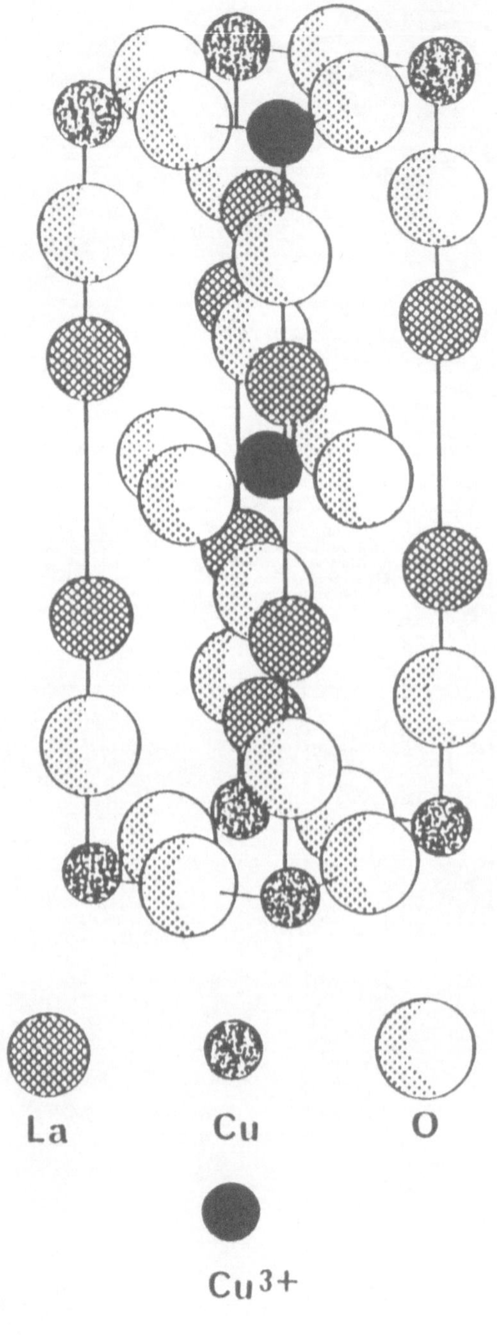

Table (5)

Charge disproportionation and bipolaron formation energies

Reaction or Process	Energy/eV Thermal	Optical
$Cu^{2+} + Cu^{2+} \rightarrow Cu^{3+} + Cu^{+}$	4.20	9.59
$Cu^{3+} + Cu^{3+} \rightarrow Cu^{4+} + Cu^{2+}$	4.38	9.84
$O^{-} + O^{-} \rightarrow O^{o} + O^{2-}$	1.29	7.00
Peroxy Formation	-0.23	2.94
Cu^{3+} Bipolaron Formation		
Intra-layer (figure 5a)	0.46	2.07
Inter-layer (figure 5b)	0.04	0.92

SUMMARY AND CONCLUSIONS

Computer modelling techniques clearly have a major role to play in the study of high Tc materials. The techniques can be used to probe properties of the perfect and defective structures and to give insight into important questions concerning the electronic structure of these oxides. Work on the $YBa_2Cu_3O_{7-x}$ system is now in progress.

ACKNOWLEDGEMENTS:

We would like to thank Dr A.M. Stoneham for several useful discussions.

References

1. Colbourn EA, Tasker PW and Mackrodt WC: "Transport in Nonstoichiometric Compounds", eds. Simkovich and Stubican, NATO ASI Series B: Physics Vol. 129, Plenum Press, New York (1985).
2. Tomlinson SM, Catlow CRA and Harding JH: ibid.
3. Cormack AN, Saul P, Catlow CRA: ibid.
4. Grande B, Muller-Buschbaum Hk. and Schweizer M: Z. Anorg. (Allg.) Chem., 428, 120 (1977).
5. Catlow CRA and Mackrodt WC: "Computer Simulation of Solids", Springer Verlag, Berlin (1982).
6. Leslie M, SERC Daresbury Laboratory Report DL-SCI-TM31T (1982).
7. Lewis GV: PhD Thesis, University of London (1984).
8. Islam MS, Leslie M, Tomlinson SM and Catlow CRA: J. Phys. C., 21, L109 (1988).
9. Catlow CRA, Tomlinson SM, Islam MS and Leslie M: J. Phys. C., submitted.
10. Cohen RE, Pickett WE, Boyer LL and Krakauer H: Phys. Rev. Lett., 60, 817 (1988).
11. Birgeneau RJ et al.: Phys. Rev. Lett., 59, 1329 (1987).
12. Kastner MA et al.: Phys. Rev., B37, 111 (1988).
13. De Jongh LJ: Physica C, 152, 171 (1988).
14. Yu J, Freeman AJ and Xu JH: Phys. Rev. Lett., 58, 1035 (1987).

DEFECTS AND TRANSPORT IN $YBa_2Cu_3O_{7-x}$

G.M. Choi*, H.L. Tuller, and M-J. Tsai
Crystal Physics & Optical Electronics Laboratory
Department of Materials Science & Engineering
Massachusetts Institute of Technology
Cambridge, MA 02139

ABSTRACT. In an attempt to understand the role of stoichiometry in controlling the transport properties of the system $YBa_2Cu_3O_{7-x}$ ($0 < x < 1$), we have performed electrical conductivity (σ) and thermoelectric power (Q) measurements, in situ, as a careful function of temperature, oxygen partial pressure, and oxygen stoichiometry.

In the low temperature orthorhombic phase, a continuous transition from semiconducting to metallic conduction was observed as temperature decreases and Po_2 increases as a result of exothermic generation of holes. Both the magnitude of Q and the Po_2 dependence of σ and Q confirm the p-type nature of the carriers. In the high temperature tetragonal phase a p-n semiconducting transition is detected which shifts to higher temperature as the Po_2 is increased. Based on these results and results of measurements performed as a function of temperature for fixed values of x, we conclude that the n-type contribution is a result of the disproportionation of Cu^{1+} to Cu^{0} and Cu^{2+} which is characterized by an energy of 1.24 eV. A preliminary defect model is proposed which provides a framework for understanding the Po_2 dependence of σ and Q.

1. INTRODUCTION

The $YBa_2Cu_3O_{7-x}$ (YBCO) system has generated great interest due to its high transition temperature (~92K) to the superconducting state for x values close to 0. Its transport properties, e.g., electrical conductivity and oxygen diffusivity have been shown to vary dramatically due to the extensive range of nonstoichiometry exhibited by this cuprate. Nevertheless, little reliable data presently exists in which a systematic examination has been made between the transport properties and the parameters, e.g. temperature and Po_2, which control the degree of

---

*Present Address: Dept. Material Science & Engineering, Pohang Institute of Science and Technology, Pohang, Korea

J. Nowotny and W. Weppner (eds.),
Non-Stoichiometric Compounds Surfaces, Grain Boundaries and Structural Defects, 451–470.

nonstoichiometry. In the present study, we examine the electrical conductivity and thermoelectric power of polycrystalline specimens of $YBa_2Cu_3O_{7-x}$ over an extensive composition range and on the basis of these results examine preliminary defect models which are consistent with these results.

2. EXPERIMENTAL PROCEDURE

Powders of YBCO were prepared by the Pechini method(1). This technique is noted for its ability to insure accurate control of cation stoichiometry and atomic scale mixing of the cations. Pellets were made by isostatically pressing the powders at 40 Kpsi. Specimens were sintered at 930°C for 12 hours in flowing oxygen resulting in sintered densities of over 90% of theoretical density.

Standard 4 probe d.c. conductivity measurements were performed. Thermoelectric power was determined by measuring the emfs induced in samples placed in either controlled constant temperature gradients or by application of a heat pulse. In the former case, the temperature gradients were kept below 5°C to avoid inducing significant chemical gradients. In the latter case, we adopted a design similar to that used by the authors previously(2), in which a small heater next to the sample was used to generate the heat pulse. Absolute thermoelectric power values were obtained by subtracting out contributions from the platinum leads.

The first series of σ and Q measurements were taken at 25°C intervals between 300°C and 800°C at a series of fixed oxygen partial pressures ranging from 10^{-4} to 1 atm. At each temperature, specimens were allowed to "equilibrate" for one hour. In the second series of measurements which were performed isothermally as a function of Po_2 for temperatures of 600, 700, and 800°C, the specimens were maintained at a given Po_2 for 1-3 days to insure equilibration.

The third series of measurements were designed such that the stoichiometry of the specimen, fixed initially by equilibrating the specimen at 800°C in a Po_2 of $10^{-3.5}$, 10^{-3}, $10^{-2.5}$, or 10^{-2} atm, remained essentially constant during subsequent cooling and reheating. This was accomplished by sealing the specimen into a leak tight chamber together with extra powder of $YBa_2Cu_3O_{7-x}$ to buffer the oxygen content in the gas phase. A similar approach was previously applied successfully to the study of small polaron transport in CeO_{2-x}(3) by one of the authors.

3. RESULTS

Fig. 1 shows a plot of log σ vs. 1/T for YBCO at five oxygen partial pressures ranging from 10^{-4} to 1 atmosphere. In region I, where temperature is low and oxygen partial pressure is high, the conductivity is nearly temperature insensitive and of high magnitude, resembling typical metallic behavior, i.e., the conductivity is inversely proportional to temperature. However, the conductivity becomes thermally activated, with negative activation energy, at intermediate temperatures and reduced oxygen partial pressures (region II) . Further, at high temperature and low oxygen partial pressure (region III), the conductivity reverses and show a thermally activated process with positive activation enthalpy.

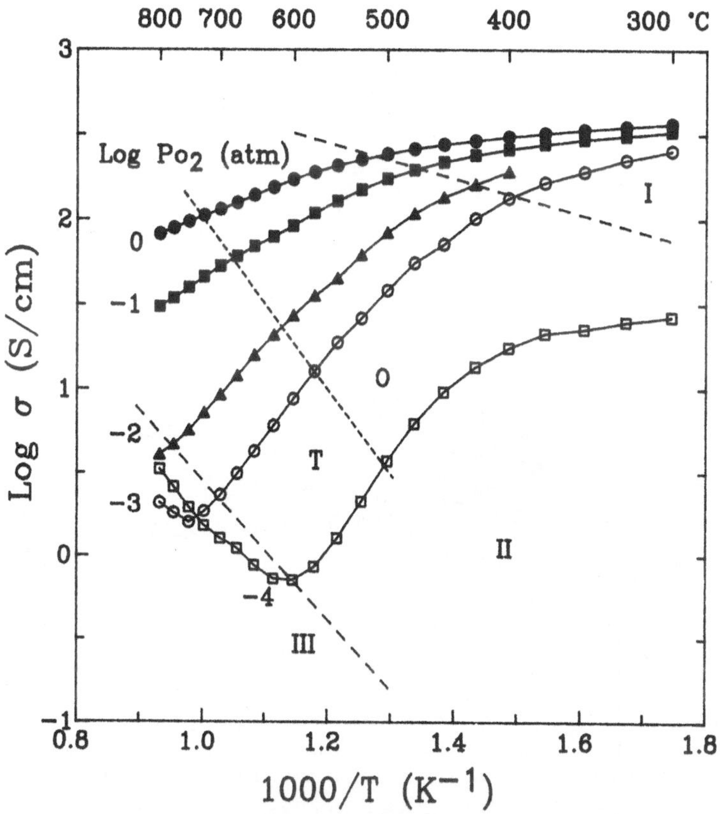

Figure 1. Log σ versus 1000/T for YBCO measured at a series of controlled Po_2's as indicated. The different conduction regimes designated by roman numerals are described in the text. The letters O and T refer to the orthorhombic and tetragonal phases respectively (from Ref. 4).

Fig. 2 shows the thermoelectric power plotted vs. 1/T. In the region corresponding to region I of Fig. 1, the thermoelectric power is positive and temperature insensitive. In region II, the thermoelectric power remains positive but increases rapidly with temperature, confirming that a) charge carriers are holes, and b) the charge carrier density decreases with increasing temperature and decreasing oxygen partial pressure. In region III, the thermoelectric power passes through a maximum and reverses sign to n-type at the highest temperatures and lowest oxygen partial pressures, suggesting a p-n transition at elevated temperature. The results shown in Fig. 1 and 2, which provide an overview of the range of electrical properties exhibited by $YBa_2Cu_3O_{7-x}$, are reproduced from an earlier publication by the authors.(4)

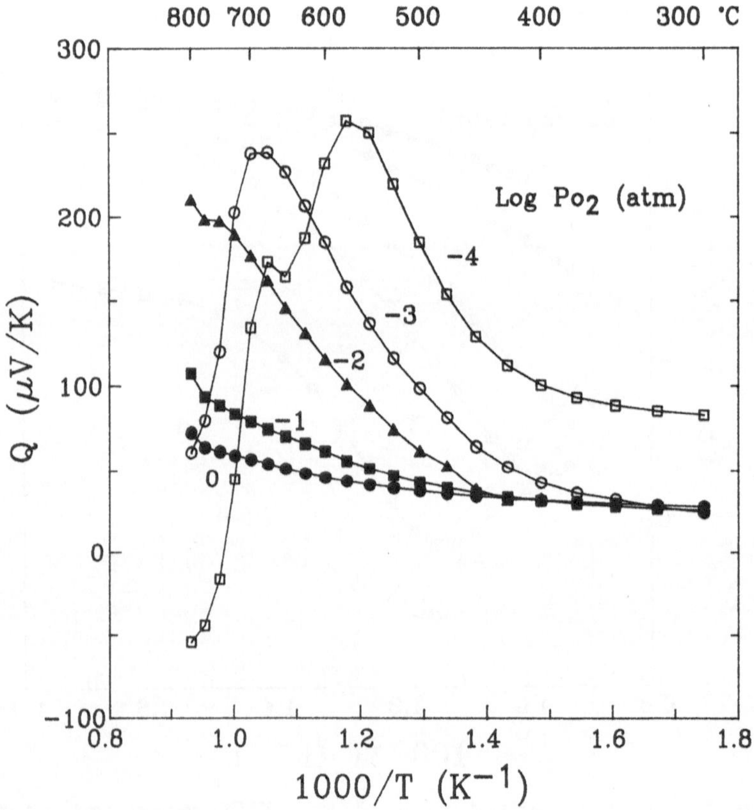

Figure 2. Thermoelectric power versus 1000/T for YBCO measured under the same conditions as in Fig. 1 (from Ref. 4).

When log σ is plotted versus log Po_2 (Fig. 3), one finds several trends. At 600°C, σ follows a $Po_2^{1/2}$ dependence closely below 10^{-2} atm followed by a weaker dependence for more oxidizing conditions. At 800°C, a transition from an approximately $Po_2^{1/2}$ to a $Po_2^{-1/2}$ dependence, characteristic of a p to n transition, is observed at low Po_2's. The 700°C isotherm exhibits both the 600 and 800°C trends at the extremes of Po_2. The same trends are confirmed by the Po_2 dependent thermoelectric power measurements shown in Fig. 4. The onset of the p-n transition is clearly observed for both the 700 and 800°C isotherms initiating at higher Po_2, the higher the temperature.

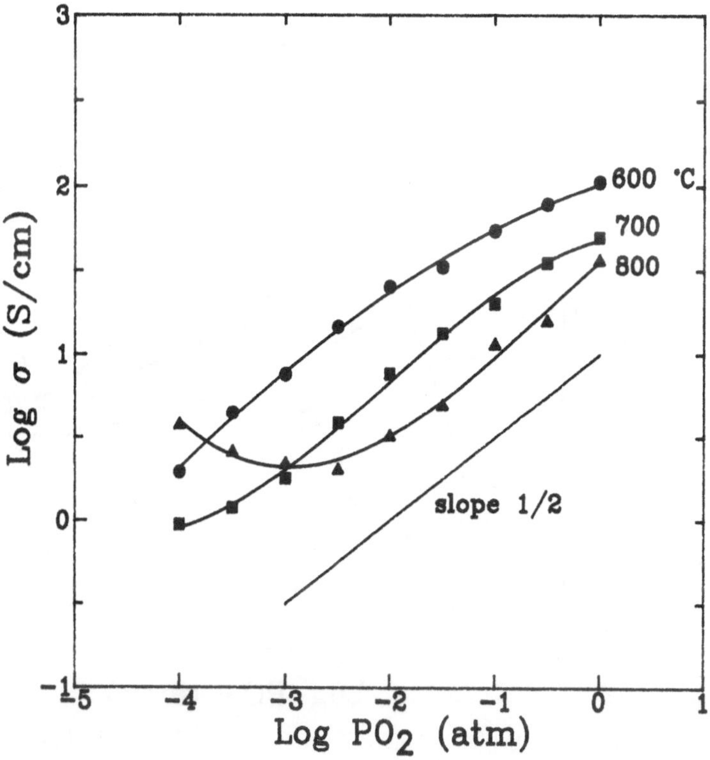

Figure 3. Log σ versus log Po_2 for 600, 700, and 800°C isotherms.

The temperature dependence of σ measured for three fixed values of x between ~0.8 and 0.9 are shown in Fig. 5. The percent oxygen indicated next to each curve corresponds to the atmosphere the sample was pre-annealed in at 800°C prior to sealing the system. See Table I for the correspondence between 7-x and the Po_2 used at 800°C according to Kishio et al (5) and O'Bryan et al (6). Several trends are noteworthy. First, in contrast to the results of Fig. 1, taken at constant Po_2, the conductivity decreases monotonically with decreasing temperature. Second, the temperature dependence, particularly at the lower temperatures, decreases rapidly with decreasing x. At the highest temperatures, all three curves appear to approach a common curve.

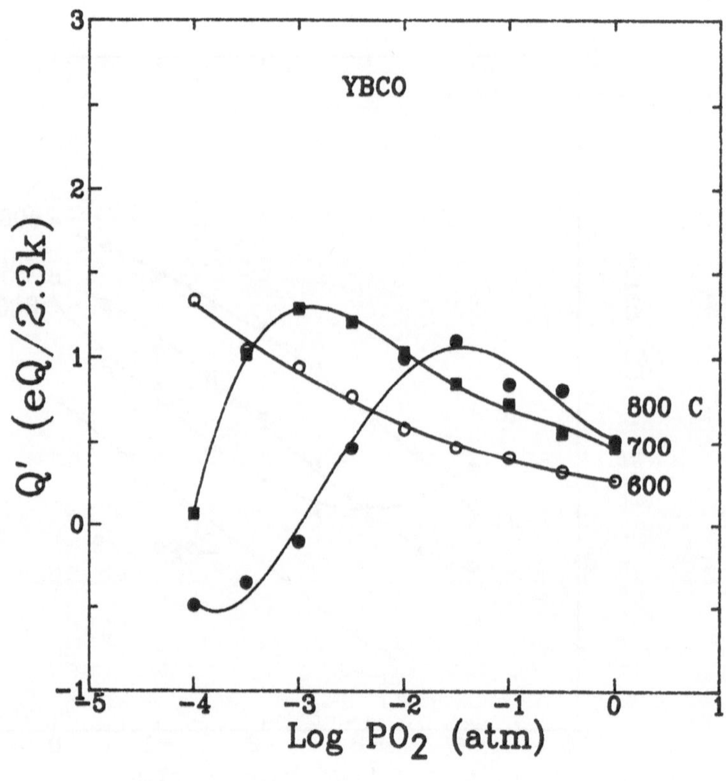

Figure 4. Normalized thermoelectric power versus log Po_2 for the conditions described in Fig. 3.

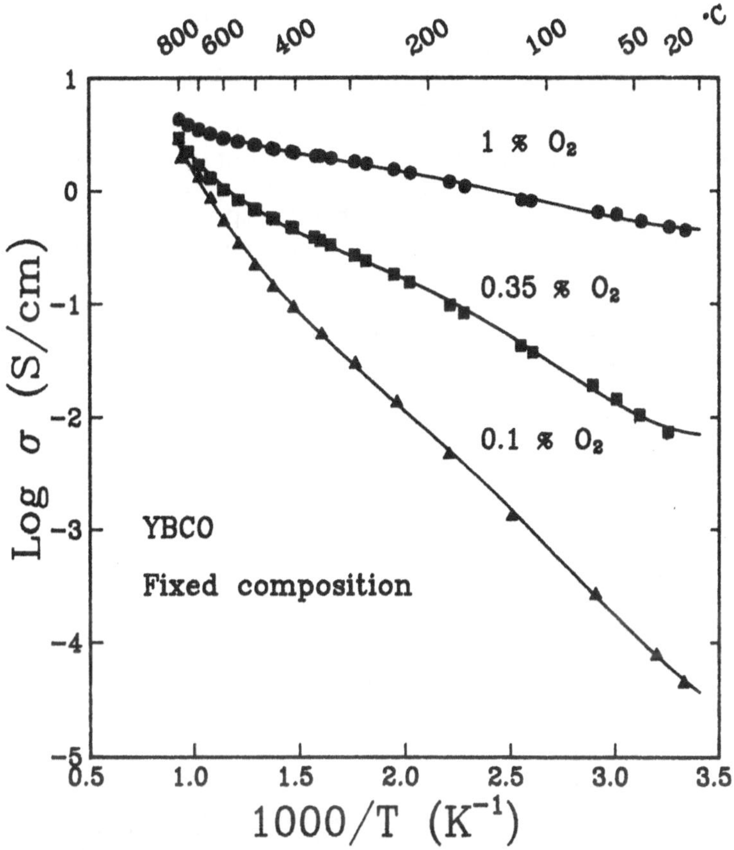

Figure 5. Log σ versus 1000/T for YBCO with fixed values of x. See text for explanation of percent oxygen designation.

TABLE I	Fixed Stoichiometry		
Temp.	Po_2	7-x	
		Kishio(5)	O´Bryan(6)
800°C	0.1%	6.08	6.14
	0.35%	6.13	6.20
	1.0%	6.20	6.24

The thermoelectric power, normalized by e/2.3 k, is shown plotted in Fig. 6 as a function of reciprocal temperature for conditions identical to those described in Fig. 5. At reduced temperatures, Q is positive for all values of x, consistent with the expected p-type nature of the carriers. Further, the decreasing magnitude of Q with decreasing x suggests that the corresponding increase in σ shown in Fig. 5 is at least in large part due to an increase in carrier density. At elevated temperatures, a clear transition from p to n type conduction is observed with increasing temperature with the onset of the transition occurring at lower temperatures the larger the value of x. As we show later, this p-n transition is likely connected with intrinsic electron-hole generation at elevated temperatures.

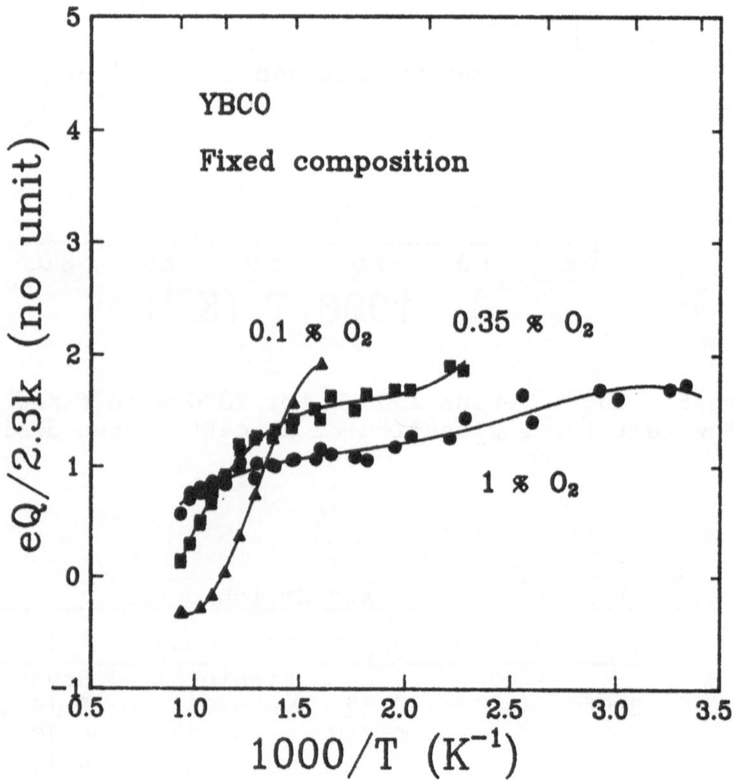

Figure 6. Normalized thermoelectric power versus 1000/T for conditions identical to those in Figure 5.

4. DISCUSSION

4.1 Variable Stoichiometry

A continuous transition from metallic to apparent p-type semiconducting behavior is observed as temperature increases and oxygen partial pressure decreases. This transition from metallic to semiconducting behavior is supported by the large increase in the magnitude of thermoelectric power observed at the higher temperatures. Further, the increase in σ with Po_2 further confirms the p-type nature of the material at intermediate Po_2 and temperature. Of particular note in this system is the unusual exponentially decreasing nature of the p-type conductivity with temperature characterized by an apparent activation energy of approximately 1 eV.

The orthorhombic to tetragonal transition is reported to occur near 630°C(6) in oxygen, and at lower temperatures in more reducing atmospheres. Thermogravimetric studies(5) show that the orthorhombic-tetragonal transition occurs at x values close to 0.5.

The conductivity and thermoelectric power both point to a p-n transition which shifts to lower temperature as the oxygen partial pressure is decreased.

Of special interest is the source of the near $Po_2^{\pm 1/2}$ dependence of σ observed in Fig. 3 at higher temperatures and lower Po_2's. Let us begin by designating the cuprate of interest by $Y_1Ba_2(Cu_a^{1+}Cu_b^{2+}Cu_c^{3+})O_{7(1-\delta)}$ where a,b, and c designate the number of Cu ions which are of 1+, 2+, and 3+ valence respectively. Since the total number of formula units of copper must add to three, we write the following mass balance relation:

$$a + b + c = 3 \quad (1)$$

To maintain charge balance, we further require,

$$a + 2b + 3c = 7(1-2\delta) \quad (2)$$

Assuming further that electron-hole generation can be a result of Cu^{2+} disproportionation we obtain,

$$2Cu^{2+} \rightarrow Cu^{1+} + Cu^{3+} \quad (3)$$

for which we may write the mass action relation,

$$ac/b^2 = K_D \quad (4)$$

in which K_D is the equilibrium constant appropriate to this reaction.

Equilibration of the cuprate with the gas phase may be approached in a number of ways. We choose to view the cuprate with oxygen content less than "O_7" as oxygen deficient and therefore write the oxidation reactions in the following manner:

$$Cu_{Cu}^{2+} + V_o^{\cdot} + 1/2O_2 = Cu_{Cu}^{3+} + O_o^{x} \qquad (5)$$

and

$$2Cu_{Cu}^{2+} + V_o^{\cdot\cdot} + 1/2O_2 = 2Cu_{Cu}^{3+} + O_o^{x} \qquad (6)$$

The first of the two equations assumes that the oxygen defect is singly ionized while the second assumes full ionization. The corresponding mass action relations are given by

$$\frac{c}{b} \frac{(1-\delta)}{\delta} = K_R \; Po_2^{1/2} \qquad (7)$$

and

$$\frac{c^2}{b^2} \frac{(1-\delta)}{\delta} = K_R' \; Po_2^{1/2} \qquad (8)$$

respectively.

In order to simplify the solution of these simultaneous defect equations, it is useful to select experimental conditions for which some of the species in the above reactions become negligible. Since we find experimentally that the conductivity and thermoelectric power are, for the most part, p-type, it is reasonable to assume that the Cu^{1+} contribution is negligible under most circumstances. This condition implies that a << b, c and therefore Eq´s 1 and 2 become

$$b + c = 3 \qquad (9)$$

and

$$2b + 3c = 7(1-2\delta) \qquad (10)$$

Solving, we find

$$\delta = (1-c)/14 = (b-2)/14 \qquad (11)$$

Combining this result with Eq. 7 and solving for c, one obtains the following solution for the quadratic equation in c:

$$c = \frac{13 + 4n}{2(1-n)} \left\{ -1 \pm \left[\frac{1 + 12n\,(1-n)}{(13 + 4n)^2} \right]^{1/2} \right\} \qquad (12)$$

where $n \equiv K_R Po_2^{1/2}$. Note, that due to a minor error in their calculations, Yoo et al(7) obtained this same solution for the doubly ionized case described in Eq. (8).

In Figure 7, we plot the solution for c (Eq. 12) as a function of Po_2. The curve has been fit to the 600°C experimental data of Fig. 3 thereby obtaining values for $K_R = 5.6$ atm$^{-1/2}$ and $\mu_h = 6.8\times10^{-2}$ cm^2/V.s. A $Po_2^{1/2}$ dependence is indeed predicted at low values of Po_2 followed by a saturation in c at high Po_2 which corresponds to the condition at which c approaches 1, i.e, when the $YBa_2Cu_3O_7$ stoichiometry is reached beyond which no further oxidation is believed to occur.

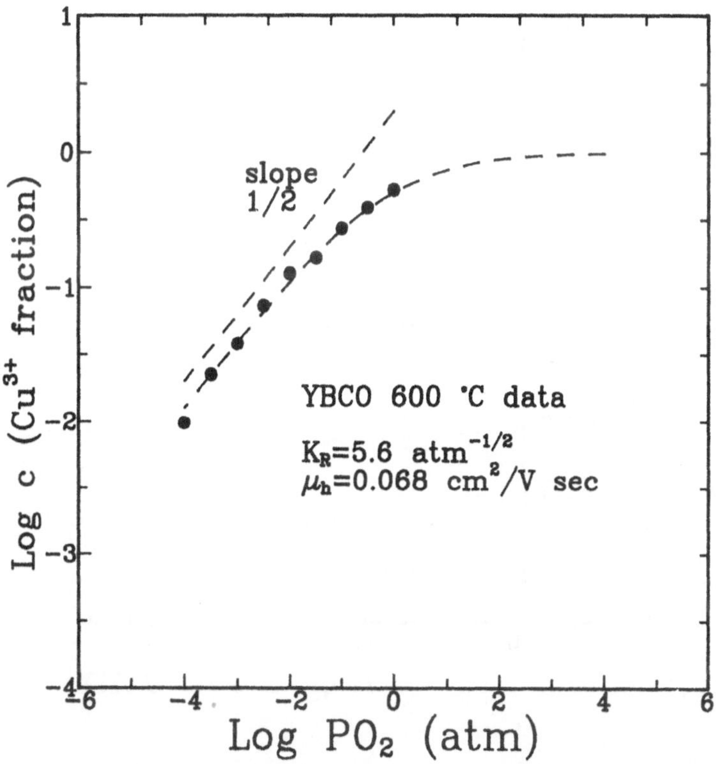

Figure 7. Log σ versus log Po_2 as obtained from Eq. 12 and fitted to the 600°C experimental data (solid circles) of Fig. 3.

Since the magnitude of the derived hole mobility is low, i.e., ~10^{-2} $cm^2/V.s.$, the small polaron transport model is implied(2,7). Under these circumstances, the thermoelectric power is given by:

$$Q = k/e \ln(b/c) \quad (13)$$

Using the above expression, the normalized thermopower is plotted in Fig. 8 as a function of Po_2. Also plotted are our experimental data. Although the general shape of the curves are similar, the experimental data are noticeably lower in magnitude than the predicted curve. This discrepancy is not entirely unexpected given the observed transition to n-type thermopower at reduced Po_2's in Fig .4.

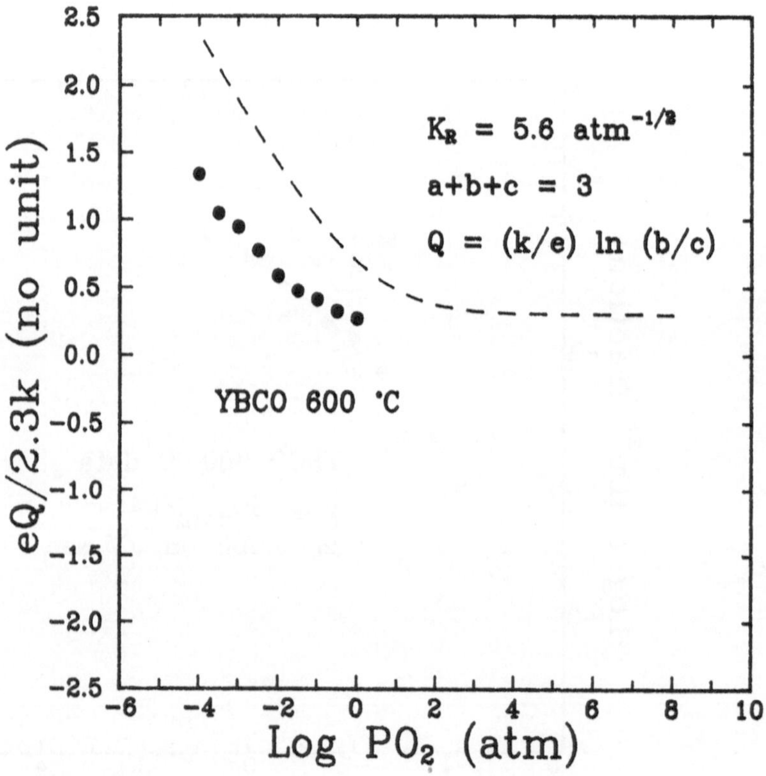

Figure 8. Normalized thermoelectric power versus log Po_2 as obtained from Eq. 13. The 600°C experimental data (solid circles) taken from Fig. 4 are plotted for comparison.

Note that the thermopower begins to show the influence of electrons at considerably higher Po_2's than does the electrical conductivity (compare Fig. 4 with Fig. 3). Eq. 13, on the other hand, ignores contributions from Cu^{1+}, i.e., a is assumed to be equal to zero under all conditions.

A second issue which deserves attention in the above model and which can affect the degree of agreement between the experimental and predicted thermopower data is the character of the Cu mass balance expressions, Eqs. 1 and 9. If one compares the unit cells of $YBa_2Cu_3O_7$ and $YBa_2Cu_3O_6$ as in Fig. 9, one finds two types of Cu sites designated as Cu(1) and Cu(2). Only the Cu(1) coordination changes during oxidation from the "O_6" to the "O_7" state. It is therefore reasonable to assume, as have others(8,9), that Cu(2) remains divalent at all times while only Cu(1) changes from Cu^{1+} at "O_6" to Cu^{2+} at "$O_{6.5}$" to Cu^{3+} at "O_7". Thus, one could rewrite the cuprate designation as $Y_1Ba_2Cu_2^{2+}(Cu_a^{1+}Cu_b^{2+}Cu_c^{3+})O_{7(1-\delta)}$ where the new Cu mass balance relation becomes

$$a + b + c = 1 \qquad (14)$$

Again making the assumption that $a \sim 0$ for high Po_2 and that Q is given by Eq. 13, one obtains a new curve as shown in Fig. 10. Here, the agreement with the experimental data is better but still not entirely satisfactory. We are presently investigating several other models which we hope will provide improved agreement with experiment. In fact, data which we have most recently obtained and which are presented in the next section, cause us to reconsider several key assumptions used in deriving the above model.

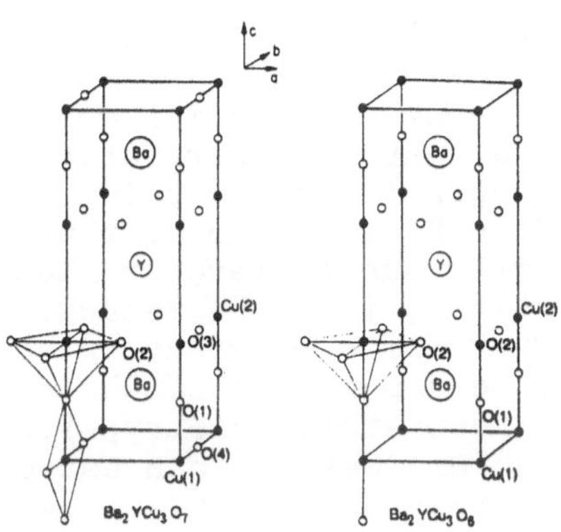

Figure 9. Schematic drawings of the unit cells of $YBa_2Cu_3O_7$ and $YBa_2Cu_3O_6$ (from Ref. 9).

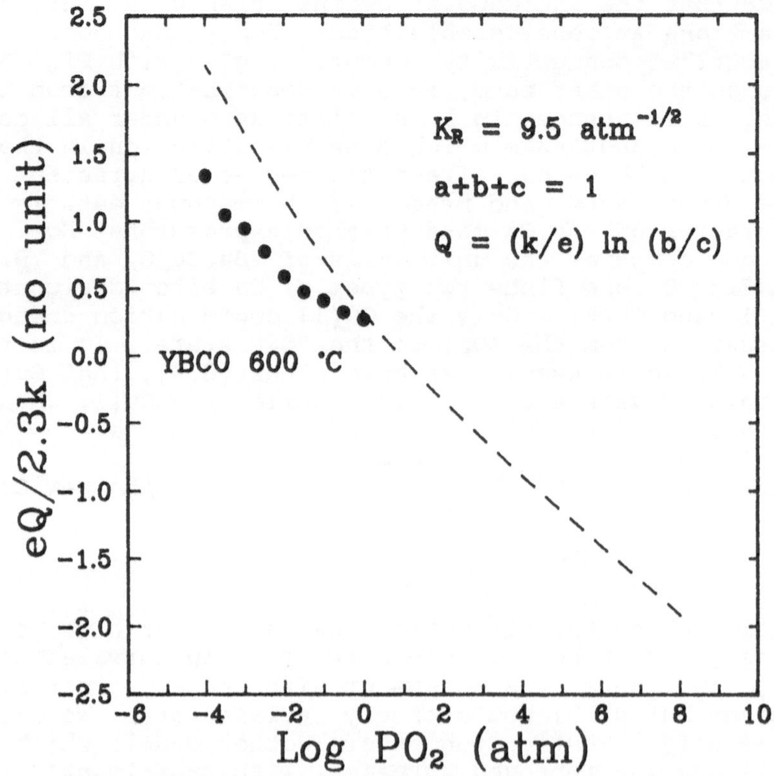

Figure 10. Normalized thermoelectric power versus log Po_2 as in Fig. 8 with the Cu mass balance expression changed to $a + b + c \approx b + c = 1$.

4.2. Fixed Stoichiometry

The data in Figs. 1 and 2 show the carrier concentration to be strongly activated at elevated temperatures. We know both from published thermogravimetric data(5,6) and from our own results as presented in Figs. 3 and 4, that these variations are substantially due to large variations in oxygen stoichiometry resulting from the reduction of the cuprate upon heating. It remains of great interest to investigate the temperature dependence of the electrical properties for a given oxygen stoichiometry, i.e. for constant values of δ. In this way, we can establish whether the carrier density and/or mobility are thermally activated and thereby address issues regarding the energy band structure and mobility mechanisms operative in these cuprates.

There are several approaches one can take for obtaining constant δ data. The most common approach is to cross correlate $\delta(T,Po_2)$ and $\sigma(T,Po_2)$ data obtained by thermogravimetric analysis and conductivity measurements respectively. Fig. 11 shows our attempt to do this by correlating Kishio´s TGA data(5) and our own conductivity data. One notes several features, i.e. (a) the conductivity increases with increasing temperature at constant δ or x in

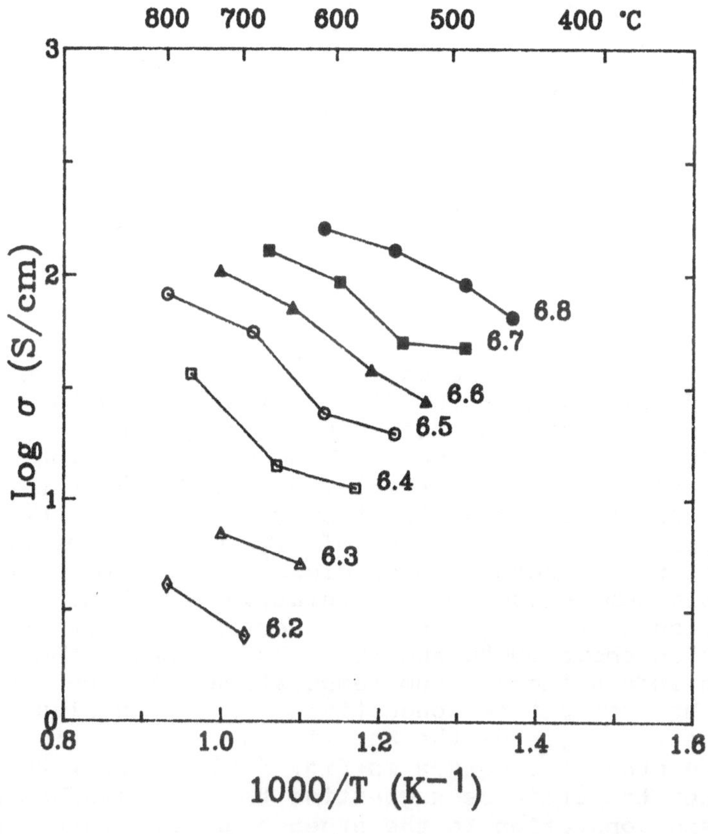

Figure 11. Log σ versus 1000/T obtained for fixed values of (7-x) by cross correlation of TGA data(4) and the data of Fig. 1.

contrast to the results obtained at constant Po_2; (b) the results do not follow an arrhenius dependence on T; and (c) the data appear to show a good deal of scatter. Similar correlations made by Kishio et al(4) and Sageev Grader et al (10) show substantially different results. We suspect that this is due to the sluggish kinetics exhibited by these materials. Consequently, the respective TGA and conductivity specimens may not have reached identical values of δ for all T and Po_2.

An alternative approach, which we favor and have successfully used elsewhere(2), is to fix the value of δ at elevated temperature (e.g. 800°C) by control of Po_2, and maintain this value of δ essentially fixed by sealing the specimen into a leak tight chamber with sufficiently low volume to insure negligible oxygen exchange with the gas phase in the chamber. The advantage of following this approach is obvious from Figures 5 and 6 which show that one can readily obtain near continuous data as a function of temperature. Data obtained upon heating and cooling at different rates showed little or no hysteresis suggesting that negligible exchange of oxygen occurred with the atmosphere during the heating and cooling cycles.

It is interesting to note that compositions with (7-x) values as low as 6.2 still retain high conductivities with weak temperature dependence at all temperatures. This composition, although it contains negligible Cu^{3+} concentrations (i.e., c ~ 0) is, according to Fig. 6, p-type with a high hole density. This observation stands in conflict with earlier suggestions(7) that the Cu disproportionation reaction given by Eq. 3 is important in YBCO. If this were the case we would expect a minimum in the conductivity near the "6.5" composition with n-type conduction predominant below "6.5" and p-type for values above "6.5." Instead, we find the p-n transition occurring close to "6.0" and only then at elevated temperatures.

If we examine the lowest curve in Fig. 5, the one which corresponds to a composition of ~$YBa_2Cu_3O_{6.1}$, we find (a) a temperature dependence which closely resembles that of an extrinsic semiconductor with relatively high acceptor ionization energy at lower T converting to intrinsic conduction above 500°C and (b) a YBCO composition which becomes insulating at room temperature. Further, Fig. 6 shows that the p-type conductivity in the extrinsic regime converts to n-type as the material goes intrinsic. Note that the other two curves in Fig. 5 also appear to converge at higher temperatures suggesting that the assignment of intrinsic conduction to the steep high temperature portion of the lower curve is reasonable.

Assuming this model of extrinsic p-type semiconductivity at lower T is appropriate, then one would conclude that the upper two curves in Fig. 5 possess successively higher acceptor levels with the decreasing slopes corresponding to the approach of degeneracy. The thermoelectric power data of Fig. 6 support this view. Note that the p-n transition shifts to higher T as x or δ decreases.

In order to explain these data, we propose some essential modifications to the earlier models proposed by ourselves and others(6,7). First, we suggest that the correct Cu disproportionation reaction be given by

$$Cu^{1+} \rightarrow Cu^{o} + Cu^{2+} \tag{15}$$

with a corresponding mass action relation given by

$$[Cu^{o}][Cu^{2+}]/[Cu^{1+}] = K_{Cu} \tag{16}$$

where $K_{Cu} = N_c N_v \exp(-E_g/kT)$. With the knowledge that Cu^{1+} has a $3d^{10}$ closed electronic shell, one can reasonably assume that Eq. (16) corresponds to the energy gap between the full Cu $3d^{10}$ derived valence band and the empty Cu $4s^{o}$ derived conduction band, assuming little or no hybridization with other Cu or O derived levels. If this were the case, E_g, as derived from Fig. 5 is twice the high temperature slope of 0.62eV or E_g = 1.24eV.

The low temperature slopes for the curves in Fig. 5 were found to be 0.33, 0.18, and 0.08 eV respectively with the slope increasing as the composition approaches closer to "O_6." For the non-degenerate case, this slope should correspond to either the acceptor ionization energy E_A or one-half its value, the former case corresponding to partial acceptor-donor compensation(11).

In Fig. 12, we compare the temperature dependence of the conductivity and thermoelectric power of the cuprate in Figs. 5 and 6 with the highest oxygen content. It is obvious from this plot that both quantities share nearly the same activation energy of 0.08eV. This demonstrates that, at least for YBCO with 7-x values equal to or greater than 6.2-6.25, the mobility is non-activated. This brings into question the use of the small polaron model(8) to explain data obtained for YBCO over its entire oxidation range. We are presently examining additional modeling approaches which take these most recent results obtained for fixed values of x into consideration.

5. SUMMARY

A defect model was initially proposed which exhibited many of the key experimental observations made by us and others(8,12) concerning the high temperature electrical properties of YBCO. This included a nominal $Po_2^{1/2}$

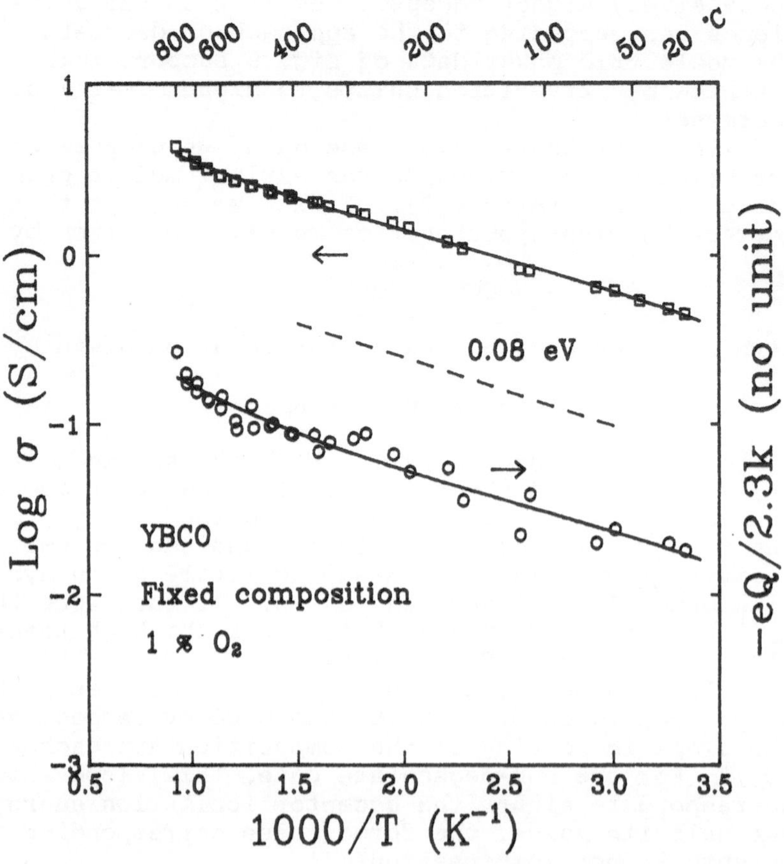

Figure 12. Log σ and eQ/2.3k are plotted versus 1000/T for YBCO with fixed composition annealed initially in $Po_2 = 10^{-2}$ atm and 800°C.

dependence of the conductivity observed at lower Po_2 which tended to saturate at high Po_2 and a p-n transition at lower Po_2 confirmed both by a slope change in conductivity towards $Po_2^{-1/2}$ and a change in sign of the thermoelectric power. The observed Po_2 dependence was replicated with a model which assumed disproportionation of Cu^{2+} to Cu^{1+} and Cu^{3+} and an oxidation reaction involving singly ionized oxygen defects in which Cu^{2+} is oxidized to Cu^{3+}. Following this model, the equilibrium constant for the oxidation reaction was found to be equal to 5.6 $atm^{-1/2}$ while the hole mobility was found to be 0.068 $cm^2/V.s.$ both at 600°C. The fit to the Po_2 dependence of the thermoelectric power was however less successful.

Measurements performed on YBCO specimens with constant values of x and compositions close to $YBa_2Cu_3O_6$ demonstrated that carrier generation is thermally activated and not solely dependent on x in contrast to earlier claims(10). Log σ and Q vs. 1/T plots resemble closely those of acceptor doped semiconductors. Analysis of the high temperature intrinsic region enables us to derive a band gap of 1.24eV which we propose corresponds to the disproportionation of Cu^{1+} to Cu^{0} + Cu^{2+}. For the least oxidized specimens we derive an acceptor ionization energy of 0.33 or 0.66eV depending on whether we have partial donor compensation or not. The more oxidized specimens exhibit increasingly higher degrees of degeneracy. Comparison of the slopes of log σ and Q at reduced temperatures suggest a non-activated mobility at least for x values less than 0.75-0.80.

A number of unresolved issues remain including the appropriateness of a single defect model for the full range of nonstoichiometry particularly given the semiconductor-metal transition observed in this material upon oxidation and the high concentrations of defects involved.

6. ACKNOWLEDGEMENT

The authors thank the MIT Center of Materials Science and Engineering for support of this work under Award #87 19217 DMR. We also appreciate receiving the preprint of H.I. Yoo et al´s manuscript prior to publication.

7. REFERENCES

1. M.P. Pechini, U.S. Patent 3,330,697 (11 July 1967).

2. G.M. Choi and H. L. Tuller, J. Am. Ceram. Soc. 71 201 (1988).

3. H.L. Tuller and A.S. Nowick, J. Phys. Chem. Sol. **38** 859 (1977).

4. G.M. Choi, H.L. Tuller, and M.-J. Tsai, in Proc. Symp. High Temperature Superconductors, (MRS Symp. Proc. V.99), eds., M.W. Brodsky, R.C. Dynes, K. Kitizawa, and H.L. Tuller, Mat. Res. Soc., Pittsburgh, PA (1988), p. 141.

5. K. Kishio, J. Shimoyama, T. Hasegawa, K. Kitazawa, and K. Fueki, Jpn. J. Appl. Phys. **26** L1054 (1987).

6. H.M. O´Bryan and P.K. Gallagher, Adv. Cer. Mat. 2 [3B] 640 (1987).

7. H.I. Yoo, `High Temperature Transport Properties of the High T_c Superconductor $Y_1Ba_2Cu_3O_x$,´ J. Mat. Res., recently submitted.

8. D. Ahuja, S.E. Dorris, M.Y. Su, Q. Robinson, D.L. Johnson, and T.O. Mason in Proc. Symp. High-Temperature Superconductors (MRS Symp. Proc. Vol. 99), eds. M.W. Brodsky, R.C. Dynes, K. Kitizawa, and H.L. Tuller, Mat. Res. Soc., Pittsburgh, PA (1988) p. 467.

9. D.W. Murphy, S.A. Sunshine, P.K. Gallagher, H.M. O´Bryan, R.J. Cava, B. Batlogg, R.B. van Dover, L.F. Schneemeyer, and J.M. Zahurak, in Proc. Symp. Chemistry of High-Temperature Superconductors (ACS Symp. Sec. 351), eds., D.L. Nelson, M.S. Whittingham, and T.F. George, Am. Chem. Soc., Washington D.C. (1987), p.181.

10. S. Sageev Grader, P.K. Gallagher, and E.M. Gyorgy, Appl. Phys. Lett. **51** 5 (1987).

11. J.S. Blakemore, Solid State Physics, 2nd Edition, Cambridge Univ. Press, Cambridge, England, 1985, p.319.

12. E.K. Chang, D.J.L. Hong, A. Mehta, and D.M. Smyth, Mat. Lett. **6** 251 (1988).

OXYGEN NONSTOICHIOMETRY AND CHEMICAL DIFFUSION COEFFICIENT OF OXYGEN IN $Ba_2YCu_3O_{7-x}$

Tsuneo Matsui[1)], Keiji Naito[1)] and Sadaaki Hagino[2)]
1) Department of Nuclear Engineering, Faculty of Engineering, Nagoya University, Furo-cho, Chikusa-ku, Nagoya 464-01, Japan
2) Superconducting Materials Development Team, Central Research Institute, Mitsubishi Metal Corporation, Kitabukuro-cho, Omiya, Saitama 330, Japan

ABSTRACT. The oxygen nonstoichiometry x and the chemical diffusion coefficient of oxygen in $Ba_2YCu_3O_{7-x}$ were measured by means of thermogravimetry as a function of oxygen partial pressure and temperature over the ranges $10^2 \leq Po_2/Pa \leq 10^5$ and $698 \leq T/K \leq 1173$. The oxygen partial pressure dependence of x in $Ba_2YCu_3O_{7-x}$ obtained in this study was in good agreement with those previously reported by other researchers. The chemical diffusion coefficients of oxygen in $Ba_2YCu_3O_{7-x}$ with various oxygen nonstoichiometry x=0.063-0.77 were measured, and they increased with decreasing oxygen nonstoichiometry x at constant temperature. The compositional dependence of the chemical diffusion coefficient changed at about x=0.35, indicating the occurrence of the phase transition from the tetragonal to the orthorhombic structure. The activation energies of the chemical diffusion coefficients of oxygen in $Ba_2YCu_3O_{7-x}$ with various x were observed to be about $115 \cdot kJ\ mol^{-1}$, irrespective of oxygen nonstochiometry x. The self-diffusion coefficient of oxygen and the diffusion coefficient of oxygen vacancy were calculated from the chemical diffusion coefficient, and were compared with those of another superconductor $La_{2-y}Sr_yCuO_{4-x}$.

1. INTRODUCTION

The usefulness of the new superconductor with perovskite structure $Ba_2YCu_3O_{7-x}$ relies on the proper high-temperature treatment. The critical temperature, the critical current and the mechanical strength vary drastically with oxygen content. The oxygen content of the sample changes dependently on temperature, oxygen partial pressure and annealing time. Accordingly, the study on the oxygen nonstoichiometry and chemical diffusion coefficient is necessary in order to optimize the annealing condition for obtaining the better superconducting properties.

The measurement of the oxygen nonstoichiometry x in $Ba_2YCu_3O_{7-x}$ has been carried out by Kishio et al. [1] and Gallagher [2]. Their data

J. Nowotny and W. Weppner (eds.),
Non-Stoichiometric Compounds Surfaces, Grain Boundaries and Structural Defects, 471–484.

agree well each other, but no clear change was found in the slope of the oxygen nonstoichiometry x vs. the oxygen partial pressure relating to the occurrence of the phase transition from the tetragonal to the orthorhombic structure reported by means of high temperature x-ray diffractometry [4,5] and electrical conductivity measurement [6]. The chemical diffusion coefficient of oxygen in $Ba_2YCu_3O_{7-x}$ has not been determined until the present experiment was finished. But just after finishing the present experiment, the study on the chemical diffusion coefficients of oxygen in $Ba_2YCu_3O_{7-x}$ with various oxygen nonstoichiometry (x=0.15-0.73) in the range $739 \leq T/K \leq 1173$ was reported by Shimoyama et al. [3]. However their chemical diffusion coefficients of oxygen are mainly in the high temperature $Ba_2YCu_3O_{7-x}$ phase with the tetragonal structure ($T \geq 925$ K) and the dependence of the chemical diffusion coefficients of oxygen upon the oxygen composition was reported only for $Ba_2YCu_3O_{7-x}$ with large oxygen nonstoichiometry $x \geq 0.48$. Accordingly the measurement of the chemical diffusion coefficient of oxygen in the orthorhombic $Ba_2YCu_3O_{7-x}$ ($x \leq 0.4$) at low temperature ($T \leq 873$ K) is still needed.

In the present study, the oxygen nonstoichiometry x and the chemical diffusion coefficient of oxygen in $Ba_2YCu_3O_{7-x}$ were measured by means of thermogravimetry as a function of oxygen partial pressure and temperature over the ranges $10^2 \leq Po_2/Pa \leq 10^5$ and $698 \leq T/K \leq 1173$ in order to reconfirm the occurrence of the structural phase transition from the oxygen partial pressure dependence of oxygen nonstoichiometry and to determine the compositional dependence of the chemical diffusion coefficients of oxygen in $Ba_2YCu_3O_{7-x}$ with both the tetragonal and the orthorhombic structure ($0.0625 \leq x \leq 0.774$). The chemical diffusion coefficients of oxygen in $Ba_2YCu_3O_{7-x}$ with various x values obtained in the present study were discussed in comparison with those in $Ba_2YCu_3O_{7-x}$ recently reported by Shimoyama et al. [3]. The diffusion coefficient of oxygen vacancy and the self diffusion coefficient of oxygen in $Ba_2YCu_3O_{7-x}$ were also determined from the chemical diffusion coefficient and were compared with those of another superconductor $La_{2-x}Sr_xCuO_{4-y}$ [7].

2. EXPERIMENTAL

Samples of $Ba_2YCu_3O_{7-x}$ were prepared either by a powder mixing method or a coprecipitation technique. For the powder mixing method, high purity Y_2O_3, $BaCO_3$ and CuO powders were mixed in an appropriate ratio, subsequently heated at 1193 K for 12 h in air. The mixture was reground and pressed into pellets which were again heated at 1193 K for 12 h in air. The pellets were reground to be fine powder by jet milling method. The powder was again pressed into pellets, which were then finally sintered at 1223 K for 8 h in air.

For the coprecipitation method, three nitric acid solutions of each constituent metal were mixed in an appropriate ratio. The ammonium oxalate solution was then used to form the precipitates from the mixed solution. The precipitates were then filtrated and dried, subsequently calcined at 1189 K for 8 h in air. They were reground, finally heated at 1189 K for 8 h in air and then ground to make fine powder. The powder, thus obtained, was pressed into pellet and then sintered at 1223 K for 8 h in air.

The pellets thus obtained were approximately 1.0 g in weight, 9.7-10.2 mm in diameter and 2.2-2.3 mm in thickness. The densities were about 93-95 % of the theoretical density. The atomic compositions of Ba and Y in $Ba_2YCu_3O_{7-x}$ were chemically analyzed by precipitating Ba as $BaSO_4$ and Y as Y_2O_3, respectively. The atomic composition of Cu was determined by titration using $Na_2S_2O_3$ solution. X-ray diffraction analysis at room temperature confirmed that the samples were single phase with the orthorhombic perovskite structure. No difference was observed in the values of the oxygen nonstoichiometry and the chemical diffusion coefficient determined with both samples prepared by a powder mixing method and a coprecipitation technique.

The measurement of the oxygen nonstoichiometry and the chemical diffusion coefficient of oxygen in $Ba_2YCu_3O_{7-x}$ was carried out with a Cahn 1000 electro-microbalance. The chemical diffusion coefficient of oxygen was determined from the weight change of the sample induced by a stepwise change in oxygen partial pressure. Before and after the stepwise change in oxygen partial pressure, the sample was equilibrated with an oxygen partial pressure for a time, ranging from 12 to 48 h, depending on temperature and oxygen partial pressure.

The oxygen partial pressure was controlled with O_2/Air or Air/Ar mixed gas. The oxygen partial pressure was determined from the mixing ratio of the gases or by measuring the electrical conductivity of cobalt oxide calibrated before use as an oxygen sensor in our laboratory.

3. RESULTS AND DISCUSION

3.1. Oxygen Nonstoichiometry

In the present study, the oxygen nonstoichiometry x in $Ba_2YCu_3O_{7-x}$ at 1173 K and $Po_2=1.0\times10^5$ Pa was assumed to be 0.600 as the standard point, since both Kishio et al. [1] amd Gallagher [2] have reported x at 1173 K in $Po_2=1.0\times10^5$ Pa to be 0.60. The variation of x with temperature and Po_2 was determined by the weight change from that assumed standard value.

The values of the oxygen nonstoichiometry x in $Ba_2YCu_3O_{7-x}$ measured at temperatures of 698-1173 K in this study are shown in Fig. 1 against oxygen partial pressure together with those previously reported by Kishio et al. [1] and Gallagher [2]. It is seen from this figure that the present result of x is fairly in good agreement with the previous results [1,2]. The values of the oxygen nonstoichiometry x at temperatures 698-1173 K are summarized in Table I.

The occurrence of the phase transition from the tetragonal to the orthorhombic structure at about x=0.25-0.42 has been reported by high temperature X-ray diffractometry [4,5] and electrical conductivity measurement [6], but no clear sign of the occurrence of the phase transition has been observed in the variation of x with oxygen partial pressure [1,2]. As seen in Fig. 1, no clear change in the slope of x

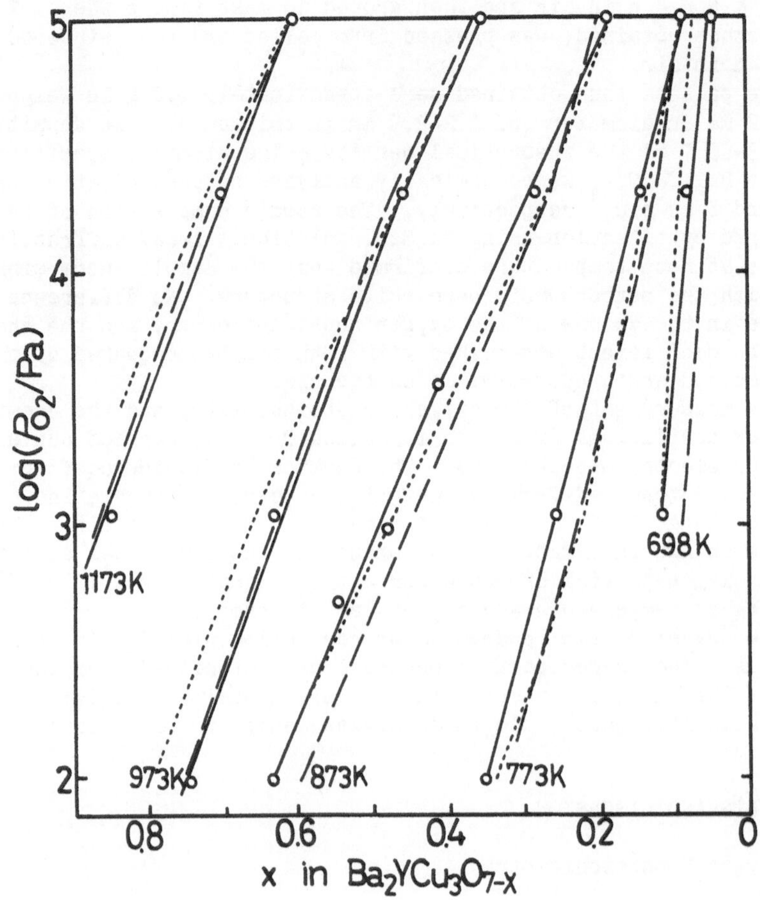

Fig. 1 Variation of the oxygen nonstoichiometry x with the oxygen partial pressure.
o—o This study, ... Kishio et al. (1),
\- - Gallagher (2)

vs. log Po_2 around x=0.2-0.4 was also found in the present study. The relation between log Po_2 and log x in $Ba_2YCu_3O_{7-x}$ obtained in this study is shown in Fig. 2. In this figure, two regions with different dependences of x upon Po_2 are seen. In the compositional region with small x values ($x \leq 0.3$), the slope values of $\partial \log Po_2/\partial \log x$ are about 5, irrespective of temperatures from 698 to 873 K. In the compositional region with large x values ($x \geq 0.4$), the slope values are about 10, irrespective of temperatures from 873 to 1173 K. At 873 K the slope value changed from 5 to 10 around x=0.4. The difference in the slope values from 5 to 10 observed in the present study is thought to be indicative of the occurrence of the phase transition from the orthorhombic to the tetragonal structure. Although the slope value

Table I Oxygen nonstoichiometry x in $Ba_2YCu_3O_{7-x}$

T/K	Po_2/Pa	x
1173	$1.0x10^5$	0.600±0.001
	$2.1x10^4$	0.699±0.002
	$1.1x10^3$	0.848±0.002
973	$1.0x10^5$	0.353±0.001
	$2.1x10^4$	0.456±0.003
	$1.1x10^3$	0.631±0.003
	$9.7x10^1$	0.745±0.003
873	$1.0x10^5$	0.184±0.001
	$2.1x10^4$	0.279±0.001
	$3.7x10^3$	0.418±0.001
	$9.7x10^2$	0.481±0.004
	$5.0x10^2$	0.547±0.004
	$9.9x10^1$	0.635±0.002
773	$1.0x10^5$	0.086±0.002
	$2.1x10^4$	0.139±0.002
	$1.1x10^3$	0.257±0.002
	$9.9x10^1$	0.353±0.001
698	$1.0x10^5$	0.045±0.002
	$2.1x10^4$	0.080±0.002
	$1.1x10^3$	0.116±0.002

$\partial \log Po_2 / \partial \log x \simeq 5$ in the orthorhombic phase seems to be simply interpreted by assuming the co-presence of the singly and doubly charged oxygen vacancies as the predominant defects, a simple defect may not be applied for $Ba_2YCu_3O_{7-x}$, since the distribution of the oxygen vacancies is not random. According to the structural studies [8-10], it was concluded that (1) the one-dimensional -O-Cu-O-Cu-O- chains along the a-axis on the plane sandwiched between the two adjacent Ba planes lose their oxygen atom and produced the ordering of oxygen vacancies in the

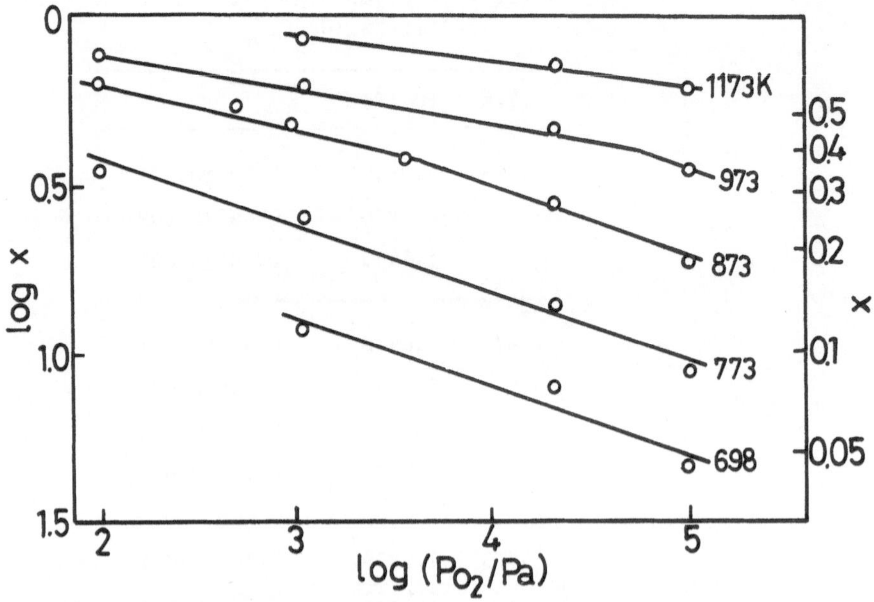

Fig. 2 Variation of log x in $Ba_2YCu_3O_{7-x}$ with log Po_2.

low temperature (low x-value) orthorhombic phase, and (2) the disordering of the oxygen vacancies relates to the orthorhombic-to-tetragonal phase transition. The defect structure including the distribution (ordering) of the oxygen vacancies and/or interaction between oxgen vacancy and copper ion may be more appropriate.

3.2. Chemical Diffusion Coefficient of Oxygen

If the relaxation phenomenon is based on the diffusion controlled process, the weight of a cylindrical sample with a size 2a in diameter and 2b in thickness changes with time t in accordance with the following equation [11]:

$$\frac{W(t)-W(o)}{W(\infty)-W(o)} = 1 - \frac{32}{\Pi_a^2} \times \left(\sum_{n=0}^{\infty} \frac{\exp[-\tilde{D}t(2n+1)^2\Pi^2/4b^2]}{(2n+1)^2} \right)$$

$$\times \left(\sum_{m=1}^{\infty} \frac{\exp(-\tilde{D}t\alpha_m^2)}{\alpha_m^2} \right) , \qquad (1)$$

where $\tilde{D}$ is the chemical diffusion coefficient, W(o), W(∞) and W(t) are the initial weight, the final equilibrium weight and the weight at t, respectively, and α_m is the mth positive root of the Bessel function of the zeroth order $Jo(a\cdot\alpha_m)=0$. After a sufficiently long time ([W(t)-W(o)]/[W(∞)-W(o)]$\geq$0.5), equation (1) becomes

$$\log\left(\frac{W(\infty)-W(t)}{W(\infty)-W(o)}\right) = -0.252 - \frac{1}{2.303}\left(\frac{(2.405)^2}{a^2} + \frac{2}{4b^2}\right)\tilde{D}t \quad . \qquad (2)$$

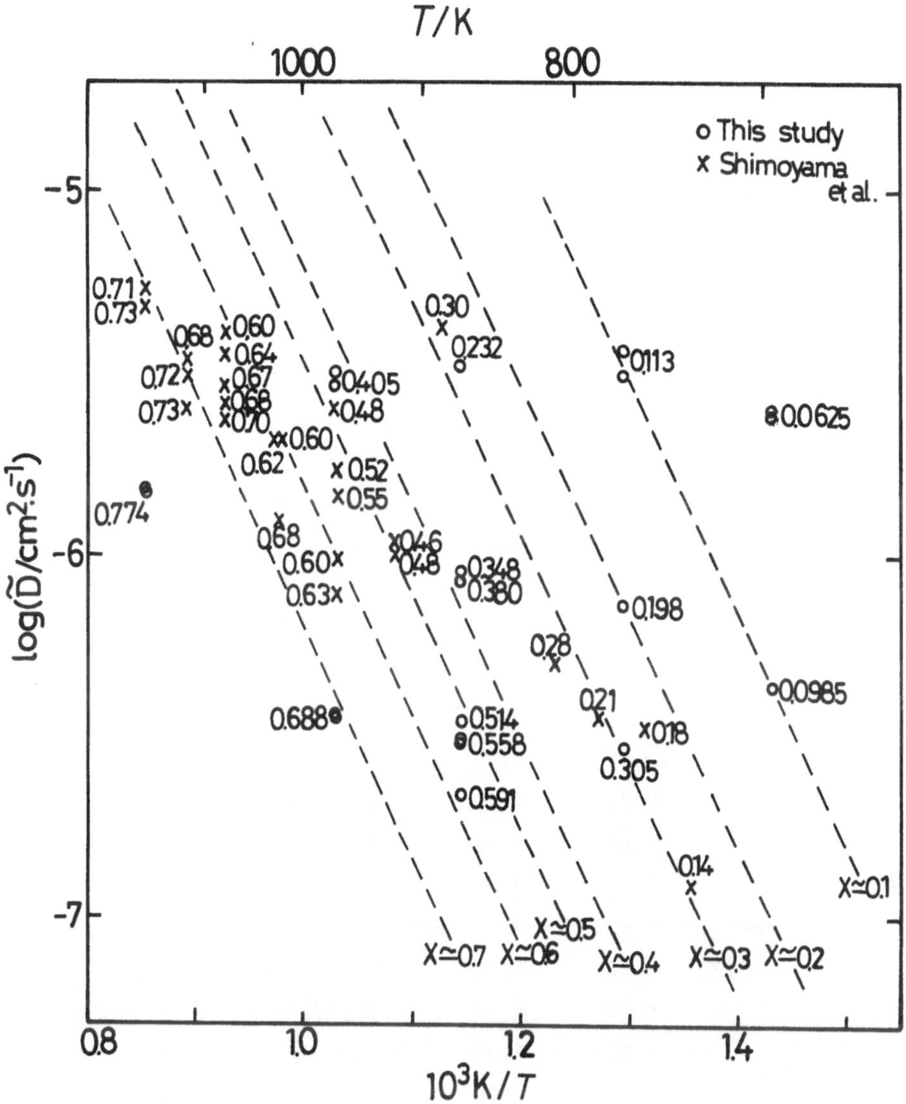

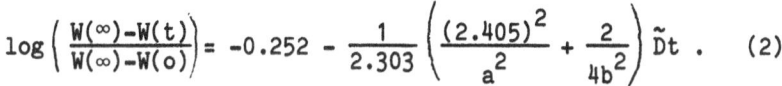

Fig. 3 Variation of the chemical diffusion coefficients of oxygen with the reciprocal temperatures.
The numbers in the figure indicate the oxygen nonstoichiometry x in $Ba_2YCu_3O_{7-x}$.

From the values of the slope of the logarithmic plot of weight change against time, the chemical diffusion coefficient can be calculated. Since the diffusion coefficient of oxygen in the perovskite oxide is generally much higher than that of cation [12], the chemical diffusion in $Ba_2YCu_3O_{7-x}$ for oxidation and reduction is mainly due to the oxygen ion rather than the cation.

Table II Chemical diffusion coefficient of oxygen in $Ba_2YCu_3O_{7-x}$

Temp. T/K	log (Po_2/Pa) initial	log (Po_2/Pa) final	Diff. Coeff. $\tilde{D}/cm^2 \cdot s^{-1}$	Averaged x
1173	4.32	3.03	$(1.50 \pm 0.04) \times 10^{-6}$	0.774
	3.03	4.32	$(1.54 \pm 0.04) \times 10^{-6}$	0.774
973	4.32	5.00	$(3.18 \pm 0.04) \times 10^{-6}$	0.405
	5.00	4.32	$(2.87 \pm 0.02) \times 10^{-6}$	0.405
	3.05	1.99	$(3.59 \pm 0.04) \times 10^{-6}$	0.688
	1.99	3.05	$(3.64 \pm 0.04) \times 10^{-6}$	0.688
873	5.00	4.32	$(3.30 \pm 0.02) \times 10^{-6}$	0.232
	4.32	3.56	$(9.04 \pm 0.02) \times 10^{-7}$	0.348
	4.32	2.99	$(8.45 \pm 0.01) \times 10^{-7}$	0.380
	2.99	2.70	$(3.51 \pm 0.05) \times 10^{-7}$	0.514
	2.70	2.00	$(2.19 \pm 0.08) \times 10^{-7}$	0.591
	2.00	3.02	$(3.15 \pm 0.04) \times 10^{-7}$	0.558
	3.02	2.00	$(3.11 \pm 0.05) \times 10^{-7}$	0.558
773	4.32	5.00	$(3.69 \pm 0.06) \times 10^{-6}$	0.113
	5.00	4.32	$(3.11 \pm 0.03) \times 10^{-6}$	0.113
	4.32	3.02	$(7.33 \pm 0.06) \times 10^{-7}$	0.198
	3.02	2.00	$(2.98 \pm 0.02) \times 10^{-7}$	0.305
698	4.32	5.00	$(2.38 \pm 0.06) \times 10^{-6}$	0.0625
	5.00	4.32	$(2.47 \pm 0.07) \times 10^{-6}$	0.0625
	4.32	3.02	$(4.40 \pm 0.04) \times 10^{-6}$	0.0985

The chemical diffusion coefficients of oxygen in $Ba_2YCu_3O_{7-x}$ thus obtained are shown in Fig. 3 as a function of the reciprocal temperature together with those recently reported by Shimoyama et al. [3] and are summarized in Table II. The absolute value and the temperature dependence of the chemical diffusion coefficients of oxygen in $Ba_2YCu_3O_{7-x}$ at high temperatures obtained in this study are seen to agree well with those reported by Shimoyama et al. [3]. It is concluded from both the present result mainly at low temperature and the previous result mainly at high temperature by Simoyama et al. shown in Fig. 3 that (1) the chemical diffusion coefficient of oxygen increased with decreasing x value at constant temperature and (2) the activation

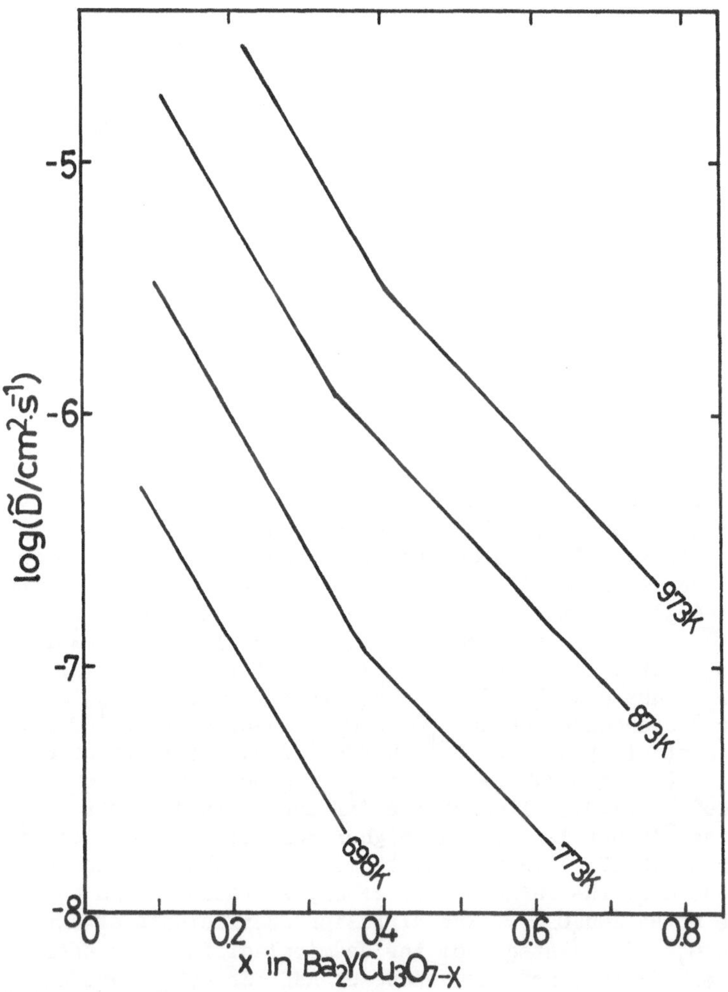

Fig. 4 Variation of the chemical diffusion coefficients of oxygen with the oxygen nonstoichiometry x.

energies of the chemical diffusion coefficients of oxygen in $Ba_2YCu_3O_{7-x}$ with various x were about 115 kJ mol^{-1}, irrespective of x values (x=0.1-0.7) and the difference in the crystal structure (orthorhombic or tetragonal). The temperature dependences of the chemical diffusion coefficients of oxygen in $Ba_2YCu_3O_{7-x}$ at constant compositions from x≃0.1 to x≃0.7 estimated from both the present result and the previous result by Shimoyama et al. are shown by the broken lines in Fig. 3. The composional dependences of the chemical diffusion coefficient of oxygen in $Ba_2YCu_3O_{7-x}$ at constant temperatures from 698 to 973 K obtained from the broken lines in Fig. 3 are shown in Fig. 4. The change in the slope of D vs. x seen at about x=0.35 in the figure is thought to be indicative of the occurrence of the phase transition from the orthorhombic to the tetragonal structure.

The chemical diffusion coefficients of oxygen $\tilde{D}$ can be related to the diffusion coefficient of oxygen vacancy Dv and the self-diffusion coefficient of oxygen Do according to the Wagner's theory [13]. In the case of $Ba_2YCu_3O_{7-x}$ in which electronic conduction prevails, the chemical diffusion coefficients $\tilde{D}$ is expressed in terms of Dv and Do as follows:

$$\tilde{D} = \frac{1}{2} Dv \left| \frac{\partial(\ln Po_2)}{\partial(\ln x)} \right| \tag{3}$$

$$= \frac{7Do}{2x} \left| \frac{\partial(\ln Po_2)}{\partial(\ln x)} \right|$$

As seen from the equations (3) and (4), both the diffusion coefficient of oxygen vacancy and the self-diffusion diffusion coefficient of oxygen can be given by the product of $\tilde{D}$ and the thermodynamic factor $\partial(\ln Po_2)/\partial(\ln x)$. The thermodynamic factor can be calculated from the slope of the plots in Fig. 2. In the compositional region with small x values ($x \leq 0.3$), the thermodynamic factor is about 5. In the compositional ragon with large x values ($x \geq 0.4$), the thermodynamic factor is about 10. The self-diffusion coefficients of oxygen in $Ba_2YCu_3O_{7-x}$ thus calculated are shown in Fig. 5 in comparison with those in another superconductor $La_{2-y}Sr_yCuO_{4-x}$ measured with ^{18}O using SIMS by Smedskjaer et al. [7]. The self-diffusion coefficients of oxygen in $Ba_2YCu_3O_{7-x}$ (x=0.1-0.7) are seen to be higher by 6-7 order than those in $La_{2-y}Sr_yCuO_{4-x}$ (x≃0), although the difference in the activation energy between them is not large. The higher self-diffusion coefficient of oxygen in $Ba_2YCu_3O_{7-x}$ is thought to be due to the presence of the open path produced by the large number of oxygen vacancies in the crystal structure of $Ba_2YCu_3O_{7-x}$. The diffusion coefficients of oxygen vacancy in $Ba_2YCu_3O_{7-x}$ calculated from the chemical diffusion coefficients are shown in Fig. 6 together with those of $La_{2-y}Sr_yCuO_{4-x}$ and $La_{1-y}Sr_yCuO_{3-x}$ with the same perovskite structure [14]. The values of the diffusion coefficient of oxygen vacancy in $Ba_2YCu_3O_{7-x}$ are nearly the same order as those of $La_{1-y}Sr_yCuO_{3-x}$ which are reported to be in dependent of x values and y value of 0 and 0.1[14], and of $La_{2-y}Sr_yCuO_{4-x}$ calculated from the self-diffusion coefficient of oxygen on the assumption of $x \simeq 1x10^{-5}$ by the present authors.

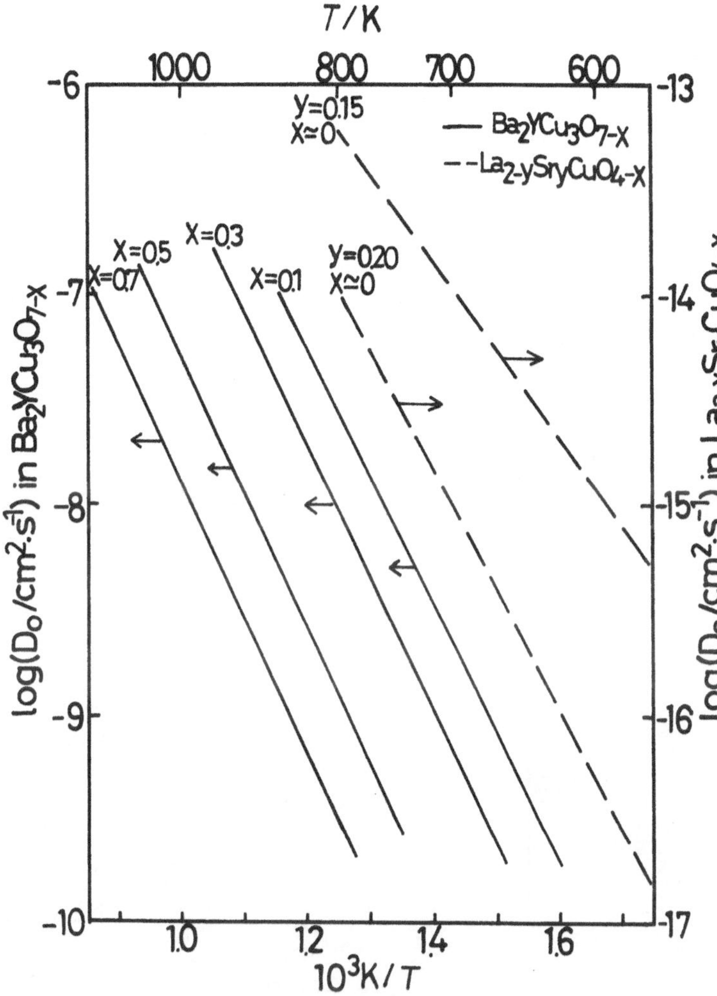

Fig. 5 Variation of the self-diffusion coefficient of oxygen in $Ba_2YCu_3O_{7-x}$ and $La_{2-y}Sr_yCuO_{4-x}$.

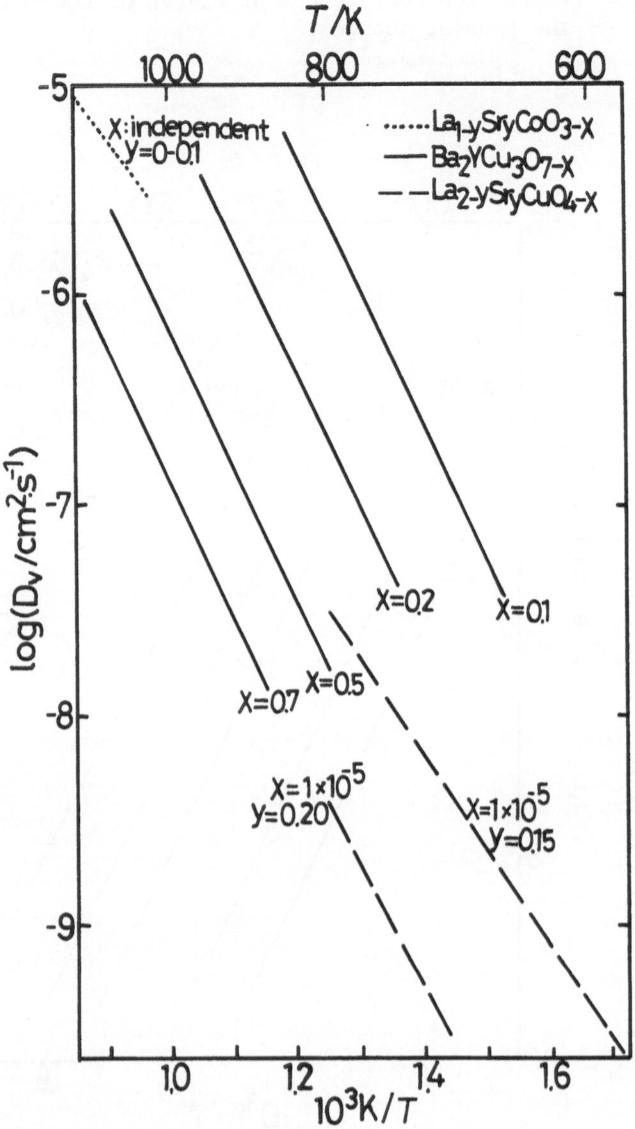

Fig. 6 Variation of the diffusion coefficients of oxygen vacancy in $Ba_2YCu_3O_{7-x}$, $La_{2-y}Sr_yCuO_{4-x}$ and $La_{1-y}Sr_yCoO_{3-x}$.

4. CONCLUSION

The occurrence of the phase transition from the tetragonal to the orthorhombic structure has not been clearly shown yet by means of thermogravimetry. The change in the slope of log x vs. log Po_2 found around $x \simeq 0.3$-0.4 in the present study was thought to be the clear indication of the occurrence of the phase transition.

The chemical diffusion coefficient of oxygen in $Ba_2YCu_3O_{7-x}$ was measured mainly in the region of the small oxygen nonstoichiometry ($x \leq 0.4$) at low temperature ($T \leq 873$ K) in this study. The dependences of the chemical diffusion coefficients of oxygen upon the oxygen nonstoichiometry x at constant temperature and upon the reciprocal temperature at constant oxygen nonstoichiometry x were first determined in the low temperature superconducting $Ba_2YCu_3O_{7-x}$ phase with the orthorhombic structure.

The compositional dependences of the chemical diffusion coefficients in $Ba_2YCu_3O_{7-x}$ with various oxygen nonstoichiometries ($0.0625 \leq x \leq 0.774$) at constant temperature determined in the present study changed at about x=0.35, indicating the occurrence of the phase transition, as was suggested from the slope change in the plot of log x vs. log Po_2.

The diffusion coefficient of oxygen vacancy and the self-diffusion coefficient of oxygen in $Ba_2YCu_3O_{7-x}$ were first calculated form the chemical diffusion coefficient and were compared with those of another superconductor $La_{2-y}Sr_yCuO_{4-x}$. The higher self-diffusion coefficient of oxygen in $Ba_2YCu_3O_{7-x}$ is thought to be due to the presence of the open path produced by the large number of oxygen vacancies in the crystal structure of $Ba_2YCu_3O_{7-x}$. The diffusion coefficient of oxygen vacancy in $Ba_2YCu_3O_{7-x}$ was nearly the same order as that of $La_{2-y}Sr_yCuO_{4-x}$ calculated from the self-diffusion coefficient of oxygen on the assumption of x 1×10^{-5} by the present authors.

REFERENCES

[1] K. Kishio, J. Shimoyama, T. Hasegawa, K. Kitazawa and K. Fueki, Jpn. J. Appl. Phys. 26 (1987)L1228.
[2] P. K. Gallagher, Adv. Ceram. Mater. 2 (1987)632.
[3] J. Shimoyama, T. Hasegawa and K. Kishio, Sankabutsuchodendotai no kagaku, p.74, Kodansha Scientific Co. (1988).
[4] E. T. Muromachi, Y. Uchida, K. Yukino, T. Tanaka and K. Kato, Jpn. J. Appl. Phys. 26 (1987)L665.
[5] K. Yukino, T. Sato, S. Ooba, M. Ohta, F. P. Okamura and A. Ono, Jpn. J. Appl. Phys. 26 (1987)L869.
[6] S. G. Grader, P. K. Gallagher and E. M. Gyorgy, Appl. Phys. Lett. 51 (1987)1115.
[7] L. C. Smedskjaer, J. L. Routbort, B. K. Flandermeyer, S. J. Rothman and D. G. Legnini, Phys. Rev. B36 (1987)3903.
[8] F. Izumi, H. Asano, T. Ishigaki, E. Takayama-Muromachi, Y. Uchida, N. Watanabe and T. Nishikawa, Jpn. J. Appl. Phys. 26 (1987)L1193.
[9] M. A. Beno, L. Soderholm, D. W. Capone, D. G. Hinks, J. D.

Jorgensen, I. K. Schuller, Appl. Phys. Lett. 51 (1987)57.
[10] J. O. Jorgensen, M. A. Beno, D. G. Hinks, L. Soderholm, K. J. Volin, R. L. Hitterman, J. D. Grace and I. K. Schuller, Phys. Rev. B36 (1987)3608.
[11] H. S. Carslaw and J. C. Jaeter, Conduction of Heat in Solids, p227, Oxford press, Oxford (1959).
[12] K. Fueki, K. Kitazawa adn K. Matsukawa, Yogyo-Kyokai-Shi 87 (1979)37.
[13] C. Wagner, Z. Phys. Chem. B11 (1930)139 and B32 (1936)447.
[14] K. Fueki, J. Mizusaki, S. Yamauchi, T. Ishigaki and Y. Mima, Reactivity of Solids, (Proceedings of the 10th International Symposium) p.339 (1984).

POINT DEFECTS AND TRANSPORT IN SMALL POLARON SYSTEMS

G. P. Sykora, M. -Y. Su and T. O. Mason
Northwestern University
Department of Materials Science and Engineering and
Materials Research Center
The Technological Institute
Evanston, IL 60208

ABSTRACT. A general defect equation has been developed which can describe defect phenomena in systems with variable valence cations (e.g. transition metals) which exhibit small polaron conduction. The types of defect phenomena considered include isolated point defects (vacancies), charged defect clusters (vacancy/interstitial), neutral clusters which link and produce percolation networks, and oxygenation (intercalation) where oxygen insertion does not result in the formation of cation vacancies but is still charge compensated by the variable valence cations. The bulk cation valence ratio (isolated defects, charged clusters, oxygen intercalated structures) and the local cation valence ratio (neutral cluster networks) can be studied via the electrical properties (conductivity, thermopower.) Examples are taken from the systems: $Co_{1-\delta}O$, $La_{2-x}Ba_xCuO_4$ and $YBa_2Cu_3O_{6+Y}$.

1. INTRODUCTION

Small polarons consist of charge carriers with large effective mass and low mobility due to the lattice polarization which causes their localization on discrete sites. In variable valence systems such as transition metal oxides this results in a mixture of valence states. Verwey [1] was the first to describe such "mixed valence compounds" and the corresponding electrical properties. The seminal work by Bosmann and van Daal [2] describes the physical properties unique to small polaron systems. Chen et al.[3] discuss the characteristic features of small polaron conduction and how to establish that this mechanism predominates in a given material.

Beginning with the well known Nernst-Einstein equation it can be shown that the mobility for small polaron conductors is [4,5]:

J. Nowotny and W. Weppner (eds.),
Non-Stoichiometric Compounds Surfaces, Grain Boundaries and Structural Defects, 485–497.

$$\mu = \frac{g(1-x)ea^2\nu_o}{kT} \exp\left\{\frac{-E_H}{kT}\right\} \qquad (1)$$

where g is a geometric factor on the order of unity, (1-x) is the probability that an adjacent equivalent site is of the opposing valence and is therefore available for hopping, e is the charge of an electron, a is the jump distance, ν_o is the lattice vibrational frequency responsible for hopping, k is Boltzmann's constant, E_H is the hopping energy and T is absolute temperature. Since the conductivity is given by the product of the mobility, electronic charge and carrier concentration, the result is:

$$\sigma = Nx(1-x)\ \sigma_o\ T^{-1} \exp\,[-E_H/kT] \qquad (2)$$

where N is the density of transition metal cations involved in the conduction system, a fraction x of which are small polarons (i.e. possess charge carriers) and σ_o is a preexponential constant incorporating most of the constants of Eq. 1.

The distiguishing features of small polaron conduction are evident in Eqs. 1 and 2. By letting x be negligibly small and setting E_H=0 an upper limit for mobility can be obtained. For $10^{13}s^{-1} \leq \nu_o \leq 10^{14}s^{-1}$ values of $0.1cm^2V^{-1}s^{-1} \leq \mu \leq 1.0cm^2V^{-1}s^{-1}$ are obtained assuming reasonable jump distances. The T^{-1} term can be diagnostic if x is temperature-independent over a large temperature range [6]. The most recognized feature is the activated mobility. Since the hopping energy is typically small, a few tenths of an eV, often the T^{-1} and exponential term cancel. Where x is approximately 0.5 as in mixed valence compounds like magnetite [5] carrier densities (Nx) can be large, on the order of $10^{22}cm^{-3}$. Combining this value with the maximum mobilities results in an upper limit for small polaron conductivity in the 10^2 to 10^3 $(\Omega\text{-cm})^{-1}$ range. Perhaps the most telling feature of small polaron behavior is the x(1-x) factor in Eq. 2. When x is significantly large, small polaron behavior will be evident by the (1-x) dependence of the mobility which translates into the x(1-x) dependence of the conductivity.

This factor will be discussed at length below.

Two expressions are possible for thermopower in small polaron systems, depending upon whether the carriers are electrons or electron holes [7]. In the former case:

$$Q = -\frac{k}{e}\left\{ \ln 2 \frac{[M^{(n+1)+}]}{[M^{n+}]} + A \right\} \tag{3}$$

whereas in the latter case:

$$Q = \frac{k}{e}\left\{ \ln 2 \frac{[M^{(n-1)+}]}{[M^{n+}]} + A \right\} \tag{4}$$

where M^{n+} is either the lower or higher valence state of the transition metal and k and e have been defined previously. The entropy of transport term, A, is negligibly small in most instances with a few noteworthy exceptions [8,9]. All iron-based transition metal oxide work in our laboratory has been interpretable with negligible A terms. See [10] for a discussion of the ferrite spinels.

By combining Eqs. 1 through 4 much can be learned about the valence distributions in small polaron systems via high temperature electrical property measurements. For example, the distribution of Fe^{2+} and Fe^{3+} (and often that of the doping species) between octahdral and tetrahedral sites in ferrospinels can be established. Systems studied to date include Fe_3O_4 and Fe_3O_4 doped with divalent cations (Co, Mg, Mn, Ni), trivalent cations (Al, Cr) and tetravalent cations (Ti). See [10] for a thorough review. In the case of $MnFe_2O_4$ the $Mn^{2+}/Fe^{3+} \Leftrightarrow Mn^{3+}/Fe^{2+}$ redox equilibrium was studied via the high temperature electrical properties [11]. Finally, the degree of octahedral site diproportionation of Mn^{3+} to Mn^{2+} and Mn^{4+} was established via high temperature electrical property measurements in Mn_3O_4 [7,11].

The present study will show that a similar approach can be taken to the study of defect structures in transition metal oxide systems which exhibit small polaron conduction. Following the development of a general defect equation, this will be used to analyze the defect behavior in systems with isolated charged defects and defect clusters (e.g. $Co_{1-\delta}O$), systems with neutral clusters linked into networks (e.g. $La_{2-x}Ba_xCuO_4$), and oxygen intercalation compounds (e.g. $YBa_2Cu_3O_{6+Y}$).

2. A GENERAL DEFECT MODEL

We previously presented a generalized equation describing the formation of clusters in transition metal monoxides [12]. We have since extended this equation to other types of oxides with other point defect phenomena. The equation is:

$$\{2(v-i) + \alpha\,[i(1+n-m) + b]\}\, M_M^{n+} + \frac{(v-i)}{2}\, O_2(g) \xrightarrow{M_\alpha O} (v-i)\, O_O^{=}$$

$$+ \alpha\,(vV_M iM_i^{m+} aM_M^{(n+1)+} bM_M^{n+}) + \{2(v-i) + \alpha\,[i(n-m) - a]\, M_M^{(n+1)+} \qquad (5)$$

where α is the cation/anion ratio of the oxide in question, v and i are the number of vacancies and interstitials in the defect or cluster being formed, n is the initial valence of the transition metal cation prior to oxidation, and m is the valence of the interstitials. As will be shown, the negative effective charge on a defect or cluster can be locally neutralized by trapping small polarons at adjacent positions. Of these a+b equivalent sites, either a are occupied by holes or b are occupied by electrons. A simplification of Eq. 5 can be written as:

$$\beta\, M^{n+} + (\gamma/2)\, O_2(g) \Leftrightarrow \gamma\, O_O^{=} + \delta\,[\text{defect/cluster}] + \epsilon\, M^{(n+1)+} \qquad (6)$$

Given the formation of small polarons, the electrical properties become a powerful tool to investigate the overall valence ratio (ϵ/β) in the case of isolated defects and clusters or intercalation, or to register the local valence ratio associated with a cluster network (a/b in Eq. 5). A transition from isolated defect behavior to cluster link-up can also be detected.

3. APPLICATIONS

3.1 ISOLATED DEFECTS/CLUSTERS: CoO

With isolated charged defects or clusters, $M^{(n+1)+}$ species are formed for charge compensation of the defects at the expense of M^{n+} species. By studying the $M^{(n+1)+}/M^{n+}$ ratio thermoelectrically or the $M^{(n+1)+}M^{n+}$ product via conductivity, the defect equilibrium can be investigated. An example of this is our previous work in magnetite, Fe_3O_4 [5]. By letting α=3/4 (spinel stoichiometry), i=a=b=0, and v=1 (isolated vacancies) in Eq. 5, we obtain for n=2:

$$2\ M_M^{2+} + 1/2\ O_2(g) \Leftrightarrow O_O^{=} + 3/4\ V_M + 2\ M^{3+} \qquad (7)$$

In [5] the experimental conductivity was modelled by varying the equilibrium constant of this reaction until fit was achieved. The resulting equilibrium constants agreed well with those obtained thermogravimetrically [13].

To further demonstrate the utility of Eq. 5 we set i=1 and v=0 in Eq. 5 (interstitials instead of vacancies), and obtain for m=3:

$$O_O^{=} + 11/4\ M_M^{3+} \Leftrightarrow 1/2\ O_2(g) + 3/4\ M_i^{3+} + 2\ M_M^{2+} \qquad (8)$$

(Species appear on the opposite side of the reaction due to negative coefficients resulting from the input data.) Since at small defect populations, Fe^{2+} and Fe^{3+} concentrations are nearly independent of defect concentration, it can be shown that vacancies and interstitials should have oxgyen activity exponents of 2/3 and -2/3, respectively, which is consistent with experimental results [13].

We have recently completed an electrical study of cobaltous oxide at large defect contents [9]. Deviations from stoichiometry as large as 2.5 percent have been achieved at 1200°C in a pressurized vessel capable of 1 to 100 atm. oxygen operation. What is surprising about the results in the 1 to 10 atm. range is the difference in oxygen pressure dependence between the various properties. The oxygen exponents for the deviation from stoichiometry, conductivity, and thermopower are 0.37, 0.27, and 0.43, respectively. To the best of our knowledge, the divergence of the two electrical properties can only be explained on the basis of small polaron theory. Furthermore, oxygen exponents for the electrical properties in excess of 0.25 (1/4) would require some kind of charged cluster. We have fit the literature data for CoO (α=1, n=2) with a model incorporating doubly charged vacancies (i=a=b=0, v=1), singly charged vacancies (i=b=0, v=a=1), and quadruply charged 4:1 vacancy/interstitial clusters (v=4, i=1, a=1, m=3). In the latter case the defect reaction would be:

$$6\ M_M^{2+} + 3/2\ O_2(g) \Leftrightarrow\ 3\ O_O^{=} + \{4\ V_M\ 1\ M_i^{3+}\ 1\ M_M^{3+}\} + 4\ M_M^{3+} \qquad (9)$$

Strictly on the basis of this reaction, the oxgyen pressure dependence of the holes (M^{3+}) would be 3/10 or 0.3. A number of different clusters were considered, however this selection gave the best fit to the data. The results are shown in Fig. 1. (Over the range displayed, the doubly charged vacancies are negligible in

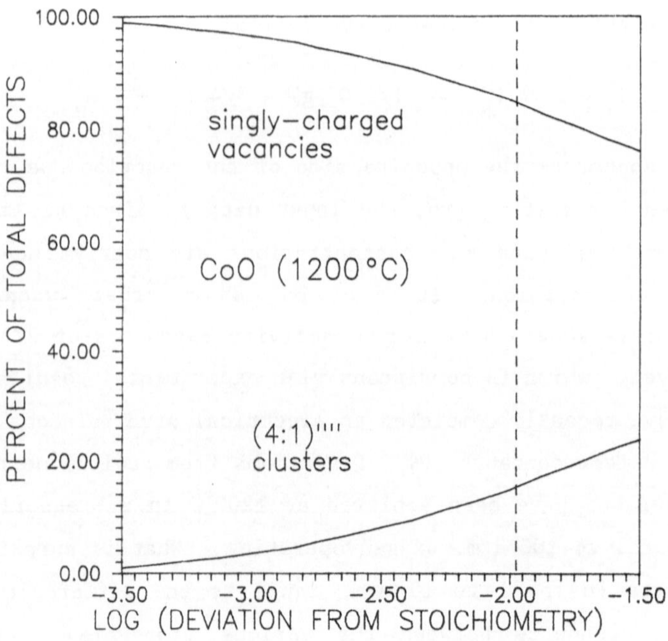

Fig. 1. Calculated vacancy and cluster concentrations in $Co_{1-\delta}O$. See [9] for details.

concentration.) The vertical dashed line at approximately 1 percent deviation from stoichiometry corresponds to the nonstoichiometry at 1 atm. oxygen; higher defect contents can only be achieved under pressurized oxygen as in [9]. It should be noted that theorists have calculated that $(vV_M iM_i)$ clusters should be stable species in the monoxides. Catlow [14] favors the 6:2 cluster whereas Khowash and Ellis [15] favor the 4:1 cluster in CoO.

The most significant feature of the high pressure work in CoO is that it allows the conduction mechanism to be unambiguously established. This has not been done to date given the negligibly small hopping energy [3] and hole mobilities which are bordering on itinerant character. The fact that the electrical properties diverge in their oxygen pressure dependencies is consistent with small polaron theory. In Eqs. 1 and 2 it can be seen that conductivity varies with the product whereas thermopower varies with the ratio of M^{3+} and M^{2+}. When defect populations become large enough to significantly reduce the M^{2+} concentration (clusters also use up M^{2+}) the two properties should exhibit divergent oxygen pressure behavior.

3.2 NEUTRAL CLUSTERS AND CLUSTER NETWORKS: $La_{2-x}Ba_xCuO_4$

Above some threshold population of clusters, each cluster will have at least one other cluster as a nearest neighbor. Furthermore, neutral clusters can be formed in small polaron systems by localizing holes immediately adjacent to the vacancies of the cluster. For example, the neutral 5:2 cluster can be formed in MO (α=1) by setting n=2, v=5, i=2, m=3, and a=4 in Eq. 5:

$$(6 + b)\ M_M^{2+} + 3/2\ O_2(g) \Longleftrightarrow\ 3\ O_O^{=}\ + \{5\ V_M\ 2\ M_i^{3+}\ 4\ M_M^{3+}\ b\ M_M^{2+}\} \qquad (10)$$

Here no "free" small polaron holes are formed; all holes are immediatly "bound" to the cluster. If hopping can take place between the M^{3+} and the equivalent b M^{2+} sites about the $vV_M iM_i$ cluster and also between adjacent clusters, a percolation network will exist through the material. We have successfully modeled the electrical properties (conductivity and thermopower) in FeO incorporating 5:2 and 13:4 clusters in the model [16,17]. In this system the minimum devation from stoichiometry is above the threshold for cluster-to-cluster percolation.

We have recently reported a similar phenomenon in the superconductor, $La_{2-x}Ba_xCuO_4$ [18]. This is a doping reaction as opposed to an oxygen incorporation reaction, but the result is analagous:

$$2\ La^{3+}_{La} + 2\ Cu^{2+}_{Cu} + 2\ BaO + 1/2\ O_2(g) \Leftrightarrow La_2O_3 + 2\ Ba^{2+}_{La} + 2\ Cu^{3+}_{Cu} \qquad (11)$$

The high temperature electrical properties (650-850°C) clearly exhibit small polaron character [18]. However, the thermopower obeys Eq. 3 only at small doping levels. At higher doping contents the thermopower achieves an asymptotic value of 5-10 μV/K beyond a doping level of x=0.25, the same composition at which a knee is observed in the electrical conductivity vs. doping content. This suggested to us a transition between isolated small polaron hopping between Cu^{3+} and Cu^{2+} and neutral cluster network formation as in FeO. An example cluster might involve 2 Ba^{2+}, 2 Cu^{3+}, and 5 Cu^{2+}:

$$7\ Cu^{2+}_{Cu} + 2\ La^{3+}_{La} + 2\ BaO + 1/2\ O_2(g) \Leftrightarrow$$

$$La_2O_3 + \{2\ Ba^{2+}_{La}\ 2\ Cu^{3+}_{Cu}\ 5\ Cu^{2+}_{Cu}\} \qquad (12)$$

This cluster can be visualized as involving the four copper sites immediately adjacent to the Ba_{La} which are in square planar arrangement with four oxygens. Two such squares can be corner-shared, leading to 7 total copper sites in the cluster, of which two are occupied by holes (Cu^{3+}). The local Cu^{3+}/Cu^{2+} ratio would yield a thermopower of 19 μV/K according to Eq. 3. A cluster with edge-shared Cu-O squares would have a thermopower of 0 μV/K. A mixture of these two possibilities is also conceivable to arrive at the 5-10 μV/K range.

We have reasonably well modeled both the electrical conductivity and thermopower vs. doping concentration in $La_{2-x}Ba_xCuO_4$ in terms of a gradual transition from "free" small polaron holes to "bound" holes in a cluster network [18]. The percolation threshold can be adjusted to correspond with the experimental value of x=0.25. The results are shown in Fig. 2. "Free" holes predominate at small doping levels; clusters play an increasingly important role as doping increases toward the threshold. We are currently conducting kinetic studies of cluster formation and high temperature diffraction experiments are planned. Smedskjaer et al. [19] have proposed a similar cluster model to explain the oxygen diffusion vs. doping behavior in Ba- and Sr-doped La_2CuO_4.

3.3 OXYGEN INTERCALATION: $YBa_2Cu_3O_{6+Y}$

Raveau was the first to suggest that the $RE_{1+x}Ba_{2-x}Cu_3O_{6+Y}$ (RE=rare earth or yttrium) series of compounds are actually oxygen intercalation compounds [20]. Depending upon the value of x, y can range from 6 to well over 7 in these unusual materials. The oxygen content is known to play a dominant role in governing the

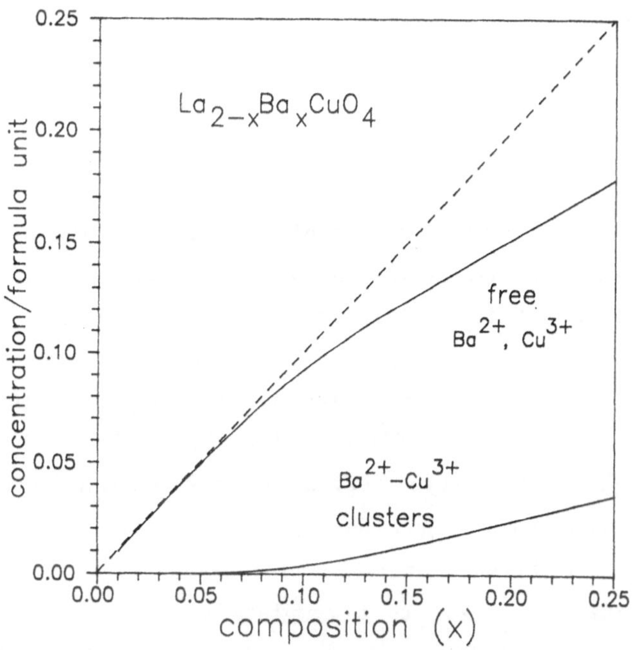

Fig. 2. Calculated cation and cluster concentrations per unit of $La_{2-x}Ba_xCuO_4$. Reprinted by permission from [18].

superconductivity of these important new materials. By setting $\alpha=0$ (no cation vacancies are required for site ratio balance), $(v-i)=1$ (we desire the reaction per additional oxygen anion), and $n=2$, Eq. 5 yields:

$$2\ Cu_{Cu}^{2+} + 1/2\ O_2(g) \Leftrightarrow O_O^{=} + 2\ Cu_{Cu}^{3+} \quad (13)$$

In previous work on $YBa_2Cu_3O_{6+Y}$ we showed that the equilibrium constant for the disproportionation reaction on the Cu(1) site:

$$2\ Cu_{Cu}^{2+} \Leftrightarrow Cu_{Cu}^{+} + Cu_{Cu}^{3+} \quad (14)$$

was approximately independent of oxygen content, and that the reaction resulting from the subtraction of Eq. 14 from Eq. 13 was similarly oxygen-independent:

$$Cu_{Cu}^{+} + 1/2\ O_2(g) \Leftrightarrow O_O^{=} + Cu_{Cu}^{3+} \quad (15)$$

From Eqs. 14 and 15, the oxygen content can be calculated. The results are given in Fig. 3 and compare well with literature data [21,22]. In this calculation, the activity of $O_O^{=}$ is assumed to be $(6+Y)/7$. This illustrates one unique feature of intercalation. The anions are treated as additional structural units as oppposed to defects, i.e. interstitials or occupied vacancies. The excess charge is compensated by adjusting the ratio of copper valence states.

On the basis of the Cu(1) site valence distribution in Fig. 3 p-type small polaron character (Cu^{3+}/Cu^{2+}) is predicted at high oxygen activities and n-type small polaron character (Cu^{+}/Cu^{2+}) is predicted at low oxygen activities. This is precisely what has been observed [23]. In the p-type regime good agreement has been achieved between a model based upon Eqs. 1 and 4 and the experimental conductivity and thermopower [18]. The n-type regime is limited in extent for $YBa_2Cu_3O_{6+Y}$ [23] so a complete analysis is not possible. Current studies are being carried out in the $La_{1+x}Ba_{2-x}Cu_3O_{6+Y}$ which, for $x=0.1$, exhibits a much more extensive n-type regime.

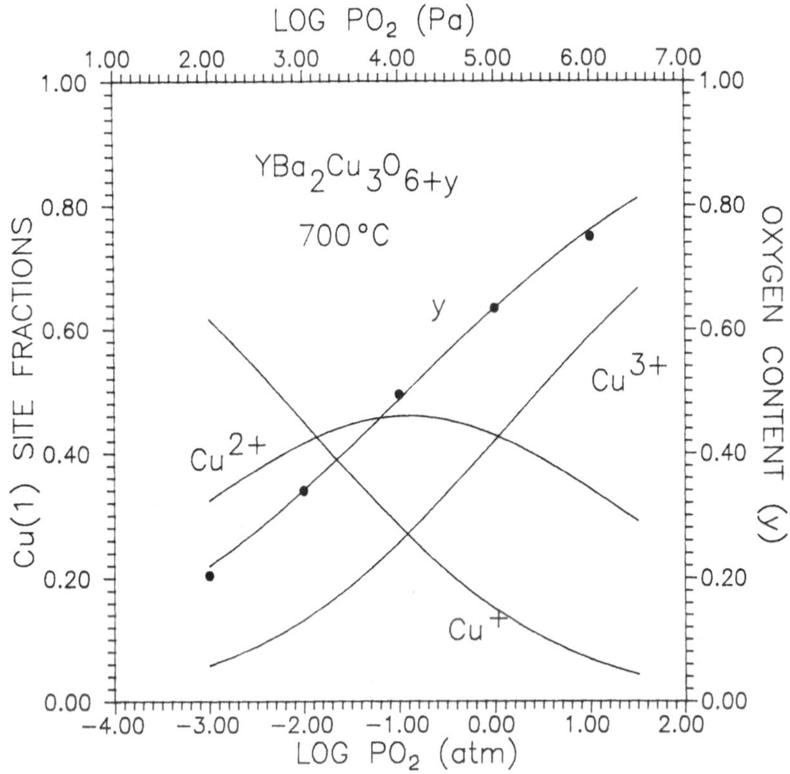

Fig. 3. Calculated (solid lines) vs. measured (solid points [21,22] Cu(1) site valence distribution and oxygen content in $YBa_2Cu_3O_{6+y}$. K(Eq. 15)=2.66, K(Eq. 14)=0.34. Reprinted by permission from [18].

4. CONCLUDING REMARKS

Charge localization leading to small polaron formation and mixed cation valence is not unique to transition metal oxides. An important early piece of work involving small polarons and mixed valence was that of Tuller and Nowick on CeO_{2-x} [4]. This indicates that a similar approach can be taken to the study of f-electron variable valence cations as we have taken to the study of d-electron variable valence cations. In both cases high temperature electrical properties represent useful means of probing the in situ valence distributions in these oxides. Similar behavior is reasonable to expect in non-oxides as well.

Variable valence oxides have long been recognized as important electrode and interconnect candidates for high temperature, oxidizing environments. The recent advent of high T_c superconducting oxides appears also to be related to the issue of mixed valence behavior. High temperature characterization of the defect and valence structures of these materials is therefore of practical as well as fundamental importance.

ACKNOWLEDGMENTS

The monoxide and cluster work was supported by the Department of Energy under Grant. No. DE-FG02-84ER45097.A005. The superconductor work was supported by the National Science Foundation through the Materials Research Center of Northwestern University under Grant. No. DMR 8520280.

REFERENCES

1. E. J. W. Verwey and P. W. Haayman, Physica, 8, 979 (1941).
2. A. J. Bosman and H. J. van Daal, Adv. Phys., 19, 1 (1970).
3. H. -C. Chen, E. Gartstein and T. O. Mason, J. Phys. Chem. Solids, 43 [10] 991 (1982).
4. H. L. Tuller and A. S. Nowick, J. Phys. Chem. Solids, 38, 859 (1977).
5. R. Dieckmann, C. A. Witt and T. O. Mason, Ber. Bunsenges. Phys. Chem., 87 [6] 495 (1983).
6. D. P. Karim and A. T. Aldred, Phys. Rev. B20, 2255 (1979).
7. S. E. Dorris and T. O. Mason, J. Am. Ceram. Soc., 71, 379 (1988).
8. W. J. Weber, C. W. Griffin and J. L. Bates, J. Am. Ceram. Soc., 70, 265 (1987).
9. G. P. Sykora, "Electrical Properties and Defect Structure of Highly Defective Manganese, Cobalt, and Nickel Oxides," Ph.D. thesis, Northwestern University, June, 1988.
10. T. O. Mason, J. Phys. Chem. Minerals, 14, 156 (1987).
11. S. E. Dorris, "The Electrical Properties and Cation Distributions of the Fe_3O_4-Mn_3O_4 Solid Solution," Ph.D. thesis, Northwestern University, June, 1986.
12. G. P. Sykora and T. O. Mason, in Adv. Ceram. Vol. 23: Nonstoichiometric Compounds, The American Ceramic Society (1987) p. 45.
13. R. Dieckmann, Ber. Bunsenges. Phys. Chem., 86, 112 (1982).
14. C. R. A. Catlow and A. M. Stoneham, J. Am. Ceram. Soc., 64, 234 (1981).
15. P. K. Khowash and D. E. Ellis, Phys. Rev., B36, 3394 (1987).
16. E. Gartstein, T. O. Mason and J. B. Cohen, J. Phys. Chem. Solids, 47, 759 (1986).

17. E. Gartstein, J. B. Cohen and T. O. Mason, J. Phys. Chem. Solids, 47, 775 (1986).
18. M. -Y. Su, K. Sujata and T. O. Mason, "High Temperature Defect Structure and Transport in Rare Earth-Alkaline Earth-Copper Oxide Superconductors," Adv. Ceram. Mater., in press.
19. L. C. Smedskjaer, J. L. Routbort, B. K. Flandermeyer, S. J. Rothman, D. G. Legnini and J. E. Baker, Phys. Rev. B36, 3903 (1987).
20. C. Michel and B. Raveau, Rev. Chim. Miner., 21, 407 (1984).
21. P. K. Gallagher, Adv. Ceram. Mater. 2 (3B) 632 (1987).
22. K. Fueki, K. Kitazawa, K. Kishio, T. Hasegawa, S. -I. Ichida, H. Takagi and S. Tanaka, in "Chemistry of High Temperature Superconductors," p. 38, Am. Chem. Soc., Washington, D.C. (1987).
23. M. -Y. Su, S. E. Dorris and T. O. Mason, J. Solid State Chem., in press.

ELECTRONIC STRUCTURE OF HIGH T_c CUPRATES BY ELECTRON SPECTROSCOPIES

D.D. SARMA
Solid State and Structural Chemistry Unit
Indian Institute of Science
Bangalore 560 012, India

ABSTRACT. Employing electron spectroscopic techniques, we show that the ground state of these high T_C cuprates is primarily described in terms of mixed valency between $3d^9$ and $3d^{10}$-related configurations, with very little $Cu^{3+}(3d^8)$ configuration. Temperature dependent changes in the spectral features are interpreted in terms of formation of dimerized hole species in primarily oxygen p-derived band, while Cu concomitantly assumes a larger proportion of $3d^{10}$ related configuration. It is also shown that the formation of these dimerized hole species leads to a drastic decrease in the frequency dependent surface resistivity of the samples.

1. Introduction

Ever since the discovery /1/ of high temperature superconductivity in La_2CuO_4 doped with Ba (or Sr), there have been an explosion in research activities all over the world to understand the electronic structure of these compounds. In a short span of about two years, several other oxides have been discovered with still higher transition temperatures (T_c). All these oxides share the common property of containing two-dimensional network of CuO_4 corner shared planes and of having excess oxygen stoichiometry. In order to satisfy the oxygen valency of 2-, many proposed that the excess oxygen converts Cu^{2+} partially into Cu^{3+} in these compounds. Alternatively, it is also possible that some of the oxygen ions remain in the lower valence state of O^{1-}. Since the electronic structure of solids can be probed directly using electron spectroscopic techniques, we have investigated two of these high T_c compounds, $La_{1.8}Sr_{0.2}CuO_4$ and $YBa_2Cu_3O_{7-\delta}$,

J. Nowotny and W. Weppner (eds.),
Non-Stoichiometric Compounds Surfaces, Grain Boundaries and Structural Defects, 499–507.

employing these techniques. Thus we are able to establish that there is almost no ionic Cu^{3+} ($3d^8$) component in these oxides, but instead Cu atoms exhibit mixed valency between $3d^9$ and $3d^{10}$-related configurations. The excess holes in the system are found to convert O^{2-} into O^{1-} and O_2^{2-} species. The relative proportions of these configurations are found to depend strongly on temperature. Moreover, we show that concomitantly with these changes, the high frequency surface condcuctivity of these compounds vary systematically. While in the following text we discuss primarily the results obtained in our laboratory, it should be noted that substantial amount of electron spectroscopic investigations of these compounds has been carried out by other researchers as well. Some of these works are cited in our original papers referenced in this text.

2. EXPERIMENTAL

$La_{1.8}Sr_{0.2}CuO_4$ was prepared by heating an appropriate mixture of the component oxides at 1300 K in air for 48 hours, followed by treatment in O_2 at 1100 K. The sample of $YBa_2Cu_3O_{7-\delta}$ was prepared by heating the mixture of Y_2O_3, $BaCO_3$ and CuO at 1200 K for 12 hours, followed by heating in oxygen atmosphere at 1100 K for 36 hours in the form of a pellet. Following the oxygen annealing, the sample was slowly cooled in flowing oxygen. The samples were characterized by X-ray diffraction and measurement of resistivity by the four-probe method.

Photoemission, Auger electron and electron energy loss spectra were recorded with commercial VG ESCA3 and ESCALAB spectrometers. For photoemission and Auger electron spectra, we used a MgK_α source. In the photoemission experiments, the total resolution was ~ 0.8 eV and in the Auger electron experiments, a total resolution of ~ 0.4 eV was achieved. For electron energy loss spectra, we used an incident energy of 10 eV for the primary electron beam. The resolution in this technique was ~ 25 meV. The vacuum in the spectrometers was in the range of $4\text{-}9 \times 10^{-10}$ Torr. The sample surface was scraped in situ with a stainless steel blade in order to clean the sample surface.

3. RESULTS AND DISCUSSION

In Fig. 1 we show the $Cu(2p_{3/2})$ and $Cu(L_3VV)$ regions in $La_{1.8}Sr_{0.2}CuO_4$ recorded at 300 K and 80 K from ref. 2. The peak at ~ 933 eV binding energy in $Cu(2p_{3/2})$ spectra is primarily due to the well-screened $3d^{10}\underline{L}^1$ state ($\underline{L}^1$ representing a hole in the ligand band), while the peak at ~ 942 eV is due to the poorly screened $3d^9$ state. While the

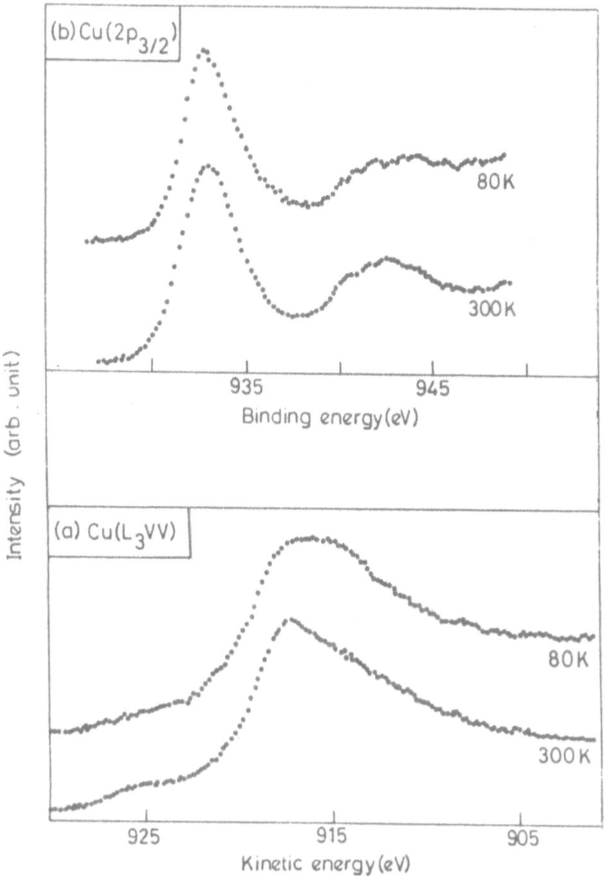

Fig 1. (a) X-ray photoelectron spectra of $Cu(2p_{3/2})$ region, and (b) Auger electron spectra of $Cu(L_3VV)$ region in $La_{1.8}Sr_{0.2}CuO_4$ at 300 K and 80 K.

above assignment is true for the semiconducting stoichiometric La_2CuO_4, in the case of superconducting $La_{1.8}Sr_{0.2}CuO_4$ the signal around 933 eV would also have components of $3d^{10}\underline{L}^2$ configurations due to the presence of excess holes. Likewise, the poorly screened signal at ~ 942 eV has partial admixture of $3d^9\underline{L}^1$ configurations in the superconductor. It should also be noted that in the spectra we do not find any evidence of Cu^{3+} ($3d^8$) signal which is expected to be ~ 23 eV above the main peak /3/. Thus, it appears that copper in this oxide is best described in terms of valence mixing between $3d^{10}$ and $3d^9$ related configurations, with negligible admixture of $3d^8$ configuration.

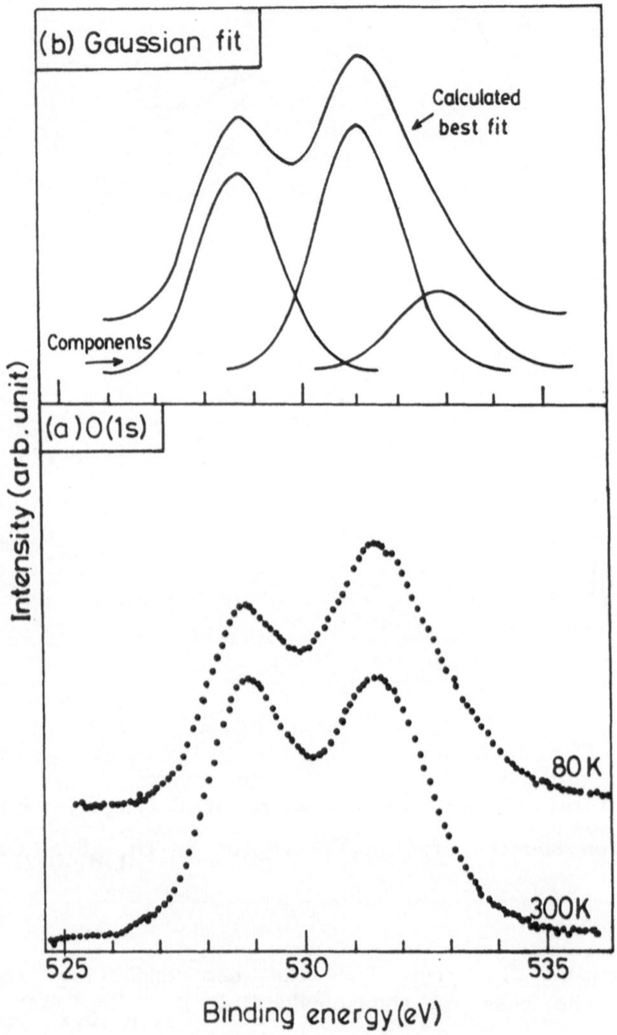

Fig. 2: (a) X-ray photoelectron spectra of O(1s) region in $La_{1.8}Sr_{0.2}CuO_4$ at 300 K and 80 K. (b) Decomposition of the 80 K spectrum in(1) in terms of three Gaussians.

In cooling, the $Cu(2p_{3/2})$ spectrum exhibits a strong reduction in the $3d^9$ related poorly screened peak (Fig. 1). This clearly indicates that the proportions of $3d^{10}$ related configurations in the ground state increase on lowering the temperature. We have further analyzed the $Cu(2p_{3/2})$ spectra employing configuration interaction calculations /2,3/. From a simulation of the experimental spectra, we obtain the total number of d-electrons (n_d) as 9.6 and 9.8 for each Cu at 300 K and 80 K, respectively. Further support for the increase in the proportion of $3d^{10}$ configurations on lowering the temperature comes from the $Cu(L_3VV)$ spectra in Fig. 1. Comparing the spectra recorded at 300 K and 80 K, we see an increase in the intensity at ~ 915 eV kinetic energy in the spectrum at 80 K. Comparing the Auger spectra with those of Cu_2O and CuO, we identify this increase with an increase in the proportion of $3d^{10}$ configurations.

In Fig. 2 we show the corresponding photoemission spectra in the O(1s) region of $La_{1.8}Sr_{0.2}CuO_4$. The spectrum at 300 K exhibits three different features with peak positions at ~ 529, 531.5 and 533 eV, as shown in the least-squared-error fit in Fig. 2. We attribute the 529 eV feature to O^{2-} species, while the peak at 531.5 eV can be either due to O^{1-} species or due to impurities like $(CO_3)^{2-}$ and $(OH)^{1-}$. We interpret the peak at ~ 533 eV as a signal from dimerized hole species in the oxygen band. We believe that this particular species plays an important role in the mechanism of high temperature superconductivity of these compounds. On lowering the temperature to 80 K, we find clear evidence of increase in the proportion of the 533 eV peak, indicating an increase of the dimerized hole species.

Our observations on the other superconducting oxide, $YBa_2Cu_3O_{7-\delta}$, are very similar /4,5/ to those described for $La_{1.8}Sr_{0.2}CuO_4$. In this case also we find a decrease in the weakly screened $3d^9$ configuration related feature (at ~ 942 eV) in Cu(2p) spectra with decreasing temperature. The temperature dependence of this peak intensity relative to the well-screened peak (at ~ 933 eV) is shown in Fig. 3. The increase in the 533 eV peak (due to the dimerized hole species) in the O(1s) region is plotted in Fig. 4 as a function of temperature. We believe that the decrease of the $3d^9$ related peak in Cu(2p) spectra and the increase in the intensity of the dimerized hole species in O(1s) spectra with decreasing temperature may have important bearing on the mechanism of superconductivity in these oxides.

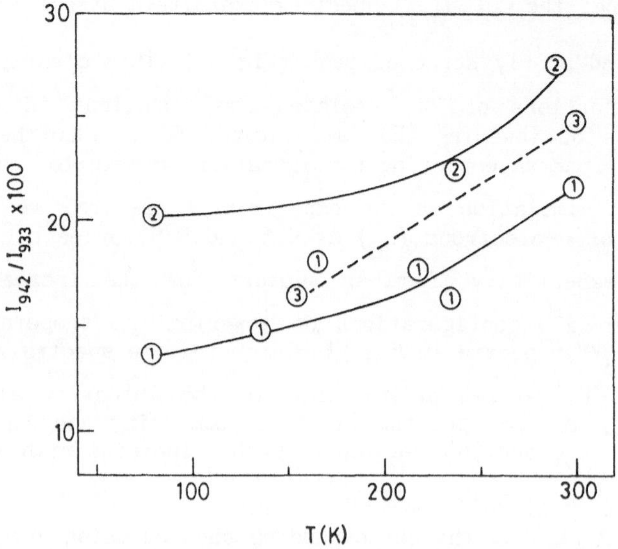

Fig. 3: The temperature dependence of the intensity of the $3d^9$-related peak relative to the $3d^{10}$-related peak in the $Cu(2p_{3/2})$ spectrum of $YBa_2Cu_3O_{7-\delta}$.

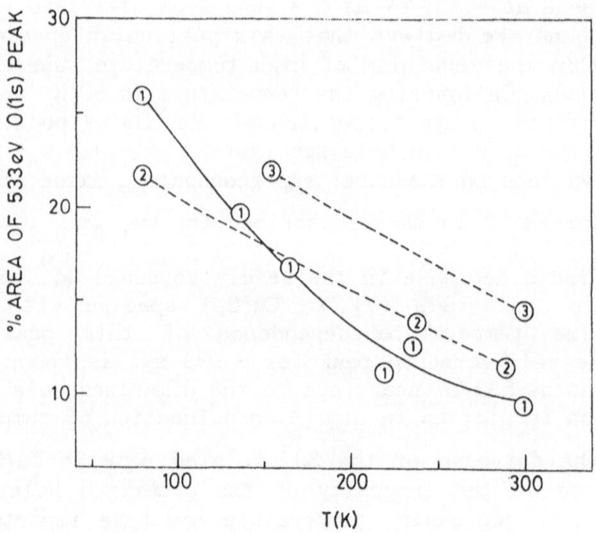

Fig. 4: The temperature dependence of the intensity of the dimerized hole species in O(1s) spectrum of $YBa_2Cu_3O_{7-\delta}$.

Having established the above-mentioned changes in the electronic structure of these compounds employing highly surface sensitive (~ 20 Å) electron spectroscopies, we have attempted to monitor the conductivity behaviour at the surface of these compounds. While it is not possible to measure the dc resistivity of the surface with similar surface sensitivity, we have monitored the frequency dependent resistivity of the surface, as reflected in the intensity of inelastically scattered electrons from the surface in an EELS experiment /6-8/. In Fig. 5 we show typical EELS spectra recorded from superconducting $YBa_2Cu_3O_{7-\delta}$ surface at two temperatures. The prominent decrease of the loss intensity with temperature is a clear indication of decreasing surface ac resistivity on cooling. We plot in Fig. 6 the intensity, I_{in}, of inelastically scattered electrons (corresponding to E_{Loss} = 120 meV) from the surface of a $YBa_2Cu_3O_{7-\delta}$ as a function of

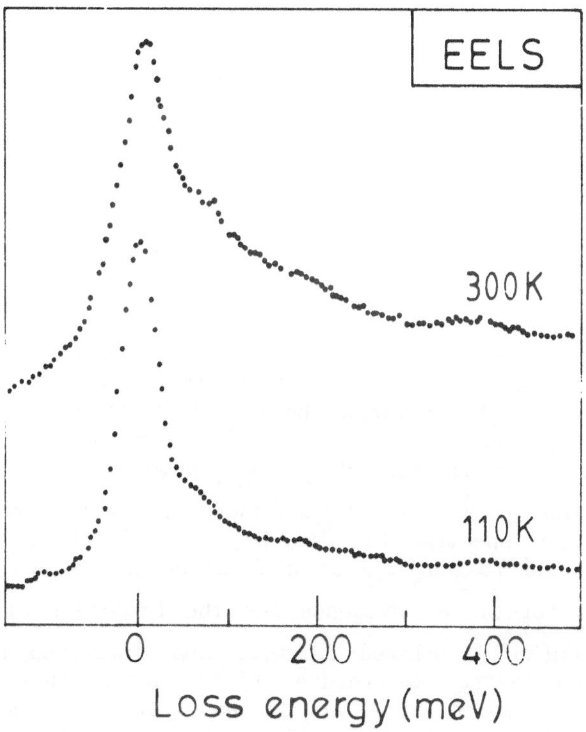

Fig. 5: Electron energy loss spectra of $YBa_2Cu_3O_{7-\delta}$ at 300 K and 110 K, showing the reduction of the loss intensity on cooling.

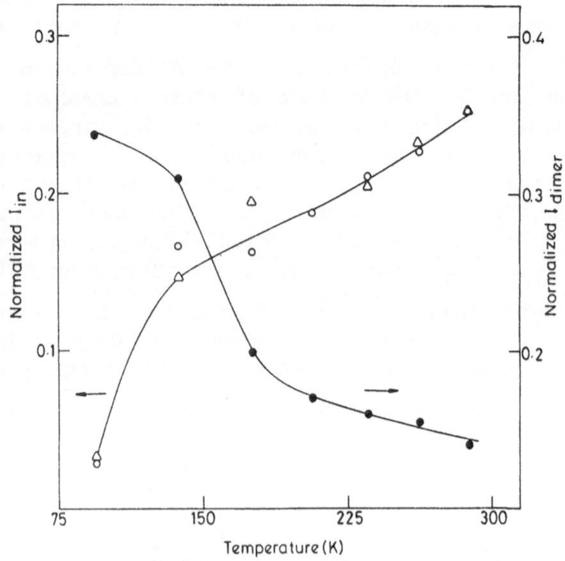

Fig. 6: The temperature dependences of the intensities of the loss spectra, I_{in} and the dimerized hole species in O(1s) spectra, I_{dimer} of $YBa_2Cu_3O_{7-\delta}$.

temperature. Since I_{in} is expected to be proportional to the frequency dependent resistivity, we find a rapid decrease in surface resistivity of the sample with decreasing temperature below ~ 140 K. Plot of the intensity of dimerized hole species in O(1s) spectra from the same surface as a function of temperature (also shown in Fig. 6) establishes a close link between the formation of dimer species and the surface conductivity. We point out that similar changes are not observed on non-superconducting $YBa_2Cu_3O_{6.2}$ samples.

In conclusion, we have shown that there are systematic changes in the photoemission spectra in the Cu(2p) and O(1s) regions of $La_{1.85}Sr_{0.2}CuO_4$ and $YBa_2Cu_3O_{7-\delta}$ as a function of temperature. These changes are interpreted as evidence for the formation of increasing proportion of $Cu(3d^{10})$ related features and dimerized oxygen hole species, with decreasing temperature. It is further shown that these changes are accompanied with an increase in frequency-dependent conductivity of the sample surface. Thus it appears that the observed changes may have important bearings on the mechanism of superconductivity in these compounds.

ACKNOWLEDGEMENT

The author gratefully acknowledges extensive collaboration with Prof. C.N.R. Rao on these experiments described here. The author is also thankful to Profs. P. Ganguly, K. Jacobi, G. Kaindl, and T.V. Ramakrishnan for many helpful discussions and collaboration.

REFERENCES

1. J.G. Bednorz and K.A. Müller, Z. Phys. B 64, 189 (1986)
2. D.D. Sarma and C.N.R. Rao, J. Phys. C 20, L659 (1987)
3. D.D. Sarma, Phys. Rev. B 37, 7948 (1988)
4. D.D. Sarma, K. Sreedhar, P. Ganguly, and C.N.R. Rao, Phys. Rev. B 36, 2371 (1987)
5. D.D. Sarma and C.N.R. Rao, Solid State Commun. 65, 47 (1988)
6. K. Jacobi, D.D. Sarma, P. Geng, T. Simmons, and G. Kaindl, Phys. Rev. B (in press)
7. D.D. Sarma, K. Prabhakaran, and C.N.R. Rao, Proceedings of Interlaken Conference, Physica B+C (1988)
8. D.D. Sarma, K. Prabhakaran, and C.N.R. Rao, Solid State Commun. (to be published)

NONSTOICHIOMETRY AND DEFECT EQUILIBRIA IN $YBa_2Cu_3O_x$

A. Mehta and D. M. Smyth
Materials Research Center #32
Lehigh University
Bethlehem, PA 18015
U.S.A.

ABSTRACT. The value of x in $YBa_2Cu_3O_x$ varies from just under 6 to nearly 7 as a function of temperature and oxygen activity, $P(O_2)$. Published data on the oxygen content, determined by thermogravimetry, have been fit by a simple equilibrium oxidation reaction that involves the filling of vacancies in the reference structure, $YBa_2Cu_3O_7$. A single thermally-activated mass-action constant, with a single enthalpy of oxidation, fits the experimental data quantitatively over the range $450 < T < 850°C$, $-3 < \log P(O_2) < 0$, and $6.2 < x < 6.9$.

1. INTRODUCTION

The high temperature superconducting system, $YBa_2Cu_3O_x$, is notable for its wide range of oxygen content. The value of x appears to vary continuously from slightly less than 6 to nearly 7. It is of interest to understand the chemical and structural behavior of the system as its oxygen content is varied by oxidation or reduction, and to find a model that can lead to an expression that links the composition to the experimental variables, i.e. the temperature and the ambient oxygen activity of equilibration. For that purpose, we have analyzed the thermogravimetric data of Kishio et al.(1) and of Gallagher (2), and have found that a surprisingly simple mass-action treatment of the oxidation reaction is in good agreement with the experimental results. The use of a mass-action approach for these very large deviations from stoichiometry is justified by the observed constancy of the apparent enthalpy of the oxidation reaction over a wide range of temperature and oxygen content. The form of the mass-action expression that is required to fit the data suggests some interesting structural features.

2. THE STOICHIOMETRIC COMPOSITION

Nonstoichiometry cannot be discussed quantitatively unless the stoichiometric composition is known. In most simple compounds this is

J. Nowotny and W. Weppner (eds.),
Non-Stoichiometric Compounds Surfaces, Grain Boundaries and Structural Defects, 509–520.

obvious, but for a complex material such as $YBa_2Cu_3O_x$, it is not immediately apparent. At the stoichiometric composition, the compound is electronicly balanced and all atomic species are in fixed formal oxidation states. The electron and hole concentrations are equal, and result solely from ionization from a filled valence band to an empty conduction band. Thus the electron and hole contributions to the electrical conductivity are equal, neglecting any difference in their mobilities. Reduction from this composition will result in an increasing n-type conductivity, while oxidation will result in an increasing p-type conductivity. Therefore the stoichiometric composition corresponds closely to a minimum value of the conductivity measured under equilibrium conditions as a function of oxygen partial pressure, $P(O_2)$, at constant temperature. The equilibrium conductivity of $YBa_2Cu_3O_x$, shown in Fig. 1, clearly shows conductivity minima at low pressures at the higher temperatures (3), and these should correspond to the stoichiometric composition. Comparison with the equilibrium compositional data shows that the minima correspond quite closely with the oxygen content x = 6. Thus the stoichiometric composition is clearly $YBa_2Cu_3O_6$, and the more oxidized versions, including the superconducting compositions near x = 6.9, should be described as $YBa_2Cu_3O_{6+y}$. The appearance of a narrow n-type region at values of $P(O_2)$ just below the minima indicates that a small degree of

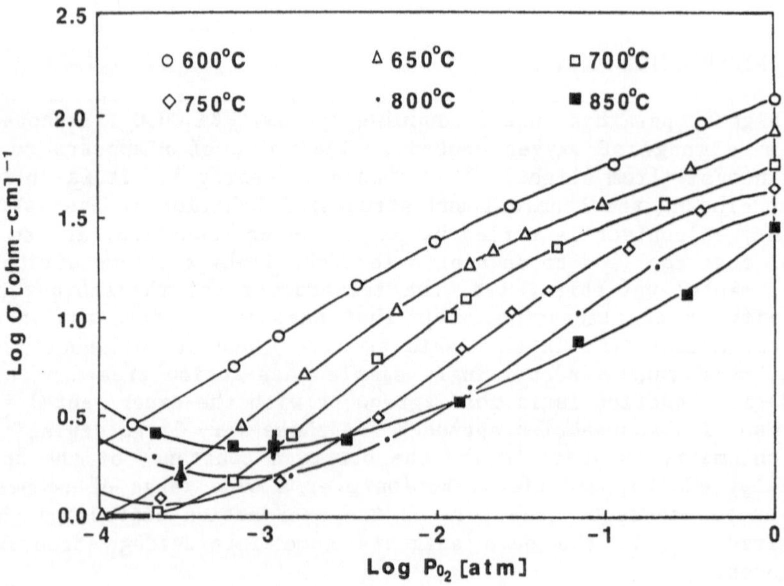

Fig. 1 The equilibrium electrical conductivity of $YBa_2Cu_3O_{6+y}$ (3). The vertical marks near the conductivity minima at 800 and 850°C correspond to the location of the composition y = 0, according to the data of Kishio et al. (1).

oxygen deficiency, i.e. $x < 6$, is tolerated before the structure collapses. From the conductivity data shown in Fig. 1, it is apparent that the defect chemistry of $YBa_2Cu_3O_6$ is characterized primarily by a broad region of oxygen-excess, p-type material as x increases from 6 toward 7. This is to be expected for a system that contains an easily oxidizable cation (Cu^{+1}) and that has the necessary space to accommodate the extra oxygen.

3. THE MASS-ACTION MODEL

The defect models will be discussed in terms of the structure of $YBa_2Cu_3O_6$ (4), as shown in Fig. 2. Within the triple perovskite-like unit cell, all of the oxygen in the plane of the yttrium, and in the basal planes at the top and the bottom of the cell, are missing relative to a filled perovskite structure. The missing oxygen in the yttrium plane will be considered to be an unchangeable part of the structure, while the empty sites in the basal planes can be nearly half-filled to approach $x = 7$. Those sites in the basal planes that are filled in the orthorhombic structure at $x = 7$ to give the Cu-O chains are conventionally designated as O(4) sites, while those that remain empty are designated as O(5). It is assumed that reduction of $YBa_2Cu_3O_6$ will result in the formation of oxygen vacancies in the tetragonal structure, but this region is of little practical interest. For the important oxygen-excess direction, it is necessary to consider whether the oxidation process is filling empty interstitial sites in a tetragonal structure, resulting in an increase in the defect concentration with oxidation, or whether oxygen vacancies are being filled with a resulting decrease in the defect concentration. In the reference structure of $YBa_2Cu_3O_6$, the available oxygen sites must clearly be

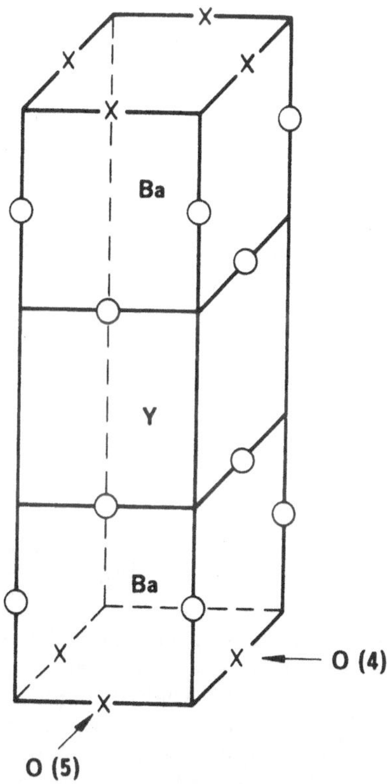

Fig. 2 The structure of the stoichiometric reference composition $YBa_2Cu_3O_6$ (4). The oxygen sites that are occupied in the ideal orthorhombic structure of $YBa_2Cu_3O_7$ are designated as O(4), while those that remain unoccupied are designated as O(5).

treated as empty interstitial sites, but whether this picture of random occupation of the interstitial sites can be maintained as a significant fraction of the sites is filled remains to be tested.

If the oxidation process involves the filling of empty interstitial sites, i.e. the O(4) and O(5) sites which are indistinguishable in the tetragonal structure, the reaction can be written as

$$V_I + 1/2\ O_2 \rightleftharpoons O_I'' + 2h^{\cdot} \tag{1}$$

which has the corresponding mass-action expression

$$\frac{[O_I''] \ p^2}{[V_I]} = K_{ox} \exp(-\Delta H_{ox}/kT)\ P(O_2)^{1/2} \tag{2}$$

where $p = [h^{\cdot}]$, $K_{ox}\exp(-\Delta H_{ox}/kT)$ is the mass-action constant, and ΔH_{ox} is the enthalpy of the oxidation reaction as written, i.e. the enthalpy per oxygen atom. If the sites to be filled are treated as vacancies, the oxidation reaction is

$$V_o^{\cdot\cdot} + 1/2\ O_2 \rightleftharpoons O_o + 2h^{\cdot} \tag{3}$$

with the mass-action expression

$$\frac{[O_o] \ p^2}{[V_o^{\cdot\cdot}]} = K_{ox} \exp(-\Delta H_{ox}/kT)\ P(O_2)^{1/2} \tag{4}$$

In either case, the left-hand-side of the mass-action expression will be a constant at constant composition, i.e. at a fixed value of x. Thus a plot of $\log P(O_2)$ vs. $1/T$ at constant x will have the slope $-2\Delta H_{ox}/k$, which should be independent of x if the mass-action expressions are valid. The thermogravimetric data of Gallagher are presented in this form as a family of parallel lines for various values of x, indicating that this condition is fulfilled, and that the enthalpy of oxidation is independent of the composition (2).

Since the available oxygen sites are clearly interstitial sites at x = 6, Eq. (2) will first be tested for compatibility with the experimental data of Kishio et al. (1), which has been replotted in a more convenient form in Fig. 3. It is seen that the pressure dependence, expressed as the value of n in $P(O_2)^{1/n}$, varies upward from a value of 4. In this model, the excess oxygen must be treated as a defect, and the interstitial concentration is taken as y in $YBa_2Cu_3O_{6+y}$. p is then 2y if the oxidation reaction is the major

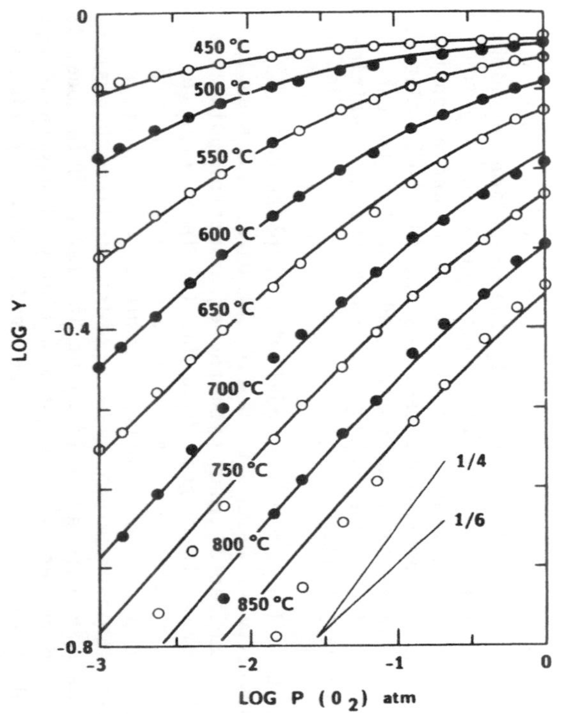

Fig. 3 The data of Kishio et al. (1), plotted as log y (in $YBa_2Cu_3O_{6+y}$) vs. log $P(O_2)$ at the indicated temperatures. The points are from the experimental data (1); the lines are calculated from Eq. (8).

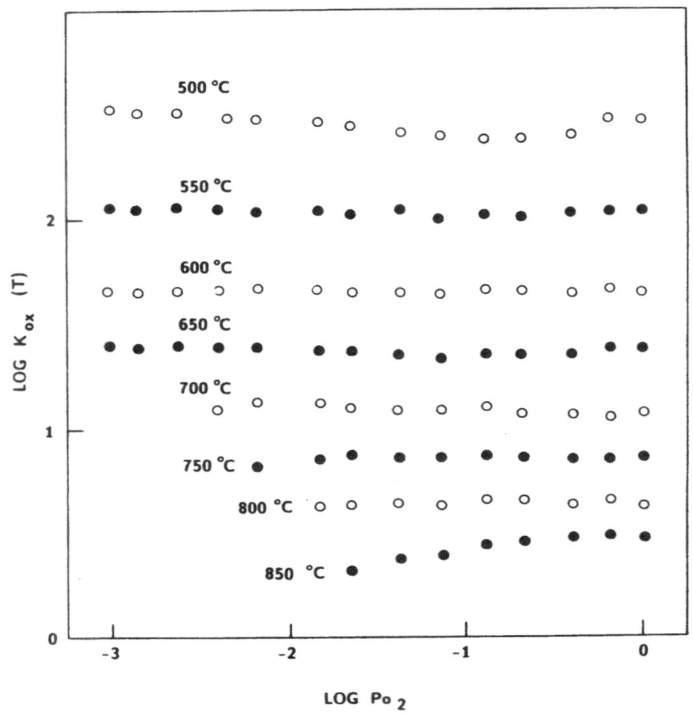

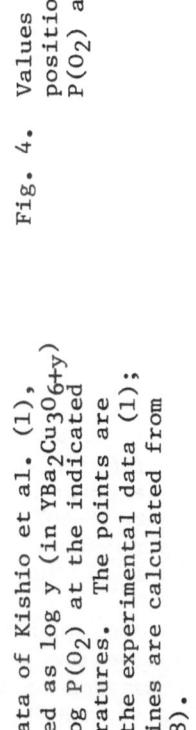

Fig. 4. Values of $K_{ox}(T)$ obtained from the composition data at different values of $P(O_2)$ at each temperature.

source of defects. From the experimental data it is clear that the concentration of sites for excess oxygen is slightly less than y = 1. The concentration of available sites will thus be taken to be A-y, where A is the total fraction of sites available for excess oxygen. The mass-action expression then becomes

$$\frac{4y^3}{A-y} = K_{ox}(T)\ P(O_2)^{1/2} \tag{5}$$

where $K_{ox}(T)$ is the mass-action constant and A is the concentration of all available interstitial sites. The maximum achievable amount of oxidation has been commonly observed to saturate at about x = 6.93, and this is apparent in Fig. 3. In fact, Kishio et al.(1) found that this value was not exceeded even when the material was subjected to 80 atm. of oxygen pressure at 400°C. Thus the value of A is taken as 0.93. The compositional data can be fit quite well by this expression for high values of y, but there is a major problem for small y. This can be seen in the range of possible values of n in the pressure dependence, $P(O_2)^{1/n}$,

$$n = d \log P(O_2)/d \log y \tag{6}$$

From Eq. (5), n is

$$n = 6 + \frac{2y}{A-y} \tag{7}$$

and the minimum value for n is 6, which disagrees with the value of 4 seen at low values of y. A possible way out of this is to invoke a large amount of intrinsic disorder so that the concentration of occupied interstitial sites is unaffected by the oxidation reaction at small values of y. This changes the left-hand side of Eq. (5) to $4y^2I/(A-y)$, with I being constant at low values of y, and n can then range upwards from 4. However, this requires intrinsic disorder of about 30-40%, which seems very unlikely, and the fit to the experimental data is still unsatisfactory. Thus the interstitial model does not work very well.

The justification for a vacancy model will be postponed while we explore the possibility of using it to fit the data. As seen by comparison of the calculated lines with the experimental points in Fig. 3, the fit is surprisingly good when Eq. (4) is interpreted as follows. The available vacancies are still taken to be A-y, with $A \sim 0.9$, as indicated by the experimental results. The concentration of occupied oxygen sites is given by 6+y, which implies that all oxygen sites in the structure are equivalent insofar as this model is concerned. $[O_o]$ then represents the total activity of oxygen in the material, independent of local crystalline environment. A similar assumption was used by Su, Dorris, and Mason in a slightly different interpretation of the oxidation reaction (5). With these assumptions,

Eq. (4) becomes

$$\frac{(6+y)\ y^2}{A-y} = K_{ox}(T)\ P(O_2)^{1/2} \tag{8}$$

and the pressure dependence is

$$n = 4 + \frac{2y}{6+y} + \frac{2y}{A-y} \tag{9}$$

and can have values of 4 and greater.

The adjustable parameters in the fitting process are A and $K_{ox}(T)$ at each temperature; however, the possible values of A are severely limited by the experimental results. After the optimum value of A has been determined at each temperature, $K_{ox}(T)$ can be obtained at each temperature by a one-point fit. Fig. 4 shows the values of $K_{ox}(T)$ obtained from the experimental data taken at different values of $P(O_2)$, and hence at different values of y. $K_{ox}(T)$ is seen to be satisfactorily independent of the data point chosen for the fitting process, and these values were averaged to obtain the mass-action constant used at each temperature. Since the values of $K_{ox}(T)$ are seen to form a family of parallel, horizontal lines, both the enthalpy and entropy of oxidation are independent of composition, and that is the necessary and sufficient condition for the application of the law of mass-action. The values of $K_{ox}(T)$ set the scale of the calculated lines, in Fig. 3, while the curvature is essentially fixed by the form of the equation. Thus it is not a matter of having so many freely adjustable parameters that any arbitrarily convoluted set of data can be reproduced. The values of A and $K_{ox}(T)$ used at each temperature are tabulated in Table I. An Arrhenius plot of the mass-action constants are shown in Fig. 5, and the slopes correspond to a value of -0.99 eV (-95 kJ/mole) per oxygen atom for the enthalpy of the oxidation reaction, Eq. (3). This may be compared with the values of -1.04-1.15 eV reported by Gallagher (2), and with the value of -0.99 eV obtained by calorimetry by Morss, Sonnenberger, and Thorn (6). To summarize the results of the fitting process, the values of y in the formula $YBa_2Cu_3O_{6+y}$ can be reproduced over the experimental range 450 T 850°C, -3 log $P(O_2)$ 0, and 0.2 y 0.9 by Eq. (8) with the tabulated values of A and $K_{ox}(T)$ given in Table I.

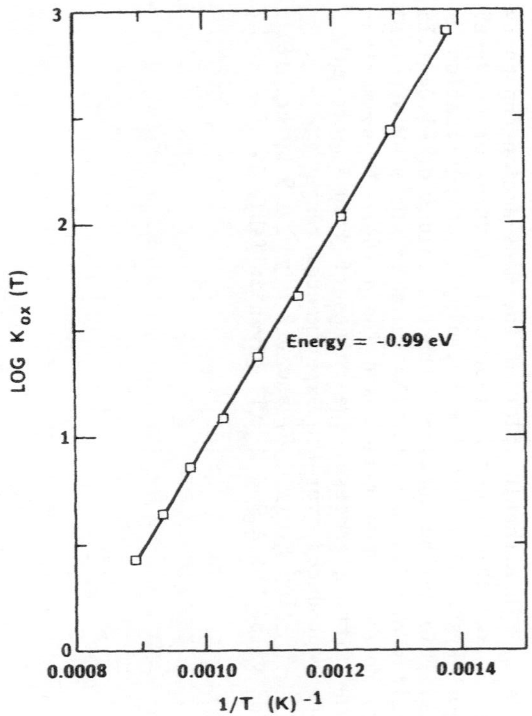

Fig. 5 An Arrhenius plot of the mass-action constants for the oxidation reaction, Eq. (3).

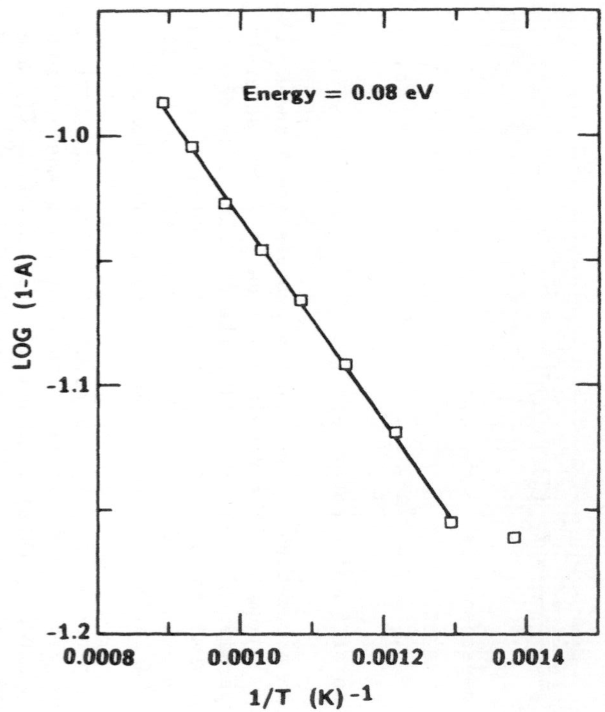

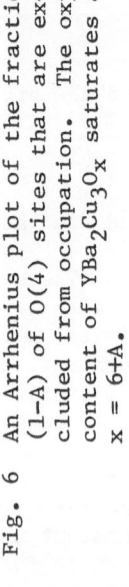

Fig. 6 An Arrhenius plot of the fraction (1-A) of O(4) sites that are excluded from occupation. The oxygen content of $YBa_2Cu_3O_x$ saturates at $x = 6+A$.

Table I. Values of A and $K_{ox}(T)$ in Eq. (8).

T°C	A	$K_{ox}(T)$
450	0.931	797
500	0.930	270
550	0.924	107
600	0.919	45.1
650	0.916	23.4
700	0.910	12.0
750	0.906	7.17
800	0.901	4.39
850	0.897	2.68

4. THE VACANCY MODEL

Since the collective O(4) and O(5) sites in the basal planes of $YBa_2Cu_3O_6$ are clearly structural features, they should be treated as interstitial sites for small deviations from stoichiometry. Over the range of compositions $0 < y < 1$, however, the deviations are far from small, and, in fact, include virtually complete occupation of the available sites. Somewhere in this range it will become impossible to identify a predictable set of normally unoccupied sites, since the pattern of occupied sites will be essentially random. The unoccupied sites must then be treated as vacancies relative to a reference structure, in this case, the structure of $YBa_2Cu_3O_7$. The successful fitting of the experimental results by the vacancy model suggests that this change in structural viewpoint occurs at rather modest values of y, perhaps 0.1-0.2.

An oxygen vacancy bears a formal charge relative to the reference structure, i.e. it corresponds to two missing negative charges, and this charge must be balanced by an equivalent number of formal positive charges. As $YBa_2Cu_3O_7$ is reduced, electrons enter Cu-derived states, and the average oxidation state of Cu is reduced. In a chemical sense, Cu^{+++} is being reduced to Cu^{++}, or even to Cu^{+}. In a formal sense, then, the oxygen vacancies that are created by reduction of $YBa_2Cu_3O_7$ are charge-compensated by the formation of an equivalent number of Cu^{++} or Cu^{+} ions on sites normally occupied by Cu^{+++}. The picture is not dependent on the specific location of the various charge states of the Cu, or even on whether the charges are delocalized. In essence, $YBa_2Cu_3O_7$ becomes acceptor doped on reduction by the presence of Cu with smaller than the normal oxidation state of +3, and the charges of the Cu "acceptors" will always just balance the charges on the oxygen vacancies. This situation can be summarized by the charge-neutrality condition $[Cu''] = [V_o^{\cdot\cdot}]$, assuming that the reduced state is Cu^+. Thus $YBa_2Cu_3O_6$, or $YBa_2(Cu^{++})_2(Cu^+)_1O_6$ is viewed structurally as an acceptor doped version of $YBa_2Cu_3O_7$, or $YBa_2(Cu^{++})_2(Cu^{+++})_1O_7$. $YBa_2Cu_3O_6$ remains the reference composition for deviations from stoichiometry, however, because it fulfills the thermodynamic criteria for that definition, particularly in that it represents the dividing line between regions of excess electrons and excess holes, and thus corresponds to the presence of electronic states that are formally either completely filled or completely empty. A distinction must be made between the reference structure and the stoichiometric reference composition. The situation is analogous to that for any acceptor-doped oxide, e.g. TiO_2 doped with Al_2O_3. The stoichiometric reference composition is $(1-x)TiO_2 + (x/2)Al_2O_3$, and this will be the composition at the minimum in the equilibrium conductivity measured as a function of $P(O_2)$. However, the oxidation reaction involves the filling of vacancies in the rutile structure, which is thus the reference structure.

It is well-known that the oxygen content of this system saturates at about x = 6.93, and it is thus necessary to limit the number of available sites accordingly in order to reproduce the experimental data. In our case, we have selected the value of A at each temperature that gives the best fit. These are listed in Table I, and it is seen that the range of values is restricted to 0.90 to 0.93. While small, the variation seems to be meaningful, because an Arrhenius plot of 1-A, the fraction of excluded oxygen sites, is quite linear, as seen in Fig. 6, and has an activation energy of 0.08 eV. The physical meaning of a thermally activated concentration of unavailable sites is not yet clear.

For compositions corresponding to the upper range of y values, the structure is known to be the ordered orthorhombic version. For smaller values of y, the system has been treated variously as a disordered tetragonal structure, or as a sequence of short-range ordered systems that are not detected by x-ray diffraction (7). We find a single enthalpy for the filling of oxygen sites across the entire compositional range. If there is an order-disorder transition, then the energy difference between the two states must be small compared with the precision of our fit to the data, and may well be too small to account for the formation of an orthorhombic structure from a tetragonal precursor. This lends credence to the viewpoint that there is some form of ordering at all oxygen contents.

Our analysis of the oxidation reaction differs somewhat from those of Su, Dorris, and Mason (5), and of Verweij (8). A relatively minor difference is our limitation of the available oxygen sites to slightly less than that of $YBa_2Cu_3O_7$. More importantly, we differ in the number of hole-type species that are formed per added oxygen. The other authors propose oxidation reactions that yield only one hole-type species per added oxygen atom. The concentration of this species then appears in the mass-action expression to the first power.

As seen in Eq. (3), we propose that two singly-charged hole species are formed per added oxygen, and their concentration then appears as the square in the mass-action expression, Eq. (4). We find that the squared term is required to fit the experimental data. We have not specified the exact nature of these hole species; they may very well be in the form of small polarons as has been proposed (5). Finally, in order to maintain a crystallographic site balance, the concentration of the sites that can be occupied during oxidation must be included in the oxidation reaction and in the associated mass-action expression.

5. CONCLUSIONS

The oxygen content of $YBa_2Cu_3O_{6+y}$ can be quantitatively related to the temperature and $P(O_2)$ of equilibration by a single equilibrium oxidation reaction that forms two hole-like species per added oxygen atom. A single thermally-activated mass-action constant, with a single enthalpy of oxidation, fits the entire range of compositions determined thermogravimetricly by Kishio el al.(1), except for the smallest values of y at the highest temperatures and lowest $P(O_2)$, where the experimental data show a distinct shift from the overall pattern. All occupied oxygen sites are treated as equivalent in the model. 7-10% of the O(4) sites are unavailable for occupation, and the fraction of excluded sites is thermally activated with an activation energy of 0.08 eV. The enthalpy of oxidation shows no variation that could be attributed to energetic differences related to vacancy disordering, or to a tetragonal-orthorhombic transition.

6. ACKNOWLEDGEMENT

The authors are grateful for support from the Division of Materials Research of the National Science Foundation, and by the Lehigh University Consortium for Superconducting Ceramics.

REFERENCES

1. K. Kishio, J. Shimoyama, T. Hasegawa, K. Kitazawa, and K. Fueki, Jap. J. Appl. Phys. 26, L1228(1987).

2. P. K. Gallagher, Adv. Ceramic Mat. 2, No. 3B, Special Issue, 632 1987).

3. E. K. Chang, D. J. L. Hong, A. Mehta, and D. M. Smyth, Mat. Letters 6(8,9), 251 (1988).

4. A. Santoro, S. Miraglia, F. Beech, S. A. Sunshine, D. W. Murphy, L. F. Schneemeyer, and J. V. Waszczak, Mat. Res. Bull. 22, 1007 (1987).

5. M.-Y. Su, S. E. Dorris, and T. O. Mason, submitted to J. Solid State Chem.

6. L. R. Morss, D. C. Sonnenberger, and R. J. Thorn, Inorg. Chem. 27, 2106 (1988).

7. C. Chaillout, M. A. Alario-Franco, J. J. Capponi, J. Chenavas, P. Strobel, and M. Marezio, Solid State Commun. 65, 283 (1988).

8. H. Verweij, Solid State Commun. 64, 1213 (1987).

V. Redox Processes

MODEL CALCULATIONS OF METAL OXIDATION AT AMBIENT TEMPERATURES

E. Fromm
Max-Planck-Institut für Metallforschung, Institut für Werkstoffwissenschaften, Seestraße 92, 7000 Stuttgart-1, B.R.D.

ABSTRACT Oxidation of clean metal surfaces at ambient temperatures is a heterogenious reaction composed of several individual partial steps such as adsorption of oxygen on the oxide surface, formation and transport of lattice defects in the oxide layer and formation of the oxide at the reaction front. Reaction models describing limiting cases, e.g., very thin layers according to the Cabrera-Mott mechanism or transport-controlled layer growth for constant defect concentrations at the phase boundaries may be found already in the literature. The aim of the model calculations presented is a straight forward description of oxidation kinetics including the limiting cases as well as interface reactions at the phase boundaries gas/ oxide and oxide/metal. A set of non-linear algebraic equations are solved numerically. Results are discussed for systems with only one lattice defect being responsible for the transport of atomic species inside the oxide. They are found to be in good agreement with oxidation experiments performed with the quartz microbalance technique using initially gas-free metal films and for exposure times ranging from 10^2 to 10^5s.

1. INTRODUCTION

The formation of oxide layers on uncovered metal surfaces is a complex phenomenon at ambient or relatively low temperatures. These tarnishing layers with a thickness in the range of 1 to 10 nm affect properties of metals like weldability, friction and wear of construction materials or contact potentials, resistivity and electron emission of components in electrical circuits.

The reaction mechanisms occuring are discussed in the literature and numerous time laws are derived from experiments /1,2,3/. In contrast to high temperature oxidation a large scatter in experimental data is the usual situation and the reaction models available consider only limiting cases, i.e., only part of the real oxidation process. This unfavorable condition for a quantitative treatment of the problem arises from the difficulty beginning an experiment with the sample surface in a well defined initial state and from the small amount of oxygen consumed in the process. In modeling one has to keep in mind that the effect of individual parameters on the reaction rate is strongly dependent on the layer thickness. For this reason, the role of various mechanisms governing the overall process is quite different in different stages of the oxidation reaction starting with extremely high reaction rates in the first seconds, passing a transiton period of several hours and ending with very low growth rates producing measurable changes only after weeks of air exposure.

J. Nowotny and W. Weppner (eds.),
Non-Stoichiometric Compounds Surfaces, Grain Boundaries and Structural Defects, 523–534.

This paper does not present any new aspect of a specific reaction partial step. It is intended only to demonstrate how a reaction model must look in order to avoid assumptions inconsistent with basic physical principles. These principles are: the reaction rate is limited by the impinging rate of gas molecules at low pressure; ionic and electronic species are produced from neutral atoms in equal numbers by interface reactions; defect reactions at interfaces must obey kinetic or equilibrium laws; after a short incubation period necessary for establishing a specific concentration distribution of defects and electric charges steady state conditions for atoms, ionic, and electronic defects must hold. This means that the sum of ionic and electronic currents must be zero with respect to change transfer.

In contrast to high temperature oxidation these conditions cannot be satisfied simultaneously by merely introducing some simplifying conditions if a realistic description of the total process is intended at R.T. Therefore, even for the most simple model conceivable, numerical methods are needed for the solution of a set of non-linear equations containing numerous parameters which are usually not known for a specific system. Consequently, the aim of such models cannot be fitting of experimental data, but rather a better understanding of the influence of experimental parameters such as pressure, temperature, or oxide layer thickness on oxidation kinetics in various stages of the overall reaction for a specific oxidation system under consideration.

2. PROBLEM DEFINITION

2.1 Reaction Partial Steps

In systems forming adherent oxide scales the overall reaction

$$xMe + \frac{y}{2} O_2 \longrightarrow Me_xO_y \qquad (1)$$

is impeded since metal and oxygen atoms are separated by the oxide film. Oxygen or metal atoms have to be transferred through the oxide scale as lattice defects and the reaction front can be either the oxide/gas or the oxide/metal interface depending on the defect structure of the oxide.

In a system with metal interstitial ions, e.g., Me^+, and O_2^--chemisorption species on the surface the following partial reactions have to be considered:

a) O_2-physisorption on the oxide
   O_2 (gas) $\longrightarrow$ O_2 (phys).
b) O_2^--chemisorption
   O_2 (phys) + e $\longrightarrow$ O_2^-.
c) electron transport to the surface
   e (me) $\longrightarrow$ e (surf).
d) Me^+ interstitial formation
   Me $\longrightarrow$ Me^+ (ox) + e (me).
e) Me^+ transport in the oxide
   Me^+ (ox/me) $\longrightarrow$ Me^+ (ox/gas).
f) oxide formation
   Me^+ (surf) + O_2^- (surf) $\longrightarrow$ MeO_2.

Thus, the overall reaction eqn. (1) summarizes six partial reactions comprising surface reactions, defect formation, transport processes and finally the oxide formation as a reaction of charged metal atoms and O_2^- molecules on the oxide surface. In Fig. 1 the mechanism is schematically shown for metal ion (1a) and oxygen ion transport through the oxide (1b). In the system (1b) the reaction front is at the interface oxide/metal.

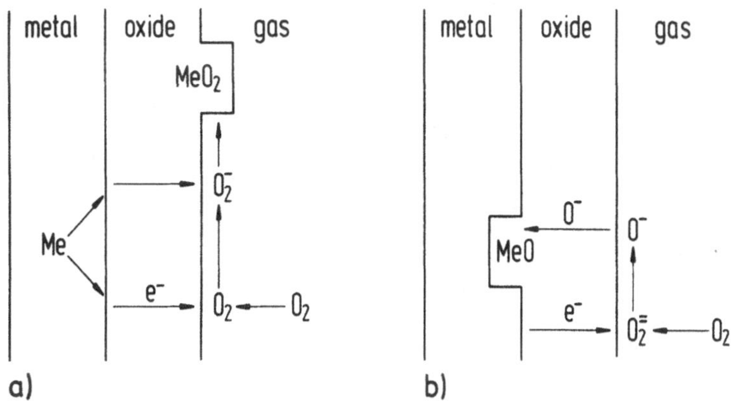

Figure 1. Partial steps in oxidation;
a) metal ion interstitials, b) oxygen ion interstitials

Vacancies as well as interstitials can also transport metal or oxygen ions through the oxide scale. Furthermore, electron acceptors at the surface can be O_2^-, $O_2^=$, O^- $O^=$ or other sites. Therefore a large number of surface and defect species are candidates for reactants in partial reactions and must be specified for each reaction model under consideration.

In order to keep the numbers of free parameters in model calculations small, one should restrict consideration to models with only one chemisorption, one ionic and one electronic species. Such relatively simple models describe oxidation in systems where the concentrations of other species are small, a preasumption justified for many real systems within a limited pressure and temperature range.

2.2 The Steady State Approach

The condition of overall charge neutrality in the system requires that the charge transfer by ionic defects and electrons or holes must cancel in steady state /1,2/. The steady state conception is a highly effective approach in reaction kinetics which minimizes difficulties in the mathematical treatment without violating essential principles of physical chemistry.

It is assumed that at each point of the system the sum of fluxes of the species i and the sum of reaction rates producing or consuming this species is zero or $\left.\frac{dn_i}{dt}\right|_x = 0$.

In this approach the slow local variation of defect concentrations as a consequence of oxide layer growth is disregarded. This omission does not result in large errors as long as the change of defect concentrations over the oxide layer consumes less ions than the production of new oxide at the reaction front in the same period of time.

2.3 The Mott-Field

Transport equations of charged particles depend on the electric field in the oxide. Such a field in the oxide can be produced by surface charges. Space charges can be neclected for small defect concentrations. Cabrera and Mott /4/ were first to demonstrate that the large field present in very thin oxide films is responsible for the high initial reaction rates at ambient temperatures. If the energy of electrons in chemisorption species on the oxide surface is in the range of 1 to 5 eV below the Fermi level in the metal, the electric Mott-field over thin oxide layers with a thickness in the range of 1 nm is 10^7 to 10^8 V cm^{-1} and it strongly reduces the activation barrier for ion diffusion. This gives rise to very fast initial oxidation rates as long as the Mott-field can be maintained by tunneling electrons.

2.4 Transport Mechanisms

2.4.1 *Electron transport.* Electron tunneling through thin films is very fast for a layer thickness below 3nm /1,2/. However, it decreases exponentially with layer thickness (Fig.2a) according to the equation

$$I_{e,t} \sim \exp\left(-d \sqrt{U-V}\right) \tag{2}$$

For thicker layers, electrons have to jump by thermal activation into the conduction band of the oxide, a process which is orders of magnitude slower than the initial tunneling process.

$$I_{e,th} \sim \exp\left(- \frac{U+V}{kT}\right) \tag{3}$$

However, this second mechanism has to be considered in a reaction model if growth rates for extended exposure times or for slightly elevated temperatures are included.

2.4.2 *Defect Transport.* Ionic lattice defects in the oxide are transported by chemical diffusion as a result of the concentration gradient and of electrical conductivity caused by the field. Furthermore, for very thin layers the activation energy in the diffusion coefficient is drastically reduced by the field (Fig. 2b).

$$I_{ion} = \frac{D^\circ}{d} \cdot \exp\left(\frac{-A + kV/d}{kT}\right) \cdot F(c(o), c(d), V, d) \tag{4}$$

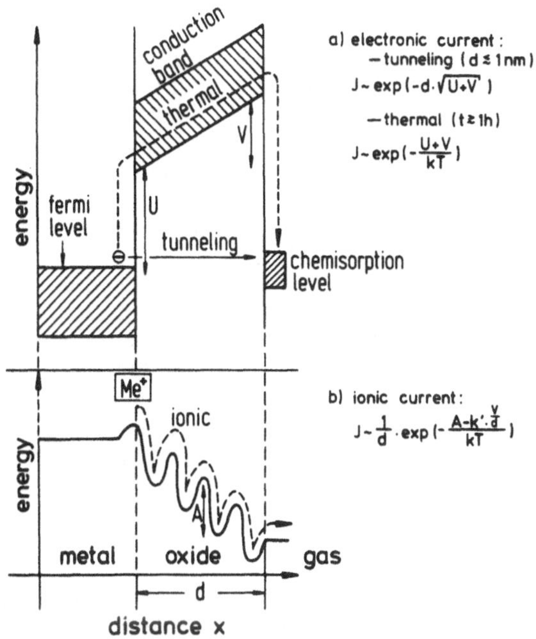

Figure 2: Electronic (a) and ionic (b) transport mechanisms, schematic.

A parabolic rate law is obtained as long as $I_i \sim 1/d$ holds and the ionic current is rate determining. The second term in Eq. (4) represents the reduction of the activation barrier A by the field V/d and is the reason for the very large growth rates observed in the initial stage of oxidation at ambient and low temperatures. Other terms have been comprised by a function F containing diffusion and conductivity terms as derived from the "hopping model". In the books of Fromhold /2/ this problem has been treated in detail and he presents the laws for the limiting case of small space charge and numerical solutions for problems including space charge effects. In the first models of our treatment we use the formula given by Fromhold for ionic currents in the low space charge limit.

2.4.3 *Surface Reactions.* Rate laws can be formulated for the forward and backward reaction of interface processes such as physisorption, chemisorption or oxide formation from defects according to standard procedures used in reaction kinetics. The terms usually have a structure like

$$v \sim \Sigma a_i \Sigma a_j K \cdot \exp\left(\frac{-A - \Delta\varphi}{kT}\right) \qquad (5)$$

with a_i the activity of reacting species, a_j the activity of empty sites and $\Delta\varphi$ the change of the electrical potential between both sides of the activation barrier.

2.5 Model Calculation Procedure

The model calculations have been performed with the aim of obtaining oxidation curves comparable to experimental curves in a plot of thickness vs. time with O_2-pressure and temperature as parameters /7,8/. The problem can be solved if atomistic parameters of the system are given or if reasonable values are assumed for each evaluation run of a model system under consideration.

The calculation is subdivided into the following steps:

a) Determination of potentials in the system, e.g., U and distance W of the chemisorption level from the Fermi level in the metal.
b) Definiton of the electronic and ionic defects in the system responsible for the transport mechanisms in the oxide.
c) Replacement of the defect concentration c(o) and c(d) at the interface x = o and x = d in eqn. (4) by p_{O2} and $\Delta\varphi$-values at the interfaces (e.g., by $\Delta\varphi = aV/d$, a = distance of atomic planes in the oxide).
d) Establishment of the formula for the steady state condition as a function of V (see Fig. 2), p, T and layer thickness $d = N \cdot a$. N is the number of atomic layers in the oxide.

$$\Sigma z \cdot J_{ions} + \Sigma z \cdot J_{electrons} = 0 \tag{6}$$

This equation for the variable V must be solved by numerical methods for given constant values of p and T and for a specific oxide scale thickness $d = N \cdot a$.

The ion current J_{ion} multiplied by a constant geometric factor gives the growth rate $\frac{dN}{dt}$.

e) Finally by numeric intergration methods, the oxidation curve

$$d(t) = a\int_0^t \frac{dN}{dt} \cdot dt$$

is obtained with p and T as experimental parameters. Further parameters of the curves are inherent constants of the model system considered such as rate constants or equilibrium constants of defect reactions and potential differences between electronic states.

With this final step the problem is solved. Other quantities of interest such as electric potentials, the electric field, concentrations of defect species or currents of ionic and electronic particles can be plotted from the data compiled during the computer run for each model as a function of the parameters under consideration, i.e., position inside or on interfaces of the oxide, time, or layer thickness d.

3. RESULTS

Fig. 3 shows a typical curve Δm vs. log t obtained by model calculations for the system shown in Fig. 1a with reasonable parameters for the reaction equations and the electronic structure of the oxide. The three branches of an oxidation curve as observed, e.g. in the experimental curve for Fe, are obtained also by the model. The data points for individual concentrations, currents and potentials indicate that in the stage I, electronic equilibrium is established between chemisorbed O_2^- species on the surface and the metal due to a very fast tunneling process (Fig. 4). Across the oxide layer, a more or less constant Mott potential V is established which strongly

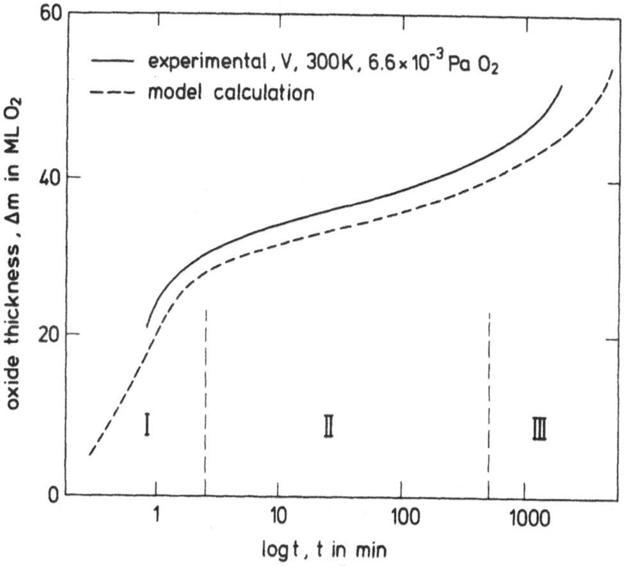

Figure 3. Experimental and calculated oxidation curves at 300 K.

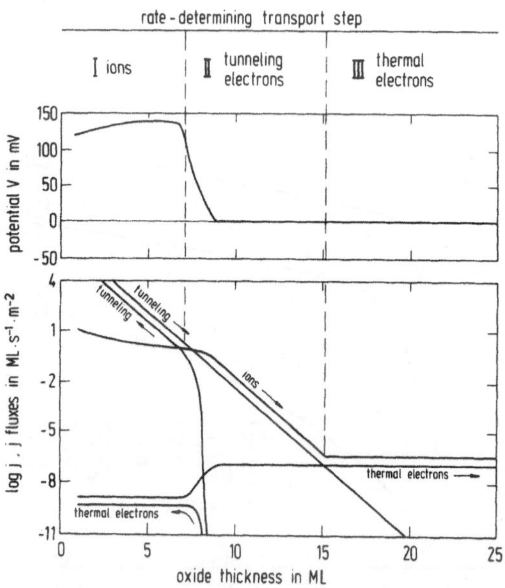

Figure 4. Ionic and electronic currents and potential V as a function of oxide layer thickness d. (1 ML = 10^{15} O_2 molecules / cm^2)

increases the ionic current (see Fig. 4). This branch corresponds to the rate law of the Cabrera-Mott theory /3,4/.

With increasing layer thickness the electronic tunneling current decreases and the tunneling forward current and ionic current are equilibrated by the value of the potential V (see Fig. 4). The rate determining step in this stage II is the tunneling current and thus, the time law becomes logarithmic. Finally, in stage III the thermal electronic current, which is not dependent on layer thickness, becomes larger than the tunneling current and is now rate determining.

The general shape of the thickness vs. time curves depends remarkably on parameters like pressure, temperature, charge of the defects, electronic band structure, and electronic energy levels on the surface of the oxide.

4. COMPARISON WITH EXPERIMENTAL CURVES

4.1 Pressure Dependence

Figs. 5 and 6 show experimental curves for vanadium and iron oxidation at 300K for various O_2 pressures obtained with film samples by the quartz microbalance technique /5-7/. Model calculations with metal interstitial defects and O_2^- chemisorption species simulate the vanadium curves quite well. Iron oxidation obviously needs another model system for simulation (Fig. 7).

4.2 Temperature Dependence

Fig. 8 shows two curves for chromium oxidation at 35 and 60 °C. A model calculation with temperature changes simulates the effects observed. It is interesting to note that temperature changes affects only the upset point to stage III since electron tunneling does not strongly depend on temperature (Fig. 9).

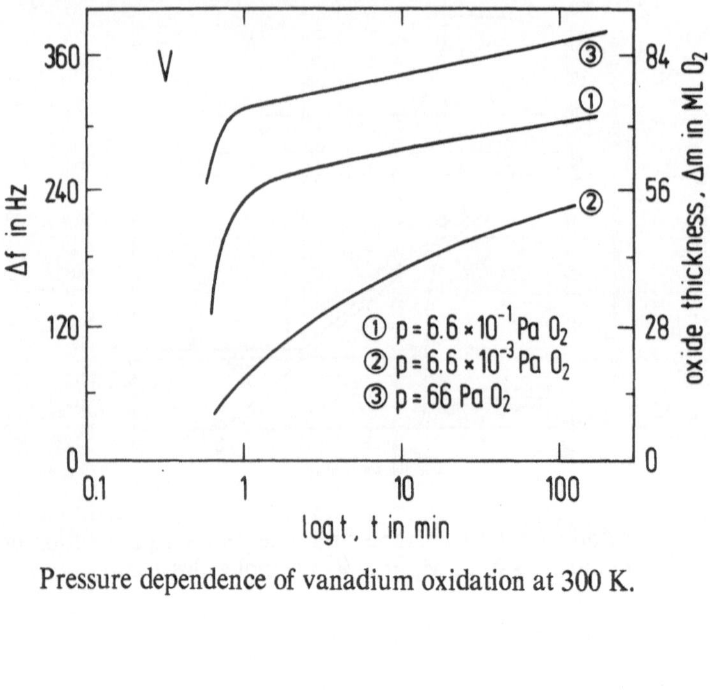

Figure 5. Pressure dependence of vanadium oxidation at 300 K.

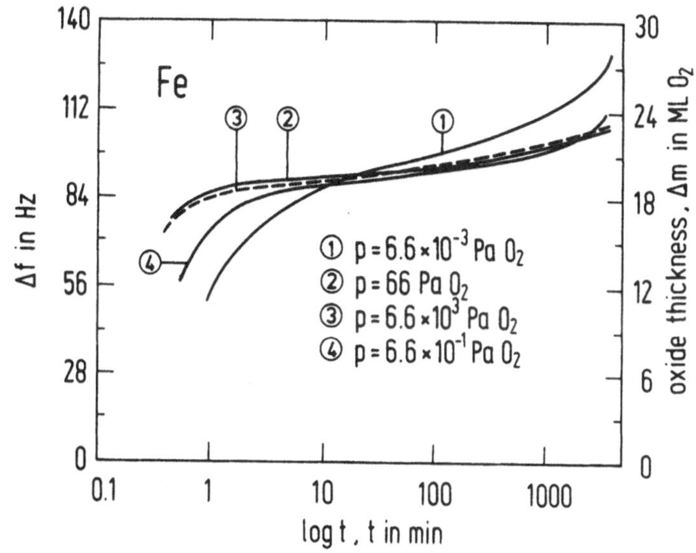

Figure 6. Pressure dependence of iron oxidation at 300 K.

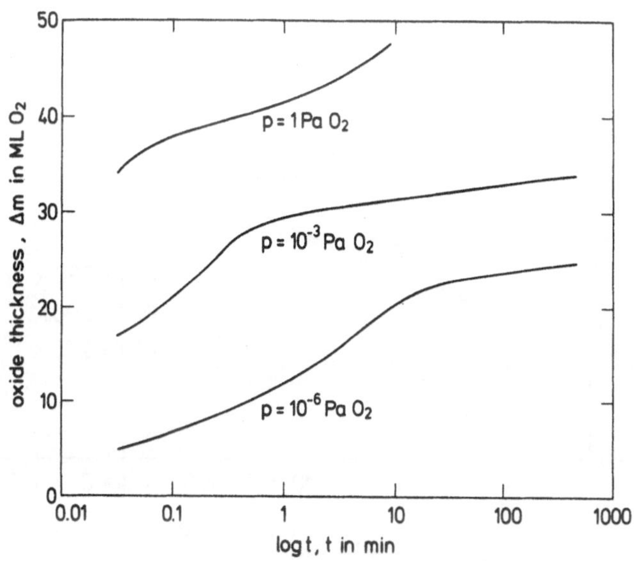

Figure 7. Model calculation of a system with metal ion interstitials as defect species at different pressures.

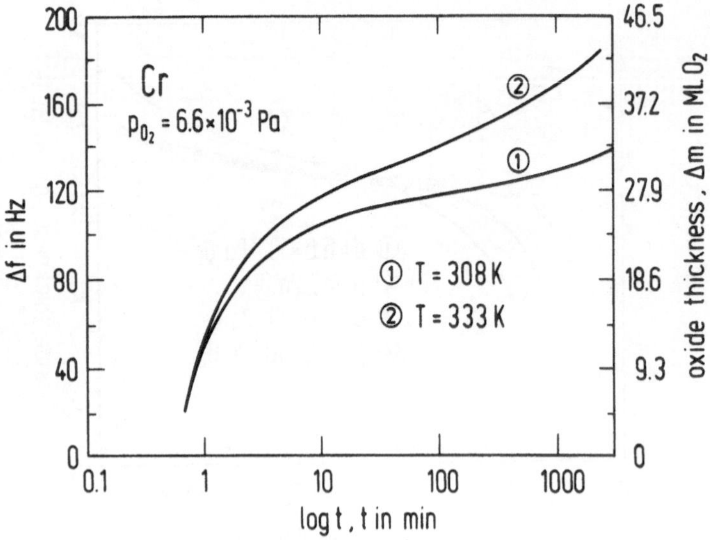

Figure 8. Temperature dependence of chromium oxidation at $6.6 \cdot 10^{-3}$ Pa.

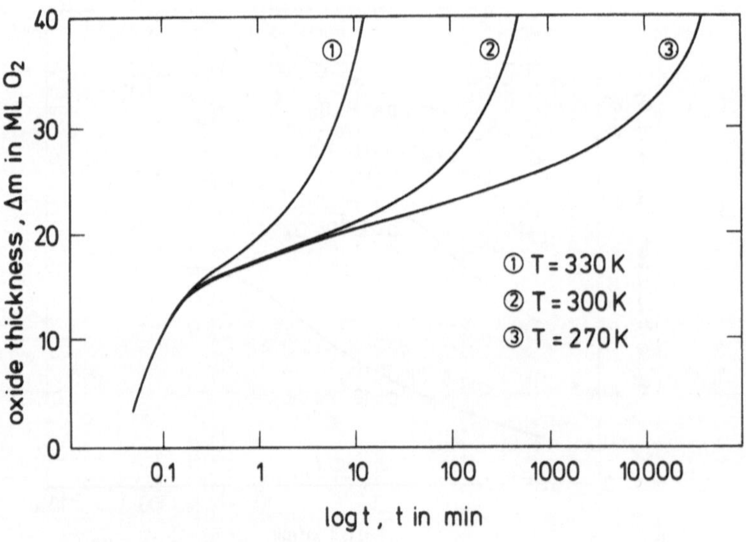

Figure 9. Model calculation of a system with metal ion interstitials as defect species at different temperatures.

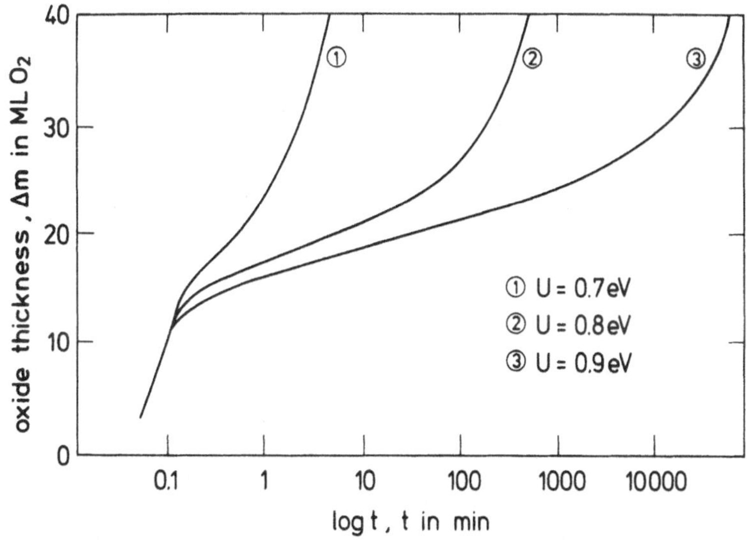

Figure 10. Effect of the electronic structure of the system on model calculations. U is the distance between Fermi level in the metal and the conduction band in the oxide.

4.3 Electronic Structure of the System

In Fig. 10 the distance U between the Fermi level in the metal and the bottom of the conduction band in the oxide has been varied in the model calculation. This parameter affects the shape of the curves in a similar way as temperature changes since thermal electrons become active in both cases at an earlier stage of the oxidation process if T or U become smaller.

5. CONCLUSIONS

Experimental curves as well as the model calculations demonstrate that the room temperature oxidation of metals cannot be described adequately by mechanisms considering only one rate-determining partial step. The main features of oxidation curves are simulated correctly only by models which consider at least the following points:

defect structure of the oxide
band structure of the metal and the oxide
chemisorption states and/or electronic states on the oxide surface
transport mechanisms of atomic, ionic, and electronic species
kinetic and/or equilibrium constants of interface reactions and adsorption reactions.

The presence of several partial reaction steps in the models introduces a relatively large number of parameters which are usually not exactly known. Fitting of experimental curves by arbitrary parameter variation becomes, therefore, very easy but it does not yield much detailed information on the real mechanism.

The advantage of modeling is less an exact fitting of experimental oxidation curves but instead a better understanding of the influence of individual parameters and partial steps on the shape of oxidation curves. Depending on the defect and electronic structure of the system, the effect of the experimental parameters pressure and temperature on the shape of the curves is quite different. Frequently, only one of the three stages of an oxidation curve shows marked changes. The information obtained from various model types and additional detailed analysis of the changes of potential, local concentration of species, and the contribution of individual transport mechanisms to the electronic and ionic currents are valuable data for a semi-quantitative interpretation of complex phenomena observed in room temperature oxidation. For example, the smooth transition from parabolic time laws in stages I and III to the logarithmic time law in stage II in Fig. 2 explains the occurrence of laws like $d \sim t^{1/3}$ sometimes observed experimentally under specific conditions but without introducing additional details in the reaction models.

The programs developed so far cover systems with only one electronic, one ionic and one chemisorption species and disregard space charge effects inside the oxide. They run with tolerable speed on PC and AT personal computers.

An extension of the method to systems with more species or describing also space charges involves no principal difficulties. This is only a question of computer capacity and the maximum number of parameters desired for the interpretation of results.

Acknowledgement
This study has been supported by the Deutsche Forschungsgemeinschaft.

References

/1/ Hauffe, K.: Reaktionen in und an festen Stoffen, Springer, 1966.
/2/ Fromhold, A.T.: Theory of metal oxidation, Vol. I, North Holland Publishing Company, Amsterdam-New York-Oxford 1976.
/3/ Fehlner, F.P.: Low-temperature oxidation. The role of vitreous oxides, Academic Press, New York 1986.
/4/ Cabrera, N., Mott, N.F.: Rept. Phys. 12 (1949)
/5/ Lu, C., Czanderna, A.W.: Applications of piezoelectric quartz crystal microbalances, Elsevier, Amsterdam, 1984.
/6/ Cichy, H., Fromm, E.: Oxidation kinetics of metal films at 300K studied by the piezoelectric quartz microbalance technique. Interconnection Technology in Electronics, Lectures of the 3rd Int. Conf. in Fellbach 1986. DVS, Düsseldorf.
/7/ Cichy, H.: Doctoral thesis, Universität Stuttgart, 1987.
/8/ Cichy, H., Fromm, E., Grajewski, V.: Model calculations on the kinetics of metal film oxidations at 300 K. Interconnection Technology in Electronics. Lectures of the 4th Int. Conf. in Fellbach 1988. VDS, Düsseldorf.

POINT DEFECTS AND THE MECHANISMS INVOLVED IN THE REDUCTION OF HEMATITE INTO MAGNETITE

C. GLEITZER
Laboratoire de Chimie du Solide Minéral, C.N.R.S. U.A. 158
Université de Nancy I
B.P. 239 - 54506 Vandoeuvre-lès-Nancy Cedex
France

This process presents several unusual features. We shall focus here on the strong cracking of hematite crystals when they are reduced at low temperature and low CO/CO_2 ratio, typically 350°C and CO/CO_2= 20/80.

The investigated parameters are : temperature, CO/CO_2 ratio, thermal pretreatments and gas impurities (H_2O,COS). The kinetics presents an induction period in relation to cracking.

Point defects are shown to play a role in several steps of the detailed process :

- during saturation (and super saturation), interstitial iron should prevail over vacancies ; then it may increase the yield strength
- nucleation involves a high local volume increase if interstitials are dominant ; it is also facilitated by transient γFe_2O_3, thanks to dislocations-interstitials interaction
- in the fracture process, point defects increase the brittleness, as demonstrated by substitution Al-Fe
- OH groups, from adsorbed water, help dislocations to glide and then decrease breaking and induction period (we show that they are connected).

1. INTRODUCTION

The reduction of hematite into magnetite presents 2 unusual features :

i - at low temperature, the hematite crystals are severely fractured, even for low conversion degree (typically at 400°C with CO/CO_2 = 20/80) ; cf. fig. 1

ii - at high temperature, magnetite grows inside the matrix, in a semi-coherent way, along given crystallographic orientations

We deal here with the first phenomenon. It is of interest for the ironmaking industry, because the degradation of iron sinters, in the upper part of the blast furnace, is a consequence of that cracking.

J. Nowotny and W. Weppner (eds.),
Non-Stoichiometric Compounds Surfaces, Grain Boundaries and Structural Defects, 535–545.

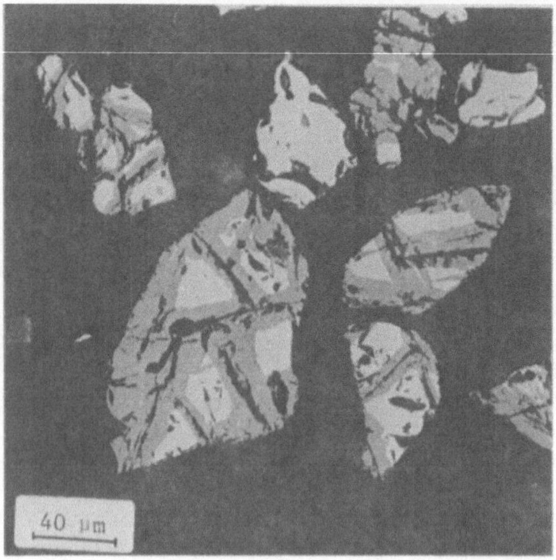

Figure 1. Example of cracking at 400°C with CO/CO_2 = 20/80

2. EXPERIMENTAL

An iron hematite ore, from Itabira (Brazil), has been provided and analyzed by the Institut de Recherche de la Sidérurgie. It was treated with HF (5 %), ultrasounds, decantation in bromoform, and sieving. This yields single crystals particles in the range 100-125 μm with a few agglomerates of smaller particles. The crystals purity was checked with an electron micropobe : 0.1 % Al was the only impurity detected.

In some experiments we also used the natural ore, as received, and also, for comparison, synthetic crystals prepared by transport. In the reduction runs, the samples (∿ 50 mg) were placed on a flat gold crucible in a flow of 100 Nl/h $CO-CO_2$ with, when needed, additions of H_2O or COS.

3. RESULTS

A typical example of cracking is shown on fig.1, for T = 350°C, CO/CO_2= 20/80,. Cracking clearly occurs from the first nuclei. We note the following :

i - The influence of T and CO/CO_2 is given on fig.2

ii - The kinetic curve $\alpha = f(t)$ has a sigmoide shape indicating an induction period

iii - Modification of the surface properties through doping with H_2O or COS traces has given striking effects :

- a few Torrs of H_2O suppress the induction period, significantly decrease the cracking, slow down the overall reduction rate, and make the magnetite nuclei more uniformly distributed

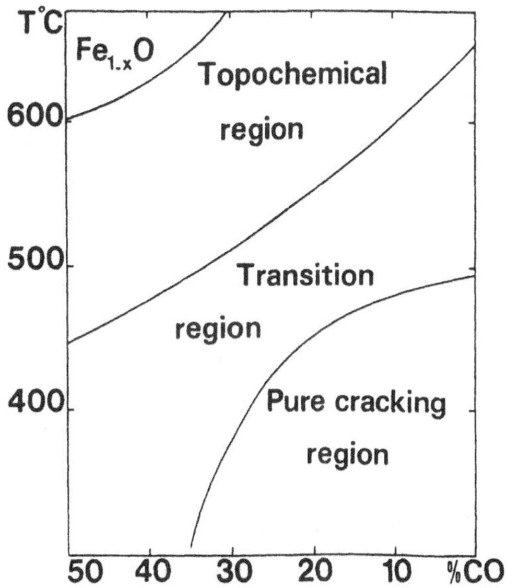

Figure 2. Influence of temperature and gas composition

- a few vpm of COS severely slow down the reaction ; cracking remains, but magnetite now has 2 forms : a porous one inside the particles, and a dense one covers the surface.

iv - In brittle solids cracking often starts from surface microcracks. In order to heal such possible microcracks, a sample has been annealed at 1300°C ($T/T_m \sim 0.84$) for 4 d. in oxygen and slowly cooled. But no change was observed for cracking, when reduced as usual. This means that the reaction itself is the cause.

v - The kinetic curves have been analyzed. The classical Shrinking Core Model should not apply, and the Cracking Core Model (1) should be more appropriate. This analysis will be reported elsewhere (2).

4. INTERPRETATION

It seems clear that the hematite cracking comes from the first stages of the reaction. So let us consider the magnetite nucleation and, before, the saturation and supersaturation of hematite.

4.1. (Super)saturation

During this step, surface reaction occurs and stores point defects near the surface (or deeper if diffusion is fast). Actually the hematite defect structure is not completely elucidated. It is admitted that both oxygen vacancies and interstitial (mainly divalent) iron are present. In 1986, we have suggested (3) that interstitial iron should predominate at low oxygen pressure, due to its greater oxygen dependence. This was confirmed by Catlow et al (4) who have also been able to calculate the formation energies of these point defects (PD).

A question is : how do these PD influence the hematite mechanical properties ? Couzin et al (5) have recently measured the Rutile deformation vs. p_{O_2} at 1030°C : when p_{O_2} decreases, the deformation first increases and then sharply decreases. They proposed the relationship :

$$\Delta \sim \alpha[V_O^{\bullet\bullet}] - \beta[T_i^{\bullet\bullet\bullet\bullet}]$$

leading to a good simulation when α/β varies. It is no difficult to see that the same trend would be obtained for hematite, since low p_{O_2} favours interstitials. However in the present range of temperatures, plastic deformation of hematite more relies on dislocations than on PD : at 400°C, for instance, $T/T_m \sim 0.33$, these dislocations are able to glide, not to climb, because diffusion is slow ; it is known that PD tend to pin the dislocations, as impurities, and especially precipitates do : in first approximation, the yield strength is given (6) by :

$$\sigma_y \propto ([V_O^{\bullet\bullet}] + [Fe_i^{\bullet\bullet}] + [M])^{1/2} \quad (M : \text{substitution})$$

At low p_{O_2}, intrinsic regime can prevails, then :

$$[M] << [V_O^{\bullet\bullet}],\ [Fe_i^{\bullet\bullet}]$$

$$\sigma_y \propto p_{O_2}^{-1/12} \text{ or } p_{O_2}^{-1/8} \text{ if } V_O^{\bullet\bullet} \text{ or } Fe_i^{\bullet\bullet} \text{ dominates}$$

Consequently hematite is more brittle at low p_{O_2} : from 1 to 10^{-28} atm (equilibrium hematite-magnetite), σ_y is multiplied by $\sim$ 200 or 2000 respectively. However condensation of PD also occurs, and may bring dislocation loops, which, when generating new dislocations, operate inversely.

4.2. Nucleation

Semi-coherent nucleation is expected, since the transformation hcp-cfc allows epitaxy along $(0001)_H//(111)_M$, with a misfit of $\sim$ 2 % between the oxygen rows (7, 8), and with interfacial dislocations (8). Previous authors indeed observed oriented nucleation, followed by rapid breakaway at 600°C (9).

The classical treatment for nucleation reads for the situation of Fig. 3

$$\Delta G = \gamma l^2 + (\varepsilon + \Delta G_V) e l^2 \qquad [1]$$

γ : interface energy (hematite-magnetite)
ε : elastic energy at the interface
ΔG_V : free enthalpie change in the reaction per unit volume

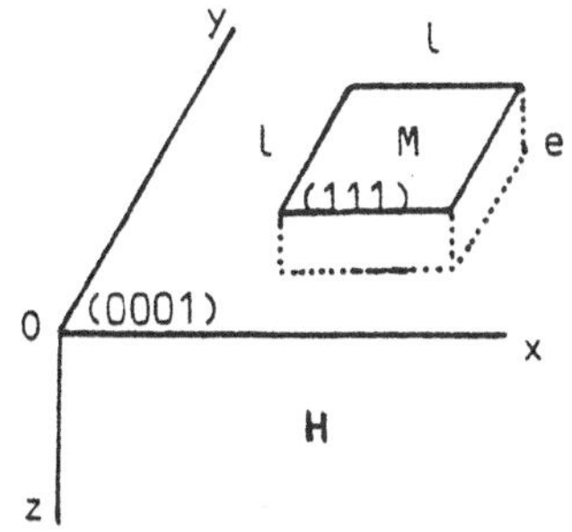

Figure 3. Coherent nucleus of magnetite on (0001) hematite

4.2.1. The hematite-magnetite interfacial energy is given by (11)

$$\gamma_{int} = \gamma_{adh} + \gamma_{Fe_2O_3} + \gamma_{Fe_3O_4}$$

$\gamma_{Fe_2O_3}$, the hematite surface energy, has been calculated by Mackrodt et al (10) ; is is 1.53 J/m^2 for (0001). The other terms are unknown. However we can compare with an oxide-metal interface because the dielectric constants are very different : 1.4 and 21 for hematite and magnetite(12). In the case of oxide-metal epitaxy $\gamma_{adhesion}$ is comparable to γ_{oxide} ; then here $\gamma_{int} \sim \gamma_{Fe_3O_4}$ (111) which should be somewhat lower than $\gamma_{Fe_2O_3}$ (0001).

4.2.2. The elastic energy includes 2 terms

i - the part stored in interface dislocations

$$\sigma_D = \frac{2\sqrt{3}}{D} \cdot \frac{Gb^2}{4\pi(1-\nu)} \ln \frac{\alpha R}{b} \sim 0.55 \ J/m^2$$

D : dislocations periodicity : $\sim$ 240 Å
G : hematite shear modulus = 6.5×10^4 MPa
b : Burger vector $\sim$ 3 Å (13)
ν : hematite Poisson ratio $\sim$ 0.38 (13)
α : core coefficient $\sim$ 4 (14)
R : stress field radius $\sim$ D/2

ii - The true elastic component, besides the dislocations net, is given by (15)

$$\Delta g = \frac{Gel^2}{1-\nu}[2\varepsilon_1^2(1-\nu) - \frac{\pi e}{4l}\{\varepsilon_1^2(3+4\nu) - \varepsilon_2^2 - 2\varepsilon_1\varepsilon_2(1+2\nu)\}]$$

ε_1 and ε_2 : parallel and perpendicular components of the deformation tensor

4.2.3. Equation [1] can now be written, the driving force ΔG_v being

$\Delta G_v = -1.83 \times 10^9 J/m^3$ at 350°C with CO/CO_2= 20/80 for instance.

In this conditions, the critical nucleus size may be calculated by derivation of :

$$\Delta G = \gamma(l^2+4el) + el^2\Delta g - el^2\Delta G_v \qquad \rho = e/l \qquad [2]$$

$$\partial\Delta G/\partial\rho = [2\gamma(1+4\rho)+3.15.10^8(1.2+4\rho)\rho l-4.10^9\rho l]l$$

$$l_c = 20\gamma(1+4\rho)/\rho(3.6-1.25\rho) \qquad \text{cf. Fig. 4}$$

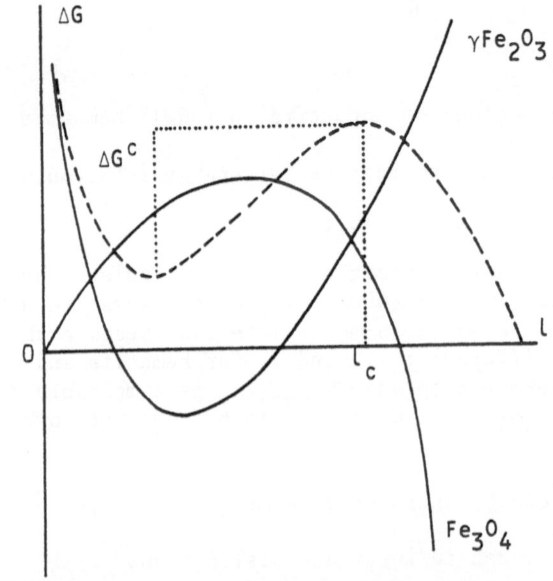

Figure 4. Free enthalpie variation during nucleation versus size of the nucleus. The dashed curve corresponds to the path $\alpha Fe_2O_3 \rightarrow \gamma Fe_2O_3 \rightarrow Fe_3O_4$

If, for instance, $\gamma = 1\ J/m^2$, then, for a cubic nucleus ($\rho = 1$) :

$$l_c = e_c = 43\ \mathring{A}$$

whereas for a plate ($\rho = 0.1$)

$$l_c = 80\ \mathring{A} \qquad e_c = 8\ \mathring{A} \qquad \text{(the width of a cell)}$$

The nucleation frequency is given by :

$$I = (kTN/h)\exp-(\Delta G^c+v)/kT \qquad [3]$$

N : potential sites (1 per cell of hematite yields $N=4.5\times10^8 m^{-2}$)
v : Fe^{2+} transfer energy through the interface ; $v \sim 10$ kT (15)

If we want $I > 1$ per crystal and per hour $\rightarrow I > 6.10^3 m^{-2}h^{-1}$; this needs $\Delta G^c < 3.0$ ev and therefore $\gamma < 0.25\ J.m^{-2}$. Otherwise nucleation uses another path, which could be γFe_2O_3 (13). Then PD play a double role.

i - As the α-γ transition has a positive free enthalpy change, the nucleation of γ needs to be assisted by dislocations, but, in the expanded region of an edge dislocation the interstitials concentration is c_i^D given by :

$$c_i^D/c_i = \exp-\left(\frac{Gbv_i}{3\pi r} \cdot \frac{1+\nu}{1-\nu}/kT\right)$$

c_i : interstitials concentration far from a dislocation
r : distance to the dislocation line, v_i : volume for an interstitial

For Fe^{2+} at 10 Å from the dislocation line $c_i^D/c_i \sim 6$ at 350°C.

ii - When a γFe_2O_3 nucleus is obtained, it is indispensable, so that it can grow, that interstitials fill the cation vacancies ; this is indeed the condition for obtaining a favourable free enthalpy change.

4.2.5. Point defects and volume change during nucleation :

The local volume change, when the accumulated PD condensate into a nucleus, is not the same when major PD are vacancies or interstitials :

i - with oxygen vacancies :

storage of defects : $3Fe_2O_3 + CO = 6Fe_{Fe}^x + V_O^{\cdot\cdot} + 2e' + CO_2 + 8\ O_O^x$

precipitation : $V_O^{\cdot\cdot} + 2e' + 6Fe_{Fe}^x + 8\ O_O^x = 2Fe_3O_4$

local volume change : $3Fe_2O_3 \rightarrow 2Fe_3O_4$ $\quad \Delta V/V = -0.02$

ii - with interstitials :

storage : $9Fe_2O_3 + 3CO = 2Fe_i^{\cdot\cdot} + 4e' + 16Fe_{Fe}^x + 24\ O_O^x + 3CO_2$

precipitation : $2Fe_i^{\cdot\cdot} + 4e' + 16Fe_{Fe}^x + 24\ O_O^x = 6Fe_3O_4$

local volume change : $4Fe_2O_3(+Fe) = 3Fe_3O_4$ $\quad \Delta V/V = 0.015$

Therefore in the first case the vacancies condensate into micropores, and, in the second, the excess of matter yields an internal stress which may explain the observed cracking, in the same way as in polymorphic transformations (16).

4.3. Cracking

There are at least 2 possible causes for the hematite cracking : the above-mentionned volume change, and the shear stress due to the interfacial misfit.

4.3.1. The volume change consequences

If the matrix were uncompressible, the internal stress inside a magnetite nucleus would be :

$$\sigma = K\Delta V/V = 1.9\times10^4 \text{MPa}$$

(K : magnetite bulk modulus : 1.83×10^5MPa, according to ref. 17)

Actually the stress is distributed over both nucleus and matrix. If, in agreement with observation of cross-sections (13), we compare the nucleus to a wedge (Fig. 5), the stress distribution may be calculated as when a pressure is applied inside a crack, and this is equivalent(18) to a tensile stress applied on the crystal in mode I (opening mode). According to the known mechanical properties of hematite (19), the yield stress can be reached, at least in the <0001> direction, hence the stress distribution is alike that given on Fig. 6.

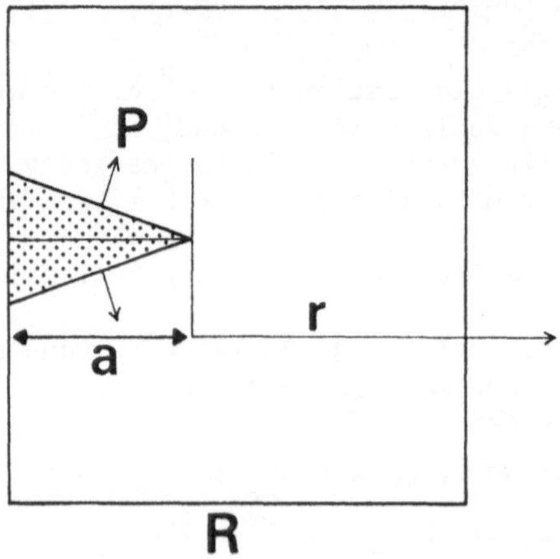

Figure 5. Magnetite nucleus acting as a wedge on the matrix

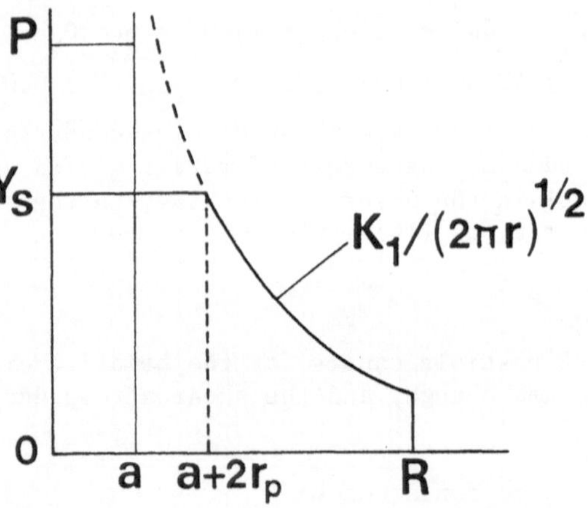

Figure 6. Stress distribution ahead of the nucleus

Now the question is : do we fulfil the conditions for crack propagation ? We can use the Griffith criterion for plane strain :

$$\sigma_f = [2E(\gamma+p)/a(1-\nu^2)]^{1/2}$$

σ_f : critical stress for fracture
γ : hematite surface energy
p : energy stored in the plastic deformation

The major problem is the evaluation of p, because it relates to the non linear strain-stress dependence ; this needs a finite elements calculation, which is in progress (13). Here again PD play a role : as already mentionned, they lower the plastic deformation, so making the solid more brittle. Then, it is of interest to control the PD amount, for instance through substitution of Fe by Al or Cr : that is what we did, and indeed with 2 % Al (At) for instance, we observed a significant increase of cracking (20).

4.3.2. The interfacial shear-stress

Due to the misfit (between 2 and 5 % according to orientation), it is partly relaxed by interfacial dislocations, and may be calculated as for thermal stress in composites materials (6) as :

$$\tau \sim [V_M E_M/(1-\nu_M)]\varepsilon \; 1/e$$

V_M, E_M, ν_M : volume fraction, Young modulus and Poisson ratio for magnetite

$V_M \sim 0.1$, $E_M \sim 1.5\times10^5$MPa (21), $\nu_M \sim 0.3$, $1/e \sim 1$, yields $\tau \sim$ 400 MPa

This is a low value compared to σ.

4.4. The role of water vapour

It is known that water strongly adsorbs on hematite, and among others, takes the place of CO (22). Actually it has been recently shown that H_2O adsorption is favoured by surface oxygen vacancies (23) ; we then write this process :

$$V_{OS}^{\cdot\cdot} + O_{OS}^{x} + H_2O = 2\ OH_{OS}^{\cdot} \quad \text{(S : on the surface)}$$

One consequence is that the effective positive charge, carried by the vacancies, is now trapped on the surface, because OH groups have not been observed inside the hematite lattice. Consequently this positive charge may trap electrons in the near surface layers and repel positive PD. However it may be assumed that the mobility is much higher for electrons than for other PD. Then, if we need both negative electronic and positive atomic PD, as in the nucleation processes described above, a positive surface charge is useful, otherwise the highly mobile electrons would spread in the lattice.

Secondly OH groups are known to diffuse inside quartz crystals along dislocations and to increase its ability to plastic deformation, because the glide of dislocations only breaks Si-OH bonds which are weaker than Si-O (24). If the same effect works here, it can explain the cracking lessening, and also the lack of induction period because the 2 facts are connected :

i - the crack advance meets an energy barrier (6)

$$\Delta G^{*} = \Delta G - \sigma V^{*} + \gamma' v/\rho \qquad (\gamma' = \gamma + p)$$

ΔG : stress free activation energy
V^{*} : activation volume
v : hematite molar volume
ρ : crack tip radius

This barrier is obviously lowered when σ increases and vice-versa.

ii - the nucleation frequency is connected to the mechanical stress through the volume change ; obviously if the matrix strength is lowered there is less resistance to nucleation ; in other words, the shear modulus G, which enters in the free enthalpy change for nucleation ΔG^{c}, is lowered (cf. eq. [2] and [3]).

4.5. The role of COS

The two observed features : reaction strong slowing down, and twofold magnetite morphology, may be inter-related as follows : Nicolle and Rist (25) have shown that the iron morphology, in the reduction of wustite, depends on the ratio $\alpha = kr/D$, where k is the rate constant, r the particle size and D the iron diffusion coefficient ; when α decreases the iron density increases up to whisker formation. If sulphur poisons active surface size, α decreases and the magnetite is more dense, but this is not usually (25) uniform, because where sulphur activity is lower reaction is fast and magnetite grows inside the hematite crystals in a porous form. However other mechanisms, such as near-surface effects should also be taken into account (26).

ACKNOWLEDGEMENT

The author is grateful to Dr. P. Riboud, head of the Physico-chemistry Analysis and Structures Department of the Institut de Recherche de la Sidérurgie (IRSID-CIREP) and to his collaborators, for numerous valuable discussions. This work has also been supported by the Centre National de la Recherche Scientifique.

REFERENCES

1 - J. PARK et O. LEVENSPIEL - Chem. Eng. Sci., 30 (1975) 1207
2 - F. ADAM, F. JEANNOT, B. DUPRE et C. GLEITZER - Journées de Cinétique Hétérogène, DIJON 13-14/9/1988
3 - M. ETTABIROU, B. DUPRE and C. GLEITZER - React. Solids 1 (1986) 329
4 - C.R.A. CATLOW, J. CORRISH, J. HENNESSY et W.C. MACKRODT - J. Amer. Ceram. Soc. 71 (1988) 42
5 - J. COUZIN, J.J. OEHLIG et A. DUQUESNOY - CR Acad. Sci. 303, Ser. II, (1986) 1549
6 - W.D. KINGERY - Introduction to Ceramics - Publ. J. WILEY 1960
7 - P. BECKER, J.J. HEIZMANN and R. BARO - J. Appl. Cryst. 10 (1977) 77
8 - L.A. BURSILL et R.L. WITHERS - J. Appl. Cryst. 12 (1979) 287

9 - P. HAYES and P. GRIEVESON - Met. Trans. 12B (1981) 579
10 - W. MACKRODT, R. DAVEY, S. BLACK et R. DOCHERTY - J. Cryst. Growth 80 (1987) 441
11 - A. STONEHAM et P. TASKER - in "Ceramic microstructures 86 : role of interfaces", Ed. J.A. Pask, Publ. Plenum Press
12 - E. McCAFFERTY et A. ZETLLEMOYER - Disc. Faraday Soc. 52 (1971) 239
13 - F. ADAM - Thesis Nancy - september 1988
14 - J.P. HIRTH et J. LOTHE - "Theory of dislocations", Publ. McGraw Hill (1968)
15 - J.W. CHRISTIAN - "The Theory of transformations in metals and alloys", Pergamon Press (1965)
16 - A. CHUPAKHIN, A. SIDELNIKOV, V. BOLDYREV - React. Solids 3 (1987) 1
17 - L. FINGER, R. HAZEN and A. HOFMEISTER - Phys. Chem. Miner. 13 (1986) 215
N. NAGAKIRI, M. MANGHNANI, L. MING and S. KIMURA - ibid. 238
18 - R. LABBENS - "Introduction à la mécanique de la rupture". Ed. Pluralis (1980)
19 - C. HENNIG-MICHAELI and H. SIEMES - "High Pressure Research in Geoscience", ed. W. SCHREYER, p.133-150, publ. Schweizerbart Verlag, Stuttgart (1982)
20 - K. GHARIBI - Thesis Nancy, to be submitted in 1989
21 - M.E. FINE and N.T. KENNEY - Phys. Rev. 94 (1954) 1573
22 - A. OZAKI and K. TANAKA - J. Cat. 20 (1971) 422
23 - M. HENDEWERK, M. SALMERON and G. SOMORJAI - Surf. Sci. 172 (1986) 544
24 - J.C. DOUKHAN - in "Dislocations et Déformation Plastique" ; Ecole d'Eté d'Yravals (1979), publ. P. GROH et al.
25 - R. NICOLLE and A. RIST - Met. Trans. 10B (1979) 429
26 - Z. ADAMCZYK and J. NOWOTNY - React. Solids 4 (1987) 139

JOINING OF METALLIC GRAINS BY THERMAL OXIDATION

Teodor Werber
Dept. of Materials Engineering, Technion
32000 Haifa, Israel

ABSTRACT. The subject of this work is the mechanism of formation of intergranular oxide joints (IJ) in uncompacted nickel and iron powder layers oxidized in air at 400 - 800°C. Four kinds of nickel powders of different size and morphology of grains and one iron powder were investigated. The strength of IJ depends on the size of contact area formed between oxide layers of adjacent grains. It is a function of time and temperature of oxidation and the size, shape and morphology of the grains. On the interface between both contacting oxide layers is formed a grain-boundary-like zone, which is responsible for the cohesion in the IJ.

1. INTRODUCTION

Metallic powders have a tendency to form agglomerates of interconnected grains during thermal oxidation. The intergranular joints (IJ) are often strong. In a majority of cases the agglomeration is detrimental but in the past there was no special interest in studying and explaining this phenomenon. A few studies were made on the oxidation of partially compacted metallic powders [1-3]. These investigations concerned the kinetic aspects of the oxidation of porous metallic materials and the closing of pores by the growing oxidation products. The mechanism of bond formation between the oxidized grains was not a subject of discussion.

The goal of present work was to determine the factors causing the intergranular bonding during the oxidation of metallic powders. This work has a preliminary character and will be continued.

J. Nowotny and W. Weppner (eds.),
Non-Stoichiometric Compounds Surfaces, Grain Boundaries and Structural Defects, 547–556.

2. EXPERIMENTAL AND RESULTS

The experiments were made on four kinds of nickel powders and one iron powder. The powder parameters are listed in Table 1. Before oxidation the powders were put

Table 1: Powder Characteristic

Powder Grade	Abbreviation in the text	Producer	Grain Sizes, μm	Grain Shape
Nickel Powder	Ni I	unknown	20-40	nodular
INCO HDNP (high density nickel powder)	Ni II	INCO	5-10	nodular
Nickel Powder	Ni M	Merck	<10	dendritic
Nickel Powder	Ni 123	INCO	3-7	spike particle
Iron Powder	-	unknown	2-5	nodular

via a sieve on rectangular supports made of nickel sheet (1.5 - 2.5 cm^2, 0.15 mm thick). Excess of powder was removed by passing the samples through a 1.0 mm gap. The powder layer prepared in this manner had a 50-60% porosity*.

The nickel and iron samples were placed on refractory plates and oxidized in air in a chamber furnace. The nickel powder samples were than oxidized in isothermal conditions in the temperature range 400-800 °C (temperature stability ± 5 deg) for exposure times of 0.5-18 hrs. The iron powder samples were oxidized only at 400 °C for 0.5-1 hr. After being taken out of the furnace the samples were cooled in air.

The intergranular joining in the samples after oxidation were tested only qualitatively. The criterion

---

* The porosity was determined by measuring the weight of the powder layer 'w' on a given surface 's'

$$\% = 100 - \frac{w}{d.l.s}$$

where -
d - metal density; l - thickness of the layer.

of IJ formation was the stability of the sample surface while being rubbed against a typewriter sheet. The IJ was distributed in three groups:

1. no joining - without rubbing the grains spill from the sample surface of their own accord;

2. weak joining (W) - during the rubbing grains separate from the surface;

3. strong joining (S) - during the rubbing the grains do not separate from the sample.

Table 2 shows the results of the IJ test, from which it follows that the strong joints were already formed

Table 2: Strength of IJ formed at 600°C

Oxidation time, hr	powder grade			
	Ni I	Ni II	Ni M	Ni 123
1.5	W	S	W	W
6.0	W	-	W	W

between Ni II powder grains in one of the shorter reaction time (1.5 h). For both Ni 123 and Ni M powders the bond fastness probably did not increase with oxidation time, because the quantity of grains separated during the rubbing were similar after 1.5 and after 5 hrs of heating. On the other hand, the bond strength between the grains of Ni I powder increased after 6 h of oxidation, but really strong joints for this powder were obtained only after 4 h heating at 700°C.

Fig.1 shows grains of precursor and of oxidized nickel powders for three types of powders (Ni II, Ni M and Ni 123). The nickel grains before oxidation are generally nodular shaped and have a rough surface. In the precursor powders the quantity of agglomerates* rose with decreasing grain size (highest quantity of agglomerates was found in Ni 123). The surface of oxidized grains is smoother. In the case of Ni II powder (Fig.1d) IJ and the broken intergranular oxide bridges are seen. They are also seen with powders smaller grain sizes (Fig.1e,f)

* The agglomerates in percursor powder are caused by fabrication.

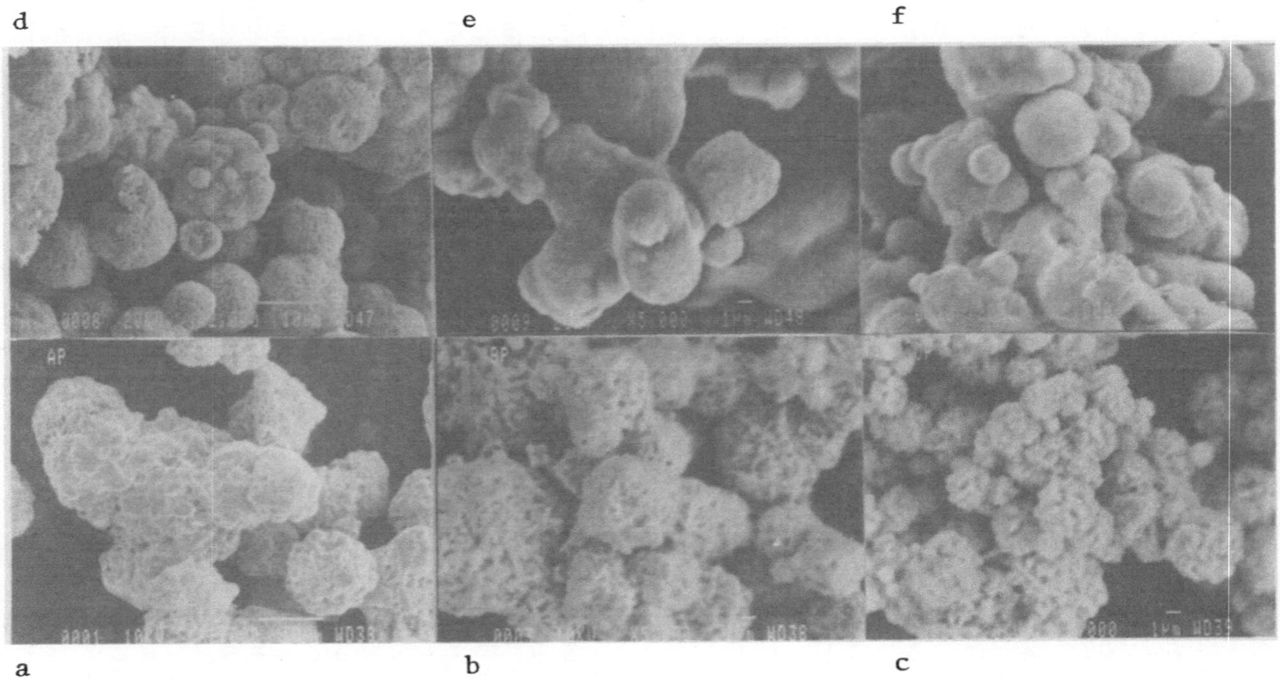

Fig.1. SEM pictures of unoxidized (a,b and c) and oxidized (d,e and f) grains of nickel powders (oxidation at 600°C of 1.5 h) a, d - Ni II; b, e - Ni M and c, f - Ni 123.

The IJ created during oxidation of iron powder were strong and the formed intergranular oxide bridges are seen in Fig.2a.

a

b

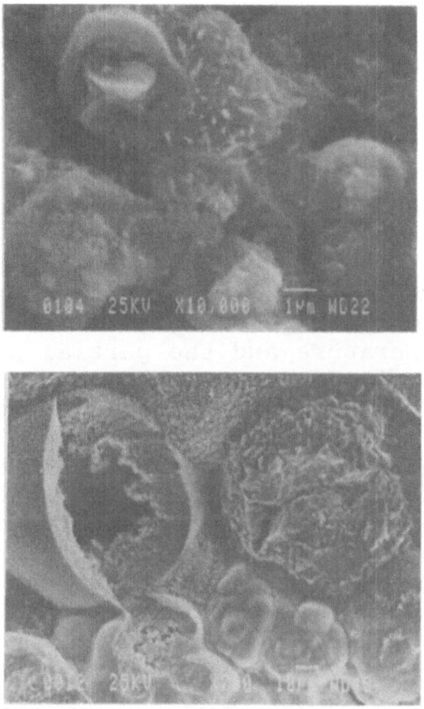

Fig.2. SEM pictures of: a - oxidized iron powder grains at 400°C of 0.5 h; b - oxidized Ni I powder grains at 800°C of 18 h.

3. DISCUSSION

During thermal oxidation the metallic powder grains "swell", because the oxide layer formed has a lower density than the consumed metal. The grains swelling lead to an increase in the contact area between adjacent grains already in contact at the moment reaction began or it is formed later during the reaction as a consequence of the grain swelling. The interface formed between the contacting growing oxide layers is responsible for the creation of IJ.

In the case of oxide layers with prevailing cationic transport of reagents as occurs during the oxidation of nickel and iron [4], the oxide growth take place on the external oxide surface and in the places of consumed metal empty spaces are created (Fig.2). This leads to a greater swelling than would be expected from the common relation between the volumes of consumed metal and the oxide formed from it. The formation of new lattice planes on the external surface of the oxide stops it from further growth, when the surfaces of adjacent grains come into contact. The contacting external surfaces adapt their shape to each other and also repelling of swelling grains is excluded.

The strength of the IJ is related to size of the contact area formed between the adjacent grains during the oxidation process. The size of the contact area is a function of the rate of the oxidation process and the size and shape of the contacting grains. The first factor depends on the temperature and the partial pressure of oxygen, which commonly determines the rate of oxide formation. The influence of grain size on the size of the contact area is shown in Fig.3. It follows from the drawing that the contact area between spherical grains is a function of the grain diameter and the thickness of the oxide formed. The contact area between two grains of the same diameter increased with increasing diameter (Figs.3b,d). When two grains with different diameters come into contact, the size of the contact area is determined mainly by the diameter of the lower grain, and this leads to a reduction of the contact area (Fig.3c). The influence of the grain shape on the size of the contact area is related to the radius of curvature of the surfaces coming into contact. Surfaces with a smaller radius of curvature form smaller contact areas in otherwise identical reaction conditions, as is seen in Fig.3, which illustrates this effect for spherical grains.

In a majority of cases the grains of metallic powders are not monolithic, but have an open and/or closed porosity. The roughness of the grain surfaces is a particular case of this effect. Both morphological features - open pores and surface roughness - have a negative influence on IJ formation, because the oxide growth in the places where these features are present does not lead to a significant grain swelling and increase of the surface area between oxide layers forming on adjacent grains. Fig.4 shows a dendritic grain of Ni M powder, in which the nickel oxide can grow in the direction of the grain radius and laterally to it. The volume of empty spaces between nickel platelets is sufficient to contain the NiO formed during the oxidation of the grain. The lateral growth of oxide does not increase the grain

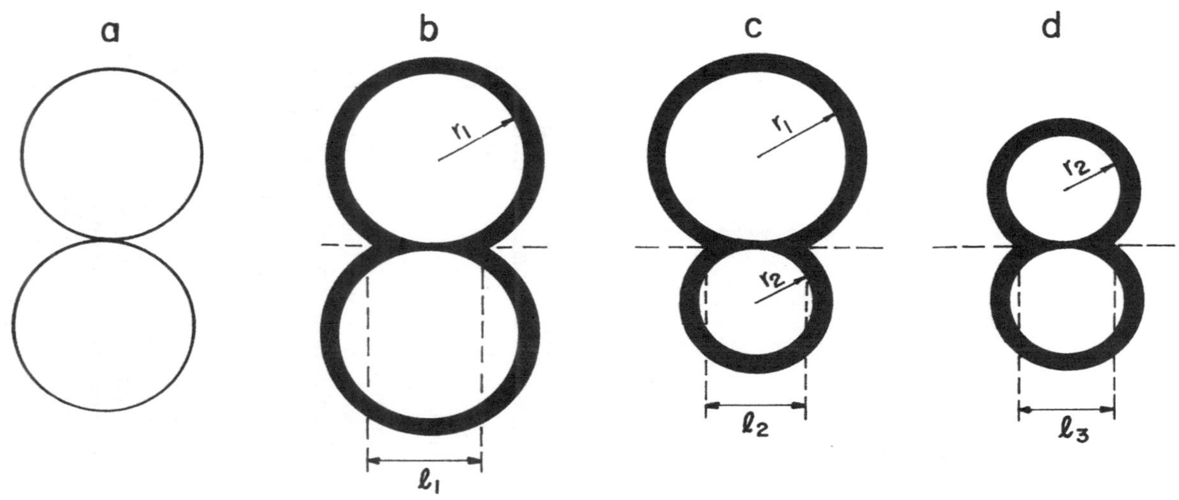

Fig.3. Effect of grain diameter on the size of contact area; a - initial contact between adjacent unoxidized grains; b, c, and d - contact area formed between oxidized grains of different radiuses ($r_1>r_2$) relation between contact area diameters: $l_1>l_3>l_2$; (black rim - oxide layer formed).

dimensions, but only changes the surface morphology, which becomes much smoother (Fig.1e,f). The increase of the contact area between initially contacting grains is determined only by the growth in the direction of the grain radius.

Fig.4. Grain of Ni M powder

An adhesion effect was found by the Author when growing (oxide, sulfide) scales came in contact with an adjacent solid substance that does not react with the oxidizer [5]. The adhesion forces arise on the interface scale/solid substance only when the growing scale expands on the solid surface and the scale forming process takes place on the outer surface of the scale (cationic transport in the scale). It was assumed that the newly produced lattice planes are not completely ordered, and this can lead to physical and/or chemical interactions with the contacting surface to a formation of adhesion bonds.

In the case of oxidation of metallic grains of uniform composition oxide layers of identical chemical compositions (eg. NiO) exist on both sides of the contact interface. Because the contact on the interface between growing oxide surfaces is very close and the external (last formed) lattice planes are only partially ordered, an ordering process takes on the interface. This atomic rearrangement leads to the formation of a grain boundary between crystallites consisting of the same oxide. Therefore the IJ is a cohesive bond created by grain boundary formation. The oxide located between the remainders of both initial metallic grains can be treated as a monolithic layer common to both grains. The locations of the cracks in different places of oxidized grains (Fig.3) confirm this supposition.

The differences in the strength of the IJ created during oxidation in different types of nickel powder (Table 2) can be explained by factors mentioned in the foregoing discussion. These factors are: temperature and time of reaction, and the size and morphology of powder grains. The powders with a higher specific surface (lower grain diameter) oxidize faster, and because the oxidation reaction is exothermic, this can lead to an overheating of the sample and an acceleration of the reaction. The difference in the rates of IJ formation between the Ni I and Ni II powders can be caused by an overheating effect. The grains of Ni I have diameters a few times bigger than Ni II grains and this leads to a significant drop in specific surface of the powder and to the reduction of overheating during oxidation. The powder Ni I barely forms a strong IJ at 700 C after 4 hrs of heating.

The weak IJ of both Ni M and Ni 123 powders (Table 2) is caused both by the small sizes and by the morphology of the metallic grains (Figs.1 and 4). The small grain diameters potentially lead to the formation of narrow contact areas between oxidized grains. Additionally, the open porosity and the significant surface roughness reduce the grain swelling during oxidation. All these factors limit the size of the contact areas and decreases the strength of IJ created.

4. CONCLUSIONS

1. During thermal oxidation of uncompacted nickel and iron powders intergranular joints are formed.

2. The strength of the intergranular joints depends on the time, the temperature of oxidation and the size, shape, and morphology of the powder grains.

3. The creation of intergranular joints is connected with formation of a grain-boundary-like zone between intimately contacting oxide layers formed on adjacent grains.

5. ACKNOWLEDGEMENT

The Author is grateful to Mr. Guy Proactor for carryhing out the SEM studies and to INCO Europe Ltd. for supplying free of charge INCO HDPN and INCO 123 nickel powders.

6. REFERENCES

1. I.M. Fedortshenko, A.P. Lyapunov and W.V. Shorochod, Powder Metallurgy No.12, 27 (1963).

2. E.T. Denisenko, Poroschovaya Metallurgiya, Nr 7, 4 (1966).

3. K. Słoczyńska and T. Werber, Archiwum Hutnictwa, 13, 323 (1968).

4. A. Atkinson, Reviews of Modern Physics, 57, 437 (1985).

5. T. Werber, Materials Science and Engineering, 88, 283 (1987).

IMPEDANCE AND VOLTAGE RELAXATION STUDIES OF THE OXYGEN SENSOR SYSTEMS $Pt/O_2/YSZ$, $Pt/O_2/TiO_2$ and $Pt/O_2/\delta\text{-}Bi_2O_3$

B. LEIBOLD and N. NICOLOSO
MAX-PLANCK-INSTITUT FÜR FESTKÖRPERFORSCHUNG
HEISENBERGSTR. 1
D-7000 STUTTGART 80
FRG

ABSTRACT. The oxygen sensor systems $Pt/O_2/YSZ$, $Pt/O_2/TiO_2$ and $Pt/O_2/\delta\text{-}Bi_2O_3$ have been investigated by impedance and voltage relaxation measurements. Mainly single crystal and thin film oxide materials were used.

In the case of single crystal YSZ, the oxygen exchange reaction involves at least two interrelated reaction steps at the Pt or YSZ interface. Dissociative adsorption and incorporation of O or O^- are believed to account for this behaviour. However, no final reaction model can be given on the basis of the electrochemical data. Polycrystalline TZP can be described by a barrier layer model with grain boundary layers $\ll$ 100 Å. Current fractal models are not suited to describe the $Pt/O_2/YSZ$ system. In this context, recent results on the CPE behaviour of the impedance and its possible fractal origin are briefly reviewed. In addition, a short introduction to impedance spectroscopy is given.

For single crystal TiO_2 the surface excess conductivity follows a $p_{O_2}^{1/2}$ dependence which indicates adsorption and incorporation of monoatomic oxygen species like O or O^-. Thin polycrystalline films ($300 \le d \le 10^4$ Å) show no significant surface contribution. Probably, the surface and bulk defect structure is dominated by the same kind of defects ($V_O^{\cdot\cdot}$, $Ti_i^{\cdot\cdot\cdot\cdot}$), whose concentrations or mobilities appear to differ within only one order of magnitude.

In the case of $Pt/O_2/\delta\text{-}Bi_2O_3$, preliminary voltage relaxation studies indicate that the diffusion coefficient of the minority carriers exceeds the ionic diffusion coefficient by several orders of magnitude. This is the same behaviour as seen in the YSZ system.

J. Nowotny and W. Weppner (eds.),
Non-Stoichiometric Compounds Surfaces, Grain Boundaries and Structural Defects, 557–579.

1. IMPEDANCE SPECTROSCOPY (IS) - ELEMENTARY ANALYSIS OF SPECTRA

Electrochemical processes are often represented by their electrical analogs or "equivalent circuits", whose various elements (such as R, C, L) or distributed elements like the Warburg impedance W, may be obtained from impedance measurements. Usually, a small monochromatic signal ($\Delta U_{pp} \leq 10$ mV) is applied and the amplitude and phase shift of the resulting current are taken over a frequency range of 10^{-4} - 10^{7} Hz. In the low frequency range (10^{-4} - 10 Hz) the 'Fast Fourier Transform' technique is advantageous because of the reduction of measuring time. For a detailed description of measurement techniques the reader is referred to Ref. 1.

1.1 BASIC DEFINITIONS

Impedance may be expressed in complex numbers:

$$Z = Z' + j\,Z'' \tag{1}$$

with Z' and Z'' being the real and imaginary components, respectively.

Its magnitude is $|Z| = [(Z')^2 + (Z'')^2]^{1/2}$. The phase shift between voltage and current is represented by an angle φ:

$$\varphi = \arctan\,(Z''/\,Z'). \tag{2}$$

Similarly, **admittance,** the inverse of impedance, can be written as

$$Y = Z^{-1} = Y' + j\,Y'' \tag{3}$$

Z and Y may be also expressed in terms of resistive and capacitance components:

$$Z = R_s(\omega) - j\,X_s(\omega) \tag{4}$$

and

$$Y = G_p(\omega) - j\,B_p(\omega), \tag{5}$$

where $X_s(\omega) = [\omega C_s(\omega)]^{-1}$ is the **reactance**, $B_p(\omega) = \omega C_p(\omega)$ the **susceptance** and $G = R^{-1}$. The subscripts s and p stand for "series" and "parallel".

In case of a series network, Z will be just the sum of the impedances of the individual elements. Similarly, in a parallel network Z can be evaluated from the sum of admittances.

1.2 SIMPLE RC CIRCUITS

The most basic network unit is (RC), the parallel combination of a resistor R and a capacitor C. Its impedance is:

$$Z = (R - j\omega CR)\,(1+\omega^2C^2R^2)^{-1} \tag{6}$$

As can be seen from Eq. 6, the dc-resistance value of the circuit is recovered in the low frequency-limit $\omega \longrightarrow 0$. For $\omega \longrightarrow \infty$, Z is zero and, since Eq. 6 describes a semicircle, a maximum occurs at $\omega CR = 1$ in the complex plane representation $-Z''$ vs Z' ("impedance spectrum"). In the time domain the process has a characteristic time $\tau = RC = 1/\omega_{max}$. Time constants may vary between some μs and seconds in electrochemical systems. Hence several processes may be easily separated, if the time constants differ by some orders of magnitude. Generally, the high frequency part of an impedance diagram corresponds to the dielectric contribution of the bulk material (C_b typically in the pF range for usual sample dimensions). Accordingly, the slowest step, usually an electrode process, produces low frequency features. A typical example of a simple (RC) circuit is given in Fig. 1a, where the impedance of single crystal TiO_2 is shown. The apparent dielectric constant (≈150) derived from $C=(\omega_{max}R_b)^{-1}$ and the known sample geometry agrees well with literature[2].

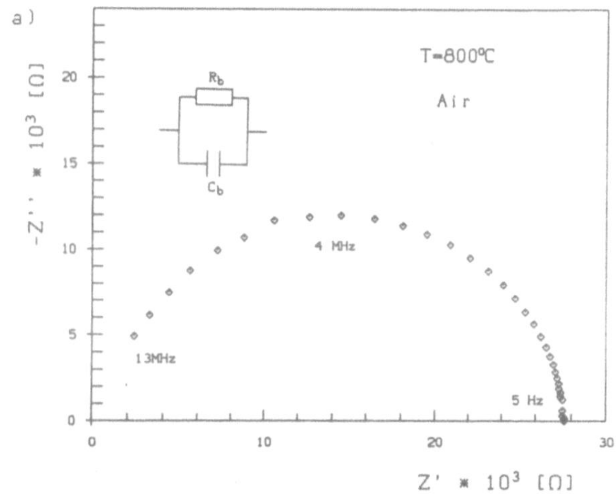

Fig. 1a: Impedance spectrum and equivalent circuit of single crystal TiO_2 at T = 800 °C under reversible conditions.

A more complex impedance spectrum arises if non-reversible electrodes are attached. In addition to the high frequency arc of bulk TiO_2, further arcs with C-values typical for electrode processes (μF- range) are discernible in the low frequency part of the spectrum (Fig. 1b).

A second example which illustrates the use of IS for a first qualitative understanding of chemical processes is the reaction of O_2 at the interface Pt/polycrystalline stabilized zirconia (SZ). According to Bauerle[3] three distinct contributions to the overall cell resistance can be separated (see Fig. 2): the bulk electrolyte conductance (G_3), the electrode conductance (low frequency part, arc 1) and a contribution from the grain boundaries (arc 2).

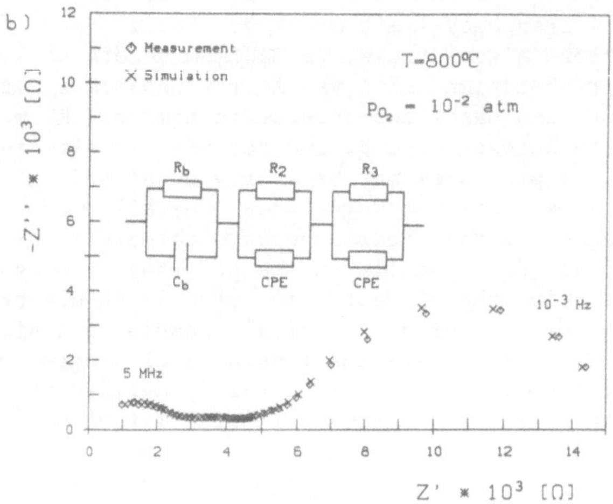

Fig. 1b: Impedance spectrum and equivalent circuit of single crystal TiO_2 at T = 800 °C under blocking conditions.

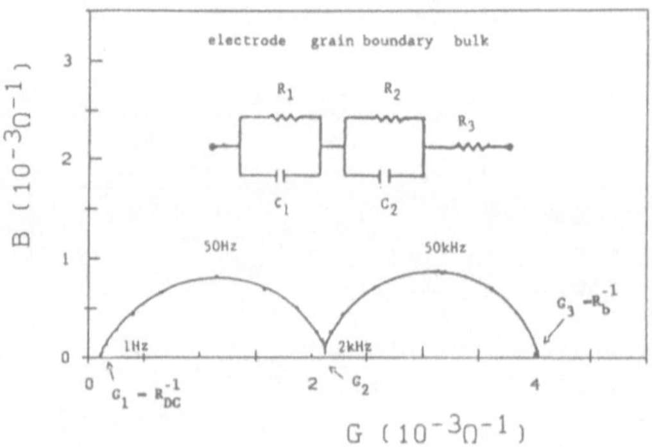

Fig. 2: Admittance spectrum and equivalent circuit of polycrystalline YSZ/porous Pt (Bauerle [1969]).

However, care must be taken in the use of idealized circuit elements and in establishing the correct equivalent circuit as these are inherent sources of error. For the series arrangement of the three (RC)-elements mentioned above, two other circuit descriptions may be used to give the same impedance (see Fig. 3). The R and C values in these circuits will naturally be altered. By varying the experimental conditions (T, p_{O_2},..) an optimal choice can be made. In the case of stabilized zirconia, arguments in favour of the series arrangement have been given by Bonanos et al[4].

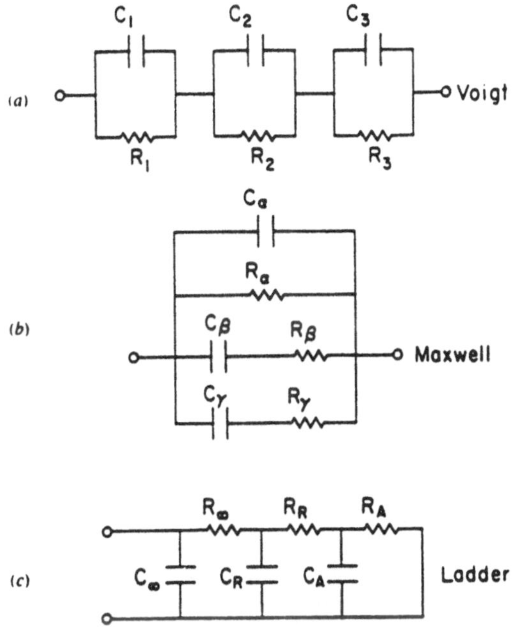

Fig. 3: Ambiguous circuits: Three equivalent circuits, which can have the same impedance at all frequencies, when the circuit elements are properly interrelated (after Ref. 1, p. 96).

1.3 CIRCUITS WITH DISTRIBUTED ELEMENTS

1.3a DIFFUSION RELATED ELEMENTS - WARBURG IMPEDANCE

Carrier diffusion at electrolyte/electrode interfaces may be described more adequately by distributed elements than by a finite combination of ideal circuit elements. For example, the RC transmission line shown in Fig 4, is the analog of the finite length Warburg diffusion impedance Z_W generally abbreviated by the symbol W. The case of one-dimensional semi-infinite particle diffusion was first treated by E. Warburg and F. Krüger[5]. Two impedance features are typical of such a diffusion controlled process (see eqns. 16 and 18 of Appendix 1):

1) A capacitive phase shift of $\pi/4$ (easily detectable in the low frequency part of an impedance spectrum by a straight line with an angle of 45°) and

2) A linear dependence of Z on $\omega^{-1/2}$ (the slope is determined by the diffusion coefficient D_i and the equilibrium concentration of diffusing particles).

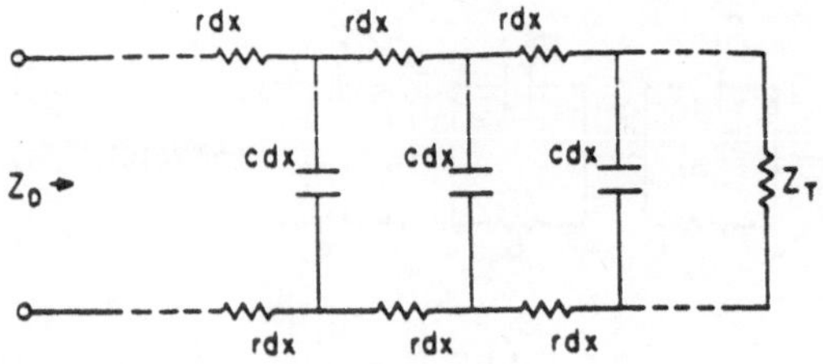

Fig. 4: Uniform continuous transmission line, involving series resistance of 2r per unit length and shunt capacitance of c per unit length, terminated by an impedance Z_T. When $Z_T = 0$, $Z_D = Z_W$ and when $Z_T = \infty$, $Z_D = Z_{DOC}$ (after Ref. 1, p. 89).

Warburg-type diffusion was first observed in liquid systems. This will not be discussed here, but the readers attention is drawn to the works of Randles and Somerton[6], Sluyters[7], Sluyters and Ohmen[8] and the reviews of Sluyters and Rehbach[9], and Armstrong et al.[10]. In electrochemical systems, with solid electrolytes, the effect of diffusion is often seen in depressed or skewed semicircles. Fig. 5 illustrates some IS-spectra and equivalent circuits depicting situations of this type. An outline of the evaluation of the important parameters of the symmetrically depressed semicircle is given by J.Ross Macdonald and W.B. Johnson in Ref. 1.

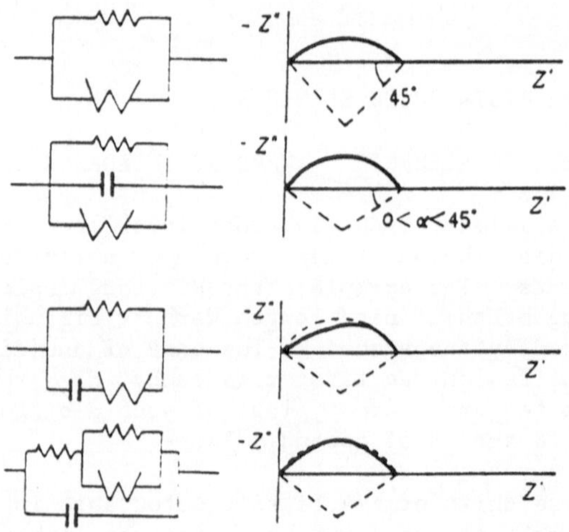

Fig. 5: Impedance spectra and equivalent circuits, including a Warburg impedance W, in different combinations with R and C elements.

1.3b CPE - BEHAVIOUR OF IMPEDANCE

Particularly under blocking conditions, where one introduces electrodes not permeable to the majority carriers, a constant phase element (CPE) is often found in the impedance, and $Z(\omega)$ follows a power law

$$Z(\omega) \propto (j\omega)^{-\alpha} . \tag{7}$$

The exponent α varies between 0 and 1. By plotting log $|Z|$ vs log ω and φ vs log ω ('Bode-plot') the CPE-behaviour of a system can be demonstrated. Fig. 6 illustrates the two limiting cases $\alpha = 0$ and 1 (pure resistance and pure capacitance). The corresponding phase angles of 0 and 90° are constant over several decades in frequency. In fact, this spectrum corresponds to a simple resistor with $R = 10^9\ \Omega$ and illustrates the nuisance of cable and stray capacities when measuring highly insulating materials. The third limiting case, the Warburg diffusion impedance with its exponent $\alpha = 1/2$ has been discussed previously (see above).

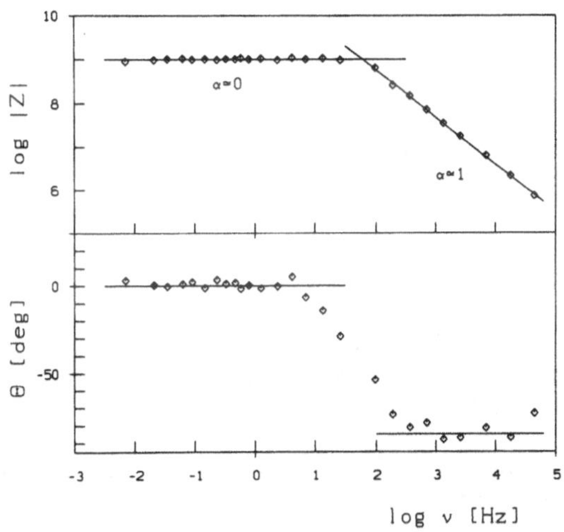

Fig. 6: Impedance of a simple resistor with $R = 10^9\ \Omega$ (Bode-plot). The two limiting cases of CPE-behaviour, $\alpha = 0$ and $\alpha = 1$ (pure resistance and pure capacitance), and the corresponding phase angles 0° and 90° are seen in the low and high frequency parts of the spectrum. The deviation from the ideal resistive behaviour arises from cable and stray capacities.

Except in these three well defined cases, the CPE-behaviour - a phenomenon ubiquitous in IS-spectroscopy - is still not understood. However, some promising models have been suggested. These include:

a) a distribution of relaxation times,

b) an inhomogeneous current flow in the contact region,

c) the fractal nature of the interface.

The first two probable causes are discussed in Ref. 1. The third one is currently a matter of considerable debate, mainly because the various theories differ substantially and experimental evidence is scarce. Notably the approaches of Liu et. al.[11] and of Nyikos and Pajkossy[12] lead to rather different dependences of the CPE exponent α on the fractal dimension d of the interface - see part III of this work.

Besides the fractal approach on topologically disordered interfaces, several pore models should be mentioned. De Levie[13] has treated the case of a V-shaped groove in an electrode surface. The corresponding transmission line is shown in Fig. 7.

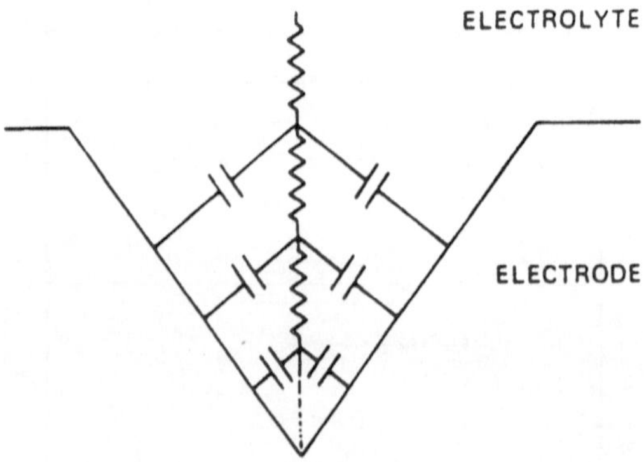

Fig. 7: Transmission line of a V-shaped groove in an electrode surface (after de Levie [1965]).

Its most important impedance feature is a gradual phase change from $\pi/2$ to $\pi/4$ (low and high frequency limit - see curve 3 in Fig. 8). The branched transmission line (discussed by Scheider[14]) leads to an impedance with a constant φ of 3/4 ($\pi/2$). Wang[15] extended de Levie's approach using position dependent resistors and capacitors. The important feature of all the differently constructed transmission lines is that they produce CPE-behaviour with an angle not restricted to $\pi/4$, the Warburg diffusion value. Fig. 8 illustrates schematically the impedance spectra expected for different kind of pores.

Evidently, elucidation of the nature of the CPE-behaviour will have a large impact on impedance spectroscopy and the interpretation of electrochemical processes not yet understood.

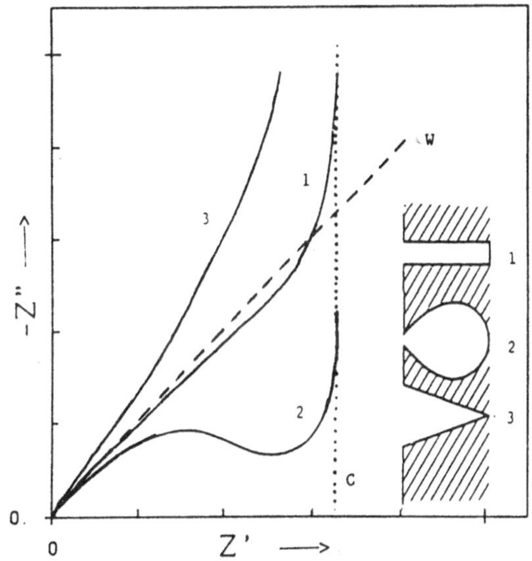

Fig. 8: Impedance spectra expected for differently shaped pores. The limiting cases of the Warburg impedance W and a pure capacitance C (broken lines) are included for comparison.

2. APPLICATION OF IMPEDANCE SPECTROSCOPY TO OXYGEN SENSOR SYSTEMS

2.1 STABILIZED ZrO_2

2.1a YTTRIUM STABILIZED CUBIC SINGLE CRYSTALS

A typical series of IS-spectra on single crystal yttrium stabilized ZrO_2 (YSZ, 9.4 mol% Y_2O_3) with blocking Pt-electrodes attached to the (111) planes is shown in Fig. 9. The following observations can be made from this data:

1. Two interface processes with well separated τ's and CPA-exponents are discernible in the whole temperature range studied.

2. Until approximately 750 ^oC both processes have the same activation energy of $\approx$ 1 eV. Above this temperature, the faster process is not activated, or only weakly activated ($E_A \leq 0.1$ eV).

3. With increasing temperature the spectra become more uniform and a distribution in τ about a single value is expected for T $\gg$ 1000 ^oC.

4. The slower, rate determining step may be affected by a change of the gas flow[33].

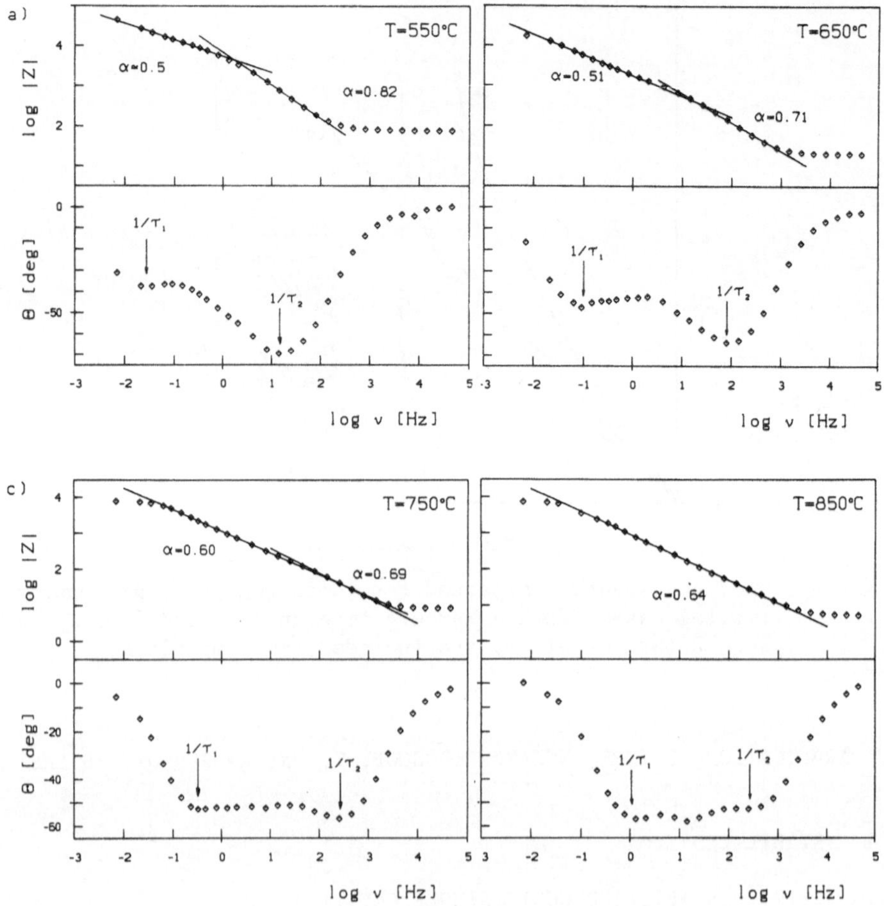

Fig. 9: Impedance of single crystal cubic YSZ (9.4 mol% Y_2O_3) under blocking conditions (dense sputtered Pt-electrodes, $p_{O_2} = 10^{-3}$ atm) at different temperatures. The slopes of the Bode-plots correspond to the CPE-exponents α, see text.

Similar spectra have been obtained for crystals contacted on the (100) and (110) planes. The impedance depends rather strongly on the morphology of the Pt deposited on the crystal surface and possible effects from the crystal orientation can therefore not be discerned. This behaviour also suggests that a characterization of the $Pt/O_2/YSZ$ system by the fractal approach might be useful (see below). A detailed description of the experimental conditions is given in Ref. 16.

2.1b TETRAGONAL POLYCRYSTALLINE YSZ (TZP)

Tetragonal YSZ is a potentially interesting sensor material, because of its favourable mechanical and electrical properties. Fig. 10 illustrates typical impedance spectra of ZrO_2, doped with with 2 mol% Y_2O_3. The polycrystalline sample has been pressed and sintered to a pellet of approximately 1 cm in diameter and a thickness of 0.1 cm. The arc in the centre of the spectrum represents the grain boundary contribution, the semicircle to the left represents the bulk contribution and the incomplete arc at the right side represents the electrode contribution (see Fig. 10a).

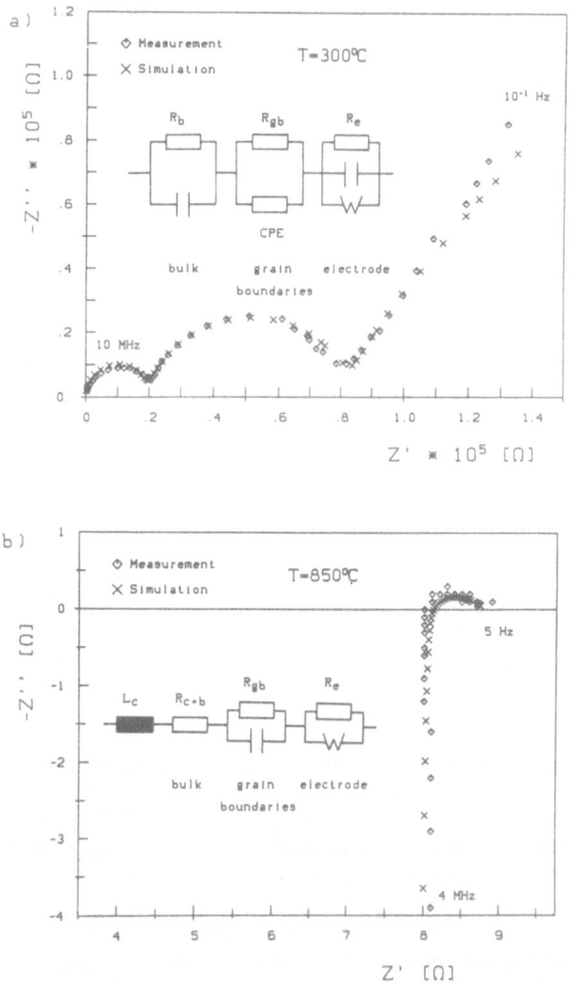

Fig. 10: Impedance spectra and equivalent circuits of polycrystalline Y-TZP (2 mol% Y_2O_3) with porous Pt-electrodes at a) T = 300 °C and b) T = 850 °C in air.

Fig. 10 b shows the impedance of the sample at a higher temperature. Such spectra, mainly distorted by the inductance of the connecting wires, may be analysed by NLLS-fitting procedures[17] provided that the correct equivalent circuit has been established at lower temperatures. In our case, the improvement in data analysis and simulation leads to a revision of the published Arrhenius behaviour of tetragonal YSZ[18] (and also of cubic YSZ[16]): Up to temperatures of 1000 ^{o}C the Arrhenius plots show no bending (or only marginal bending) and thus, no unusual kind of electrical transport, or defect association has to be invoked. Fig. 11 shows the Arrhenius diagram of the various conductivity contributions σ_b, σ_{gb} and σ_e for the sample with 2 mol% Y_2O_3. Activation energies of 0.74 and 1.13 eV are found for the bulk and grain boundary transport. Above ≈ 700 ^{o}C the transport within the electrode/electrolyte interface region and grain boundaries is similar. The large change in the electrode part at ≈ 750 ^{o}C demonstrates that the oxygen exchange is rate limited at lower temperatures mainly by kinetic processes at the Pt/TZP interface.

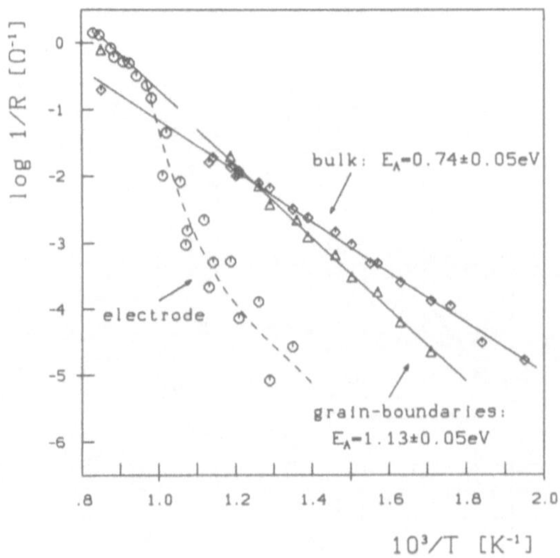

Fig. 11: Arrhenius diagram of the bulk, grain boundary and electrode interface conductance of TZP (2 mol% Y_2O_3).

The impedance analysis also indicates that the apparent grain boundary capacity is nearly constant in the temperature range investigated (300-900 ^{o}C) with a value of about 10^{-9} F. As has been shown by Nowotny and Sloma[19], segregation of Y^{3+} and other impurities such as Al^{3+} occurs in the near surface region and, as in varistors, a thin barrier layer may be generated. In Fig. 12 the bulk and grain boundary activation energies $E_{A,b}$ and $E_{A,gb}$, evaluated in the temperature range 300 ≤ T ≤ 600 ^{o}C, are shown as a function of the dopant concentration. Within the bounds of experimental error, $E_{A,b}$ increases linearly with Y_2O_3 addition. Such behaviour agrees well with the single vacancy pair model derived from neutron diffraction data[20], and is not in accord with extensive defect association or trapping. $E_{A,gb}$ remains constant within $2 \le c \le 9$ mol% dopant which implies that the grain boundary composition and structure is kept fixed. Possible explanations are the freezing in of an eutectic mixture during sintering, or the formation of a kinetically preferred "second phase". According to the experimental $E_{A,gb}$ of 1.08 ± 0.07 eV grain boundaries with cubic structure and a composition of about 11 mol% Y_2O_3 may coexist with the tetragonal grain core. It is interesting to note that in this concentration range the conductivity of YSZ has a maximum which is intimately connected with an oxygen vacancy distribution providing the highest jump efficiency ($\approx$ 5% oxygen vacancies). At higher concentrations, the vacancy jumps become less effective and hence, the mass transport will also decrease. The saturation tendency of the grain boundary composition, observed by Burggraaf et. al.[34], is in accord with such a picture. The shift of some of the $E_{A,b}$ values by a constant amount towards the grain boundary value (see Fig. 12) is probably caused by a substantial contribution of a thin layer type grain boundary. The enrichment of Y^{3+} in the near surface region, reported in Ref. 19, and the constant C_{gb} (see above) yield further evidence for a grain boundary layer model. A rough estimate of the layer thickness δ may be obtained from a thin barrier layer model. In this model, the effective layer capacitance C_{eff} exceeds the bulk capacitance C_b by a factor d/δ. An upper limit of δ = 100 Å results from the experimental value $C_{eff}/C_b \ge 100$ and a typical grain size of some microns.

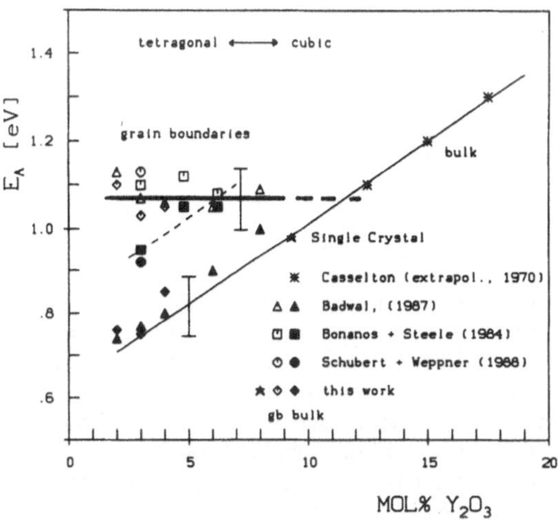

Fig. 12: Bulk and grain boundary activation energies $E_{A,b}$ and $E_{A,gb}$ of polycrystalline yttrium stabilized zirconia as a funtion of the dopant concentration.

2.2 TiO_2

It is still an open question as to what extent the oxygen exchange reaction in TiO_2 is determined by either surface or bulk properties. Separation of the surface and bulk conductivity contribution should yield the first qualitative results about this problem. Two kinds of approaches have been pursued. Firstly,

a) experiments on the surface excess conductivity of single crystal material and secondly

b) experiments on thin films from which the thickness dependence of the conductivity may be derived.

2.2a TiO_2 SINGLE CRYSTALS

Three single crystals, with different surface to volume ratios, were contacted on the (110) planes with reversible Pt-electrodes and impedance data were then made as a function of the oxygen partial pressure ($1 \leq p_{O_2} \leq 10^{-2}$ atm). From this data the excess conductivity of the crystal surface exposed to the ambient gas was determined. The following results have been obtained[16]:

1. The surface excess conductivity follows a $p_{O_2}^{1/2}$ dependence which indicates chemisorption or incorporation of monoatomic oxygen species such as O or O^- in the boundary layer.

2. For partial pressures $p_{O_2} \geq 10^{-1}$ atm, the surface conductivity appears to be enhanced with respect to the bulk. Whereas for lower partial pressures the situation is reversed.

2.2b THIN POLYCRYSTALLINE TiO_2 FILMS

Thin films prepared by electron beam evaporation, rf-sputtering and the Ion Cluster Beam technique are currently investigated in a joint project under UHV and under normal pressure, with different gas mixtures as oxygen partial pressure sources. Typically, the film thickness varies between several hundred Å and one micron. Fig. 13 shows the conductivity results obtained to date by Van der Pauw[21] measurements. As can be seen by comparison with the data of Marruco et al.[22] on thick slabs, the thin films behave like bulk material. Note the good agreement with this work although a different kind of gas mixture was used to provide the oxygen partial pressure (the small difference is probably due to H_2 acting as a donor). However, a rather large discrepancy is found between the measurements under UHV-conditions and those under normal pressure. In UHV the conductivity is higher and the activation energies are distinctly reduced (see Fig. 13, values in brackets). A more consistent picture, especially with respect to the activation energies, is obtained if one shifts the UHV data to lower partial pressures by roughly 6 orders of magnitude. The most probable cause of this deviation is a different surface composition and structure in both experiments. A different segregation behaviour of impurities and a change in the defect structure of the near surface region by, e.g., incorporation or loss of OH^-, may also contribute. Concerning the p_{O_2} dependence, the exponent of the log σ vs -log p_{O_2} plot is close to 1/6. This suggests that double ionized oxygen vacancies $V_O^{\cdot\cdot}$ and Ti^{4+} interstitials are the dominant defects up to the limit of Magneli phase formation.

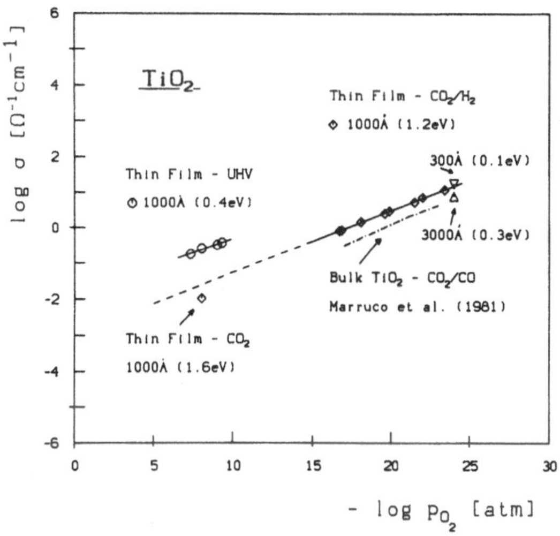

Fig. 13: DC-conductivity of thin TiO_2 films prepared by rf-sputtering and electron beam evaporation as a function of the oxygen partial pressure at T = 750 °C. The values in brackets correspond to the activation energies E_A of the conductivity.

A weak deviation of the surface conductivity from the bulk conductivity is found in both the experiment on the surface excess conductivity of single crystal material (see above) and the experiment on the thickness dependence of thin films. A similar result has been obtained for the samples under UHV conditions[23]. Here a slight reduction in the carrier concentration was deduced. Further investigations are in progress and a more complete report will be published at a later date.

2.3 δ-Bi_2O_3

Pressed pellets and thin rf-sputtered films of δ-Bi_2O_3 have been investigated by impedance and voltage relaxation measurements. An introduction to the latter type of experiment is given in Ref. 24. The diffusion coefficient of the minority carriers deduced from such an experiment - see Fig. 14 for a typical example - is in the range of 10^{-4} cm^2/s to 10^{-3} cm^2/s at T = 800 °C and $p_{O_2} = 10^{-3}$ atm. It should be noted, however, that there is a large uncertainty in this value, mainly because the blocking electrode, prepared by covering the Pt-electrode with a glass, may have not been working perfectly at high temperatures. As a first indication, the electronic conduction exceeds the ionic conduction by about two orders of magnitude. Further experiments on the voltage relaxation, the current- voltage characteristics and the impedance of δ-Bi_2O_3[25] are currently under way. These will allow a comparison with the stabilized zirconia system for which more data is available[26].

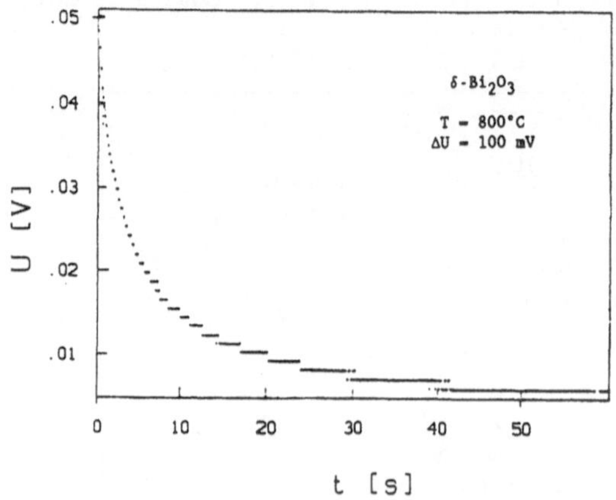

Fig. 14: Voltage relaxation in δ - Bi_2O_3 at T = 800 °C and $p_{O_2} = 10^{-3}$ atm after excitation with a short 100 mV pulse.

3. IMPEDANCE OF FRACTAL INTERFACES

As mentioned in Section 1.3b, CPE-behaviour might arise from the fractal nature of an interface. Various model structures have been taken as the starting point for the theoretical description of topographically disordered interfaces. Fig. 15 shows the two models (Koch curve and Cantor bar) and their respective electrical equivalents used by Liu et. al[11] and Nyikos and Pajkossy[12]. A further approach, dealing with the so called Sierpinsky electrode[27], yields essentially the same dependence of the CPE-exponent on the fractal dimension as the model of Liu. Two other works should also be mentioned, that of LeMehaute[28] and that of Keddam and Takeneouti[29]. With respect to the dependence of the impedance exponent α on the fractal dimension d, three rather different results emerge:

$$\alpha = (d-1)^{-1} \quad \text{(Nyikos and Pajkossy, LeMehaute)} \tag{8}$$

$$\alpha = 3-d \quad \text{(Liu, Sapoval)} \tag{9}$$

$$\text{none} \quad \text{(Keddam and Takeneouti)} \tag{10}$$

To our knowledge none of these theories has been hitherto satisfactorily verified by experiment. Nyikos and Pajkossy[12] claimed good agreement of their theory with a model experiment where they used large fractal Cu electrodes exposed to a liquid electrolyte. However, for convincing experimental proof, independent information is needed on d so that the value obtained from impedance can be checked. The work of Nyikos and

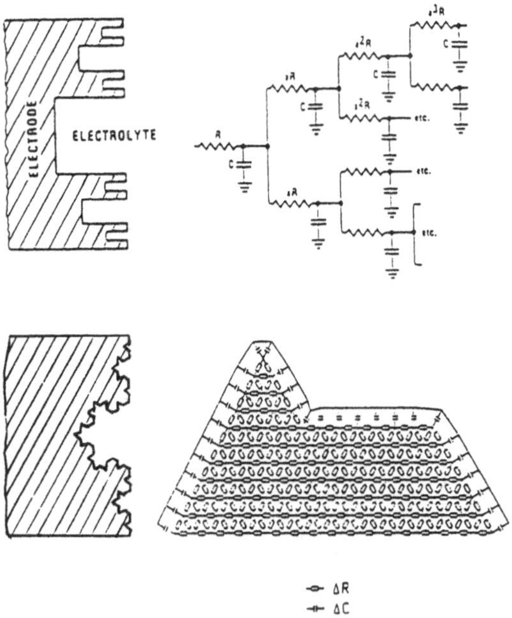

Fig. 15: Models for fractal interfaces (based on the Cantor bar and the Koch curve) and their electrical analogs (after Liu [1985] and Nyikos and Pajkossy [1985]).

Pajkossy has been questioned by Bates and Chu[30] who measured the impedance of blocking Pt-electrodes in contact with liquid H_2SO_4. In their experiment, the fractal dimension of the electrodes was determined independently by scanning the surface profile with a tiny diamond tip. From this type of experiment, the structure function $S(\delta) = \Delta h^2(\delta)$ may be obtained. $\Delta h(\delta)$ is the difference in height h of the profile at x and $x+\delta$. The fractal dimension d of the surface is related to the slope of the log S vs log δ plot by $d = 3 - \text{slope}/2$. No correlation with the value from impedance was found. A similar experiment has been carried out by the authors on blocking Pt/YSZ interfaces[16]. In this experiment the fractal dimension of the interface obtained from impedance was checked by voltage relaxation measurements. According to Nyikos and Pajkossy[31], carrier diffusion to a fractal boundary leads to a power law dependence of the diffusion current I with the time t:

$$I \propto t^{-\beta} \qquad \text{(Nyikos and Pajkossy)} \qquad (11)$$

with $\beta = (d-1)/2$. Thus, by taking the exponents α and β from impedance and voltage relaxation studies the Nyikos and Pajkossy approach may be checked in situ. As demonstrated in Figs. 9d and 16, the exponents α and β obtained from the slopes of log Z vs log ω and log I vs log t -plots are approximately equal at a given temperature. Therefore, this result is not in agreement with the Nyikos and Pajkossy model (unless d is assumed to be fixed at 2.41 (see eqns. 8 and 11)). Within the temperature range $550 \leq T \leq 1050$ °C, the exponents α and β vary between 0.5 and 0.8. This is also not explained by the temperature independent Nyikos and Pajkossy model. Serious doubt about the theoretical validity of the Nyikos and Pajkossy approach on impedance has been raised by Wang[32].

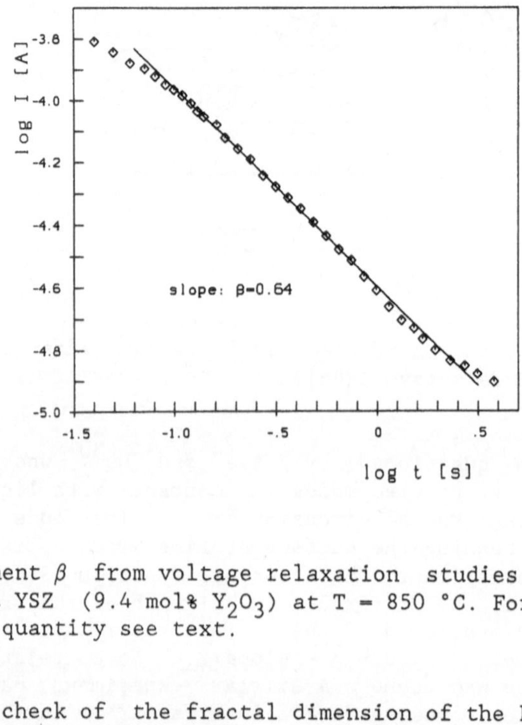

Fig. 16: Exponent β from voltage relaxation studies of single crystal cubic YSZ (9.4 mol% Y_2O_3) at T = 850 °C. For a description of this quantity see text.

An additional check of the fractal dimension of the Pt/YSZ interface at room temperature by the method used in Ref. 30 confirmed our result obtained at higher temperatures. As can be seen from Fig. 17, the fractal dimension $d = 3$ - slope/2 is close to 2 and not $\gg 2$ as indicated by the impedance. Again, no clear relation exists between the experiment and the above mentioned theories.

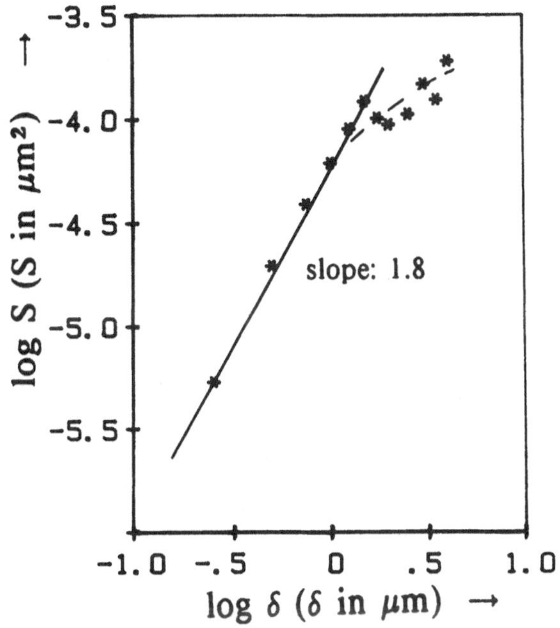

Fig. 17: Graph of log S versus log δ. $S(\delta) = \Delta h^2(\delta)$ is the structure function and $\Delta h\ (\delta) = h(x+\delta) - h(x)$ the difference in height of the profile at x and $x+\delta$. The fractal dimension d of the Pt/YSZ-interface at RT can be obtained from d = 3 - slope/2.

SUMMARY AND CONCLUSIONS

Despite the information available, a final description of the oxygen exchange reaction in oxide materials such as stabilized zirconia and bismuth oxide can not be given. This is due to the fact that fundamental microscopic properties, such as the types of defect states involved in the process, or the structure of the adsorption layer on the platinum, or the electrolyte surface, are still unknown. Some qualitative results about the oxygen exchange can be drawn, however, from electrochemical investigations. The impedance data on single crystal YSZ indicate that at least two interrelated processes occur at the electrolyte/electrode interface in the complete temperature range studied (350-1050 °C). The slower, rate determining step, depends on the concentration of O_2 in the very first adsorption layer, which can be affected - inter alia - by the diffusion in the gas phase[33]. The rather long relaxation times observed at $T \leq 700\ ^{\circ}C$ ($\tau_1 \gg 1$ sec) may therefore be caused by an adsorption step impeded by strongly bound surface entities like hydroxyl groups. As suggested in Ref. 30, diffusion of ions into the deeper region of the electrode, across locally screened tips of a rough surface, is another possible explanation. The second process is not, or only slightly, dependent on the gas flow and presumably involves surface diffusion and incorporation of O or charged oxygen species like O^- or O_2^- in the first atomic layers of the electrolyte. At T ≥ 750 °C this process is not activated (or only rather weakly activated) which suggests tunnelling or leakage - perhaps at chemical imperfections - through a thin barrier. On the other hand, the apparent temperature independence can simply be mimicked by a rather small change in the adsorption kinetic of a coupled reaction sequence like, for example, dissociative adsorption of O_2 and transport and incorporation of the decisive oxygen species. The $p_{O_2}^{1/2}$ dependence of the work function of single crystal YSZ (9.4 mol % Y_2O_3) reported in ref. 19, fits well with a dissociative adsorption step. The change from a $p_{O_2}^{1/4}$ to the $p_{O_2}^{1/2}$ dependence observed in this experiment during the first increase in temperature most likely indicates a segregation induced turnover, from bulk to surface control, of the oxygen exchange rate.

In polycrystalline material, here tetragonal YSZ, the contribution from grain boundaries is not suppressed. As in the case of single crystal material[16], the interface defect properties appear to be determined by the defect structure of the material in closest contact with the metal. As discussed in Section 2.1b, TZP can be described by a layer model with a grain boundary barrier layer less than 100 Å.

Concerning the question of whether platinum, with its topographically non-uniform surface structure, determines the interface properties, the following answer can be given: the first experimental tests of current fractal models are unsatisfactory and the morphology of the Pt does not seem to be the decisive parameter. Alternatively, the observed CPE-behaviour of the impedance may originate from a distribution of relaxation times. This is implied by the following observations: the CPE exponents are temperature dependent and therefore activated; at T ≫ 750 °C the relaxation times τ_1 and τ_2 converge and a uniform distribution seems to be established at high temperatures.

In the case of single crystal TiO_2, the surface excess conductivity follows a $p_{O_2}^{1/2}$ dependence which indicates adsorption and incorporation of monoatomic oxygen species like O or O^- in the boundary layer. However, no such surface contribution has been observed in polycrystalline thin films (300 ≤ d ≤ 10^4 Å). From the p_{O_2} dependence they behave like bulk material. This suggests that the surface and bulk defect structures are dominated by the same defects ($V_O^{\cdot\cdot}$, $Ti_i^{\cdot\cdot\cdot\cdot}$) whose concentrations may of course be different in the bulk and surface region. Preliminary measurements of the thickness dependence of the conductivity show that the concentrations (or mobilities) differ within only one order of magnitude.

Polarization experiments on δ-Bi_2O_3 yield a first estimate of the diffusion coefficient of the minority carriers. As in stabilized zirconia, the electronic diffusion is much faster than the ionic diffusion (by approximately two orders of magnitude). Electronic carriers in the sensoric active interface region are expected to be of crucial importance and further investigations of the type and number of the minority charge carriers are in progress.

Acknowledgement

We like to thank D. Fischer, Batelle Inst., Frankfurt, FRG, and U. Kirner, University of Tübingen, FRG, for the preparation of thin TiO_2 films. We are also grateful to H. Schubert for placing samples of tetragonal zirconia at our disposal. It is a pleasure to thank J. Nowotny for a critical reading of the manuscript and W. Weppner and A. Rabenau for their continuous interest and support. Financial support of the Bundesministerium für Forschung und Technologie and the Max-Planck-Gesellschaft is gratefully acknowledged.

References

1) Impedance Spectroscopy, ed. by J.Ross Macdonald, J. Wiley, New York 1987

2) D.C. Cronemeyer, Phys. Rev. **87**, 876 (1952)

3) J.E. Bauerle, J.Phys.Chem.Solids **30**, 2657 (1969)

4) N. Bonanos, E.P. Butler and B.C.H. Steele, see ref.1, chapter 4

5) E. Warburg, Wied.,Ann. **67**, 493 (1899) and F. Krüger, Z.physikal. Chemie **45**, 1 (1903)

6) J.E.B. Randles and K.W. Somerton, Trans.Farad.Soc. **48**, 937 (1952)

7) J.H. Sluyters, Rec.Trav.Chim. **79**, 1092 (1960)

8) J.H. Sluyters and J.C. Ohmen, Rec.Trav.Chim. **79**, 1101 (1960)

9) M. Sluyters-Rehbach and J.H. Sluyters, in Electroanalytical Chemistry, vol.4, ed. A.J. Bard, Marcel Dekker, New York 1970, pp. 1-127

10) R.D. Armstrong, M.F. Bell and A.A. Metcalfe, in Electrochemistry, Chemical Society Specialist Periodical Report **6**, 98 (1978)

11) S.H. Liu, Phys.Rev.Lett. **55**, 529 (1985); see also T. Kaplan, L.J. Gray and S.H. Liu, Phys.Rev. **B 34**, 4870 (1986)

12) L. Nyikos and T. Pajkossy, Electrochim.Acta **30**, 1533 (1985) and J. Electrochem.Soc. **133**, 2061 (1986)

13) R. de Levie, Electrochim.Acta **8**, 751 (1963) and Electrochim.Acta **10**, 113 (1965)

14) W. Scheider, J.Phys.Chem. **79**, 127 (1975)

15) J.C. Wang, J.Electrochem.Soc. **134**, 1915 (1987) and Solid State Ionics **18&19**, 224 (1986)

16) N. Nicoloso et. al., 6th Int. Conf. Sol. State Ionics, Garmisch-Partenkirchen, FRG, 1987, to be published in Solid State Ionics

17) "Equivcrt", a Non Linear Least Square Fit program written by B.A. Boukamp, University of Twente, Enschede, The Netherlands

18) W. Weppner and H. Schubert, to be published in Advances in Ceramics **24**, Science and Technology of Zirconia III

19) J. Nowotny and M. Sloma, in Surface and Near Surface Chemistry of Oxide Materials, eds. J.Nowotny and L.C. Dufour, Elsevier, Amsterdam 1988

20) N.H. Andersen et al., Physica **136B**, 315 (1986)

21) L.J. Van der Pauw, Philips Res. Repts. **13**, 1 (1958)

22) J.F. Marruco, J. Gautron and P. Lemasson, J.Phys.Chem.Solids **42**, 363 (1981)

23) W. Göpel, U. Kirner, G. Rocker and H.D. Wiemhöfer, 6th Int. Conf. Sol. State Ionics, Garmisch-Partenkirchen, FRG, 1987, to be published in Solid State Ionics

24) W. Weppner and R.A. Huggins, Ann.Rev.Mat.Sci. **8**, 269 (1978)

25) H.A. Harwig and A.G. Gerards, J.Sol.State Chem. **26**, 265 (1978)

26) W. Weppner, Z.Naturforsch. **31a**, 1336 (1976) and Electrochim. Acta **22**, 721 (1977)

27) B. Sapoval, Solid State Ionics **23**, 253 (1987); see also Y.T. Chu, Solid State Ionics **26**, 299 (1988)

28) LeMehaute, J.Stat.Phys. **36**, 665 (1984) and Solid State Ionics **25**, 99 (1987)

29) M. Keddam and H. Takeneouti, C.R.Acad.Sci. **302** (Ser.II), 288 (1986)

30) J.B. Bates and Y.T. Chu, 6th Int. Conf. Sol. State Ionics, Garmisch-Partenkirchen, FRG, 1987, to be published in Solid State Ionics

31) L. Nyikos and T. Pajkossy, Electrochim.Acta **31**, 1347 (1986)

32) J.C. Wang, 6th Int. Conf. Sol. State Ionics, Garmisch-Partenkirchen, FRG, 1987, to be published in Solid State Ionics

33) A. Löbert and N. Nicoloso, "Oxygen Exchange Reaction of Zirconia", (review), to be published.

34) A.J. Burggraaf et. al., Mater. Sci. Monographs 28, P. Barret and L.C. Dufour, Eds., Elsevier, Amsterdam, 1985

Appendix 1

In the following a concise derivation of the Warburg impedance is given for the simplest case of one-dimensional semi-infinite diffusion. Extensions are discussed in Ref. 1.

The space and time dependent change in carrier concentration c_i at the interface obeys Fick's second law:

$$\partial c_i/\partial t = D_i \; \partial^2 c_i/\partial x^2 \tag{12}$$

where t = time, x = distance from the electrode surface and D_i the diffusion constant. For a sine wave modulated current density $j = I \sin \omega t$ (usually an ac-voltage of that wave type is applied) and the boundary condition

$$(\partial c_i/\partial x)_{x=0} = - \text{ const } \sin \omega t \tag{13}$$

the following solution for Eq. 12 may be found:

$$c_i(x,t) = \langle c_i \rangle + \text{const}' \exp(-x/x_o) \sin(\omega t - 2\pi x/\lambda + \varphi) \tag{14}$$

where $\langle c_i \rangle$ is the equilibrium concentration of the diffusing species, λ the wavelength of the concentration wave and φ its phase shift. Partial differentiation of Eq. 14, insertion into Eq. 12 and coefficient comparison yields

$$x_o = (2D_i/\omega)^{1/2} \quad \text{and} \quad \lambda = 2\pi x_o \tag{15}$$

The first partial derivative with respect to x and coefficient comparison with the boundary condition, Eq. 13, yields for x=0:

$$\text{const}' = I/nF \, (D_i\omega)^{-1/2} \quad \text{and} \quad \varphi = -\pi/4 \tag{16}$$

A concentration gradient $\Delta c_i = c_i(0,t) - \langle c_i \rangle$ will lead to an overvoltage

$$\begin{aligned} \eta &= (RT/nF) \ln c_i/\langle c_i \rangle \\ &= (IRT/n^2F^2) \langle c_i \rangle^{-1} (D_i\omega)^{-1/2} \sin(\omega t - \pi/4) \end{aligned} \tag{17}$$

or a diffusion impedance

$$Z_W = \eta/I = (RT/n^2F^2) \langle c_i \rangle^{-1} D_i^{-1/2} \; \omega^{-1/2} \tag{18}$$

where n,R,T and F have their usual meanings.

SUBJECT INDEX

A

B

C

H

I

L

M

S

T

U